PHYSICAL SCIENCE, ENERGY, AND OUR ENVIRONMENT

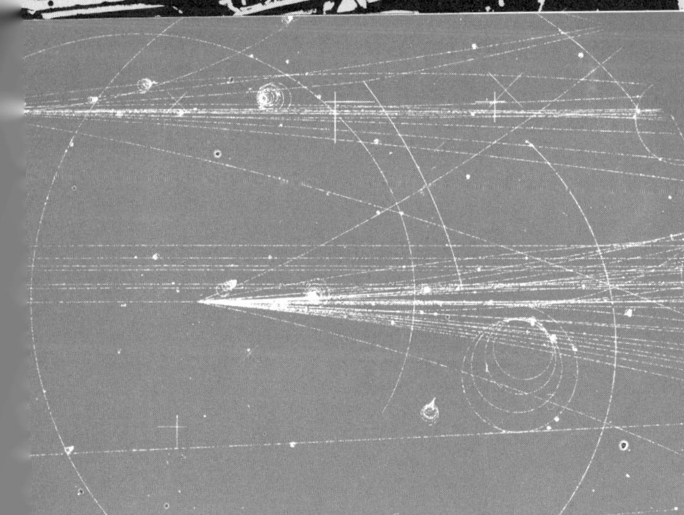

PHYSICAL SCIENCE, ENERGY, AND OUR ENVIRONMENT

Peter Fong

Professor of Physics, Emory University

Macmillan Publishing Co., Inc.
NEW YORK

Collier Macmillan Publishers
LONDON

Printed in the United States of America

Macmillan Publishing Co., Inc.
866 Third Avenue, New York, New York 10022

Collier Macmillan Canada, Ltd.

Library of Congress Cataloging in Publication Data

Fong, Peter P
Physical science, energy, and our environment.

Includes index.
1. Science. 2. Force and energy. 3. Pollution.
I. Title.
Q161.2.F65 1976 500.2 74-33094
ISBN 0-02-338660-6

Printing: 1 2 3 4 5 6 7 8 Year: 6 7 8 9 0 1 2

PREFACE

This physical science text for liberal arts students is written to help students gain a better understanding of the energy and environmental problems that currently beset our society. In a one-semester course of physical science or physics for liberal arts students this objective can be accomplished by omitting the starred sections and chapters. The complete book is designed for a one-year course, in which more complete coverage of topics in physics, chemistry, astronomy, and geology is possible. The mathematical knowledge necessary for understanding is elementary; only simple algebra is used.

Because this book is an introductory text for students without much scientific and mathematical background, we present the physical science material by centering on two basic concepts, matter and energy. All natural phenomena are considered as transformations of matter and energy. This viewpoint can be easily grasped by beginning students. Most of the material in physics and chemistry for the course under consideration can be organized into such a presentation—as studies of matter and energy on the bulk, molecular, atomic, and nuclear levels. This body of information, which can be considered as natural philosophy, can be applied to the study of natural history—the material usually contained in astronomy and geology. We concentrate on the problems of the origin of the universe, the chemical elements, the solar system, the earth, and life in Chapters 21 to 25, which are starred. These are fascinating problems full of intellectual challenge. Considerable progress has been made in recent years in attempting to understand them; therefore much of the material presented here is relatively new and has not appeared in general textbooks before. These discussions also serve to illustrate the basic principles of science and to demonstrate how physics, chemistry, astronomy, geology, and biology can be integrated to solve some of these fascinating problems. This approach is particularly advantageous in discussing energy and the environment, which are special features of the book. Energy is first dealt with as a central concept in basic science; its practical applications and environmental consequences follow naturally. Moreover, environmental modifications are a natural extension of the natural history.

The environmental crisis has attracted great public attention, and now we are faced with another crisis, that of energy. Many of the issues involved in these crises encompass a broad range of knowledge in physics and chemistry. It is thus desirable to teach physical science by emphasizing energy and environmental issues, which is demonstrated by the many original homework problems included at the end of each chapter based on energy and environmental problems (complete solutions are given in the instructor's manual). In recent years many college courses have been criticized as "irrelevant." In this presentation such a charge can be readily dismissed.

From the point of view of physical science, all environmental problems originate from economic processes involving transformations of matter and energy. The energy and environmental topics discussed in this book—such as the earth's energy resources, the utilization of exotic forms of energy, the energy industry, the automobile problem, the environmental consequences of energy utilization, air pollution, water pollution, land pollution, thermal pollution, noise pollution, radiation pollution, resource conservation, waste disposal, and so on—are developed as illustrations and applications of the basic scientific principles and thus are integrated into the presentation of physical science.

Contrary to popular conception, many environmental problems are problems of physical science. They are inadequately discussed in conventional environmental texts, which is particularly true of the energy- and radiation-related topics. This book fulfills the needs of an environmental discussion from the physical science point of view and may also be used in courses studying the environment and energy.

Because of the limitation of space we cannot develop every concept in rigorous detail, as is possible in a book for science students. Elementary concepts are used without further discussion, which permits the coverage of more interesting and significant topics.

This book went through several stages of development, and substantial changes were made along the way. The author was helped to a significant extent by comments from colleagues and friends, whom he wishes to acknowledge, including in particular Professors R. J. Ruch of Kent State University; Gordon E. Jones of Mississippi State University; Gene F. Graven of Oregon State University; Francis E. Dart of the University of Oregon; Roland A. Hultsch of the University of Missouri, Columbia; M. Lynn James of the University of Northern Colorado; Joel R. Sherman of Brevard Community College; and Richard A. Stagliano of Cuyahoga Community College.

It is hoped that this book will contribute to understanding some of the pressing societal problems of our time.

P. F.

CONTENTS

PHYSICAL SCIENCE, ENERGY, AND OUR ENVIRONMENT

1 Science, Technology, and Society

1.1 Introduction

Modern science, beginning with Galileo and Newton, emerged from the medieval world in the seventeenth century. Since then, in the short span of three centuries, science has transformed the world profoundly. The most visible change is the rapid development of technology, not only in industry but also in agriculture and medicine. The current material affluence of technologically developed societies, exemplified by abundant food supply, adequate medical service, and the variety of useful inventions, can largely be traced to the advancement of science. The technological changes that have occurred since the industrial revolution have led to profound social changes that have forged our present social structure. Not as visible but no less important are the changes in our concepts of nature, man, and society.

In all these developments science has been regarded as a positive factor contributing to the progress of society. From time immemorial mankind had been plagued by poverty, famine, and pestilence. In the struggle for subsistence and survival science and technology were the main allies. In the past century, for the first time, many of these problems were adequately dealt with and the developed nations entered a new era of affluence. The preoccupation with subsistence has conditioned social thought and ideology. Progress, expansion, higher standards of living, better health care, and so on, were considered undisputed social goals by everybody. Personal relations, ethics, law, religion, and so on, are all oriented accordingly.

However, a drastic change has taken place in the past decade. Rapid industrial development, while improving the standard of living, has created new problems that not only have detrimental effects on our present well-being but also endanger our future existence. These include the degradation of the environment, which impoverishes our life and endangers our health, and the depletion of natural resources, which disrupts and may eventually bring to a halt our economic activities. The current energy crisis is the beginning of one of our resource problems.

As these new problems began to claim our attention, attitudes toward science and technology changed. Suddenly, in the environmental movement, science and technology were seen as villains, because of the degradation of the environment, in spite of the fact that without science and technology we would still be struggling for subsistence and any talk of a desirable environment would be impractical and irrelevant. Installations of power plants were delayed, hindered, or even denied because of possible environmental effects (and no one seemed to be concerned with the resulting increase in unemployment and decrease in gross national product). Space exploration was ridiculed. Expansion and progress were frowned upon. There was much intellectual confusion. Scientists were amazed and stunned by the sudden turnaround of social values.

The very recent energy crisis adds a new dimension to the problem. We are beginning to realize how much we are dependent on technology. We also realize that environmental issues must be dealt with in relation to many other problems. The antiscience and pro-environment trend has begun to change again, as is reflected in the halting of litigation by environmental groups against the construction of the Alaska pipeline. The pressing need for oil in the current energy crisis overbalances the environmental sacrifices. At least in the immediate future we need science and technology to solve the energy crisis. And we shall also require the help of science and technology to solve our problems in the distant future. It is time to review the role of science and technology in our society and to charter a viable course. The study of energy and environment issues will provide the necessary information for such an evaluation.

In the following sections we begin by studying the origin and nature of our environmental problems. The energy problem is one aspect of the problem of depletion of natural resources that will be part of our environmental studies. Current manifestations of the environmental and energy crises are described more fully in Chapters 2 and 3. Chapters 4 to 16 contain basic physical science material concerning the transformations of matter and energy, together with some specific applications to energy and environmental problems. Chapter 16 deals with the general principles of transformations of matter and energy. These are also the general principles of environmental studies, because environmental changes are manifestations of transformations of matter and energy. These principles govern the general study of the energy and environmental problems presented in Chapters 17 to 20. These chapters conclude our energy and environmental studies. The rest of the book deals with the applications of physical science to the study of the origin and evolution of the physical world as described in the preface.

1.2 Population and the Environment

Environmental concerns generally include the expansion of population, depletion of resources, and degradation of environment; these are the three main topics of environmental study.[1] The relation between population and environment is illustrated by the following example of bacterial growth.

When a single bacterial cell is introduced into a proper nutrient medium, it reproduces by cell division and the number of bacteria increases from one to two, two to four, four to eight, and so on. The number multiplies *exponentially;* that is, the number increases as an exponential function of time.[2] However, sooner or later growth slows down, and eventually the number increases no more; the number of new births just compensates the new deaths. An *equilibrium* is established and the population is said to be *saturated.* Growth can be represented by a curve showing the number of bacteria as a function of time, and the preceding growth pattern is repre-

[1] General references: William W. Murdoch, ed., *Environment* (Stamford, Conn.: Sinauer Associates, Inc., 1971); Richard H. Wagner, *Environment and Man* (New York: W. W. Norton & Co., 1971).

[2] An example of the exponential function may be found in Sec. 14.5.

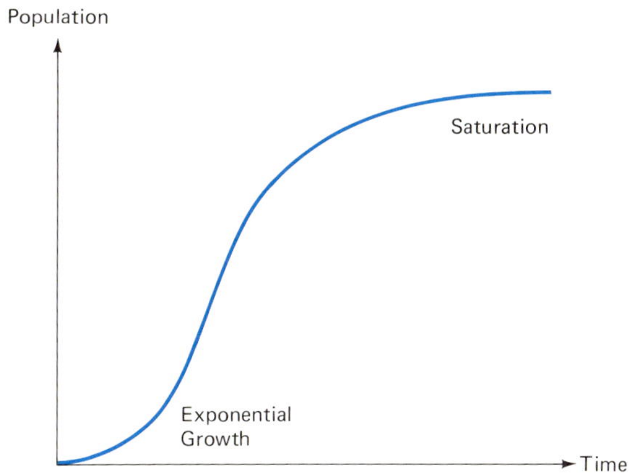

Figure 1.1 Sigmoid curve of population increase.

sented by a *sigmoid* curve (Fig. 1.1). It is characterized by an exponential growth at the beginning and a saturation at the end.

Exponential growth is a characteristic of life. Each living organism has the ability to reproduce. If the environment remains favorable all the time, the rate of reproduction is expected to be always the same. The rate of increase of population at a particular time is thus proportional to the population itself at that time. Under such conditions the growth curve is an exact exponential curve. (Mathematically this can be proved easily by the method of integral calculus. The situation is similar but opposite to that of radioactive decay.)

Exponential growth obviously cannot continue indefinitely. The total number of bacteria, and so the total mass, would increase to infinity—which is impossible, because the total amount of nutrient is finite. In fact, growth slows down long before the depletion of the nutrient. The intake of nutrient by the bacteria is accompanied by the excretion of waste products; this process of *metabolism* is another basic characteristic of life. As the population grows, the waste products accumulate. Their toxic effect slows down growth, eventually establishing an equilibrium between the population and the environment. The population stabilizes at a point at which any further increase would cause increased toxicity that would lead to more deaths than "births."

This pattern of population growth is common to other ecological systems—for instance, the population of a certain animal species in a certain region. When the environment is favorable, rapid exponential growth occurs. As the population becomes sufficiently large, the rate of growth starts to decline because of competition for food and living space with other species and within the species. Eventually an equilibrium is established and the population is stabilized.

The preceding example shows that although environmental problems are relatively new on the human scene, they previously existed and affected such simple ecological systems as the bacterial colony. The growth of the colony is slowed down by pollution of the environment (the nutrient medium) by the waste products the colony generates. Eventually the pollution becomes the principal factor that determines the size of the colony. In other words, the colony grows as far as it can tolerate its own pollution. The same may be said about pollution of the human environment. The population growth of California is slowing down. Although this is a complex social phenomenon, one of the reasons for decreased population growth is the increased air pollution (smog) of its major cities. In 1970 for the first time in the past century emigration from the Golden State equaled or perhaps surpassed immigration from the other states. In its peak years the state attracted a net immigration of 1,000 persons a day. Air pollution is thus not a mere nuisance. It plays a considerable part in social changes and historical development.

1.3 The Limit to Population Expansion

The problem of population expansion has long been the domain of biologists. Food shortage is considered the major factor limiting population growth. Now physicists have entered the field; other factors are being considered. Their conclusions will startle many biologists. The following is a view offered by John Fremlin.[3]

The population of the world is doubling every 34 years. At this rate, after 100 years the present population will grow to 30 billion. Although food shortage in the past has been a factor in limiting population, in the future this need not be so. We are already in the midst of a green revolution (high-yield miracle rice and wheat) that will boost grain production drastically. Even though progress of this kind has its limitations, the difficulties are by no means insurmountable if we are willing to take drastic and ruthless measures against the earth environment, which probably will be done when man is driven to it. These measures would have to include the destruction of all inedible plants on earth so that every inch of the surface could be used to grow food. The overriding principle would be to make maximum use of the sun's energy. There would be no more green pastures, because the utilization of solar energy through green grasses by way of milk and meat is too inefficient. Based on the total amount of solar radiation and the plant conversion ratio, we can calculate the maximum amount of food that can be produced. The amount is sufficient to support a population of 3,000 billion. Although that population is a thousand times greater than today's, it will be reached in a mere 330 years if we keep on multiplying at the present rate. In such an economy, no human activity could be permitted above ground because it would waste surface area needed for interception of solar energy by green plants. No more automobiles would be allowed, because all road surfaces would have to be used for food production. Not even houses would be permitted, because they would take away sunny spaces. All residences, and thus cities, necessarily would be built floating in the ocean, perhaps 50 ft beneath the surface, so as not to keep sunlight away from the edible algae in the sea.

Even that limitation of sunlight could be broken, according to Fremlin, and the population increased further. Perhaps giant reflectors could be installed in satellites to reflect sunlight to the night side of the earth. This might keep humans multiplying beyond the doomsday for another 60 years. More importantly, sunlight need not be the indispensable energy source of food. In the future atomic energy can do the same without geographical limitation. Man already can produce food chemically without sunlight. In Grangemouth, Scotland, a chemical plant is already producing 4,000 tons of protein a year from oil waste, using chemical reactions mediated by bacteria and yeasts. Chemical syntheses of carbohydrates and fats without the help of microorganisms can be done already. Atomic energy and chemical raw materials are sufficient to produce food for an ever-increasing population. Before they run out, another limitation would set in. In 800 years the population would reach such a great density that there would be 100 people per square yard on the earth's surface. They could be housed in 2,000-story buildings and still have plenty of room, plenty of food, and even plenty of oxygen to breathe, even though 98 per cent of the oxygen would be extracted from human excreta and only 2 per cent from fresh material. The major problem, and thus the limiting factor, would be thermal pollution resulting from the heat generated by our own bodies and machinery.

Both the human body and the machine consume energy, but this energy is not destroyed. According to the principle of conservation of energy, it is merely changed from useful energy to useless heat energy. Because a large population consumes so much energy, dumping of the equally large amount of waste heat would become a serious problem. If the entire surface of the earth were populated by people, the only way to dump heat energy would be through the rooftops into "outer space." With so great a population, the heat generated could be radiated into outer space only by raising the roof temperature to white heat (by pumping heat from living spaces below to the roof with air conditioners)—the earth would become a shining "star." A further doubling of population would cause the roof to melt. That would be the ultimate limit of human popula-

[3]John H. Fremlin, *The New York Times,* August 22, 1971, p. E13.

tion. Thermal pollution sets the limit we can never circumvent. Food is not the problem. We would not die of starvation. We would die with our bellies full.

It seems quite unlikely that such a situation will actually come to pass. Clearly the traditional argument that food will run out because of the population explosion is not invincible. However, the population cannot grow indefinitely, because the earth's environment is finite; some limitation will set in sooner or later. At the present rate of population growth, that limit will be reached in a very short time. Already environmental problems have given us the warning signal. Air pollution means that city population and present technology have loaded the atmosphere with waste products to the limit, and some reduction of city population or revision of technology is needed. Water pollution indicates that population and technology have pushed the utilization of water to the limit. In the next chapter we will learn about other environmental problems, such as thermal pollution, land pollution, and so on. All of these indicate that the stage of exponential growth, in which we need not worry about the environment, is almost or already over. From now on, like it or not, environment will be an important aspect to consider for any technological and social developments. Because it is population growth that overloads the environment, population stabilization is undoubtedly one of our most pressing issues.

1.4 Other Aspects of the Environmental Problem

Infinite and Finite Environment. The bacterial colony example illustrates another aspect of the environmental problem. Pollution of the environment is typically a problem that sets in at the late stage of population growth. In fact, it is the limitations of the environment that bring about this "late stage." In the early stage, when the population is small, the environment may be regarded as infinite—endless and inexhaustible. The small amount of resources drained from the environment does not cause depletion, and man has been tapping these resources for years. The small amount of waste dumped into a large environment becomes so dilute that it does little harm. Humans spewed smoke into the atmosphere for ages without giving rise to serious problems until recently. In an infinite environment the problem of pollution does not exist. Yet nothing on the earth is truly infinite; what made it seem infinite was the small population of the past. The smog problem afflicting modern cities indicates that the population (at least of the cities) has grown so large that the finiteness of the environment becomes manifest. It sounds the warning that we are moving into the later stage of the sigmoid curve.

Thus pollution problems occur not just because the pollutants are released, but because the release exceeds a critical amount relative to a particular environment. In economic decisions in the past, it was usual to consider efficiency or profit as the main determining factor. If a number of technical alternatives were available, usually the most efficient one was adopted, and the others were abandoned. This works well with an infinite environment. On the other hand, in a finite environment in the later stage of development, the most efficient alternative may not be the best choice if it is seriously detrimental to the environment. For a similar reason, if man used a variety of motor vehicles that each burned a different kind of fuel and released different kinds of exhaust products, then the concentration of any single pollutant would be much smaller than that resulting from all vehicles using the same fuel. Given the same amount of total traffic, diversification of technology might reduce the concentration of any particular pollutant and mitigate the pollution problem. Although this would not solve the problem basically, it is well to recognize that pollution is relative to the capacity of the environment.

Capitalism Versus Socialism. Pollution is not exclusively a capitalistic sin nor a problem restricted to populated areas. In remote reaches of Soviet Asia,

Lake Baikal is in danger of being polluted by woodpulp installations. Chimney emissions and automobile exhaust pollute the air of major Soviet cities and industrial areas. Russians claim that these sources of pollution can be controlled, though a large investment is required. They have developed a "neutralizer" to be mounted on automobiles that is claimed to reduce the carbon monoxide emission by 90 per cent and the hydrocarbon emission by 66 per cent. Filters have been installed to clean up smoke from chimneys. A new process of making nitric acid has been developed that eliminates the "foxes tails," the red-brown clouds of nitrogen oxides from plant chimneys.[4]

Social Implications. The philosophical, historical, and social significance of environmental problems perhaps has not been widely appreciated. Historically, the so-called Golden Ages are usually associated with an exponential growth, be it growth of population or wealth or cultural and scientific creations. The nearly exponential growth of the gross national product in the 1950's and 1960's has been taken by many as signifying an economic Golden Age. The exponential growth of the number of physicists and the volume of physics research publications in this period is undoubtedly taken as the sign of a Golden Age of physics. The first mongooses transplanted to a new habitat as predators to poisonous snakes must be as happy as the discoverer of a new continent. Delicious food abounds (cobras)! No adversaries! Endless opportunity to multiply! A Golden Age! Yet Golden Ages never last long. They are sometimes followed by dark ages; they sow the seeds for the dark ages. The inflation, recession, and unemployment (of even the professionals) of the early 1970's are a reflection of the headlong expansion of the early years. Exponential increase makes population multiply so rapidly that it soon reaches the limit set by the environment. The exponential curve bends down into a sigmoid curve and population soon becomes saturated. Saturation could mean stagnation. Stagnation could lead to degeneration and decay or decline and fall. The rise and fall of civilizations are the major concerns of philosophers, historians, humanists, and social scientists. The environmental problem is not merely one stirred up by a band of enthusiastic nature lovers over ugly sights and bad smells. It is a basic social problem that concerns everybody.

[4]Igor Petryanov, *The New York Times,* August 15, 1971, p. E15.

The Ideology of Progress. Modern Western social thought is dominated by the concepts of progress, expansion, innovation, and development—by a zealous pursuit of the Golden Age. To this end conservatives and radicals hold essentially the same view, and leftists and rightists are unwitting bedfellows. Yet the Golden Age is tied to exponential growth, and exponential growth is but a fleeting moment characterizing the early stage of a historical development. It soon leads to its own demise. Perhaps modern Western social thought was formulated at a time when human population and human modification of the environment were still negligibly small and the environment could be regarded as infinite, allowing unlimited exponential growth. At any rate, such a view must be regarded as shortsighted, because the finiteness of the earth cannot be denied and the speed of exponential growth is rapid indeed. Now that modification of the environment has become substantial, hindering exponential growth, the rug has been pulled out from under our basic social thought; ideologies from the left to the right will have to be reoriented. Incidentally, the traditional social thought in the East emphasizes harmony and balance without any special emphasis on innovation and expansion.

It is obvious that the future course of mankind will have to be decided through balanced consideration of all aspects involved. Expansion and development will still be pursued by man, like food and drink, but will have to be restrained for "health" reasons. New frontiers, as in cultural and artistic areas, should be developed continually. Scientific and technological potentialities are far from being exhausted. Yet their development will have to be a concerted, balanced effort instead of chaotic "gold-rush" activities. The time is not too late. Opportunities abound. The future is in our hands and will be what we want it to be.

Problems

1.1 Place one grain of wheat in the first square of a chess board, two in the second, four in the third, eight in the fourth, and so on, until all the squares are occupied. How much wheat is there altogether? This is what a courtier once asked the king for as a meager reward, which the king gladly granted. Can any king afford it? This example illustrates the rapid rate of multiplication of exponential growth.

1.2 Roughly speaking, to raise 1 lb of chicken, pork, and beef requires, respectively, 2, 4, and 8 lb of feed. What is the bearing of this fact on future population growth?

1.3 A large part of the surface of the earth (70 per cent) is covered by the oceans, which so far are little used for the production of food (except for fishing). How can they be utilized for this purpose? What is the effect on the future of the population?

1.4 Assume the availability of endless energy from nuclear fusion. How can this energy be utilized for food production? (*Hint:* Three-dimensional farming by artificial light.) With this food source how many times may the population be expanded?

2 The Environmental Crisis

The physical world around us involves the lithosphere, hydrosphere, and atmosphere of the earth, representing, respectively, the solid, liquid, and gaseous components of our planet. The biological and industrial activities of man change the constitution of the air, the water, and the earth in a detrimental way, which we call pollution. First, the material contents are changed by the use of material substances, leading to *material pollution*. The material may be airborne, waterborne, or land-bound, resulting in *air pollution, water pollution,* or *land pollution*. Second, the energy contents are changed by the use of energy, leading to *energy pollution*. This includes *thermal pollution, noise pollution,* and *radiation pollution*. The various kinds of pollution are discussed separately.

2.1 Air Pollution

The composition (volume percentage) of air is normally four-fifths nitrogen (78.03 per cent) and one-fifth oxygen (20.99 per cent), together with a trace of five rare gases—helium (0.0005 per cent), neon (0.0018 per cent), argon (0.93 per cent), krypton (0.0001 per cent), and xenon (0.000009 per cent). In addition, there is present in the air a small amount of carbon dioxide (0.03 per cent), hydrogen (0.01 per cent), and ozone (0.00006 per cent). Finally there is the water vapor; its percentage is not fixed but varies according to weather condition (humidity). Many of these gases enter and leave the atmosphere in cycles in connection with many of the physical and biological changes taking place on the surface of the earth.

After the advent of industrialization, the burning of fuel and trash resulted in dumping a large amount of gaseous products into the air in population centers (Figs. 2.1 and 2.2). In the past, when population was small, the amount of dumping was small. The pollutants were scattered and diluted in the atmosphere and therefore presented no serious problem. As population increased and industrialization intensified, the volume of dumping increased rapidly and the pollutants were not readily dispersed.

Let us consider the body of air above a metropolitan area. The chimneys and incinerators spewing pollutants into the air act as the sources of the flow of pollutants. The dispersing mechanisms, such as diffusion and wind movement, remove the pollutants from the body and scatter them. The dispersal is determined largely by natural environment and remains nearly the same, whereas the sources increase with population and industrialization. An imbalance of input and output results in an increase in the concentration of pollutants. When the concentration becomes sufficiently high, serious health effects, such as irritation of the eyes, nose, and throat, begin to appear, and the pollution problem makes itself man-

Figure 2.1 St. Louis under smog, April 10, 1966. [U.S. Environmental Protection Agency]

Figure 2.2 Jet smoke trails blacken the sky over Love Field in Dallas. [U.S. Environmental Protection Agency]

ifest. With the rapid increase of sources of pollutants, the pollution problem is destined to strike any metropolitan area. The problem emerged first in areas with bad air circulation. Industrial towns in mountain valleys (in parts of Pennsylvania, for example) have high levels of pollution because foul air stays in the valleys. Coastal cities backed by a mountain range (such as Los Angeles) are also afflicted with bad circulation. It was in Los Angeles that the air pollution problem was first recognized as a basic problem of modern urban society. The Los Angeles smog caught national attention two decades ago. Since then the smog problem has spread to all major cities.

The Air Pollutants. It has been estimated that in one year (1965) people in the United States dumped into the air 142 million tons of pollutants (roughly 1 ton per person per year), the sources of which can be broken down as follows: the automobile, the major polluter, 60 per cent; industries, 17 per cent; electrical power plants, 14 per cent; space heating, 6 per cent; and trash burning, 3 per cent. The pollutants are of many kinds. About one half of the total is carbon monoxide (CO);[1] nearly 90 per cent of it comes from automobiles. The others, in descending order, are sulfur dioxide (SO_2), largely from burning of coal by industries and electric utilities; hydrocarbons and oxides of nitrogen, largely from automobile emission; and particulates (dust and dirt), largely from coal burning by industries and utilities.[2] See Fig. 2.3 for a detailed breakdown.

Current air pollution control services monitor three pollutants: carbon monoxide, sulfur dioxide, and particulate pollutants. The services established tolerable limits of the concentration of each pollutant, above which the degree of pollution is considered to be hazardous to health. The established tolerable limit of particulate pollutants is 150

[1] C stands for carbon; O stands for oxygen. CO stands for the molecule made of one carbon atom and one oxygen atom. If more than one atom is involved, the number is written as a subscript. Thus CO_2 stands for carbon dioxide.

[2] The data are taken from *Sources of Air Pollution and Their Control,* Public Health Service Publication No. 1548 (Washington D.C.: U.S. Department of Health and Welfare, 1966).

Figure 2.3 Sources and composition of air pollutants.

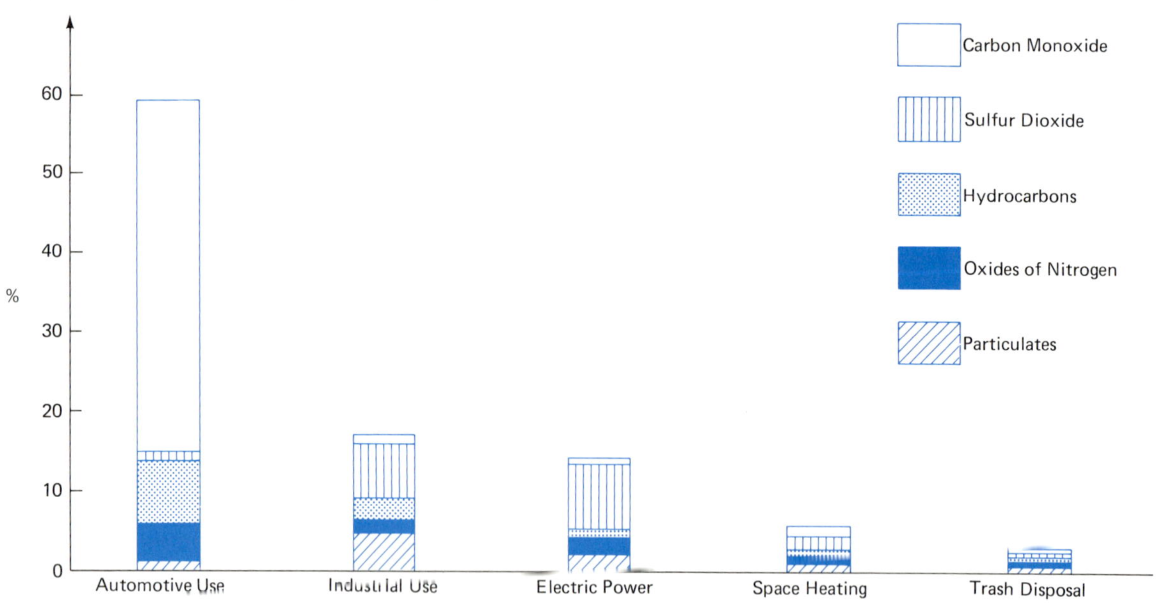

Figure 2.4 Blight of vegetation caused by smog. Pinto bean plants, 10 days old, exposed to 5 ppm PAN, a smog ingredient, for 10 minutes. Left plant, lights were on during fumigation. Right plant, lights were off during fumigation. [U.S. Environmental Protection Agency]

micrograms per cubic meter ($\mu g/m^3$) of air, that of sulfur dioxide 0.11 parts per million (ppm), and that of carbon monoxide 15 ppm. The daily concentration of pollutants is measured. Conversion to percentages of the tolerable limits gives the pollution indexes, which show how serious the pollution situation of the day is on a scale of 100. For convenience, a composite pollution index formed by averaging the three indexes is used as a general indicator (though it is less meaningful). This index has become another parameter listed in daily weather reports, together with such standard weather parameters as temperature, pressure, and humidity.

As already mentioned, the pollutant carbon monoxide comes largely from automobile exhaust. It is toxic. In fact, when exhaust fumes leak inside the automobile and are not removed by adequate circulation, the resulting carbon monoxide poisoning often causes death. Even sublethal doses are hazardous to health. Also as mentioned, the pollutant sulfur dioxide comes largely from the burning of soft coals with high sulfur content. The chimneys of power plants and industrial plants burning coal are the major sources of this pollutant. Although not seriously toxic, sulfur dioxide irritates the nose and the throat and gives rise to a number of respiratory diseases. The oxides of nitrogen and hydrocarbons coming from auto exhaust yield secondary pollutants by photochemical reactions, which will be discussed later. These pollutants, typical in smog, cause eye irritation and other health hazards, produce blight in vegetation (Fig. 2.4), and even cause deterioration of rubber.

The gas sulfur dioxide in the air may be oxidized to sulfur trioxide SO_3:

$$SO_2 + \tfrac{1}{2}O_2 \longrightarrow SO_3$$

Sulfur trioxide, when dissolved in water (moisture or rain water), becomes sulfuric acid, H_2SO_4:

$$SO_3 + H_2O \longrightarrow H_2SO_4$$

Sulfuric acid is highly corrosive, and its creation leads to another detrimental effect of air pollution: the corrosion of clothing fabric (Fig. 2.5) and building materials, such as metal, limestone, sandstone, and marble. The effect on stone material is clearly visible in historic places. In Venice, many fine buildings and bridges are made of white stone and are decorated

Figure 2.5 Sulfurous pollution causes a nylon stocking to disintegrate in St. Louis area. This is a U.S. Air Force secretary forced to change stockings twice in one day. Photo taken April, 1966. [U.S. Environmental Protection Agency]

with exquisite stone sculptures and statues. The increased industrialization of the region has increased the pollution by sulfur dioxide. Now the white stone structures and sculptures are covered with black streaks. Even the stone itself has begun to crumble. The once beautiful city proud of its art treasures has become a desperate victim of air pollution.

The particulate components of air pollution come from chimneys and incinerators. They are of a variety of chemical compositions, depending on what was burned. Hazardous elements, when burned, may remain attached to particulate pollutants and have deleterious effects. Unlike the gaseous pollutants, which are of simple chemical composition and are more amenable to control, the particulate pollutants are more complex and more difficult to handle. Numerous hazardous chemicals are present in them in trace amounts; the identification and measurement of these dangerous compounds are usually beyond standard chemical means. This makes it difficult to control the sources and to ascertain the hazards. Yet the problem may be attacked by more sophisticated technology. One approach is through *nuclear activation analysis,* in which the trace amounts of chemical elements, after exposure to nuclear radiation such as neutron bombardment, change into radioactive elements that can easily be identified and measured. The identity of the element may be established by the half-life of the radioactive decay; the concentration may be determined by the counting rate of radioactive decay. A total of 35 elements in particulate pollutants may be monitored this way—in particular, arsenic and mercury, which are toxic. The level of such pollutants in the air may be measured to safeguard public health. Their sources might be traced and proper control measures might be installed.

Exotic Air Pollutants. The U.S. Environmental Protection Agency (EPA) has recently identified asbestos, beryllium, and mercury as hazardous air pollutants and intends to set up limits on the amounts that may be tolerated. These pollutants are considered so hazardous to health that even slight exposure may cause serious illness or death. Asbestos, a fibrous mineral product used for heat insulation and fire protection in construction, may cause serious lung disorders. A connection between contact with asbestos and a high incidence of lung cancer has been established. Once inhaled into the lung, asbestos stays there permanently. It is not known why asbestos causes lung cancer. One theory is that the substance, which is highly absorptive, attracts and keeps permanently in the lung such cancer-causing chemicals as tars, which otherwise would be removed by body mechanisms. Asbestos is sprayed on steel girders of high-rise buildings. Its thermal insulation prevents heat from reaching the steel in case of fire, and thus prevents the frame from softening and the building from collapsing before people are evacuated. When walking on a New York street near a new construction site on a summer day, one may see white flakes floating down. This "July snow" is asbestos from the high-rise construction. Even though the pollution of air may be reduced by changing the method of application from spraying to pasting, there is still the more serious problem of large-scale asbestos contamination of air from demolishing old buildings. When the wrecking ball knocks down the wall, asbestos filaments burst into the air. Some foreign textile manufacturers use asbestos to make fire-resistant fabrics, which have lower import duties. Shaking a coat made of such fabric will fill the air with asbestos dust.

Beryllium is a toxic metal and may cause lung disease. It becomes an air contaminant in the processes of mining, processing, and manufacturing. One of the uses of this light metal is in the beryllium rocket engine. Mercury is known as a water pollutant. When it builds up in the human body, it can cause brain damage. The EPA is also considering the addition of other toxic substances, such as cadmium and arsenic, to the list of pollutants.

Climate and Radiation Effects of Air Pollution. Another aspect of air pollution, its long-range effect on global climate, may be mentioned here. It is a common experience that on a sunny summer day the temperature inside an automobile, if all the windows are closed, can be very much higher than outside. The reason is that the glass windows are transparent to sunlight, so energy from the sun can enter the car; but they are opaque to heat waves, so after the sunlight is absorbed by the upholstery inside and converted into heat, the heat cannot leave the interior of the car. As a result, the temperature inside becomes much higher. The same condition is found in a glass greenhouse and is referred to as the greenhouse effect.

On a global scale a similar effect occurs because of the carbon dioxide in the atmosphere. The CO_2 molecule does not absorb short-wave radiation from the sun and lets it through to the surface of the earth. Once converted into heat, the long-wave radiation is absorbed by CO_2 and does not leave the earth quickly. Therefore an increased concentration of CO_2 in the atmosphere may have a worldwide warming effect on the climate. This might have serious consequences, because it would only take a slight imbalance in heat flow (Chapter 24), an increase of a few degrees in temperature, to melt the polar ice cap over a period of time, albeit quite long. If the ice cap melts completely, the ocean water level will be raised by 400 ft and many of the world's population centers will be inundated. Before 1900, the CO_2 concentration in the atmosphere was 0.029 per cent. At present it has risen to 0.033 per cent because of the increased use of fossil fuel. However, the dust particles added to the air by polluters can scatter short-wave radiation from the sun back into outer space and thus can cool the earth. Also, dust particles act as nuclei of condensation of water vapor. An increase in dust particles means an increase in fog and low clouds, which reflect part of the sun's radiation back to outer space. It has been estimated that an increase in low cloud coverage from the present 31 per cent of the earth's surface to 36 per cent would decrease the earth's surface temperature by 4°C, which is enough to cause the advance of the glaciers and to bring forth an ice age.

Environmentalists, who oppose the supersonic transport (SST, see Fig. 2.6), argue that the combustion of jet fuel in the engines produces a large amount of water that would be dumped into the stratosphere, where the air circulates very slowly. This pollutant might have a warming effect on global climate because water vapor absorbs infrared radiation just as carbon dioxide does. Another hazard concerns the naturally present ozone (O_3) at the SST altitude. At an altitude of 15 miles, natural ozone (produced by sunlight) has a concentration of 6 ppm of air (at sea level, ozone concentration is less than 1 ppm). It is this ozone layer that shields the earth

Figure 2.6 Models of U.S. supersonic transport under study before the project was discontinued. [NASA photograph]

from a hazardous dose of ultraviolet rays of the sun and thus makes life on earth possible. But the water from jet exhaust will react with and reduce the amount of ozone. Furthermore, as will be discussed later, ozone can react with auto pollutants ($O_3 + NO \rightarrow NO_2 + O_2$, for example). The SST fuel pollutants, essentially the same as auto pollutants, therefore would react with natural ozone and might also reduce ozone concentration. Even a slight reduction in ozone might increase ultraviolet ray dosage at the earth's surface, increasing the rate of development of skin cancer. (According to one estimate, a fleet of several hundred SST's would cause 10,000 additional skin cancer cases yearly.) Advocates of the SST claim that its climatic and ultraviolet ray effects are negligible. Recently it was learned that the spray-can propellant freon has an effect on the ozone layer comparable to that of the SST's.

2.2 Water Pollution

At one time, American rivers were sparkling clean. They provided wholesome drinking water, offered decent recreational facilities, and maintained abundant riverine life. Now, many rivers, such as the Hudson River, which flows by New York City, have almost become open sewers (Figs. 2.7 and 2.8). Fish are dying or have already disappeared at places. The same is true in all industrial nations. The pollution of the Rhine River is comparable to that of the Hudson. Only in the extreme upper streams, as in

Liechtenstein, do we see the Rhine's natural beauty and purity preserved.

Supply and Use of Water. According to White,[3] the supply and use of water in the United States may be analyzed as follows. (The statistics are taken from the year 1958.) The volume of precipitation is 4,300 billion gallons (gal) per day, of which 3,040 billion is evaporated and 1,260 billion goes to the surface water system of rivers and lakes. We use water in the streams for hydropower, navigation, and recreation. In addition, we take 300 billion gal of water per day out of natural reserve for various uses, most of which comes from surface water (217.5 billion), part from groundwater storage (54 billion) and the ocean (27.8 billion). The use of this 300 billion gal per day is broken down as follows: irrigation (127.5 billion), electrical utilities (89.9 billion), industrial (56.4 billion), municipal (19.7 billion), and rural domestic (5.8 billion). After the water is used, 117.2 billion of the total goes back to the air through evaporation and the remaining 182.1 billion returns to the surface water system, carrying pollutants with it.

[3]G. F. White, *Strategies of American Water Management* (Ann Arbor, Mich.: University of Michigan Press, 1969).

Irrigation contributes little to the pollution of fresh water, except that the water may carry away with it some unused fertilizer and some of the mineral content of the soil. The electric utilities' contribution is largely thermal pollution, which will be discussed separately. Industry is a major polluter, emptying into the streams a large amount of inorganic and organic substances. Residences also contribute to pollution; their discharge, sewage, con-

Figure 2.7 Oil slicks on the Blue River, a tributary to the Missouri River. [U.S. Environmental Protection Agency]

tains mostly organic material. The metal industries (largely iron and steel) are the leading users of water, followed by chemical, paper, and petroleum industries and other lesser users in that order. The industrial users of water dump 18,000 million lb of solid waste into it per year, of which the sugar-refining industry is the leader (5,000 million), followed by the metals (4,700 million), paper (3,000 million), chemical (1,900 million), and other lesser polluters in that order. All uses, agricultural, industrial, and residential together, amount to 1,600 gal per person per day, of which less than one tenth is domestic.

Sewage and Water Pollution. The pollution of water by sewage discharge (Fig. 2.9) is an old problem. Villages built by a river pour sewage into it and let it carry the waste away. In the old days when villages were small, this was not too much of a problem. The stream "purified" itself in a few miles. What happens in this self-purification process is that the organic waste materials are broken down by bacteria with the help of oxygen from the air. The organic products are then absorbed by riverborne plants, and the bacteria are eaten by animals in a food chain ending in the riverine fish. The water is thus returned to its original pure state.

As population increases, villages grow into cities. The amount of sewage increases, but the river is the same. It takes a longer time and thus a longer distance downstream for the water to purify itself. As city population keeps on increasing, it is often the case that before the river has a chance to cleanse itself, it comes upon another city, which finds itself in the unpleasant position of having to deal with polluted water for its own use. This city in turn adds its own sewage to the river, so that in passing one metropolitan area after another the river never has a chance to purify itself and simply becomes an open sewer for the cities. A similar situation arises when cities situated by a lake pour sewage into it.

When the undesirable effects of direct sewage disposal were recognized, sewage treatment plants began to be installed. During sewage treatment, first solid waste is removed by settling. The liquid is then

Figure 2.8 Oil-coated ducks are piled up on the shores of the Mississippi River, near Spring Lake, Minnesota, victims in 1963 of the petroleum- and soybean-oil-polluted waters there. These, collected in 20 minutes, were among the many hundreds of ducks killed by the oil. [U.S. Environmental Protection Agency]

Figure 2.9 Discharge from sewage treatment plant into South Platte River, Denver, Colorado, May, 1972. [U.S. Environmental Protection Agency, Project DOCUMERICA. Photo by Bruce McAllister]

subjected to a process essentially the same as the natural process, except that it takes place in a confined volume at an accelerated rate (with air being pumped through it). Bacteria are killed by the addition of chlorine. After another settling stage to remove the sludge, the clear liquid is released into rivers or lakes. This is the general procedure carried out in a typical two-stage treatment plant (Figs. 2.10 to 2.12).

Eutrophication. Sewage treatment does not completely solve the environmental problem of sewage disposal. The treated sewage still contains a large amount of nutrients. When released into rivers and lakes, these nutrients upset the ecological balance and give rise to other environmental problems. For instance, they cause the excessive growth of algae. The large amount of algae may deplete the oxygen content of water (at night) and thus kill fish. Masses

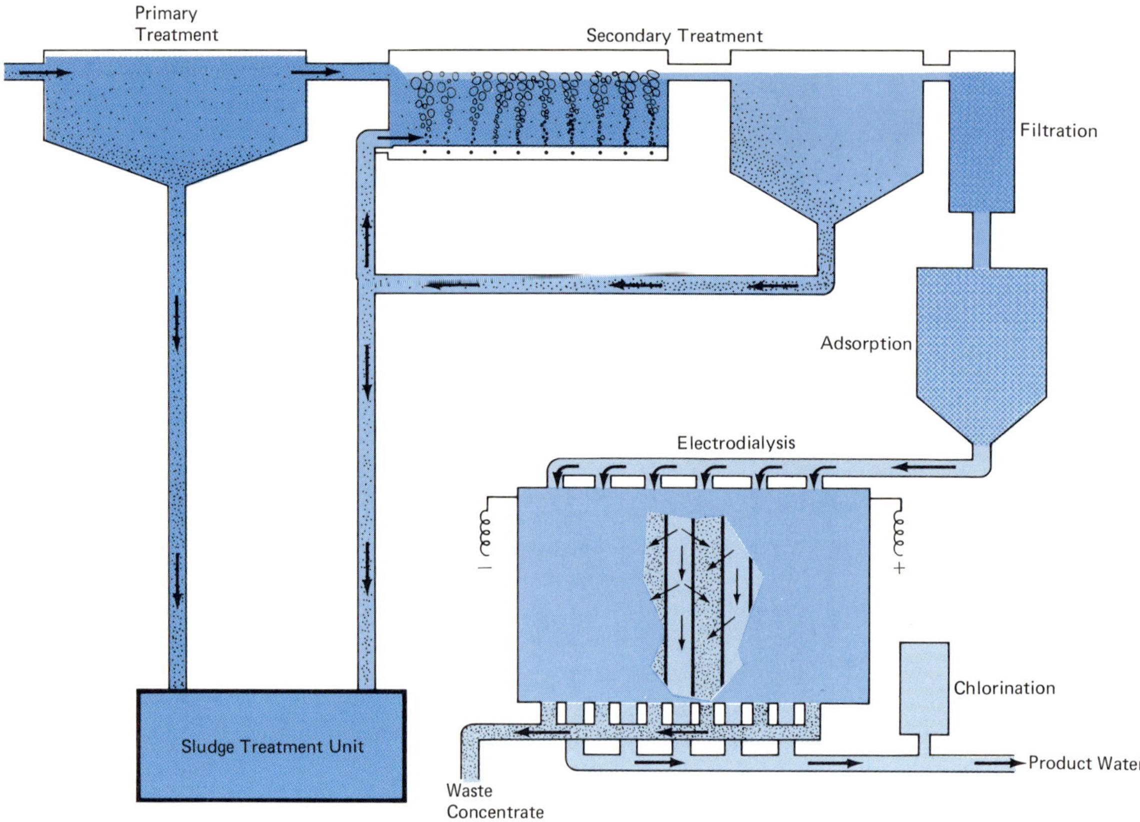

Figure 2.10 Schematic representation of a waste water reclamation plant. (1) Primary treatment. Removal of suspended solids in preliminary or primary settling tanks. (2) Secondary treatment. Removal of biodegradable organics by the activated sludge process. (3) Filtration. (4) Adsorption. (5) Electrodialysis. (6) Chlorination. [U.S. Environmental Protection Agency]

Figure 2.11 The aeration tanks of the Akron Sewage Treatment Plant, part of the secondary sewage treatment process. [U.S. Environmental Protection Agency]

Figure 2.12 Philadelphia's Southwest Sewage Treatment Plant. Shown in the photo are four primary settling tanks and eight circular sludge-digestion (by anaerobic bacteria) tanks. [U.S. Environmental Protection Agency]

of decaying algae are just as unsightly and unpleasant as raw sewage (Fig. 2.13). The increased amount of microscopic algae causes the water to appear cloudy, making it undesirable for drinking and recreation. The problem is particularly serious in lakes. Many lakes in Central Europe, such as the one in Zurich, have gone through these changes (*eutrophication*) and are undergoing "decay." Lake Washington, next to the city of Seattle, is another example, but the trend is being reversed by diverting the sewage into the ocean. Eutrophication takes place naturally but slowly; pollution greatly accelerates it.

The nutrients in sewage (largely from human waste) amount to 1.4 lb of phosphorus and 11 lb of nitrogen per person per year. Detergents add to the phosphate content of the sewage and account for about half of it. The resulting environmental prob-

Figure 2.13 The ugly scene along a bank of Lake Minnetonka, north of Minneapolis, showing dead and dying algae which prohibits any life in these waters. [U.S. Environmental Protection Agency]

lem may be reduced by controlling these nutrients, of which phosphate is the least difficult to regulate. It may be removed chemically in another stage of sewage treatment. It may also be reduced by eliminating the use of phosphate-containing detergents. That phosphate detergents caused excessive algae growth has been established. However, recently (1971) the Surgeon General of the United States advised American housewives to continue to use phosphate detergents because the available substitutes give rise to worse problems.

Industrial Pollutants. In addition to dumping solid waste, pollution of water by industry may involve dumping soluble toxic material. Mercury is an example of such a toxic material and has been recognized as a major threat to health. Many mercury compounds, organic or inorganic, are watersoluble and can be widely dispersed in the water system. Upon entering the human body, mercury compounds tend to concentrate in the nervous system, causing serious or lethal impairment of the brain. Its use in industry has greatly increased during the past 20 years. In the plastic industry it is used as a catalyst. In the pulp industry it is used as a fungicide. In electrochemical industries metallic mercury is used for electrodes. Much of the mercury put to this use is discarded in the Great Lakes. The fish from Lake Erie contain such an excessive amount of mercury (1.3 ppm, far in excess of the legal limit of 0.5 ppm) that their sale has been banned. Many other toxic materials have the property of concentrating in the bodies of certain animals. An outstanding example is DDT, which plagued many animal species on land and in water. Coho salmon from the Great Lakes were banned from sale because they were found to contain large concentrations of DDT.

Fish caught in the Hudson River near a battery factory are found to contain about 10 ppm of cadmium, 1,000 times more than normal. The cadmium-contaminated fish seem unfit to eat, though the Food and Drug Administration has not set any guidelines on cadmium pollution. The factory makes nickel-cadmium batteries and dumps cadmium- and nickel-containing waste into the Hudson River. High cadmium levels in the human body may cause liver damage, high blood pressure, hardening of the arteries, and so on. Cadmium poisoning is relatively rare in the United States, but it struck in Japan during the 1950's when more than 100 persons died after eating rice that had been irrigated by cadmium-polluted water from the Jintsu River.

The cost for cleaning up the polluted waters may be high. It is estimated that $8 billion is needed just to provide adequate sewer treatment facilities of the kind currently in use. The total cleanup cost may run up to $100 billion. Society as a whole will have to bear the burden. The rehabilitation of the environment will be one of the major social problems in the near future.

Agricultural Use of Water. In addition to pollution, there are other environmental problems related to the use of water by man. Irrigation (Fig. 2.14) provides one example. If enough water can be found for irrigation, the arid land, including most of the American west, may become fertile and produce tremendous wealth. This is exactly what happened in Colorado when Rocky Mountain water was diverted for use in the plains. However, all the mountain water had soon been allocated; the search for additional sources turned to the underground water supply, and deep wells were drilled. This brought further wealth but also gave rise to new problems. The underground water table was lowered, resulting in drought for the area. Alkali content of the surface soil increased, making the land unsuitable for agriculture. As a result, measures had to be taken to conserve water use to maintain the underground water level and to remove the alkali content of the soil.

Municipal Problems. The use of underground water may lead to other environmental problems; for example, the city of Tokyo is sinking slowly.[4] The downtown area (Ginza and Marunouchi) is sinking at the rate of 1 in. per year. In the reclaimed land area (Fukagana), the sinking rate can be as high as 8 in. per year. The sinking rate has doubled since World War II. Bridges have sunk so low that ships can no longer pass under them. The main cause to which this sinking is attributed is the large-scale use of underground water by industry (as a coolant). Over a million tons of water are pumped out every day. This large outlet upsets the balance of underground water. As the reserve of this water diminishes, the ground above "caves in" to fill up the evacuated space. The same problem besets the city of Venice, which is perhaps the city that has the worst environmental problems. If the current rate of sinking continues, Venice may become uninhabitable in less than half a century. A recently proposed program to save Venice calls for sealing the artesian wells. Again, the use of underground water is considered to be the main cause of the sinking. Another big city, Shanghai, suffers the same problem. Because the withdrawal of underground water is the main cause of sinking, an attempt to stop the trend is made by pumping water back into the underground water system. A report from Shanghai claims the measure has been successful.

Another environmental problem related to water

[4] J. P. Sterba, *The New York Times,* September 26, 1971.

Figure 2.14 Irrigation on a Montana farm is accomplished by use of a ditch and the siphoning of water into the rows of crops. [U.S. Environmental Protection Agency]

Figure 2.15 A burning dump. [U.S. Environmental Protection Agency]

and caused by the upset of the natural balance of water may be mentioned here. Long Island, New York, used to have unsalted underground water except along the coastline. In recent years the construction of residential and commercial buildings, superhighways, and parking lots caused a sizable portion of the surface area to be covered with asphalt. As a result, rainwater can no longer seep into the soil to add to the underground water reserve. Instead it is carried away by drains to the ocean. The decrease of freshwater input upsets the balance of underground water, and salt water from the ocean begins to seep in. The saltwater line is now advancing inland.

2.3 Land Pollution

Solid waste, unlike the gaseous and liquid wastes that can be freely distributed in the environment, can usually be confined in a limited physical locale and therefore can be controlled better than the other wastes. But just the same, solid waste contributes to the pollution of the environment. The waste dumps, junkyards, and scrap heaps dotting the countryside are eyesores (Fig. 2.15). Landscape pollution is just as undesirable as atmosphere pollution and water pollution. Although the health hazards can be limited, there are other long-range problems that will be discussed later in connection with the problems of solid waste production and disposal (Chapter 20).

2.4 Thermal Pollution

Water is the most common coolant used to remove unwanted heat. The unwanted heat is not destroyed by water; it is simply absorbed by water and flushed away with it (conservation of energy). The used coolant water eventually returns to the rivers or lakes and raises their temperatures. The Hanford Nuclear Plant uses Columbia River water as a coolant and has raised the temperature of the river above normal. Water from the Mahoning River in Ohio is used as a coolant by various industries to such an extent that its temperature in the area has been raised to 140°F. Thus the natural environment is significantly changed by man-made processes, and the change will lead to undesirable consequences. The current scale of heat dumping is so large that the natural environment in this respect can no longer be considered infinite and we are confronted with the new problem of thermal pollution.

Most of our present cooling needs come from the use of heat engines. A heat engine converts heat energy into mechanical energy which may be used directly in factories. It may be used indirectly in homes and plants by first converting it into electrical energy in electric power generating plants. But only a fraction of the total heat energy Q_1 supplied is converted into mechanical energy (Sec. 16.3); the rest becomes unwanted heat Q_2, which must be removed by coolants. In a steam engine this removal occurs when the spent steam (dead steam) passes through the condenser and condenses into water, releasing the latent heat and transferring it to the coolant water passing through the condenser. The ratio of Q_2 to Q_1, according to the Carnot theorem (Sec. 16.3), equals at least the ratio of the condenser temperature T_2 to the boiler temperature T_1 on the absolute scale,[5] which is a sizable fraction. For a steam power plant the ratio is about 65 per cent; thus out of one unit of heat energy supplied by burning fuel, only 35 per cent changes into electrical energy and the rest, 65 per cent, becomes waste heat to be thrown away. For a nuclear power plant this ratio Q_2/Q_1 is even higher—75 per cent. Electrical power accounts for 20 per cent of the total energy needs of the nation [10 kilowatts (kw) per person], and its consumption increases about 8 per cent per year.

We can calculate how much waste heat must be disposed of each year. Because the heat is dumped into the river system, we can predict that in a few decades, if the present rate of increase were to continue, all major rivers in the United States would be heated to the boiling point. Before such an absurdity could happen, even a small fractional increase of the river temperature would give rise to ill effects. As the temperature increases, the amount of oxygen dissolved in the water decreases. This is undesirable, because oxygen not only is needed by fish but also plays an important role in breaking down the organic waste in water. Furthermore, the power plant dumps heat in the river at an uneven rate. It operates at peak power in daytime, dumping a large amount of heat and making the water warm, and at night it operates at a fraction of peak power, which allows the water temperature to drop. It may even shut down periodically. This creates drastic temperature variations over short periods of time, so that neither cold-water fish nor warm-water fish can survive. Furthermore, the hot water coming out of the plant may flow on top of the cold water, covering it like a blanket, with very slow mixing, because hot water has less density. This prevents the cold water below from obtaining an oxygen supply from the air. Finally, the life cycle of some fish is controlled by seasonal temperature changes, and the artificial temperature change disrupts this cycle—causing spawning at the wrong time, for example.

[5]Temperature on the absolute scale has a value equal to the temperature on the centigrade scale (more properly, Celsius scale) plus 273. See Chapter 6 for further discussion.

The whole ecology of the river is upset by artificial temperature changes.

Future power plants will have to find other ways of removing unwanted heat. Three methods have been developed that seem adequate for the short-term power needs: wet cooling towers, dry cooling towers, and cooling ponds. The cooling towers (Figs. 2.16 and 2.17) remove heat from the used coolant water and dump it into the air, either by vaporization (wet) or by radiation (dry). The cooling pond also dumps the waste heat in the air.

Heat Balance of the Earth. The earth's surface receives energy input largely from the radiation of the sun; the other sources [e.g., geothermal energy and gravitational (tidal) energy] account for only 0.02 per cent of the total. Of the total solar radiation, 1.73×10^{17} watts, 30 per cent in short-wave radiation is directly reflected or scattered back into outer space. This forms the *albedo,* which is important in other discussions later. The remainder is absorbed by the earth's surface, either changing into heat of the land, water and air (47 per cent) or into energy of

Figure 2.16 A new cooling tower being built by the Bethlehem Steel Corporation plant on the Lehigh River. [U.S. Environmental Protection Agency]

Figure 2.17 Two 372-ft natural draft cooling towers and the reactor and turbine buildings of the first unit (831 megawatts) of the Metropolitan Edison Co.'s Three Mile Island Nuclear Station on the Susquehanna River, 10 miles south of Harrisburg, Pennsylvania. [U.S. Environmental Protection Agency]

the hydraulic cycle (23 per cent). Only 0.21 per cent is converted into the kinetic energy of the ocean (waves) and the air (wind). A still smaller fraction, 0.023 per cent, is captured by the green plants. But all the absorbed energies other than heat are eventually converted into heat (according to the second law of thermodynamics; see Sec. 16.3). The heat of the earth's surface gives rise to heat radiation, and energy is lost to outer space by long-wave radiation. The fact that the surface of the earth is maintained at a constant (average) temperature implies that there is a balance of heat flow (i.e., the energy absorbed from the sun equals that lost by heat radiation). Otherwise, there would be a net accumulation or loss of heat. This would mean that the surface temperature of the earth would rise or fall until the heat radiation losses, rising or falling with the surface temperature (according to the Stefan–Boltzmann law; see Sec. 13.2), caught up with the energy input and established equilibrium again.

The heat balance of the surface of the earth is delicate and is complicated by the existence of a large mass of water. At the present time, a sizable portion of the water exists in the form of ice in the polar ice cap. If the earth receives more energy than it loses by radiation, the excess energy will first cause the melting of the ice cap. As already mentioned, if the cap completely melts, the ocean level will rise 400 ft, flooding all coastal cities. On the other hand, if the earth receives less energy than it loses by radiation, then more water will be converted into ice, the glaciers will advance southward, and a new ice age will be initiated. Both eventualities would be disastrous to humans, and it would take only a change of a few degrees centigrade of the earth's surface (average) temperature to bring either about. Therefore we must make sure not to upset the natural balance of heat flow one way or the other.

In later chapters we will learn that all energy used eventually turns into heat. With the increased consumption of industrial energy, we are adding a new source of heat to the earth's surface. By burning fossil fuels we are putting back into the earth's surface within a matter of a few hundred years the solar energy trapped by green plants over hundreds of million years. The amount of heat addition at the present is still too small to upset the heat balance of the earth. But if other sources of energy (e.g., nuclear) are used in large amounts so that energy consumption keeps on doubling every 17 years, then it will take only 80 years for the energy consumption to increase 25 times and raise this artificial heat source to the earth's surface to 0.2 w/m^2. This level, according to Budyko,[6] is almost great enough to cause the melting of the polar ice cap. Thus the problem of thermal pollution by power plants is not merely one affecting the fish. It poses a serious problem for the future of mankind.

The heating of the earth's surface is affected by the change of the albedo. According to Manabe and Wetherald,[7] a 1 per cent increase in albedo will cause a 1°C decrease in the average surface temperature of the earth. Paved surfaces (urban areas, highways) reflect more sunlight than natural surfaces, and thus correspond to a higher albedo. The long-range global effect of man's modification of the earth's surface by construction, farming, and so on, has yet to be investigated. Moreover, the pollution of the ocean by oil, which leaves a thin film over the water's surface, may affect the heat exchange between ocean and atmosphere and may have a global effect on the climate. This is another aspect that should be investigated carefully in the future.

[6] M. Budyko, *Meteorologiya i Gidrologiya,* No. 11 (1967).

[7] S. Manabe and R. T. Wetherald, *J. Atmos. Sci.,* **24,** 241 (1967).

2.5 Noise Pollution

Any mechanical operation produces sound classified as noise, a form of energy wasted in any engineering process. The excessive noise constitutes a form of pollution of the environment. Noise is distributed throughout air space by a wave motion, and the only way to control it is to eliminate it at the source.

In the cities, where a large amount of human activity takes place, noise builds to a high level. Noise has long been considered an undesirable feature of city life. With the increase of population and industrialization, the noise level also increases and becomes a serious environmental problem. Moreover, more sophisticated technology usually involves the use of more powerful machinery, which emits a correspondingly higher level of noise. An outstanding example is provided by the supersonic transport (SST) plane. It is noisier than the ordinary jet plane on take-off and landing; the noise pollution that it creates around airfields is a major reason against its

Figure 2.18 Sonic boom produced by supersonic flight (right) compared with subsonic aircraft noise (left).

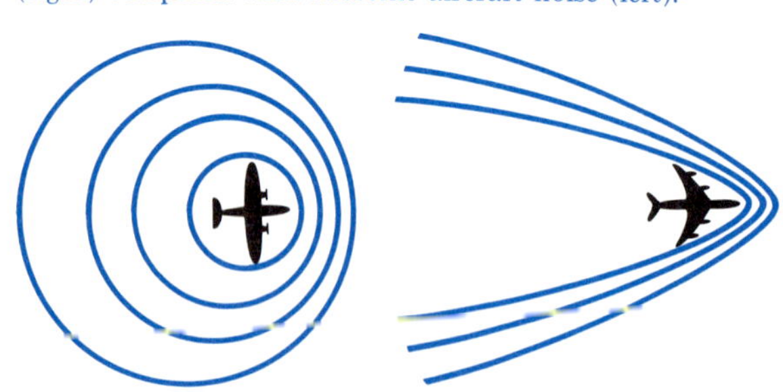

development. Furthermore, the noise pollution on flights is even more serious. Anything moving with a speed higher than that of sound produces a sonic boom—a sudden big noise with mechanical force powerful enough to shatter windowpanes. The origin of the sonic boom (Fig. 2.18) may be illustrated by the *wake* behind a speed boat on a placid lake. Sound energy is concentrated on the V-shaped *shock front*. When the front sweeps by an observer, he experiences a sonic boom. It is not only disturbing to the ear and nerves, but also likely to cause property damage. This is the major reason for the proposed ban on overland flights by the SST.

2.6 Radiation Pollution

When we speak of radiation pollution, we usually refer to ionizing radiation, which includes radiations from the nucleus (i.e., α-rays, β-rays, and γ-rays) and "hard" radiation from the atom (i.e., X rays). In a sense, the term includes both material pollution (α-rays and β-rays) and electromagnetic energy pollution (γ-rays and X rays). They are considered together because both are penetrating and both produce ionization in matter (e.g., in air and in the human body); it is ionization in body tissue that causes biological damage. In this sense nearly all nuclear radiations are hazardous, whereas among the atomic radiations only X rays are. Atoms may radiate electromagnetic energy in a wide range of wavelengths, from the long-wave infrared rays to visible light rays, ultraviolet rays, and short-wave X rays. Except for X rays, these atomic radiations do not penetrate the body tissue, but they might be hazardous in other ways, posing the dangers of sunburn, ultraviolet-induced skin cancer, microwave damage from leaks in microwave ovens, and so on.

Exposure and Dosage. The amount of *radiation exposure* (for a unit volume of material or body tissue integrated over the time of exposure) is measured by the old unit *roentgen* (that exposure producing ions of 1 electrostatic unit (esu) positive and negative charges in 1 cc of air under standard conditions). The production of ions requires energy, and a new unit, the *rad*, is used to measure *radiation dosage* or energy deposition by ionizing radiation (that dosage depositing 100 ergs of energy in 1 g of body tissue). The health hazard is gauged by the dosage. One roentgen exposure gives rise to approximately 1 rad of dosage.

Yet the same amount of energy deposition from different kinds of radiation may produce different physiological effects. The same amount of energy delivered by X rays produces fewer ion pairs than α-particles do. It is ionization that is mainly responsible for the *somatic damage* of the body tissue, because it disrupts chemical bonds between atoms in the molecules that constitute the body tissue. When the bonds are reformed, atoms may be connected in a different pattern leading to a different molecule, and the original function thus may be lost. As a result, for the same amount of energy deposition the α-particle is more harmful than the X ray. Therefore a new dosage unit called *rem* is introduced to measure the degree of harmfulness in man; it is defined by the product of the rad with a factor called relative biological effectiveness (RBE). The factor RBE for X rays is set equal to unity as a reference standard. The value of RBE for fast neutrons is around 10 and that for 1 MeV α-particle is around 25. A smaller unit, *millirem* (mrem), is defined as one one-thousandth of a rem.

To give a sense of the magnitude of the unit, one may note the dosage we receive from natural background radiation is from 90 to 200 mrem a year. (All three units—roentgen, rad, and rem—refer to totals added up over a period of time; they are not time rates.) This comes partly from cosmic rays and partly from rays of the radioactive elements in the earth's crust, the two amounts being comparable to each other. The amount varies with altitude, latitude, and local geology. In addition, there is a small fraction of 24 mrem per year emitted by the radioactive atoms, mainly ^{40}K, inside the human body itself.

Radiation Safety. For mammals, the mean lethal dosage is 400 rem. (Half of the population will die shortly after being given a dosage of 400 rem.) Whether the 400 rem is administered all at once or spread out over a short period of time does not matter. It is the total amount that matters. However, this does not apply when the radiation is spread over a very long period of time, for then the body has time to repair radiation damage continuously to offset the small amount of damage per unit time. In a short period the repair mechanism is overwhelmed and death results.

Large doses of radiation destroy bone marrow, leading to leukemia and death. Very small doses of radiation, even without somatic damage, may produce long-term *genetic damage*. The genetic information is coded in the macromolecule DNA (Sec. 10.7) composed of a string of small molecular units, i.e., nucleotides, of which there are four kinds with different bases (adenine, guanine, cytosine, and thymine). The serial sequence of the four kinds of nucleotides form a telegraphlike message that conveys the genetic information. Ionizing radiation may cause changes of molecular structure (such as formation of thymine dimers) so that the genetic information is garbled, resulting in a deleterious mutation.

Therefore the practical safe dosage level is set far below the level of physiological safety for genetic protection, and for that dosage only the exposure of the gonad cells need be an object of concern. The Federal Radiation Council set a guideline of 5 rem per generation (30 years), or about 170 mrem per year, as the maximum permissible gonadal dose for the general public from man-made radiation, except for medical irradiation. This limit corresponds roughly to a doubling of the natural background radiation level.

Concerning radiation hazards, it is well to remember that nature itself pollutes the environment with radiation. Perhaps 5 to 12 per cent of all spontaneous genetic mutations are caused by natural background radiation. Some stillbirths and forms of cancer may also be attributed to it. Because of its high altitude (higher cosmic ray intensity) and uranium deposits (higher radioactivity in the earth's crust), Colorado—including Denver—has a radiation background twice as high as the background at sea level. But this was the environment of the area before man moved in. It is just another of the natural hazards, like earthquakes and hurricanes. Because people live in Denver peacefully and vacation in Colorado happily, it appears that a doubling of the natural background radiation from that found at sea level, where most of the population resides, is not particularly hazardous.

Exactly what a safe radiation level is has long been a subject of controversy. It was hotly debated in connection with the banning of atomic bomb tests in the atmosphere. Nuclear fallout causes genetic radiation damage, including such abnormalities as stillbirth and defective offspring. Unlike the other hazardous agents, there is no *threshold* of danger in the amount of radiation below which no damage will be wrought. A small dose of a chemical poison below the danger threshold will do no harm. The minute amount taken into the body will be excreted without damage to the body. This is not so with ionizing radiation. When the radiation level is reduced, only the probability of causing damage is reduced; the possibility cannot be eliminated completely. Thus for the population as a whole, as the radiation level is reduced, the radiation-related injuries (number of stillbirths and deformities) decline but do not completely disappear.

Radiation from Nuclear Tests. In the case of global fallout from nuclear tests, the increased radiation level is smaller than that of natural background radiation and is not as excessive as the natural overdose in Denver. The amount in the peak testing years is estimated to have been 30 mrem per year. Thus according to one school of opinion, the hazard is small and negligible. On the other hand, the increased number of stillborn and deformed infants that can be attributed statistically to the increased radiation level caused by fallout is still in the thousands, even though the percentage is small.

Thousands of such genetic abnormalities are a great many, and the hazard, according to another school of opinion, is large and intolerable. The difference between *large* and *small* is necessarily subjective, and the controversy is one beyond the scope of science. Yet it was clear that endless airborne tests were definitely harmful, and the test ban was finally agreed upon. Since that time the radiation level resulting from fallout from nuclear tests has diminished greatly. (Incidentally, if all the existing nuclear weapons were exploded, even in remote unpopulated spaces, the resulting radiation pollution would be enough to make the earth uninhabitable.)

Radiation Exposure in SST's. One of the environmental considerations of the supersonic transport is the hazard of natural radiation from cosmic rays at the SST flight altitude of 60,000 to 80,000 ft. Though the normal cosmic ray intensity is much greater at this height, the exposure is still within the safety limit. However, abnormal events in the heavens may increase cosmic ray intensity temporarily. In a solar flare, the sun suddenly spews out a large stream of particles, mostly protons. The solar flare on Washington's birthday in 1956 was so large that one hour at the SST elevation over the polar region would have inflicted a dose of 40 rem—much greater than permissible (40 rem is one tenth of the mean lethal dose). Although no death would have resulted, somatic and genetic damage undoubtedly would have been inflicted. Solar flares can be monitored, and it took hours for the stream from the sun to reach earth. Future SST operations (the French, British, and Russians now have SST's) will have to be safeguarded from such flares by cosmic weather reports on atomic storms.

Radiation from Uranium. Because uranium is involved in atomic energy, the radiation hazard of uranium mining is a subject of concern. In Central Europe in the late 1930's, half of all pitchblende miners died of lung cancer. The radium extracted from pitchblende was used partly to make luminous watch dials. Most of the early painters of these luminous dials also died of cancer. Uranium mining waste (called *tailings*) contains radium, which is washed into river systems by rainwater, resulting in radium pollution of water. Uranium tailings have been used in Colorado landfills over which many houses and buildings were constructed. Recently it was discovered that these houses and buildings are exposed to a high radiation level that is hazardous to health.

X-Ray Hazards. X rays are used in medical practice both for diagnostic and therapeutic purposes. Chest X rays and dental X rays belong to the former category; breast cancer in chest X-ray patients and finger cancer in dentists have been related to these uses. The medical profession now is much more restrictive on the use of X rays. The average dose an American receives from medical irradiation is about 60 mrem a year. A chest X ray gives a dosage of 20 to 500 mrem.

A radiation problem involved in most homes results from the X rays emitted by television sets. The TV tube is essentially a cathode ray tube. When electrons are stopped on the screen, X rays are emitted. Color television uses higher voltage and thus emits more and harder X rays. A 1959 estimate[8] gives an average annual dosage of a typical viewer of 10 mrem at a distance of 200 cm. However, a 1971 report by the X radiation Ad Hoc Committee of the Electronic Industries Association claimed that the genetically significant dose resulting from color television viewing is 0.5 mrem per year.[9] Another minor contribution to radiation pollution originating from human trivialities comes from those luminous watch dials that make use of nuclear radiation (not all luminous dials do).

Radiation Hazards of Nuclear Reactors. Nuclear reactors yield effluent containing some radioactive material. As the nuclear reactor industry grows, possible radioactive pollution will have to be

[8] Braestrup and Mooney, *Science,* **130** (1959), 1071.

[9] *Radiological Health Data and Reports,* **12,** 7 (July 1971), 363 (published by the Environmental Protection Agency).

controlled carefully. Radioactive atoms, once they have escaped into the air, cannot be reconfined. They will last as long as their lifetimes, and there is no way to destroy them. The Atomic Energy Commission has recently set up a more stringent standard for nuclear power plants, requiring the exposure limit of an individual living near the plant to be less than 5 mrem a year over and above natural background radiation. Most existing reactors operate at a level below the new standard. An ordinary chest X ray delivers more dosage than this yearly limit. If this limit is observed, and if the reactor is accident-proof, living near a nuclear power plant at sea level is far less hazardous than living in Colorado (100 mrem per year over and above natural background radiation at sea level). See Section 17.3 for further discussion.

Problems

2.1 Discuss the economic and environmental impacts of the following modern practices: (a) keeping the lawn neat and green; (b) substituting bottle feeding for breast feeding; (c) using the flush toilet; (d) using the aluminum can for soft drinks.

2.2 What is common among the practices in the preceding problem? What is the verdict on these actions from a historical and humanistic point of view?

2.3 Coal made into a fine powder can be burned as a gas. Discuss the environmental impact of this mode of energy utilization.

2.4 What are the differences for heating a house by hot-air gas heating, by hot-water base-board radiation, by electric heating, and by heat pumping, from the energy and environmental points of view?

2.5 Assume the nation spends $5 billion a year for coal to generate electricity by steam power plants. How much money goes into the waste heat flushed away, which not only is unused but also causes thermal pollution?

2.6 The abundance of food we now have is largely due to the use of artificial fertilizers, yet nitrogen fertilizer, after a series of chemical changes, ultimately produces nitric oxide, NO. What environmental effect will it lead to?

3 The Energy Crisis

3.1 The Energy Crisis Begins

We have introduced the problem of resource depletion. Natural resources, mostly mineral, may be divided into two classes: material resources (metals, nonmetals) and energy resources (gas, oil, and coal). Throughout history the demand on natural resources was small compared to natural reserves. In recent years the demand has increased drastically because of the advance of industrialization. The high rate of consumption depletes the natural reserve rapidly. In the past, because of the low rate of consumption, we tended to think of a natural resource as inexhaustible. As a result we used the resources lavishly and wastefully. This, obviously, cannot go on indefinitely. Scientists sounded the alarm of the impending crisis of depletion years ago, yet no attention was paid to the warnings and no action was taken. Every part of the economy operated as usual. Consumption was as lavish and as wasteful as ever. Then, starting in late 1972, shortage of fuels began to hit various parts of our society and suddenly an energy crisis appeared. The Arab oil embargo further aggravated the problem.

How could a land of abundance, a nation of advanced technology, find itself in such a crisis of scarcity? The major oil companies claim that the rate of consumption increased much more rapidly than they anticipated and that the expansion of refinery facilities was hindered by environmentalist opposition. A gap of a few per cent between production and consumption is all that is needed to produce *spot shortages*. On the other hand, civic groups, government agencies, and particularly the independents in the oil industry charged that the big companies colluded to bring about the shortage to drive the prices up for a higher profit and to drive the independents out of business. What is the truth?

American motorists currently burn 47 million barrels of gasoline a week, whereas refineries produce only 43 million barrels a week. The increased consumption is due to many factors. More cars are on the roads; the number of motor vehicles has now reached 100 million. Cars are bigger, with more powerful engines, which consume more fuel. In addition, the car has more power equipment—automatic transmission, power steering, air conditioning, and pollution-control devices, all consuming energy. Much of this equipment has gradually become standard. In addition to automotive needs, oil consumption has increased because coal and gas users are switching to oil. The coal users switch because of environmental restrictions. Because of air pollution many municipalities prohibit the burning of coal with high sulfur content. The gas users do so because of a natural gas shortage. All such factors put an increased burden on the oil industry.

Increase in demand has been going on for years and the oil industry has been able to expand to keep up with the demand. Why did the production lag behind this time? Domestic crude oil production is at full capacity. The United States (excluding Hawaii and Alaska) has been heavily explored and new fields are desperately needed. The oil on the continental shelf is one new source. Yet its exploitation is hindered by environmental considerations. The operation in the Santa Barbara Channel resulted in a major oil spill (Fig. 3.1), spoiling the beautiful beaches of Santa Barbara. The public outcry against the industry stopped further operation in the channel.

The discovery in 1968 of the North Slope field in Alaska, one of the greatest in America, with a reserve of 10 billion barrels was a major event. But construction of the pipeline needed to transport the oil was stopped by environmentalists for four years. They claim the pipeline will destroy the tundra and strand the caribou. They are afraid the pipeline, crossing the earthquake belt, might break in a quake and cause an uncleanable spill. They want the safest possible way to build the pipeline and want to consider an alternate route through Canada. Finally they argue there is no need to overexploit the natural resources at such a fast rate for wasteful, extravagant consumption. Natural resources are irreplaceable. We should save some for future generations. Because a better environment is what everybody wants, these arguments gain easy acceptance until we are faced with a crisis. So far the new oil field has not helped crude oil production.

Not only is there a lack of crude oil production but there is a lack of refinery facilities. This perhaps is the worst bottleneck in the flow of oil to consumers. A typical 160,000-barrel-a-day refinery plant costs $250 million and takes three years to build. In 1969 five plants were in construction. This is the rate of yearly construction that is needed in the next seven years but in 1973 no plant was under construction. Again environmental opposition is a major reason. Everybody needs gasoline but no one wants a refinery in his neighborhood. It pollutes the environment and an oil spill may spoil the seashore. When such a large investment is made, one has to consider the possibility that when American crude oil is depleted the facility will have to rely on imports to continue operation and to recover cost. The import quota system in the past has made the access to import crude oil uncertain. This consideration makes oil companies hesitate to invest in refineries.

The imported crude oil from the Middle East is transported by large tankers that need a deep-sea port to anchor. The industry has found such a site in Maine in an economically depressed area where one segment of the local population is anxious to bring in the industry to revitalize the economy. Another sector of the population, particularly the retired people, opposes the project on environmental grounds. They do not want to change their life-style and want to keep the status quo. Without the Maine facility the industry is prepared to go to Nova Scotia, where the local population is anxious to welcome the new economic opportunity.

What about imports? The oil import quota system, which was abolished May 1, 1973, because of oil shortage, was adopted to discourage imports (to help domestic producers). The trade relation cannot be changed overnight. The whole world is in the process of being industrialized and oil is in great demand.

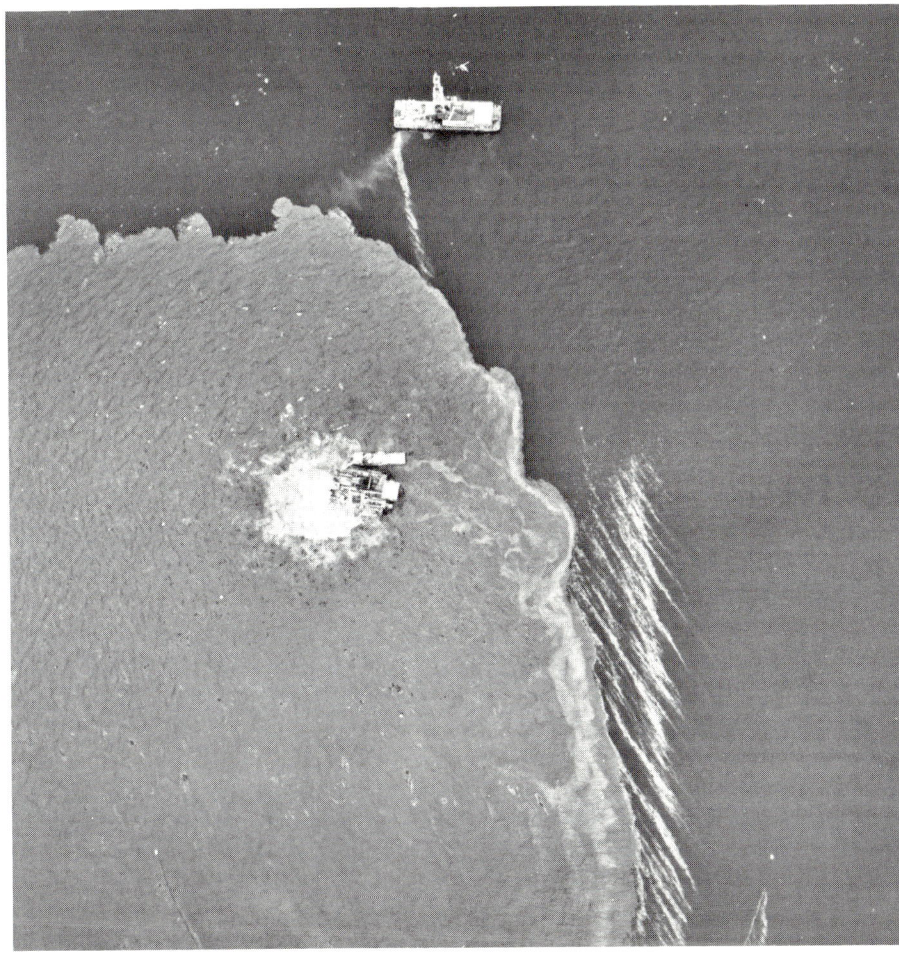

Figure 3.1 The great Santa Barbara oil spill, February, 1969. Photo covers an area 2,700 by 2,700 ft. [U.S. Environmental Protection Agency]

Production cannot be increased immediately. Moreover, there are political complications. The major oil suppliers overseas are the Middle East countries. Saudi Arabia leads all nations, with a reserve of 160 billion barrels. Kuwait follows, with a reserve of 130 billion barrels concentrated in a much smaller piece of land. Iran is third. The three together account for over half of the total world reserve. Saudi Arabia made it clear (manifested in the oil embargo) that it wants to use its oil power to influence American foreign policy regarding the Arab–Israeli conflict. Kuwait does not want to increase production, so that it has reserves for future years. The sheikdom already has a per capita income of $3,500 a year and has modern facilities. Money is no inducement. Iran likes to sell as much oil as possible, to help industrialize and modernize. Naturally it would sell to the highest bidder. Thus large-volume importing from overseas is not a simple business transaction. For national security reasons it is not desirable to be overly dependent on foreign supply for such a vital resource as oil. Already, of the 17 to 18 million barrels of oil consumed a day, only 11 million are produced domestically, out of a known reserve of 37 billion barrels. We are already heavily dependent on foreign oil. As far as refinery capacity is concerned, Europe has extra capacity and is in a position to help America.

The difference between supply and demand is still small. National gasoline stock normally is in the neighborhood of 220 million barrels. When lowered to 165 million barrels, spot shortage may occur. Because of fluctuation some stations may happen to sell more and become short in supply, whereas others may still have an adequate supply. A limitation of, say, 10 gal per customer at a time may help average out the fluctuation. However, when the gap between supply and demand becomes large, such limitation becomes meaningless and other measures, such as rationing, are necessary.

Are there conspiracy and collusion among the big oil companies to squeeze out the independents and raise the price? For an independent dealer squeezed out of business after many decades in the trade, the experience is bitter. His charge of conspiracy is only natural, particularly in view of the fact that the big oil companies are expanding their own "independent outlets." In the past, when oil supply was plentiful, the big companies had extra refinery facilities and so they sold the overproduction to the "independents" at a low price. The independents retail the products at a cut-rate price, providing feeble competition to the major brands. Over the years two gasoline markets developed: the major-brand, full-service stations (two- or three-stall service garage) and the minor-brand, no-service (pump and cash register) filling stations. As is the case with the department store and the discount store, each has its own market. Just as department stores are entering the discount store business under different names, the major brands are entering the second market under new names. When the supply becomes scarce, the "independents" are naturally the first to be shut out; they are actually the most miserable dependents.

The big oil companies laughed at the charge of collusion and price manipulation. They claim their return of 12 per cent (before 1973) is no better than that of other industries. (Oil profits in 1974 were much higher.) Yet when a few companies control nearly all the operations of an industry they must bear the blame for the shortcomings of the industry. The Federal Trade Commission filed charges against eight major companies for shutting out competition. The states of Florida and Connecticut have filed charges against a larger number of companies. The collusion and conspiracy issues are to be settled in court. Nevertheless, refinery facilities can be put to better use to increase output. Indeed, after a few months of shortage, the weekly production in the summer of 1973 was brought up to match the demand.

As if the shortage of oil is not enough, a scarcity of natural gas compounds the woes in the energy crisis. Not too long ago natural gas was burned up as useless. In the 1950's and 1960's natural gas began to be used on a large scale as a convenient and clean fuel. Now about one third of all energy needs are fulfilled by gas and one half of the population has switched to gas for home use. The rapid consumption creates scarcity. Already gas companies are not accepting new customers and the shortage threatens to close factories and schools. More serious is the possibility that there will be no gas to dry grains before they are stored after harvest. In the next 10 years billions of dollars worth of gas will be imported from the Soviet Union. The shortage is blamed on the low regulated price at the wellhead, which does not offer enough incentive for exploration of new sources.

Coal used to be the major source of energy but recently has been overshadowed by oil (45 per cent of all energy needs in 1972) and gas (33 per cent). After the depletion of gas and oil we may have to return to coal, which is more plentiful (enough to last for 300 years or so). The 1969 Mine Safety Act makes *tunnel mining* more expensive; the Clean Air

Figure 3.2 Aerial of Peabody strip coal mining operation near Nucla, Colorado. [U.S. Environmental Protection Agency, Project DOCUMERICA. Photo by Bill Gillette]

Act makes high-sulfur coal undesirable. As a result the coal mining industry turns away from the traditional coal fields in the east and focuses attention on *strip mining* (Figs. 3.2 and 3.3) in the west. In Montana a coal bed over 100 ft thick lies only 30 ft beneath the earth, whereas in the Appalachians a thinner bed lies much deeper in the earth. Such a field can be made suitable for strip mining by removing the 30 ft of earth, thus exposing the coal bed for mining. Yet strip mining is a most devastating process to the environment. To reclaim the land after mining would cost $6,000 per acre. In a region where the land cost is $20 to $30 per acre, this is not economically profitable. That region in Montana is economically depressed. Unemployment runs to 40 per cent. Young people are leaving the area. To help economic development the coal must be mined. However, the new industry is certain to change the life-style of people in the area and resistance can be expected. A rancher against selling his land remarked that if his grandfather had succumbed to the temptation to sell his land, where would he be now? In the past this attitude would have been considered peasant conservatism against progress. Currently, with the emphasis on environment and the dissatisfaction with industrial culture, this attitude has more appeal. A tug of war is emerging between economic needs and environmental preservation. The same will happen in many other situations concerning the energy problem. The energy crisis is thus closely tied to environmental conservation.

The solution of the energy crisis in the future is partly dependent on the use of nuclear energy by reactors (Chapter 15). The safety of reactors has already become an issue. It is argued that the emergency core-cooling system of the reactors has not been tested and thus there is no guarantee that it will work as expected in an actual emergency. In fact, suit has been filed by environmentalists to close down all existing reactors because of this. Besides the possibility of accidental malfunctioning, one may also consider the possibility of sabotage and natural catastrophe (earthquakes, tornadoes) that might destroy a reactor and disperse the radioactive material. In time of war, the possibility of bombing is so real that it would be very difficult to keep reactors operating safely.

World energy consumption is doubling every 16 years. The rate of increase is even higher in America, where 6 per cent of the world population consumes one third of the world's energy. As technology advances, new demands for energy emerge. The World Trade Center twin towers in New York City consume an amount of electricity equal to that of the city of Syracuse. The insatiable demand for energy cannot be met indefinitely. Sooner or later the depletion of energy resources will force us to change. But how?

For one thing, much of the energy consumption at the present time is wasteful, resulting from the business practice of maximization of profit without regard to conservation of resources. Once the finiteness of resources is realized, wasteful practices will

Figure 3.3 Waste from strip mining enters a Pennsylvania river, adding to the pollution. [U.S. Environmental Protection Agency]

have to be changed. If automobile weight is reduced from 3,300 lb to 2,500 lb the saving of gasoline amounts to 1.2 million barrels a day, comparable to the amount expected to be supplied by the Alaska North Slope oil field. If the speed of an automobile is reduced from 60 mph to 50 mph, the fuel saving amounts to 11 per cent. The use of a car pool and public transportation can save gasoline further. If a house is heated in winter 2°F lower than normal and is air-conditioned in summer 2°F higher than normal, the energy saving amounts to 10 per cent.

However, not all energy needs can be eliminated. To give up all electric facilities would be to go back to the primitive life before modern civilization, an option few would consider. Thus we are forced to look for new energy sources. This subject will be discussed in more detail in Sections 3.2 to 3.5. Some important points may be mentioned briefly here.

Because coal is more abundant but less convenient and less clean than oil and gas, the future utilization of coal depends largely on *gasification*—converting coal into gas. After all oil wells are pumped dry we may look for oil in the oil-containing rocks called *shale*. The shale in Utah, Colorado, and Wyoming contains an amount of oil six times the world oil reserve. The technology of its extraction is under development. Generally speaking, the rock is crushed and the oil is extracted by heating and condensation. But what do you do with the huge volumes of rock residue? How to dispose of them is another environmental problem. Tidal energy (50-ft tide in Nova Scotia), geothermal energy (Imperial Valley, California, experiment unsuccessful), and wind energy (Dutch windmills) cannot supply a significant amount. As to nuclear fission energy, ^{235}U[1] fuel will run out in about 40 years. The breeder reactor now under development, which uses the more abundant ^{238}U, will extend the life of the nuclear age hundreds of times. But its use in large scale will lead to a serious radioactive-waste-disposal problem. The technology of solar energy conversion is far from commercially profitable. The most abundant energy source on earth, the fusion fuel deuterium in ocean water, still cannot be tapped, in spite of intense research efforts (see Chapter 15).

[1] ^{235}U represents the uranium atom that has an *atomic weight* 235 (read U-235). There is another uranium atom that has an atomic weight of 238. These are the two *isotopes* of uranium. See Chapters 10 and 14.

In dealing with the energy crisis we come across a number of scientific problems related to energy that are vital to an understanding of our current problem and to the solution of our future problem. These we will discuss in many of the following chapters. What is the physical nature of energy (Chapter 5)? What are the kinds of energy (Chapters 7, 11, 13, 15)? How do the energies change from one to another (Chapter 16)? What is the role played by energy in the origin and evolution of the universe (Chapters 21 through 24) and of life (Chapter 25)? How is energy put to use in the industry (Chapter 8)? What are the present and future energy resources on earth (this chapter)? What are the environmental effects of energy utilization (Chapter 17)? In this book we present physical science centered around two basic concepts: matter and energy. Through the study of physical science we may gain a better understanding of the energy problem and also find its solution.

3.2 Energy from the Physical Universe

In this and the following sections we study the energy resources of the earth. This provides us with the necessary background information to discuss the current energy crisis. How to use some of the more exotic forms of energy to solve the energy crisis will be discussed in Chapter 18, after we have acquired enough scientific and technological information in the intervening chapters.

In our everyday life when we speak of energy sources we mean gasoline, coal, natural gas, and so on. This situation is a rather peculiar one associated with the particular period of time we are now in.

From a cosmic point of view, the ultimate sources of energy in the universe are nuclear and gravitational. Earth is a small part of the universe, and the energy sources on the earth are no exception. However, because energy can be transformed, some of the energy happens to have been preserved or made available to us in a variety of forms quite different from their ultimate sources. We will discuss here the various kinds of energy sources and their relationship to the ultimate sources of energy.

In astrophysics, we learn how nuclear energy and gravitational energy work to produce the heat energy of the sun (Chapter 22). Because our earth is a small part of the solar system, it is not surprising to find that most of the energy on earth is derived directly or indirectly from the heat energy of the sun. The rest, a small fraction, is essentially derived from the nuclear energy of the radioactive substances on earth.

Solar Energy. The sun radiates a tremendously large amount of energy continuously in all directions in space. The amount intercepted by earth is a tiny fraction, but even that is enormous. The amount of heat received at the earth's surface per unit time per unit area perpendicular to the sun's rays is called the *solar constant.* Its value is 1.395 kw/m^2. A more convenient figure to memorize for this important constant is 2 calories (cal) per cm^2 per minute. The interception of solar radiation by the earth is equivalent to that by a disk of 1.275×10^{14} m^2. Thus the total energy influx into the earth from the sun is 1.73×10^{17} w. This is equivalent to burning 6.5 million metric tons of coal each second.

In spite of the large amount of solar heat energy bombarding the earth, it is not easy to convert it into other forms of readily useful energy. The solar battery, which converts solar energy directly into electricity, has been used in spacecraft, which cannot afford to carry large volumes of conventional fuel. Yet this energy conversion is still too costly compared to conventional fuels, so it has not been adopted on a large scale. Nevertheless, the harnessing of solar energy is a future possibility. This source is important because it is inexhaustible as long as the sun lasts, and the sun is expected to shine for billions of years.

Biosphere and Energy. Although most solar energy is wasted, a small fraction of it is used in natural processes in such a way that it is stored in substances that can readily supply useful energy. Much of this takes place in living organisms, demonstrating the superiority of living mechanisms over our present engineering mechanisms (solar battery). (This does not imply anything mysterious about life; after all, life has had 2 billion years to experiment and select, whereas engineering is only a few hundred years old.) The green plants capture solar energy and store it in chemical compounds (carbohydrates) formed by the process of photosynthesis. These compounds supply the energy needed by the plants, and when ingested by animals they also supply the energy needed by the animals. One direct use of this energy is to power animals' muscles; this is how the domestic animals get their energy to perform their work. Another way this energy is used is in burning firewood as a fuel to power engines or to supply heat. Moreover, this energy may be preserved in fossil form to be used millions of years later. Coal is the fossil remains of plants that died millions of years ago. Petroleum is probably the fossil remains of ancient animals. When we burn coal or gasoline, the energy thus produced is actually solar energy stored in living organisms millions of years ago and preserved underground over the years. These fossil fuels, including natural gas, account for the major sources of energy consumed at the present time. Their dominance in the current energy market is due to the fact that the large-scale use of them began only recently, so their depletion is not imminent. The fact that they are economically cheap is because the major components of the cost are merely the digging and the transportation. However, their reserves are finite and limited, and from a historical point of view these fuels are just as irreplaceable as the ancient Egyptian temples. Thus the use of the fossil fuels is a lavish act. That they have been cheap is very unusual and can happen only in a very limited period of human history. We happen to be in this period.

From the chemical point of view, the photosynthetic processes by which solar energy is stored in organic compounds are very complicated. They form a complex network of pathways. Only recently have scientists begun to reveal some of the fine details. The high-energy radiation (photons) of the sun in the process of photosynthesis leads to the creation from carbon dioxide and water of compounds with a high energy level. Throughout the metabolic processes, the compounds go through a series of chemical changes in which the energy level descends; in this process the compounds release energy to carry out vital biological functions, such as muscle contraction. Eventually during the burning of firewood, dry leaves, animal remains, coal, gasoline, and so on, the remaining energy of the organic carbon compounds is released, when they oxidize into carbon dioxide and water, and the energy level drops to the bottom. The cycle is repeated when carbon dioxide is taken up by the green plants in the photosynthetic processes again. Without the photosynthetic processes to raise the energy level, the cycle will not start and there will be no life. The energy source of photosynthesis is of course the light energy of the sun.

Hydraulic Energy. Solar heat energy is responsible for another energy source, hydraulic energy, through a physical, rather than a biological, process. Water power has been utilized to operate mills for centuries. In the present time, water power is used on a large scale in hydroelectric power plants (Fig. 3.4). When water drops from a height, the potential energy (Sec. 5.2) is reduced and changes into kinetic energy (Sec. 5.1), which may be transformed into electric energy. At first glance the energy seems to come from the gravitational energy of the earth. Yet once the water at high altitude has run down the slope and thus exhausted its potential energy, it is replenished continuously by rainfall. Raindrops are formed in the sky, and the potential energy of water proceeds from the potential energy of the atmospheric moisture, which is produced by the heat of the sun over the rivers, lakes, and oceans. Thus the ultimate source of hydraulic energy is the heat energy of the sun. Earth's gravity is just an instrument for the conversion of the energy.

Tidal Energy. The sun, besides sending heat energy to the earth, also gives rise to another source of energy on earth, i.e., the tides, through gravitational interaction. The moon also contributes to this interaction. The tides are induced by the gravitational attractions of these two celestial bodies. When the water level rises, the potential energy increases and may be stored by damming up the high-tide water for release during low tide. The potential energy can be converted into kinetic energy for use in the same way as water power is used in a water mill or in a hydroelectric plant. In some coastal areas this principle has actually been used to generate electric power from the tides. Even though this energy proceeds from the gravitational attraction of the sun and the moon, most of the energy tapped this way actually comes from the rotational energy of the earth. The action of the tides slows down the rotation of the earth in a manner similar to the action of a brake shoe, and a small fraction of the rotational energy is converted into the energy of the tide.

Geothermal Energy. Besides the energy from the sun there are sources of energy from the earth itself, all ultimately nuclear in origin. The interior of the earth is molten and hot; its heat comes to the surface by conduction through the earth's crust. By measuring the temperature gradient as we dig down toward the center of the earth, it is found from the law of heat conduction (Chapter 7) that the rate of heat flow to the surface of the earth from the interior is about 0.063 w/m^2. From the total surface area of 5.10×10^{14} m^2 of the earth, the total rate of energy flow is 3.2×10^{13} w (0.02 per cent of that from the sun). The amount of heat transported from the interior of the earth to the surface by convection through volcanoes and hot springs is much smaller than that by conduction. The heat in the interior of the earth comes from the energy of the radioactive

Figure 3.4 Hiwassee Dam (307 ft high) and hydroelectric power plant (59.5 megawatts) in western North Carolina. [Tennessee Valley Authority]

atoms. When earth was first formed by accretion of earthy material (Chapter 23), it was cold. The energy of radioactive atoms released in the interior of the earth accumulated inside and eventually caused the earth to become molten. Thus geothermal energy is of nuclear origin.

Nuclear Energy. Finally, nuclear energy may be released by fission and fusion processes (Chapter 15) and those substances on earth that can be used for fission or fusion represent a potential source of energy supply. Fission occurs only in heavy elements; only in uranium and thorium, elements 92 and 90, is there the possibility of self-sustained release of fission energy. Thus the fission energy source is limited to the mineral deposits of uranium and thorium. Fusion takes place in light elements. The lightest of all, element number 1, is hydrogen. It is the rare heavy isotope of hydrogen, namely deuterium, that is considered to be the main source of fusion energy. Deuterium accounts for only 0.015 per cent of natural hydrogen but is still relatively abundant, because we have plenty of water (H_2O) in the oceans.

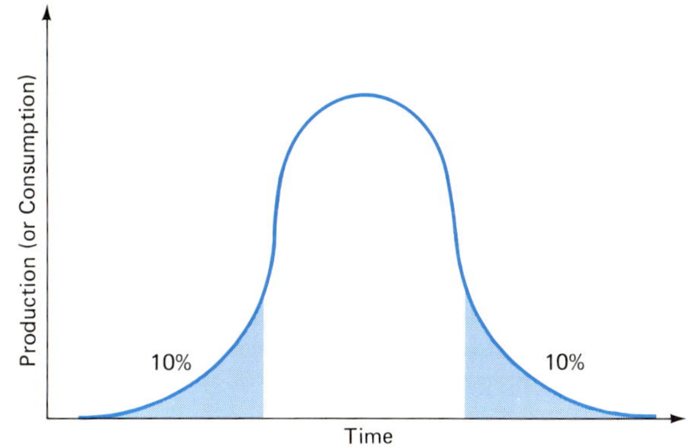

Figure 3.5 Production or consumption of a typical mineral product as a function of time.

3.3 Fossil Fuel Reserves

Although the influx of energy to the surface of the earth seems to be constant and inexhaustible, the principal sources of commercial energy at the present time are the solid, liquid, and gaseous fossil fuels (coal, oil, and natural gas), which are exhaustible. How long they will last is an important question, and the answer is disquieting.

Naively, the length of time they will last may be calculated by dividing the total amounts of reserves by the current rates of consumption. Actually, the problem is more complicated in a number of ways. First, the amount of any mineral resource on the earth can never be ascertained exactly. Some of the deposits are buried so deeply that they cannot be mined economically. Instead of considering the total physical amount, we should consider the *discoverable and recoverable amount.* The amount that has been discovered and is readily available (oil deposit with a well established on it) is called the *reserve;* the amount that has not been discovered or that is not readily available (oil deposit without a well on it) is called the *resource.* The reserve is the amount "on the shelf," which is not and cannot be very large. The resource is usually many times greater than the reserve. By past record of mineral exploration and discovery, it is possible to determine by extrapolation the total of the recoverable amount. Second, the consumption rate is not constant, and at present it is increasing. Petroleum resources have been actively exploited for only 90 years up to now. The rate of consumption (and so production) has increased as an exponential function of time in the past 80 years, doubling every 10 years (6.90 per cent annual rate of increase). The increased rate of consumption in the future must be taken into account in the estimate. Third, as mining continues, the easily accessible and high-grade resources are being depleted first; the leftovers for use in future years are less accessible and of lower quality. Therefore the cost of such fuels would be higher in future years, a factor that would tend to reduce consumption. The exponential increase of production cannot continue indefinitely and will have to be slowed down; ultimately the rate of production will be reduced to zero. Thus the production curve as a function of time would have a shape like that shown in Fig. 3.5. It has tails on either side, representing a slow start and a declining finish, and has a broad peak in the middle. The two tail regions may cover a long time period but are not significant as far as the volume of consumption is concerned. Most of the natural resource is to be consumed in a relatively short period, represented by the peak region. It has been estimated by Hubbert[2] on this basis that 80 per cent of the recoverable

[2] M. King Hubbert, *Sci. Am.*, **224,** 3 (1971) 61.

coal resource will be consumed in a period of 300 years from 2000 to 2300 A.D., although by including the tail regions of consumption of the initial 10 per cent and final 10 per cent, the total period in history in which coal plays an active role may be stretched out to about 1,300 years. The estimate is based on a total amount of 7.6 trillion metric tons of coal and a current consumption rate of 3 billion metric tons per year, roughly 1 ton per person per year. Hubbert also estimated the lifespan of oil and gas, which will be discussed as follows.

First consider the liquid fuel, oil. The estimated resource in the 48 contiguous states together with the continental shelves is about 165 billion barrels. It is estimated that 80 per cent of the resource will be consumed in a 65-year period from 1934 to 1999, with the peak of consumption at the present time. The recently discovered Alaska arctic oil field (less accessible than fields in the area of the 48 contiguous states) has an estimated reserve of 10 billion barrels. Future exploration may increase the reserve by several times, but even that is enough for no more than 10 years' consumption in the United States. The total world oil resource is estimated to be between 1,350 billion to 2,100 billion barrels, 80 per cent of which will be consumed in a period of 58 to 64 years, peaking around 2000 A.D. After oil from petroleum is used up, there still remains a substantial amount of oil available in tar sands and oil shales. The use of this oil source has just begun. The resource in the former is about 300 billion barrels, and that in the latter is about 3,100 billion barrels, of which 190 billion are recoverable by current standards. These may prolong the life of oil on the historical stage, but still the total span is a fleeting moment in human history, not to mention geological history.

The natural gas resource in the 48 contiguous states is estimated to be 1,075 trillion ft^3. It is estimated that the peak of production will come between 1975 and 1980.

The use of fuel has shifted from coal to petroleum in the past century. Before 1900, coal was the principal form of fuel. After 1900, oil use increased rapidly, and in 1970 oil consumption accounted for 41 per cent of the total energy produced. Natural gas accounted for 33 per cent of the total, and coal's contribution was reduced to 20 per cent. There are minor contributions from hydroelectricity (4 per cent) and others (2 per cent). As to the total energy resource in calories, coal accounts for 89 per cent, whereas oil and gas account for about 6 per cent and 5 per cent, respectively. Because of the limited size of the fossil fuel resource and the increased amount of energy consumption, the depletion of fuel resources is becoming a serious problem. Attention is being directed to the search for new fuel sources that can continue to supply energy after the fossil fuels have been exhausted.

3.4 Other Energy Sources

Because solar heat accounts for 99.98 per cent of the energy influx to the earth's surface and is practically inexhaustible, its harnessing would be the most desirable solution to the energy problem. Although the science and technology for this purpose are now available, economic and practical difficulties remain. The first requirement is a proper surface to intercept solar energy for conversion into other forms of energy. The area would have to be quite extensive. To provide the current United States electric power capacity of 350 million kw, the interception surface would have to be as extensive as 0.3 per cent of the country's geographical expanse. The practical and economic problems are obvious. Yet it is conceivable that in the future, roofs of houses may be specially made to intercept solar energy, and electricity may be generated to provide the energy supply for household needs.

Hydraulic power is another source of energy. The world's water power capacity is estimated to be 3 billion kw, close to the current rate of energy consumption in industry. Yet only 8.5 per cent of it is developed at the present time. Although water power

Figure 3.6 A windmill in Holland.

itself is free and self-replenished, the cost of the initial investment of building a hydroelectric plant is considerable and must be paid in full at the beginning. Besides, the developed nations that need energy most are in areas deficient in water power. Wind power is another energy source derived from solar energy. Windmills (Fig. 3.6) have been in use for centuries, but wind power is insignificant in the overall picture of world energy supply.

Tidal power amounts to 2 per cent of water power. Only one large-scale tidal power plant has been built. It is located in France and has a capacity of 240 megawatts (Mw).

Geothermal power, of the kind represented by the heat contained in volcanoes and geysers, has a capacity comparable to tidal power. A geothermal power plant making use of volcanic heat has been operating in Italy with a capacity of 370 Mw. Geothermal power from the deep interior of the earth has a much larger capacity. The exploration of this new source is now being considered (Chapter 18).

In a different sense, the earth has been used as a heat source in the operation of the heat pump used for heating and air conditioning. In the winter, heat in the earth is "pumped" into the house in an operation similar to refrigeration. The earth is being "refrigerated" and the extracted heat is released into the house to provide space heating. This operation is not without cost, and the consumer has to pay for the electric energy of refrigeration, which is also converted into heat energy for use in household heating. Because part of the heat that warms up the house actually comes from the earth free of charge, in this sense earth is an energy supplier. In this way heating costs can be greatly reduced. [Heat from the earth's surface and in the ocean is considered useless from a thermodynamic point of view (Chapter 16). This does not prevent some economic use of it. In fact such uses do have the effect of reducing fuel demand as a whole.] In the summer the house is refrigerated and the extracted heat is pumped into the earth.

3.5 Nuclear Fuels

Fission Fuel. Although none of these new energy sources has been put to wide use, the outlook for finding fuel substitutes has been altered significantly by the development of *nuclear energy* (commonly called *atomic energy*), which can now be utilized on an economical basis in competition with the fossil fuels. This possibility was first realized when a self-sustained nuclear fission chain reaction was achieved in a nuclear reactor for the first time on December 2, 1942, a landmark date in the history of science and technology. In a nuclear reactor, the atoms of the lighter uranium isotope ^{235}U undergo fission one after another and release energy continuously until all ^{235}U is used up. Present nuclear power reactors are all based on this plan. Thus the total amount of fission energy is determined by the total amount of ^{235}U in the minable uranium ores. Uranium is a relatively rare mineral, but because of the large amount of energy released per atom, the total amount of energy reserve in uranium is considerable. Because ^{235}U accounts for only 0.7 per cent of natural uranium [the rest (99.3 per cent) being ^{238}U, which is not readily fissionable], the energy made available by this process is still rather limited. At the projected rate of nuclear energy consumption in

Figure 3.7 Artist's conception of Liquid Metal Fast Breeder Demonstration Nuclear Power Plant being built at Oak Ridge. Scheduled to produce 350 to 400 megawatts of electric power in 1980, the nation's first breeder reactor using fast neutrons and liquid-metal coolant will demonstrate the technical and economic practicality of the fast breeder concept, which is expected to be commercially adopted in the mid-1980's. [U.S. Atomic Energy Commission]

the near future, the ^{235}U in high-grade ores will be used up in a few decades and the atomic age based on ^{235}U probably will last no longer than a century.

Breeder Reactor. In the nuclear reactor some ^{238}U can be converted into ^{239}Pu, which is fissionable and can be used in the same way as ^{235}U. During World War II reactors were specifically built for the purpose of producing ^{239}Pu for military use. In a reactor the ^{239}Pu produced may replenish ^{235}U as fission fuel. The possibility thus arises that a special kind of reactor, called a *breeder reactor* (Figs. 3.7 and 3.8), may be developed to breed new fuel ^{239}Pu to replenish the consumed ^{235}U. If the breeding efficiency is so high that it produces the same amount of fuel as that which is consumed, then by repeated breeding all the available ^{238}U may be converted to fissionable ^{239}Pu in the course of time, and the nuclear energy stored in all existing uranium (both ^{235}U and ^{238}U) may be released entirely. In this way the total amount of nuclear energy available is determined by the total amount of natural uranium. This amount is more than 100 times that of ^{235}U, and thus the nuclear energy resource increases by more than 100 times. One gram of natural uranium contains 8.1×10^{10} joules of energy, equivalent to 2.7 metric tons of coal or 13.7 barrels of crude oil, all of which can be released by the breeder reactor. In a similar way another heavy element, thorium (only one isotope ^{232}Th), may be converted into fissionable ^{233}U in the breeder reactor. Thus the natural resource of thorium also may be added to the resource of nuclear energy. The total uranium and thorium in high-grade ores can produce sufficient energy to maintain industrial civilization for a few thousand years.

Low-Grade Ores. A few thousand years are not a long time for the future of mankind, even though such a respite is desperately needed. On the other

Figure 3.8 A cutaway model of the fast breeder reactor. Notice the size of the man in the photograph. [U.S. Atomic Energy Commission]

hand, there is quite an abundance of low-grade ores that contain uranium or thorium. Chattanooga shale contains uranium of 6×10^{-3} per cent, and Conway granite contains thorium of 5.6×10^{-3} per cent. Though the percentages are so small (high-grade ores contain uranium of 0.4 per cent) that these are hardly ores in the conventional sense, the energy content in terms of calories per gram is still considerable. When high-grade ores are depleted, we will have to turn to low-grade ores as an energy source and will be literally burning rocks. Even though nuclear fuels from low-grade ores are more expensive, they are still economical because of the total conversion in breeder reactors. The amount of uranium and thorium in low-grade ores is so large that the total energy resource in them is enough for 100,000 years' use. If we include the trace amount of uranium and thorium in granite, the nuclear energy resource is nearly endless. These are still usable because the energy consumed in extracting them from the granite is still a small fraction of the energy that will be produced from them.

Nuclear Industry. Even though the cost of nuclear energy in terms of the material uranium is relatively small, the installation that makes the release of energy possible involves sophisticated technology and is costly. In the beginning, the cost of nuclear energy in terms of dollars per kilowatt hour was about 10 times higher than commercially available power. As technology improved during the decade after 1945, the cost was lowered continuously and finally atomic energy was made commercially competitive. The first nuclear power plant at Shippingport, Pennsylvania, began operation in 1957. A new energy industry was born and grew rapidly. By 1969 a total of 15 nuclear power stations were in operation, with a total capacity of 3.5 million kw, amounting to 1 per cent of the 1969 fossil-fuel electrical power capacity of the United States (Fig. 3.9). In 1974 nuclear power reached 9 per cent of the total. It is projected that this capacity will rise to 150 million kw by 1980, reaching about half the current fossil-fuel electric capacity.

The crucial point in lowering the cost of nuclear energy is to build bigger and bigger reactors. The energy production is proportional to the volume of the reactor, which increases to the cubic power of the linear dimension. The fixed cost of installation increases roughly in proportion to the surface area of the reactor, which increases to the square power of the linear dimension. By constructing bigger reactors, the energy produced will increase more rapidly than the fixed cost of installation will. Therefore the cost of unit energy output will be reduced.

Figure 3.9 Locations of currently operating nuclear power reactors and those firmly committed for future operation in the United States.

In the United States the fossil fuels at the present are still abundant and relatively cheap; the use of nuclear energy is not a pressing matter. On the other hand, in energy-starved countries the use of nuclear energy is not a thing of the future but is a present-day reality. At one time England had a rich reserve of coal, and this was a factor that contributed to the development of the industrial revolution. After two centuries of exploitation, British coal mines are depleted, and England is now counting on nuclear power as a major source of energy.

Fusion Fuel. Burning rock may not be easy from a technological and economical point of view, even though we know the energy is there. It is more likely that *fusion energy* will be utilized before the ultimate exhaustion of fission energy sources. Because heavy hydrogen in water is the main source of fusion energy, eventually we will literally be burning water as our ultimate source of energy supply! Harnessing fusion energy for power production has not been accomplished yet. The most promising fusion reaction, which probably will be the first to be put into operation, is the deuterium–tritium reaction. It may take place at a lower temperature than that of the deuterium–deuterium reaction. Deuterium exists abundantly in ocean water (0.015 per cent of natural hydrogen is deuterium), and its separation is a relatively cheap operation. Tritium is produced from lithium, the natural reserve of which is rather limited. Fusion energy based on the deuterium–tritium reaction (Sec. 15.4) is thus limited by the natural reserve of lithium, and it is estimated that the total amount of energy available this way is comparable to that of the fossil-fuel energy resource. On the other hand, if fusion energy from the more difficult deuterium–deuterium reaction (Sec. 15.4) is made available, the total amount will be determined by the natural reserve of deuterium or ocean water, which is abundant. The energy released per deuterium atom in this reaction is 7.94×10^{-13} joules. One m^3 of water (a metric ton) contains 10^{25} atoms of deuterium with a potential fusion energy of 7.94×10^{12} joules, comparable to that produced by 300 metric tons of coal or 1,500 barrels of crude oil. The total volume of ocean water is about 1.5 billion km^3. If only 2 per cent of the deuterium were extracted and used, the amount of fusion energy would be as much as a million times that of the total fossil-fuel energy resource. It might sustain industrial civilization for a billion (10^9) years. Fusion energy is thus the ultimate energy source on earth.

The Supply of Electric Power in the Near Future. Present technology has solved most of the problems involved in meeting short-range electric power needs up to 1985.[3] The capacity of electric power generation in the unit of 1 million kw is expected to triple from the 1970 figure of 333 to 925 in the year 1985. At present, the energy source picture is dominated by fossil fuels; of the 333 million kw in 1970, fossil fuels account for 270 million, hydroelectric power for 55 million, and nuclear power for 8 million. In the year 1985 the breakdown of the total 925 million is expected to be fossil fuels 515 million, hydroelectric power 120 million, and nuclear power 290 million. An increasingly more important part is to be played by nuclear power.

The thermal pollution problem originating from the use of river water as a coolant in the condenser of the heat engine for electric power generation can be solved by the three well-developed methods: wet cooling towers ($7 to $20/kw), dry cooling towers ($35 to $60/kw), and cooling pond ($3 to $7/kw) (Fig. 3.10). Heat from the engine condenser is considered waste energy from the thermodynamic point of view. Yet it may still be useful economically, e.g., it may be used in heating fish ponds for meat production or in heating greenhouses for growing vegetables. Any such use will help reduce the cost of pollution control.

As to SO_2 pollution from a fossil-fuel power plant, current technology is not up to removing all SO_2 produced by the burning of fuel. A future solution is the gasification of coal, making pollution-free gas out of coal.

[3] Alvin M. Weinberg, "The National Energy Problem," invited paper, American Physical Society, November 23, 1970.

Several large power plants may be clustered together to form a power park. In such an arrangement many environment-related tasks may be consolidated, such as monitoring the level of contamination, reprocessing nuclear fuel, gasifying solid fuel (coal), and so on. The concentration of power facilities in parks may be considered together with the problem of distribution of electric power. High-tension transmission lines are already considered as an environmental problem (Fig. 3.11). They are an eyesore on the landscape. Underground cryogenic cables and even superconducting cables have been under consideration as solutions. The superconducting cable may turn out to be even more economical for high-power transmission than the conventional high-tension cable.

Japan and Switzerland. In the 1971 Conference on Peaceful Uses of Atomic Energy in Geneva sponsored by the United Nations, it was learned that Japan and Switzerland are the nations most enthusiastic about developing nuclear (fission) power. In 10 years the nuclear capacity in Japan is expected to rise to 30 million kw, amounting to 20 per cent of the country's total electric power capacity. The percentage is expected to reach 40 per cent in 1990 and

Figure 3.10 Cooling tower and pressure vessel for General Electric's 1,130-megawatt pressurized water reactor at the Trojan Nuclear Power Plant near Rainier, Oregon. The plant complex will have a recreation lake, boating facilities, reflecting pool, and whistling swan lagoon. [U.S. Atomic Energy Commission]

Figure 3.11 Transmission lines near Searchlight, Nevada. [U.S. Environmental Protection Agency, Project DOCUMERICA. Photo by Charles O'Rear]

50 per cent by 2000. Switzerland expects to reach 60 per cent nuclear power by 1985; no other kind of power plant will be built in the near future. In spite of her grim memories of atomic power, Japan is now faced with the worst air pollution problem in the world, which has turned Japan's attention from fossil fuel to nuclear fuel. Besides, Japan's fossil fuels must be imported and thus are subject to political and economic influences of other nations. Notwithstanding the availability of hydraulic power in Switzerland, nuclear power is preferred because water power is too expensive and not sufficient to meet the country's energy needs. Fossil-fuel plants produce too much air and water pollution, which the Swiss want to avoid. The environmental concerns posed by the nuclear power plant are also evident in Switzerland's experience. In 1969 the reactor core at the first experimental power plant at Lucerne suffered a "meltdown." In 1970 a Swiss jetliner crashed near the Beznau Atomic Power Plant. The thermal pollution issue was raised and debated, and the result was a law requiring all future nuclear power plants to install cooling towers for heat dissipation at a cost of $6 to $10 million apiece. Even the cooling towers have elicited opposition; they are unsightly and may produce fogging that would spoil the beautiful Swiss scenery.

Problems

3.1 How should the Trans-Alaska Oil Pipeline be constructed to prevent breaks in the event of an earthquake?

3.2 From the solar constant 1.4 kw/m^2 find the total energy value the sun imparts to the earth in one day based on the electric utility rate of 3¢/kwh (kilowatt-hour).

3.3 Show that the tide energy cannot be used if the earth is completely covered with water. What does this mean concerning the origin of the tide energy?

3.4 From the solar constant and the distance from the sun to the earth 1.5×10^8 km, calculate the total amount of energy released by the sun per day.

3.5 According to the equivalence of mass and energy, $E = mc^2$, where $c = 3 \times 10^8$ m/sec in the mks system, calculate how many metric tons of weight the sun loses each day because of the energy radiated (see the preceding problem).

3.6 From the figures given in Sec. 3.2 find out how many kwh (kilowatt-hours) of heat energy are produced by burning 1 ton of coal. If all the heat could be converted into electric energy which could be sold at 3¢/kwh, how much would 1 ton of coal be worth? Compare with the actual price of coal (about $10/ton). What accounts for the difference?

3.7 When the linear dimension of a nuclear reactor is increased by a factor of 2, by approximately what factor will the unit cost of energy generated by the reactor be reduced?

4 Matter and Its Basic Laws

4.1 What Is Matter?

Matter is one of the basic concepts in physical science. *Energy* is another. In this book we shall present physical science centered on these two concepts. These two aspects will be dealt with in alternate chapters.

Most natural phenomena and many human activities involve matter and the transformation of matter. These are topics studied in physical science that enable us to gain a better understanding of the happenings that take place around us. From a scientific point of view we may say that anything that has a mass (or has a weight on earth) is matter. To put it simply, everything tangible around us is matter. Rocks, earth, water—all mineral substances are matter. Besides the inanimate material substances there are on the earth living organisms that include animals and plants. Living organisms are made of material substances and so have some properties in common with inanimate material substances. Concerning the great variety of material substances the following question arises: Are the various kinds of material substances related?

Man has contemplated this question for a long time. According to the four-element theory of the ancient Greeks, the material world is made up of four basic substances, called elements: earth, water, air, and fire. Although the theory is outdated, a number of ideas have carried over into the present scientific theory. The theory implies that all the varied material substances are related, in the sense that they are all made of a few simpler basic elements. They may be combined in a variety of ways to yield a large number of different substances. Moreover, the properties of substances may be traced to those of the elements they are made of. The few properties of the simple elements thus determine the complex properties of all substances. Finally, the very fact that there are *four* elements implies that the process of reduction to simpler entities cannot go all the way to one basic element; there are a few entities that cannot be reduced any further. These concepts are more or less carried over into modern atomic theory.

We now know that whereas water, earth, and air are material substances, fire is not. This brings out the point that the difference between matter and nonmatter is not obvious, and it took extensive investigation to settle the question completely. The recognition that fire is not a substance is worth discussing because it typifies a crucial step in the development of scientific ideas.

The Greek element of fire gradually evolved into the medieval concept of phlogiston, a hypothetical substance that is a necessary ingredient of combustible materials, the release of which in the process of burning manifests itself in the phenomenon of fire. It is a reasonable way to explain the phenomenon of fire. On the other hand, once the theory is accepted it leads to other consequences. For example, in the process of smelting iron, the ores, under the influence of fire, change into the metal iron. According to the phlogiston theory it is natural to suppose that in the process ores combine with phlogiston to make iron. If this should be the case, then the iron, having more "matter" in it than the ores because of the addition of the phlogiston, should be heavier than the ores. However, there was evidence to the contrary (rust, similar to ore, is heavier than the iron that produces it), even though no serious attention was paid to the discrepancy in the earlier times. After all, many properties of a substance can change—the volume, the color, the shape, the smell, and so on—and the change of weight did not seem to be a serious concern.

It was by the experimental work of Lavoisier, a prominent Frenchman in the period of the French Revolution, that the phenomenon of combustion was finally explained correctly. From this work a basic law of nature, the law of conservation of mass, arose. Lavoisier discovered that the process of combustion is actually one in which the combustible substance combines chemically with the substance oxygen in the air. Thus the combustion product, such as the rust (iron wire can actually burn into rust in oxygen), should be heavier than the substance before it burned. This is indeed the case, contrary to the prediction of the phlogiston theory. Thus the phlogiston theory was overthrown. The phenomenon of fire is explained as the manifestation of a process of *chemical combination* of two material substances, the combustible one and oxygen, without having to introduce a new substance, "fire" (or phlogiston). In this way and in similar developments the difference between matter and nonmatter is made clear gradually. Some of the physical entities are recognized as matter; others, such as fire, are not, but are manifestations of matter.

Lavoisier went one step further. He found that the weight of the combustion product not only is heavier than the combustible substance before burning but also equals exactly the weight of the substance and the weight of the oxygen consumed added together. Because weight (measured by a spring balance or the like) is proportional to the mass (measured by an equal arm balance), we may use the two terms interchangeably for the moment before we introduce more refined discussions later. What Lavoisier discovered is that the mass of a substance is not altered in the process of chemical combination; thus the mass of the combustible substance and the mass of the oxygen remain unchanged after they have combined and so the mass of the combustion products equals exactly the sum of the two.

This is not a trivial point. During chemical combination many of the properties of the reactants are changed—the volume, the shape, the color, the smell, and so on—the original substances seem to have disappeared completely and a new substance appears in their stead. What we learned from Lavoisier is that in spite of the drastic changes there is one aspect that is not changed, i.e., the mass. Thus out of a great number of properties of a substance we have singled out one, the mass, that is characteristic of material substance and remains always unchanged. Therefore mass may be regarded as a basic attribute of matter and we may say that anything having a mass is matter, as we stated in the beginning. We cannot use volume, color, smell, and so on, as the criterion because these attributes can all be changed or made to disappear. Air and other gases are quite different from tangible substances (solids and liquids). Yet they have mass and so are considered as material substances (in the ancient Oriental five-element theory, air was not considered an element and was not even recognized as material). Fire has no mass and is not. Matter may change its forms and properties but its mass remains always the same. This is the meaning of the *law of conservation of mass.*

The process of chemical combination may be reversed in a process called *decomposition.* A substance may be decomposed into two or more substances. Each one of them may be decomposed further and the question arises how far we can proceed in successive decompositions. The study of chemistry involving thousands of *chemical reactions* led to the conclusion that there are 92 basic substances[1] on earth, called *elements,* that cannot be decomposed any further by chemical means. Furthermore all material substances on earth are *compounds* or *mixtures* of compounds made of these 92 elements by chemical combination. Because there are many ways the elements can be combined, a large number of compounds results and this explains the great variety of things on earth. Besides, we learn by optical studies that things in heaven, e.g., the stars, are also made of the same 92 elements. Thus the whole material world is made of the 92 chemical elements. Another way to answer the question of what matter is is to say that anything made of the 92 chemical elements is matter (or material substance).

Yet there are other entities in the physical world that at first look seem not reducible to the 92 elements, e.g., electricity, magnetism, heat, light, and so on. What are they? How are they related to the material substance made of the 92 elements? Usually when we discover a new phenomenon, such as that characterized by a strange smell, we ascribe it to a new substance and carry out scientific investigations to track down this new substance (all smells are traceable to special chemical compounds). The phenomena of electricity, magnetism, heat, and light were studied in the science of physics. It was only natural that in the early days one sought to explain them in terms of various new substances, even though they might not be *material substance* related to the 92 elements. The eighteenth century was dominated by the thought of materialism after the great triumph of the materialistic Newtonian theory in the seventeenth century. Thus theories on these phenomena began by assuming the existence of a variety of new substances. For example, light was thought to be the manifestation of a substance of a corpuscular structure. Newton's theory of light was formulated along this line. Heat was thought to be due to a fluid substance called caloric that could be absorbed in material substances (like water in a sponge) and transferred from one to another. Electricity was thought to be due to one or two kinds of fluid substance that stayed or flowed in material substance. Magnetism was also ascribed to a separate magnetic entity. In all these cases apparently no weight (or mass) was seen associated with the new substances. This did not prevent them from being considered as substances. They could be so light that their mass would be hardly measurable. Moreover, at that time, mass was just one property of substances, not yet *the* property of substances, and there was no reason that one could not conceive of a substance that had no mass.

In the nineteenth century the materialistic tide was receding. The preceding theories were discarded and replaced by new ones in which no new separate substances were necessary to explain these phenomena. Heat was explained as a mechanical manifestation (molecular motion) of the material substance, not the result of a separate substance, caloric. Electricity was found to be a basic property

[1]To be exact, two of the 92, Tc and Pm, are not present in nature, and in addition a dozen new elements have been created artificially, some of which, such as Pu, are naturally present in trace amounts. ^{239}Pu is created from ^{238}U in ores by cosmic rays and ^{244}Pu, present in rare-earth ores (bastnasite), was created in pre-earth times. In spite of this, merely for convenience, we will still refer to "the 92 elements" for the chemical composition of material substance in the rest of the book, with the tacit understanding of the above exceptions.

of the ultimate building blocks of the material substance (to be elaborated shortly afterward) and thus no separate fluid substances were needed. Magnetism was explained as the manifestation of electricity (now also partly as a basic property of the ultimate building blocks of material substance) and thus no additional magnetic entity was needed. Light was explained as the manifestation of an electric and magnetic phenomenon and thus no separate light entity was necessary. All physical phenomena can be explained on the basis of one substance—the material substance of the 92 elements. The manifestations of this single entity accounts for all happenings in the physical world. The one and only substance is characterized by its unchanging mass.

If we have succeeded in reducing the great variety of material substances in the physical world to the 92 elements, then we may ask whether we can reduce the 92 elements further to a smaller number of more basic entities. This is not possible by *chemical* means, as was implied in the previous discussion. In chemistry we learn that each element has a basic unit, called the *atom,* which is immutable. Nevertheless, it was found that by *physical* means it is possible to break down the atoms of the 92 elements further into simpler building blocks. It was found that atoms of the 92 elements are all made of a few—in fact, only three—*elementary particles,* called the *electron, proton,* and *neutron.* Protons and neutrons combine to form the *nucleus* of an atom. With electrons revolving about the nucleus we have a complete atom. The differences among the 92 elements are not ultimate and absolute. They originate merely from the differences in the numbers and combinations of the three kinds of basic building blocks. In answer to the question of what matter is we may also say that anything made of the elementary particles is matter. (Later we will learn other elementary particles besides the three here mentioned.)

Each elementary particle, such as the electron, has three basic properties: *mass, electric charge,* and *magnetic moment* (it is like a tiny magnet with two equal and opposite poles; magnetic moment, defined as the product of the magnetic pole strength and the distance between the two poles, is a measure of the strength of the magnet). These attributes are the origin of the forces, gravitational, electric, and magnetic, respectively, by which elementary particles interact with each other, and, in particular, combine to form *atoms, molecules,* and *bulk material.*

Thus any material system made of elementary particles has mass, electrical properties, and magnetic properties (and therefore optical properties also). All physical properties of matter can be explained in terms of the properties and interactions of the elementary particles. Moreover, the chemical properties related to chemical reactions (the *chemical bonds*) can be explained in terms of the structure and interaction of atoms, this explanation being one of the most significant achievements of science in the twentieth century. Thus the whole physical world is reducible to the properties and interactions of elementary particles.

Whether the elementary particles can be reduced to other even more elementary entities is one problem currently at the frontier of science. Very promising speculations have been made (the quark theory, for example). Yet no definite conclusion has been established. For our purpose, as far as the understanding of everyday phenomena is concerned, such as those related to energy and environmental problems, we need to learn only the properties of the electron, proton, and neutron and their interactions, the structure of these particles themselves being largely irrelevant.

Our main concern is to learn how the material systems in the physical world are built up from elementary particles in a structural hierarchy from atoms to molecules and to bulk substances, and from these structures to learn how the material systems may be transformed physically, chemically, and subatomically. All physical and chemical phenomena in nature are to be explained on this basis. In dealing with the transformations, we come into contact with another important concept in physical science, i.e., *energy,* with the help of which much of the transformation of matter can be understood more clearly. Thus in addition to the study of matter in this chapter we follow up with a study of energy in the next chapter. The two together provide a more complete description of our physical world and thus provide a proper physical basis for discussing practical problems.

4.2 The Basic Laws of Matter

The concept of matter means little unless we know the properties of matter in a more specific way. In a scientific theory we start with a few basic properties and proceed to explain a large variety of natural phenomena. To illustrate even by the four-element theory of the ancient Greeks, earth and water are assumed to have the property of *gravity,* a tendency to move downward toward the earth, whereas air and fire are assumed to have a different property, called *levity,* a tendency to move upward toward the sky. By these properties the "natural motion" of things on earth, such as the falling of a rock and the rising of the smoke, can be explained.

In a superficial way it was recognized in ancient times that the fall of an object depends on its weight, e.g., a stone falls faster than a feather. Aristotle went as far as to say that an object 100 times heavier would fall 100 times faster. Such a statement may be regarded as a *law of nature* that describes a predictable behavior. The concern on laws of nature arose early in ancient times. In fact, the ancient Greeks had discovered a few laws of nature that remain correct to this day, e.g., the Archimedes principle (on the buoyant force) and the principle of the lever. However, Aristotle's law of a falling body turned out to be wrong.

With a great variety of material substances a large number of properties can be delineated and a great number of laws of nature can be formulated. In fact, throughout history man has been accumulating knowledge on nature in this manner. It then appears that our knowledge of nature would be a large and complex collection of properties and laws (as numerous as the contents of a mail-order catalog). However, after several hundred years of intensive study with many exciting discoveries in the process, we have finally learned that there are only a few basic properties and a few basic laws out of which all the other properties and laws (such as Archimedes' principle and the principle of the lever) can be derived by logical deduction (like the logic deduction in Euclidean geometry). This modern scientific viewpoint will be elaborated further; the details of the basic laws and their discovery will be explained shortly afterward.

Basic Theory of Physical Science—the Newtonian View. It has been mentioned already that all material substances can be reduced to elementary particles, i.e., electrons, protons, and neutrons. Each elementary particle has a number of basic properties: mass, electric charge, and magnetic moment. Besides, elementary particles can interact with one another by exerting forces. In fact each basic property is related to a basic force; thus the mass gives rise to gravitational force, the electric charge gives rise to the electric force, and so on. More important, we learn in modern science that forces are the only way particles exert influences on one another and therefore all natural phenomena, such as sound, heat, light, physical and chemical transformations, are accountable by forces. Two questions arise: "How does a force exert the influence?" and "What are the forces?"

These two questions are answered, respectively, by two sets of basic laws, the *dynamical laws* (or the *laws of motion*) and the *physical laws* (or the *laws of force*). The dynamical laws are generally and essentially provided for by Newton's laws of motion, which give a complete, unique, and self-consistent description of the way forces exert influence. The laws of force are equally simple and few. There are only a few basic forces, such as the gravitational force acting between masses and the electric force acting between electric charges. The laws of such forces are simple and well understood. The theory that uses these two kinds of laws to make a systematic study of natural phenomena is called a *dynamical theory,* of which the Newtonian theory was the first. Our present scientific knowledge is such that, based on the few basic properties of the elementary particles, the dynamical laws and the few basic forces, all other properties of matter and all natural phenomena can generally be explained. For example, we can explain why a diamond is hard and transparent and why graphite is soft and dark even though both are made of the same basic substance, the element carbon. We can explain the formation of the stars, the movements of the planets, and the changes of mountains and oceans. We can explain the burning of gasoline and the propelling of an automobile as a result of it. We can explain the Archimedes principle and the principle of the lever. And so on. Thus the elaborate catalog of properties and laws is reduced to a few basic properties and a few basic laws.

That all these can be done is truly astonishing. Science begins by studying isolated specific problems and one would expect that it ends with an elaborate catalog of detailed information. Yet general principles emerged and the complexity is vastly reduced. We have no time to narrate the whole fascinating story. What we will do here is to use this viewpoint as a basic framework to present and relate the basic knowledge of physical science and to apply this knowledge to deal with practical problems.

Modifications of the Newtonian View. The Newtonian theory, in spite of its remarkable success, is not the final, ultimate theory. It has undergone three major modifications. Still, the theory is approximately correct. Its validity has to be qualified and limited. Modifications enable the theory to be extended beyond the limitations.

The first modification is accomplished by the *field theory,* another dynamical theory. In the Newtonian view the physical world is a great void (empty space) populated here and there by material particles exerting forces directly on one another (action at a distance). Material particles are the only physical entity the world is composed of. In the field theory material particles do not exert forces directly on one another but do so through an intermediary called the field (see Sec. 4.4 for a more detailed discussion). A rocking boat causes a fishing float to bob up and down. It does not exert a force directly on the float. The action is mediated by the water wave. The field is something like the water wave that transmits an interaction. According to the field theory the empty space is filled with fields (even in a vacuum). Yet the field is a real physical entity (it has energy, for example). Thus the physical world is made of two basic entities: material particles and fields. However, the field theory has developed to such an extent that it claims that particles (matter) are not an independent entity but a manifestation of the fields. Thus the physical world is considered to be made of only one basic entity: the fields. It is on this basis that we can explain the creation and annihilation of particles in nuclear physics.

The field theory revolution (late in the nineteenth century) was not the only one in the history of physics; it was followed by two more that are even more drastic: the *relativity theory* revolution (1905) and the *quantum theory* revolution (1924 on). In each case many basic Newtonian concepts were revised. Thus in the theory of relativity space and time become relative such that the scales of length and time may change according to the state of motion. Also mass is no longer a constant but increases with speed. In the quantum theory (or rather quantum mechanics) even the concept of determinism is discarded and all physical descriptions become probabilistic. Yet in spite of these drastic changes in fundamental concepts, the Newtonian theory again turns out to be approximately correct to a very high degree of accuracy for practical problems of everyday life, if not for atomic and cosmic problems. Again the new theories include the old theory as a special case and whatever accomplishments of the old theory are preserved in the new theory (in this sense they are not revolutions at all; they are actually generalizations).

So the most sophisticated physical theory at the

present time is the relativistic quantum field theory, which replaces the Newtonian theory as the basic theory of modern natural philosophy. The modern theory has solved most problems in physical science in principle, except in *high-energy physics*. On the other hand, most applications involved in this book fall into the limited region in which Newtonian theory is accurate to a high degree. Thus we will base most of our discussions on the Newtonian theory, with elements of the other theories introduced briefly in specific problems when called for.

The preceding discussion is meant to bring out the point that the concepts and contents of the Newtonian theory are not absolute and have been subjected to revisions. In addition, it is desirable to lay down the basic elements of the complete structure of modern physical theory once and for all at the beginning, so that all subsequent applications will be confined to coherent, logical deductions; we do not want to leave the impression that science is a bag of endless, incoherent magic tricks.

Having described the limitations and qualifications of the Newtonian theory we can now return to it to give a more detailed description of the theory and its applications. As already mentioned the Newtonian theory encompasses two kinds of laws of nature: the dynamical laws, which tell how a force changes motion, and the physical laws, which tell what the force is like. We will discuss the two aspects separately.

4.3 The Dynamical Laws—Newton's Three Laws of Motion

The dynamical laws are given completely by Newton's three laws of motion. Before we state and discuss these laws a few preliminaries are in order. Because we deal with force, a quantitative measurement of force is necessary. For this purpose we may

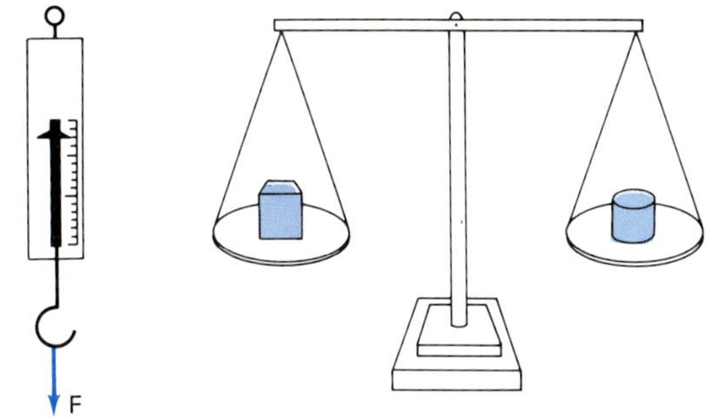

Figure 4.1 Instruments for measuring force (spring balance) and mass (equal-arm balance).

use a spring balance (Fig. 4.1). When a force F is applied to it, the spring stretches out. The indicator, showing the degree of stretch, gives us a quantitative measurement of the force F. The unit of force in the English system is the *pound*. The unit of force in the metric,[2] mks system (meter, kilogram, second) is the *newton* (equal to 0.224 lb), which will be defined later. We use mainly the mks system in this book.

Because we deal with matter, we need a quantitative measurement of the amount of matter. The measurement may be carried out by an equal arm balance (Fig. 4.1) on which we compare the amount of matter to be measured with a set of standards calibrated in the unit of *kilogram* (kg) and its subdivisions gram (g) and milligram (mg) in the mks system; the standard 1 kilogram is the one that will balance exactly 1 liter of water on an equal arm balance. The quantity so determined is called the *mass* and the unit is the kilogram.

Each mass on earth experiences a gravity force acting on it by the earth, the magnitude of which is called the *weight*. Because it is a force it may be measured by a spring balance. The weight is propor-

[2]The metric system was established as a fallout of the French Revolution. The unit of length *meter* (m) is defined as one ten-millionth of the distance from the North Pole to the equator then known. Other decimal units follow; thus 1 *centimeter* (cm) is one hundredth of 1 m. One *cubic centimeter* (cc) is a unit of volume; 1,000 cc is defined as a *liter*.

tional to the mass. It may be used as an indication of mass. When we say, "This box of cereal weighs 1 lb," it means the gravity force the earth exerts on it is 1 lb. But actually we use such a statement to convey the information on mass—how much stuff (cereal) is in the box. Thus weight is often used to indicate mass on earth. However, on the moon, where the gravity force is one-sixth smaller, a 1-lb box of cereal on the same spring balance would show a reading of $\frac{1}{6}$ lb, much smaller than on earth. However, on an equal arm balance on the moon the mass measured, 0.45 kg, of course remains the same.

Another concept necessary for presenting Newton's laws is *acceleration,* which is defined as the time rate of change of the speed of an object. The speed, obtained by dividing the distance traveled by the time, is a measure of how fast an object moves. Its unit is meter per second or foot per second (or mile per hour). Velocity means the same thing but in scientific usage speed indicates the magnitude only, whereas velocity indicates a magnitude together with a direction. The speed of an automobile may be measured by the speedometer; the indicator shows the speed of the moment on the scale. (Policemen have other means to measure automobile speed, e.g., by radar.) When speed increases (speedometer indicator sweeps up) we say that the object is being accelerated and the amount of increase of speed per second is called the *acceleration*. When an object has an acceleration a, it means that after 1 sec the speed will be increased by a m/sec, after 2 sec the speed will be increased by a total of $2a$ m/sec, . . . , and after t sec, the speed will be increased by a total of ta m/sec. If the object begins with an initial speed v_0 then the total final speed at the end of t sec will be

$$v = v_0 + at$$

The unit of acceleration is the unit of speed divided by the unit of time and thus is (m/sec)/sec or m/sec^2 or similar units in other measuring systems.

We can now state the three laws of motion of Newton. The first law states that when no force is

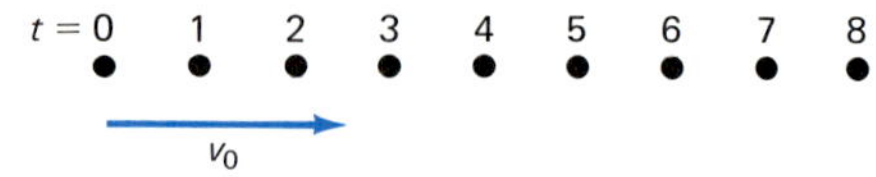

Figure 4.2 Illustration of Newton's first law.

acting, a particle (point object), once moving with an initial velocity v_0, will keep on moving with the same velocity v_0, that is, with the same speed in the same direction (Fig. 4.2). This includes the special case $v_0 = 0$; once the particle is at rest it will remain at rest.

The second law states that when a force F is acting on a particle of mass m (Fig. 4.3), the particle will have an acceleration a in the direction of the force given by

$$F = ma$$

In the mks system we use the units kilogram, meter, and second for mass, length, and time; the unit of force to be used in the preceding equation is called the *newton*. Another system using centimeter, gram, and second as the basis is called the cgs system (its unit of force is called the *dyne*). The English system uses foot, pound, and second as units of length, force, and time (its unit of mass is called the *slug,* which we will not use).

The third law states that when a particle 1 exerts a force F on another particle 2 (action), then, at the same time, particle 2 will exert a force on particle 1 (reaction) equal and opposite to F (Fig. 4.4). In other words forces always appear in pairs between two particles, and action and reaction are equal and opposite to each other.

As an example we may consider the problem of the free fall of a weight under gravity. According to the second law the gravity force gives rise to a downward acceleration. It so happens that neglecting air

Figure 4.3 Illustration of Newton's second law.

Figure 4.4 Illustration of Newton's third law.

resistance, the gravitational accelerations of all objects near earth's surface are essentially constant and are all the same. The magnitude is 9.8 m/sec^2 in the mks system and this quantity is usually designated by g. According to the previous discussion on speed and acceleration we find the speed of the object at any time t (at the end of t sec) to be

$$v = v_o + gt$$

From this result we can calculate the distance x traveled in time t, by multiplying the average speed with time. In this case, the average speed is one half of the sum of the initial speed v_o and the final speed $v_o + gt$. Thus

$$x = \tfrac{1}{2}[\, v_o + (v_o + gt)\,]\, t$$

or

$$x = v_o t + \tfrac{1}{2}gt^2$$

This equation enables us to determine the position of the object after falling t sec. In fact this equation tells us the complete future history of the motion of the object. The dynamical problem is thus solved.

When a mass of m kg is accelerated with the acceleration g, the force on it must be mg newtons according to the second law. This happens with the gravity force; thus the gravity force on a mass of m kg is mg newtons. That on 1 kg is 9.8 newtons. One newton can be calibrated on the spring balance this way.

The preceding example illustrates the general approach by which the Newtonian theory tackles the dynamical problem of a particle. The first law tells what happens when no force is acting. The second law tells what happens when a force is acting. These cover all possibilities that may happen to a particle. What results is the generation of an acceleration (including zero acceleration for zero force). From the information of acceleration a and initial velocity v_o we calculate the other information, i.e., the velocity v and the distance traveled x at any time t. These give us the complete mechanical history of a particle.

A planet moving around the sun may be treated in a similar way (Fig. 4.5). From the gravity force the sun exerts on the planet we figure out the acceleration by the second law. This time the acceleration is not constant because the gravity force changes with the distance of the planet (more details later). Moreover, the acceleration adds velocity to the planet in a direction different from the initial velocity v_o. Thus we cannot calculate v and x by the simple method used above. Besides, there are two coordinates (x,y) for the planet position. However, the calculation can be carried out by more sophisticated methods, such as calculus. Once we have calculated v and (x,y) the future history of the planet is completely determined and the problem is solved. Other information, such as the size and shape of the orbit, the period of revolution, and so on, can be derived from them readily.

Many astronomical problems are solved in this way. Once we have worked out the complete future history of the motion of the earth and the moon relative to the sun then we can predict the times in the future when the three will be along the same straight line. These are the times of the eclipses of the sun and of the moon. By the Newtonian theory these times can be predicted accurately for thousands of years to come. Eclipses were mysterious and

Figure 4.5 The motion of a planet.

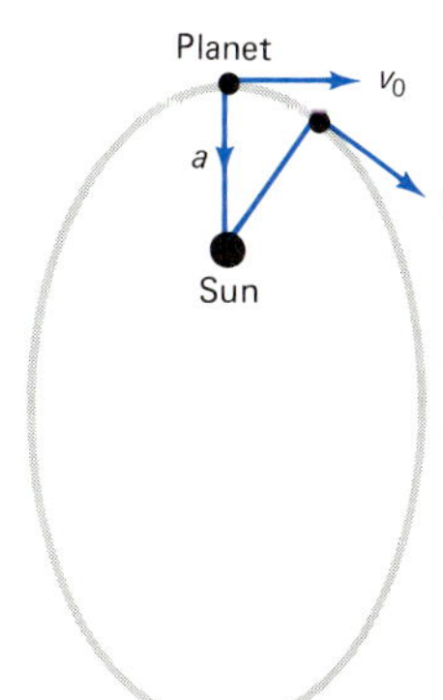

awesome natural phenomena to ancient people. From the Newtonian point of view they are reduced to a problem of the motion of three bodies. They are just a matter of space–time coincidence and are completely accounted for by the Newtonian laws. This is how science explicates the mystery of nature.

The astronomical applications are the first great triumph of the Newtonian theory. Yet the theory as stated is limited in scope because it deals only with point objects or particles. Many of the things we deal with in nature and in everyday life are extended objects. How do we deal with them?

On this point the Newtonian view asserts that any extended object may be regarded as being made up of a large number of particles. For each of the particles the three laws of Newton apply. The future behavior of each particle can thus be determined. When the results on all particles are put together the behavior and the future history of the extended body can be determined.

Whether this program can actually be carried out is not obvious. Yet, after 200 years' work we know that this can be done. Newton's laws are such that for any mechanical problem of any extended material system, be it a gas, a liquid, or a solid, the preceding process of analysis can always be carried out and it always leads to one and only one solution. In other words, Newton's laws are complete, consistent, and unique. No additional laws of motion are needed. Newton has pre-empted them all. This is the most remarkable feature of the Newtonian theory from the logical point of view. It is here that the crucial role played by the third law manifests itself. The theory would not be complete, consistent, and unique without the third law.

Much of the science of mechanics thus involves the analysis of the complex problems into elementary problems of particles that can be dealt with by the three basic laws and the integration of the results so obtained to determine the behavior of the complex material systems. Thus we have developed mechanics of rigid bodies to deal with rigid, solid objects, mechanics of elastic bodies to deal with solid but deformable objects, hydrodynamics to deal with liquids, and aerodynamics to deal with gases. All problems are uniquely solvable in principle. Whatever difficulties remain are mathematical difficulties involved in solving the equations. With the help of computers most of the remaining mathematical problems can now be solved subject to the limitations of money and time.

Many engineering problems are mechanical in nature and can be dealt with very successfully by the Newtonian theory. Sending astronauts to the moon is essentially a mechanical problem and can be solved with predictable precision. The project was mapped out 10 years ahead of time and the progress turned out generally as expected.

Still, on an overall basis, the success of the Newtonian theory as discussed so far is limited to a small area, because mechanics is just one of the many areas involved in natural phenomena. There are many others: sound, heat, light, electricity, and so on. Does the Newtonian theory have anything to do with them? It does. In fact, Newtonian theory turned out to be the foundation of all others. This makes the importance of Newtonian theory universal. It is *the* theory of physical science. First it was found that sound is exclusively a mechanical phenomenon. In spite of the fact that the hearing sensation seems beyond mechanics, sound is nothing but the mechanical vibration of the air. All Newton's laws and mechanics of the air apply and any problem of sound may be dealt with accordingly. Heat was first thought to be due to another substance different from material substance and thus was beyond the Newtonian theory. After an extensive study it was found that heat is also the manifestation of a mechanical phenomenon. It is the random motion of the molecules the material substance is made of. A mechanical theory of heat was developed and heat problems were solved accordingly. Electricity is the manifestation of a force between electric charges. Force is a Newtonian concept and electrical phenomena are analyzed on a mechanical basis according to Newton's laws. The Newtonian theory became so all-encompassing that one school of philosophy regarded everything in the universe, including life, man, and history, to be controlled by Newtonian laws.

4.4 Laws of Force

We now consider the other kind of basic laws of nature, i.e., the laws of force. In the Newtonian scheme everything of importance is determined by force and thus the essence of physics is summarized in the nature of forces. In studying forces we find that there are only a few fundamental forces and all natural phenomena may be understood as the manifestations of these few. This conclusion is arrived at a long time after Newton, even though Newton was instrumental in establishing the first fundamental force, i.e., the force of *universal gravitation.*

In our everyday life we see the manifestations of many kinds of forces. A familiar one is the gravity force of the earth, which causes heavy objects to fall downward. Newton, in an attempt to explain the movements of the planets, hypothesized the existence of an attractive force between the sun and the planets. This might well be a different kind of force from any we know on earth; after all, it deals with things in Heaven. Yet Newton pursued the problem one step further. He suspected that the force the sun exerts on the planets and the force the earth exerts on heavy objects on its surface might be of the same kind. In other words, the gravity property is not a special property of the earth but a universal property of everything in the world, including the sun. To equalize things in Heaven with those on earth was against the Aristotelian tradition as well as the religious belief of the time. Newton's idea was thus a courageous innovation. If this is true, then on the basis of the information of this force on earth (gravitational acceleration) we can make a prediction on things in Heaven. We thus calculate the period of the moon orbit and predict it to be 27 days. The observed fact is that the moon revolves about the earth once every 27.3 days. The close agreement between the prediction and the fact lent strong support to the concept of *universal* gravitation.

Thus one basic force in nature was found. The detailed nature of this force is specified by Newton's *law of universal gravitation,* which states that for any two particles of masses m_1 and m_2 separated by a distance r there is an attractive force F between them with a magnitude given by the equation

$$F = G\frac{m_1 m_2}{r^2}$$

where G is a constant, called the *gravitational constant,* the value of which is 6.67×10^{-11} in the mks system of units. The "attractive force between them" actually means two forces (Fig. 4.6): one, F_1, is that which m_2 exerts on m_1 and the other, F_2, is that which m_1 exerts on m_2. The two are action and reaction to each other and thus are equal and opposite. The particle m_1 under the action of force F_1 acting on it moves according to Newton's second law, disregarding F_2. On the other hand, F_2 affects m_2 in a similar manner. The results are that both m_1 and m_2 are accelerated toward each other and this is what we called an attraction.

Many forces can be reduced to this basic force and thus many mechanical phenomena can be explained by the same basic law underlying the force of universal gravitation. For example, buoyant force of water is traceable to its gravitational origin. Yet there are other forces of different origin. Objects moving unaided on a horizontal plane, such as a coin sliding on the floor, eventually come to a stop; they do not keep on moving indefinitely as the first law specified. This is explained by saying that there is a *friction force* acting on the coin by the floor over the surface of contact that slows down the moving coin. Human and animal muscles can exert force. Electrically charged objects exert electric forces on one another. These are forces of different origins. Each force is a manifestation of a separate physical principle and for a time it seemed that we needed a large number

Figure 4.6 Action and reaction in gravitational interaction.

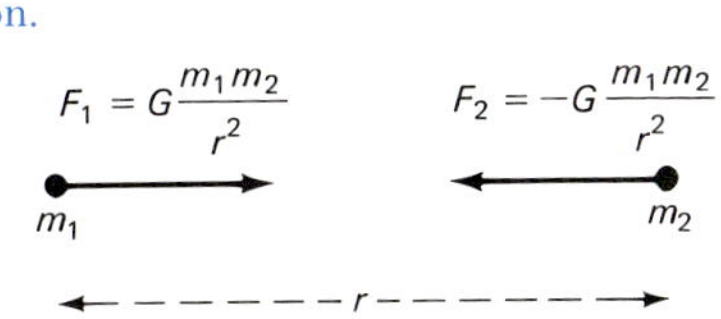

of physical principles to explain all natural phenomena in the physical world.

The situation is changed after we have learned more about the *electromagnetic force,* the second basic force. Practically every force not of gravitational origin in our everyday life can be reduced to this force and thus to one basic law. It has been mentioned that two electric charges exert force on each other. Also, two magnetic poles do the same. Thirdly, we learned that a moving electric charge (and thus a current) will exert a force on a magnetic pole; in other words, it generates a *magnetic field.* Finally, a moving magnetic pole (and a moving magnetic field) will exert a force on an electric charge; in other words, it generates an *electric field.* These four kinds of forces are so connected with one another that they are not four separate, independent forces but four manifestations of the same force, albeit a rather complex one, called electromagnetic force. Because of the complexity the law of this force cannot be written down in a simple equation as in the law of universal gravitation. Nevertheless, if we limit ourselves to the first aspect, the *electrostatic force* F between two charged particles at rest can be written down simply in an equation that turns out to be very similar to that of the gravity force,

$$F = k\frac{Q_1Q_2}{r^2}$$

where Q_1 and Q_2 are the electric charges of the two particles, r is the distance separating them, and k is a constant equal to 9×10^9. The unit of electric charge used in this equation is called the coulomb (C) and the units of the mechanical quantities F and r are those of the mks system. These will be the units used throughout this book. Like charges repel and unlike charges attract each other. This equation represents the *Coulomb law of electrostatics.* There is a similar *Coulomb law of magnetostatics* giving the magnetic force between two magnetic poles. The third and fourth aspects of the electromagnetic force are more complicated and so we do not go into their mathematical details here. We shall only state that for the third aspect there is an *Ampere law* describing the magnetic field produced by an electric current and for the fourth aspect there is a *Faraday law of electromagnetic induction* stating the electric field produced by a moving magnet or a changing magnetic field.

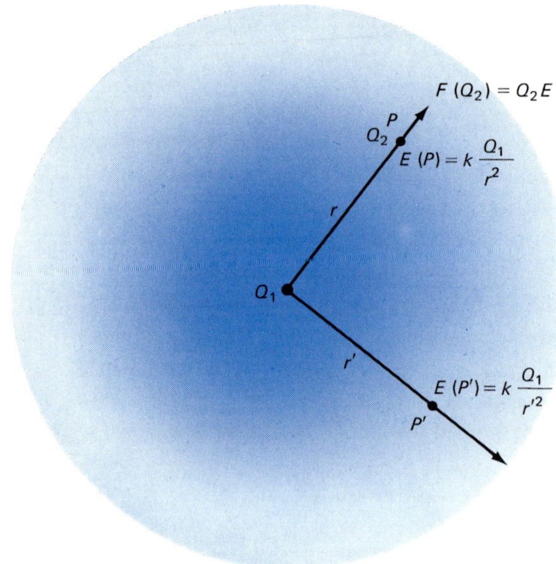

Figure 4.7 The electric field generated by a charge Q_1.

This is the proper place to introduce the concept of electric (and magnetic) *field* on a quantitative basis. According to the field theory, instead of saying that the charge Q_1 exerts a force of kQ_1Q_2/r^2 on charge Q_2 at the point P in space, we say that the charge Q_1 generates an electric field (a sort of sphere of influence) in the space surrounding charge Q_1, including the point P (Fig. 4.7). Electric field is an *attribute of the space* and it is the electric field at P (generated by Q_1) that exerts a force on Q_2. The force is exerted through an intermediary, the electric field. We define the strength or intensity of the electric field E at a point, say, P as the force the field exerts on a unit charge placed at P. Because the charge Q_2 placed at P experiences a force of $F = kQ_1Q_2/r^2$, a unit charge placed at P will experience a force F/Q_2 or kQ_1/r^2, which will be the intensity of the field at P. Thus the intensity of electric field E at P produced by Q_1 is

$$E(P) = k\left(\frac{Q_1}{r^2}\right)$$

For another point P' the field intensity can be determined in a similar manner and the magnitude is kQ_1/r'^2 when r' is the distance from Q_1 to P'. The same for all other points P'', P''', and so on. Thus for any point in the neighborhood of Q_1 we can associate a field intensity. *The totality of field intensities at all points (infinite in number) in the neighborhood of Q_1 is the quantitative definition of the electric field generated by Q_1.* There is no need to list all the numbers, because all of them follow the same formula kQ_1/r^2, and the formula $E = kQ_1/r^2$ representing the intensity as a function of the distance r is a sufficient mathematical representation of the field. It should be kept in mind that the field is a property of the space dissociated from matter, like the smile of the Cheshire cat in *Alice in Wonderland.* (The cat has disappeared but the smile is still there. To put it metaphorically: Newtonian theory deals with the cat whereas the field theory deals with the smile.) Nevertheless, the field can be visualized by the variation of the intensity over the space, just as the variation of altitude of a rolling field over the surface of the earth—the undulation of the topography—gives one a sense of the field.

Magnetic field B is introduced as an intermediary for the magnetic force. As far as the two Coulomb laws and the Ampere law are concerned the field concept is an intermediary not absolutely necessary and is dispensable. However, in the Faraday law of electromagnetic induction the field concept becomes indispensable and assumes a real significance. It enters a basic law of nature (the Faraday law) and the law cannot be stated without it. Thus it may be said that it is possessed of physical reality. That is the starting point of the field theory.

Based on the electromagnetic force all electric and magnetic phenomena can be explained in principle. Furthermore light is found to be an electromagnetic phenomenon and thus can be accounted for by the same force. Later it was found that the forces between atoms and molecules are basically electromagnetic and all their interactions may be accounted for by this force. Thus chemistry is based on electromagnetic force. Friction is a molecular effect and thus is reducible to electromagnetic force. Finally, muscle force originates from chemical reactions inside the muscle cells and thus is also reducible to electromagnetic force. In fact, most of the forces encountered in our everyday life, if not reducible to gravity force, are reducible to electromagnetic force.

Thus there seem to be just two basic forces. Together with the dynamical laws, the two forces are capable of explaining practically all physical and chemical phenomena in everyday life. Indeed, close to the end of the nineteenth century many scientists were of the opinion that all the important laws of nature were already found and that further progress could involve only the improvement of technical details. This optimism turned out to be quite premature. As mentioned before, the dynamical laws are to be modified by the field theory, the relativity theory, and the quantum theory. Also new basic forces are to be added.

Although the two basic forces account for all physical and chemical phenomena involving atoms, molecules, and bulk material, in the field of the atomic nucleus new forces are necessary. We have already mentioned that the atomic nucleus is made up of two kinds of elementary particles, protons and neutrons. How do they stick together to form a nucleus? There must be an attractive force among them. The gravity force and electromagnetic force are too weak to account for the tight binding in the nucleus. Thus a new force, the *nuclear force,* is assumed to exist among protons and neutrons. This force is taken as the basis to explain the formation, transformation, and reaction of the atomic nucleus. What the law of nuclear force is, the law corresponding to the Newton law of gravity force, is a question that has been under intensive investigation.

Unfortunately there is no simple formula known for the nuclear force like those of the gravity force and Coulomb force. Available information of this force can be expressed approximately in complicated formulas. Nevertheless, the essential features are rather clear and many phenomena in nuclear physics and elementary particle physics can generally be understood in terms of them. This force is more specifically called the *strong nuclear force* or simply the *strong interaction* to distinguish it from another nuclear force to be described presently.

There are some nuclear phenomena that cannot be understood in terms of this strong interaction. A neutron taken out of the nucleus and left by itself is unstable. After 13 minutes it has a 50 per cent probability to change itself into a proton together with the emission of an electron and another elementary particle called an *antineutrino*. This is a change not familiar in Newtonian theory, in which the particles remain always the same and only the space–time association of the particles is changed by the forces. The change we are concerned with here encompasses not only space–time but also the nature and the number of particles themselves. Yet it is a change that can be properly accounted for in the quantum field theory, and again there is a cause responsible for the change that may be considered as a force. (The term *force* is now used beyond the Newtonian sense we discussed earlier, but is a generalization of that sense.) The force so introduced is called the *weak nuclear force* or the *weak interaction*. It manifests itself in many other nuclear phenomena (beta decays and radioactive decays of other elementary particles).

Thus we have two new nuclear forces. Together with electromagnetic force they give a good account of the whole of nuclear physics. Gravity force also acts in nuclear phenomena but it is too weak to be of any significance.

With the four basic forces we can explain practically every physical phenomena from elementary particles up to bulk substances, including the whole of chemistry. Very recently investigations in high-energy physics uncovered some mysterious happenings that cannot be explained by the four basic forces. However, they can be explained by hypothesizing the existence of two additional new forces: a fifth *superstrong nuclear force* and a sixth *superweak nuclear force*.[3] These two forces are far from being firmly established and they deal with rather special situations. Their status is provisional and things may change later. With regard to everything of practical concern, it is safe to say that the four forces account for all the physical and chemical phenomena known to us.

4.5 The Modern World View

Let us first summarize the Newtonian world view, which to a large extent still dominates the modern scientific outlook.

The first element is the concept of matter as embodied in the notion of a *particle*. Newton said: "It seems probable to me that God in the beginning formed matter in solid, hard, massy, impenetrable, movable particles." This idea is carried to its ultimate completion in modern physics, in which all material substances are reduced to the combinations of three kinds of elementary particles: electrons, protons, and neutrons. It is now a fact, not merely a philosophical speculation, that ordinary material substance is indeed made of particles (and there are only three kinds). Particles of one kind are exactly the same. Each particle is always the same at all times, having a number of permanent properties that never change under any circumstances. (This part is revised in quantum field theory.) All changes in natural phenomena are considered as the manifesta-

[3] Any unexplained thing can always be explained by a new force concocted for such a purpose. Thus explanation of a new phenomenon by invoking a new force really is not an explanation. The significant point is that by assuming merely four relatively simple forces, practically all the physical world can be explained.

tions of the space–time changes of the particles; because all properties of a particle are permanent and unchanging, the only thing changeable is its space–time designation. Space–time changes are commonly called motion. They include not only movement of the particles, such as translation and rotation, but also combination and dissociation of the particles. The science that studies motion or space–time changes is *mechanics*. The Newtonian world view is thus a mechanical one, reducing all phenomena into space–time changes. Loud or soft sound sensation is reduced to strong or weak vibrations of the air molecules. Hot or cold sensation is reduced to the strong or weak random motion of the molecules. It views the whole universe as a giant clockwork that ticks according to mechanical laws.

The second element is the concept of force as embodied in a set of physical laws prescribing the nature of the forces. The fruition of this idea is manifested in modern physics by the establishment of the four basic forces that explain practically all natural phenomena. The four basic forces are gravity force, electromagnetic force, strong nuclear force, and weak nuclear force. The force is an attribute of matter (or particles) by which one particle exerts influence on another. (By influence we mean the change of space–time relations.) This attribute is related to the basic properties of the particle. Thus the mass of the electron determines the gravity force it exerts on other masses, the charge of the electron determines the electric force it exerts on other charges, and so on. How large the force is depends not only on the properties (mass, charge, and so on) of the particle exerting the force but also on the properties of the particle the force is acting upon, as well as the space–time relation between the two particles (distance, relative velocity, and so on). These points are illustrated by the Newtonian law of gravitation and the Coulomb law of electrostatics. In general, the nature of a force is specified by a physical law. At the present time we need only four basic laws for the four basic forces.

The third element is the concept of dynamics as embodied in a set of dynamical laws. The purpose of the dynamical laws is to determine how a force changes the space–time relations. Thus Newton's second law specifies that a force gives rise to an acceleration that determines space–time changes. From the space–time changes, natural phenomena may be explained. Also the future course of development may be predicted. It is obvious that the three elements are closely related and that one cannot be separated from the others. The Newtonian laws of motion are the first complete set of dynamical laws. Their establishment marked the beginning of modern science. It can be shown that the Newtonian laws of motion are complete (no other dynamical laws need to be added), deterministic (prediction of the future is definite, not probabilistic, as in dice throwing), unique (there is only one prediction, no alternatives to choose from), and consistent (the theory applied in different ways leads to the same conclusion).

As such the Newtonian theory is truly one of the most remarkable intellectual accomplishments in human history and its significance is far beyond mechanics and physics. Now we know that the theory is not absolutely correct, but it is still approximately correct. It has to be modified by the field theory, the relativity theory, and the quantum theory. These modifications comprise the modern dynamical theories. Yet many elements of the Newtonian theory remain valid.

With the three elements it appears as if we have all the necessary components of a complete theoretical scheme, but this is not the case. There is a fourth element, independent of the three and equally indispensable, that is, *the initial conditions*. Let us go back to the equation showing the velocity of a particle in a uniformly accelerated motion:

$$v = v_o + at$$

The velocity at a time t sec later is the initial velocity v_o plus t times the acceleration a. The initial velocity v_o specifies the initial condition. In the preceding equation the second part, dependent on a, comes from the three elements discussed before. The first part, v_0, comes from the initial condition having nothing to do with the previous three elements. The final answer of the problem depends on both and therefore the initial condition is an independent, indispensable part of the theory. The first three elements form a rational scheme. The initial condition is not predictable from this scheme, has to be taken as given, and has to be found out by observation. It is an irrational element. That the Newtonian theory includes an irrational element is due to the fact that the dynamical laws tell only the *rate* of change of velocity (acceleration). What the velocity was before the change begins, that is, v_0, is not told and cannot be rationally derived from the theory. It is unrelated to the rational part of the theory and has to be added from outside the theory. It has to be taken as a fact without asking why. After the Newtonian laws (of motion and of gravity) have relieved God from the daily chore of whirling the planets in the solar system, Newton still thought God indispensable; he thought it must have been God who gave the planets an initial push to endow them with an initial velocity. Once the initial condition is thus provided for, the dynamical and physical laws will take care of the motion of the planets, and the predictions of all the eclipses can be made for thousands of years to come without seeking God's consent.

Let us apply the Newtonian viewpoint to deal with scientific matters. The particle structure of matter was largely speculative in Newton's time but is now an established fact. The three kinds of elementary particles—electrons, protons, and neutrons—combine in a hierarchical structure to form first nuclei, then atoms, then molecules, and finally bulk material. The working out of this hierarchical structure is another demonstration of the fruitfulness of the Newtonian viewpoint; the combination of the particles in the hierarchy is governed by the same dynamical laws and the same basic forces mentioned above. The properties of the composite entities are thus determined by the same basic laws and no new principles need to be added. In particular their transformation from one form to another is governed by the same dynamical laws and forces. It is in transformations of material systems (Chapters 6, 10, 12, 14) that all natural phenomena manifest themselves and therefore all phenomena are explainable in terms of the same dynamical laws and forces.

Beginning with protons and neutrons, these particles combine to form atomic nuclei in the first level of the hierarchical structure. Each nucleus is formed by a definite number Z of protons and a definite number N of neutrons. The combination is due to the strong interaction. The properties of nuclei and their transformations are determined by strong, electromagnetic, and weak interactions. This part of science is called *nuclear physics* and we will discuss it in more detail in Chapter 14.

The nucleus, though a composite system, may be regarded as a particle endowed with specified forces and may enter into interaction with other particles by these forces. These forces are not new; they are the forces of the protons and neutrons left over after they combine to form the nucleus. For practical purposes the most important force a nucleus exerts is the electromagnetic force. By this force the nucleus interacts and combines with Z electrons to form an atom that represents the second level in the hierarchical structure. This part of science is called *atomic physics,* which will be discussed in more detail in Chapter 12.

An atom, though a composite system, may also be regarded as a particle and may enter into interaction with other atoms to form a molecule that represents the third level in the hierarchical structure. The force responsible for the combination of several atoms to form a molecule is again the electromagnetic force originating from the electric charges of the nuclei and the electrons.[4] This part of science

[4] In addition to the familiar electrostatic force there is a more important "exchange force" that is also of electromagnetic origin. It arises from the electronic charge distribution in the atom that is affected by a quantum effect. The exchange force is responsible for the formation of the *covalent bond.* The other bond, i.e., the *polar bond,* is explainable by electrostatic force. Thus all chemical combinations are originated in electromagnetic force. This point will be elaborated in Chapter 12.

is called *molecular physics* or *chemistry,* depending on historical background, methods of investigation, and personal preference. We will discuss it in more detail in Chapter 10.

The molecules may aggregate to form bulk material, which is the fourth level in the hierarchical structure. The force responsible for the aggregation, called *intermolecular force,* is again electromagnetic in origin.[5] The properties of bulk material and their transformations are thus explainable in terms of the properties of the molecules and intermolecular forces. These properties underlie most of the natural phenomena in our daily life and most scientific matters in our economic activities, including the scientific problems concerning environment and energy. The empirical investigation of this part of science began early in history and belongs to the part of physics now called *classical physics.* However, the complete microscopic explanation comes much later from atomic and molecular physics. We will discuss this part of science in more detail in Chapter 6.

Finally, bulk material may aggregate to form astronomical objects such as stars and planets. A planet, viewed at a large distance, may again be regarded as a particle endowed with forces and may interact with others. The forces are again those of the constituent parts (electrons, neutrons, and protons) left over after the combination processes. This time only the gravitational force remains; all other forces are either canceled out or negligibly small. Thus the movements of the planets are explained by the gravitational force. This last part, *astronomy,* ironically is actually where modern science began.

The preceding discussion on the hierarchical structure of the physical world stresses the static structure and the associated properties. It is remarkable that such static properties can be explained by the Newtonian theory, which is by nature a dynamical theory stressing the changes in the course of time. This success notwithstanding, the dynamical theory was primarily developed to deal with dynamical problems. In the macroscopic as well as the microscopic world there are problems concerning changes in the course of time, such as the motion of a planet about the sun. A large part of scientific investigations actually involves the application of the Newtonian theory to deal with such dynamical problems.

When all dynamical problems of all parts of the physical world are put together, we have collectively the problem of the whole natural history—how the whole physical universe evolves and develops in the course of time. Because the laws of nature are timeless they apply not only to the fate of the universe in the distant future but also to the origin of the universe in the distant past. The origin and evolution of the universe have always been basic problems with strong intellectual interest. The Newtonian approach is ready-made to deal with such problems. The application of modern science to solving these problems has been very successful, and at the present time we have a fairly good knowledge of the origin and evolution of the universe. We now know generally how the universe began and how the solar system was formed; this part of scientific discussion is usually found in astronomy and astrophysics. We now have a reasonably good idea of the origin and evolution of the earth; this part is usually included in geology. We now have some idea of the origin and evolution of life on earth; this part belongs to biology and is largely outside physical science. These topics will be discussed in Chapters 21 through 25, which are devoted to natural history exclusively.

After the emergence of life the important events that follow in history were the emergence of man, of society, and finally of civilization. These were dealt with by social sciences and the humanities and are beyond physical science. But historically they are parts of the whole cosmic process and are related to physical science in the historical perspective.

[5] Although the *dipole–dipole force* and *dipole–induced-dipole force* are obviously electrostatic in origin, the *London force* (*induced-dipole–induced-dipole force*) is also electromagnetic in origin, except that the charge distribution is affected by a quantum effect. Thus all intermolecular forces are electromagnetic in origin.

In recent years human activities have led to drastic modifications of the physical world and we are faced with many severe environmental problems. Many of them are related to physical science as manifestations of transformations of matter and energy. They are discussed in Chapters 2, 3, 8, 9, 17–20 of this book.

When we deal with any dynamical problem in natural history we make use of the known dynamical laws and the known force laws. In addition, we seek out an initial condition; the three elements together make prediction and explanation possible. For example, in the problem of planetary motion we not only need the laws of motion and the law of gravity force but also the initial condition on the position and velocity of the planets at the "very beginning." Now we do not believe the planets suddenly came into being on the Grand Opening day of the universe (as Newton did); we believe they were gradually formed by condensation of primordial material, which itself is a dynamical process subject to scientific analysis in the Newtonian sense. Thus the formation of planets and in particular the initial velocity by which they are formed can be explained scientifically. In this sense the initial condition of a local problem can be rationally explained by another dynamical problem preceding it. Yet from a global point of view this does not eliminate the initial condition as an irrational element. It simply pushes the initial condition to an earlier time, because the preceding problem can be solved only by a knowledge of its own initial condition. Ultimately all initial conditions of all dynamical problems are traced to the one and only initial condition at the beginning of the universe. It cannot be reduced and explained any further and has to be taken as a mere fact. This is an irrational element of science that can never be eliminated. In Chapter 21 we will show that the universe indeed has a beginning and by assuming a rather reasonable initial condition (by intelligent guess and by reconstruction from final results), we can explain reasonably well the whole of natural history in terms of the known dynamical laws and the four basic forces.

⋆4.6 Statics

Each branch of mechanics consists of two subdivisions: *statics* and *dynamics*. Statics, dealing with static problems, is actually a special case of dynamics and its emphasis is largely associated with practical applications. We will discuss two examples: the *principle of the lever* in statics of rigid bodies and the *Archimedes principle* in hydrodynamics.

The principle of the lever may be explained by the diagram in Fig. 4.8. A rod is supported at a point O. A weight W is hung at point A at one end. A force F is exerted at point B at the other end. The force F necessary to lift the weight W is found related to W as follows:

$$F = \left(\frac{n}{m}\right) W$$

where n and m are the lengths of OA and OB. This relation embodies the principle of the lever. If length OB is five times longer than OA, it needs only a force of one fifth of the weight W at B to lift the weight W at A. The principle may also be stated in the alternate form:

$$mF = nW$$

The principle is the basis of many useful devices, such as the weight-lifting machines. It is also the basic principle underlying the equal arm balance as a measuring device. The Oriental weighing device as seen in Asian market places (Fig. 4.9) is also based on this principle, but it differs from the equal arm balance by using a fixed weight and a variable arm length for the measurement of weight, which is more convenient for market place purposes.

Figure 4.8 Illustration of the principle of the lever.

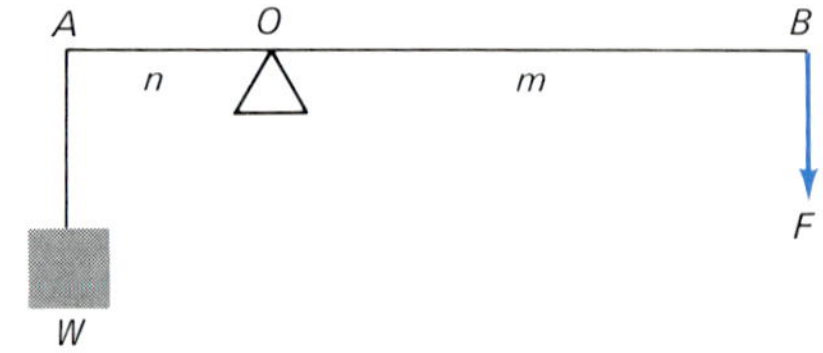

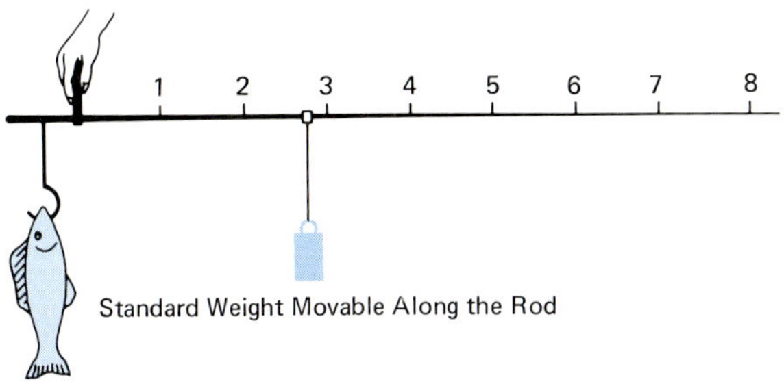

Figure 4.9 Weighing device seen in Oriental marketplaces—an unequal-arm balance called ch'ên.

The principle of the lever can be deduced mathematically in dynamics of rigid bodies, which we do not reproduce here. In this sense it is a theorem in Newtonian mechanics. The Greeks discovered it empirically 2,000 years earlier than Newton (and so it has been called a "principle" ever since, even though we now know that it is not an independent principle beyond Newton's laws). Obviously the Orientals knew it just as early, even though they did not call it a principle or put it in books. If the principle should be wrong by a few percentage points the amount of gold weighed by the weighing device would also be wrong by several per cent, a mistake they could not afford.

Are we then in a position to say that the Greeks and the Orientals contributed to the development of science significantly? It is true that the Greeks and the Orientals have discovered this piece of scientific fact, but they did not establish systematic science, which is what science is supposed to be. Only when one recognizes the significance of the fact and places it in relation to the other scientific knowledge (its derivation from the Newton Laws) does it become an integral part of science. Without recognition of its significance, it is like a piece of gold in the hand of a moron—no more than a glittering toy. Without being related to other parts of science, it remains an isolated fact, a drop in an infinite bucket, with limited significance. Such ancient discoveries are useful in a practical sense but the role they play in the development of science is incidental, not crucial. At best, they prepare for the crucial event to emerge later.

Science, being systematic knowledge, is like a mosaic art made of many pieces of stones. Each stone does not mean much. All stones together still mean little unless they are organized and related in a coherent way. The development of science began by accumulating stones. At a certain stage they were suddenly pieced together to form a grand design. This crucial step took place in the West at Newton's time. Afterward it was largely the filling, modifying, elaborating, and completing of the basic pattern. If we deal with individual drops of scientific knowledge, we can only get lost in the infinite bucket. Our purpose in this book is not to acquire the individual pieces of scientific knowledge. It is more important to grasp the overall view of the modern scientific outlook.

The Archimedes principle, attributed to the Syracusan 2,000 years before Newton, likewise is deducible from Newton's laws. It states that the buoyant force an object experiences in water is equal to the weight of the water the object displaces. The story goes that Archimedes realized this principle while taking a bath and applied it to solve the problem of determining the gold content of the king's crown.

The principle may be derived as follows: consider an object of an arbitrary shape immersed in water as shown in Fig. 4.10. Water on every side of the object exerts force on the surface of the object, as shown by the arrows in the figure. These forces can be added together to form a resultant force, many of them canceling out in the process. The resultant force is the buoyant force the water exerts on the immersed object. How large is this force? Let us imagine that the body is taken out of the water with the water outside (designated by A) undisturbed and the cavity so created filled with another body of water B. Because the body of water A is undisturbed, the forces A exerts remain the same, but this time the forces are exerted on the body of water B. The resultant, i.e., the buoyant force, will now affect B according to the second law. But B does not move; after all, A and B together form one body of water

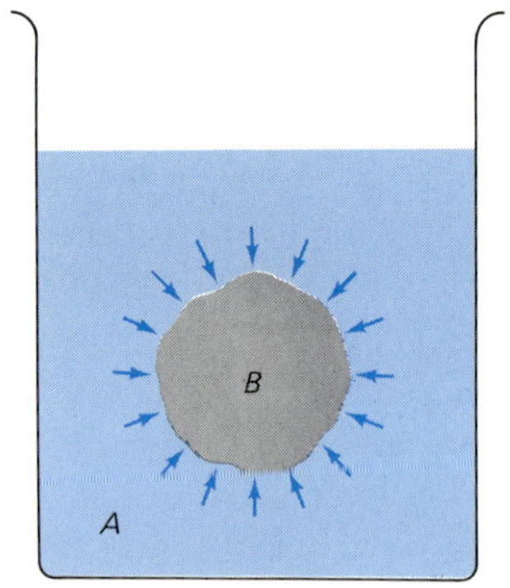

Figure 4.10 The forces water exerts on an immersed body.

Table 4.1 Densities of Materials in g/cc

water	1.0	iron	7.8
aluminum	2.7	lead	11.3
copper	8.9	platinum	21.4
gold	19.3	silver	10.5

the whole of which does not move. Therefore the net force B experiences must be zero (because acceleration is zero). Besides the buoyant force, B also experiences its own gravity force and the two must be equal and opposite. Thus the buoyant force equals the gravity force or the weight of the body of water B the object displaces.

Density. The Archimedes principle may be applied to determine the *density* of a substance. Density of a substance is defined as the mass per unit volume; its unit in the cgs system is gram per cubic centimeter designated by g/cc. When we say iron is heavier than aluminum, we mean that the same volume of iron is heavier than the same volume of aluminum. In other words, the density of iron is greater than that of aluminum. The densities of a few substances are listed in Table 4.1. The density of iron is 7.8, much greater than that of aluminum, which is 2.7. The density of water is 1 g/cc. This is the way the unit 1 g is defined.

If an object weighs 5 g in air (or more correctly, in a vacuum) but weighs 3 g in water, the loss of weight of 2 g is due to the buoyant force. According to the Archimedes principle this equals the weight of water displaced by the object. Knowing it to be 2 g and the density of water to be 1 g/cc, we figure the volume of water displaced to be $2\text{ g} \div 1\text{ g/cc} = 2\text{ cc}$. Thus the volume of the object is 2 cc. The density of the object is thus $5\text{ g} \div 2\text{ cc} = 2.5\text{ g/cc}$.

This is a convenient way to determine the density. If the crown is made of an alloy of gold and copper, the density should be somewhere between 19.3 and 8.9. In fact, from the measured density of the king's crown one can calculate the percentage of gold in the crown by simple algebraic method (Problem 4.8).

The Archimedes principle may be applied to solve the following problem. A piece of ice floats on water in a glass. When the ice melts, will the water level of the glass go up, go down, or remain the same? The answer is that it will remain the same. Why? (Problem 4.9.)

People sink in water but in the Great Salt Lake they become unsinkable. Why?

Not only for solids and liquids but also for gases we can speak of a density. The density of a gas is much smaller, usually about a thousand times smaller than that of solids or liquids. Moreover, the density of gases changes with pressure and temperature. Usually we refer to the *standard conditions,* i.e., 1 atmosphere pressure and 0°C temperature, as a guide to specify density. Thus the density of air is 1.3×10^{-3} g/cc under the standard conditions. The density at other conditions can be calculated by definite relations.

Problems

4.1 To shorten the time to accelerate a car from start to the cruising speed by one half, by how much must the propelling force be increased?

4.2 To shorten the acceleration lane at the entrance of an expressway by one half, by how much must the propelling force be increased to reach the same cruising speed?

4.3 Two lead balls of 1 kg each are placed at a distance of 1 m apart. What is the universal gravitational force between them? How does it compare with the weight of each ball itself?

4.4 In the preceding problem each ball is given 1 coulomb of electric charge. What is the electric force between them. How does it compare with the universal gravitational force?

4.5 According to Aristotle a 100-lb object would fall 100 times faster than a 1-lb object. Show that such a theory is not self-consistent by considering two identical objects falling separately and falling together. In fact Galileo began his study of gravity by such a theoretical consideration, not by an experiment on the Leaning Tower of Pisa, which is legendary.

4.6 Once the U.S. Air Force reported that one of its jet fighters fired a cannon and the shell hit that jet itself. Explain how this could happen.

4.7 Can a propeller plane fly in a vacuum (to the moon, say)? Why? Can a jet plane fly in a vacuum? Why? A space craft uses a rocket to accelerate it to outer space. How would one use the same rocket to slow it down to return to earth?

4.8 A ring is made of an alloy of gold and copper. It weighs 50 g in air and 46 g in water. What is the percentage of gold (in weight) in the ring?

4.9 An ice cube floats in a glass of water. When the ice melts and changes into water, will the water level in the glass rise, fall, or remain the same? Why?

4.10 Earthquake forecasting: Before an earthquake the ground might swell, sometimes as much as a few meters. What would be the change of the value of the gravitational acceleration g caused by this swelling? By detecting such a small change by very sensitive instruments, the impending earthquake might be predicted beforehand.

4.11 Archimedes once said, "Give me a place to stand; I can lift the earth for you." How would he do that?

5

Energy

5.1 Kinetic Energy and Work

In this chapter we will discuss another basic concept in physical science, that is, *energy,* which is also important in dealing with many practical problems. Matter and energy are two fundamental concepts in physical science. They supplement each other. Their relation will be discussed more in detail in Sec. 5.4. We choose to present the fundamentals of physical science by organizing the material around these two concepts. This also paves the way for our later discussion of the environmental problems in terms of transformations of matter and energy. Many environmental problems, such as air pollution, originate from the consumption of energy.

The notion of energy appears even in our everyday life. We use electric current in the house for many purposes: lighting, heating, cooking, and operating appliances. Actually, what we use is the energy of the electric current. This also means that energy is closely related to light, heat, and mechanics. Energy has to be paid for. This implies that energy cannot be obtained out of nothing. The notion of conservation of energy to be introduced later is thus one having a root in everyday experience.

Energy is involved in everyday life also from the biological point of view. Diet-conscious people are concerned with the amount of calories associated with various kinds of food. The *calorie* is a unit of energy. Thus food contains energy. The energy from food sustains the biological functions of an organism. Muscle energy of animals originates from food.

Although many of the manifestations of energy in our everyday life have been familiar for a long time, the scientific understanding of them in terms of a well-defined scientific concept is recent, a result of the development of modern physical and biological sciences. Even Newton, the founder of modern science, did not grasp the concept of energy. In fact, Newton emphasized the concept of force and worked out his theories without reference to the concept of energy. It took more than 100 years after Newton for the concept of energy and its principles to be developed. With these we are able to explain many seemingly unrelated phenomena in terms of a general principle. In the following we begin by introducing the concept of energy as developed in mechanics. In later chapters we will discuss other forms of energy in other branches of science.

Kinetic Energy. In mechanics we define a quantity called *kinetic energy* as follows: when a material body of mass m kg moves with a velocity of v m/sec, we say that the body possesses an amount of kinetic energy equal to $\frac{1}{2}mv^2$ joules. Why we choose to do so will be made clear a little later. The *joule* is the unit of energy in the *mks* system. From the mathematical point of view there are quantities that are proportional to two other quantities and thus to their product. Such quantities can be measured in terms of the product. Thus air traffic volume is measured by passenger-mile, workload is measured by man-hour, and area is measured by m^2 or ft^2. The quantity $\frac{1}{2}mv^2$ should be considered in the same way. This quantity is introduced through a mere definition. We can define any new quantity any way we please, but to be meaningful and useful the defined quantity must be related to basic principles and practical applications in a significant way. The quantity $\frac{1}{2}mv^2$ is related to the basic law of mechanics—Newton's second law—through a theorem to be discussed shortly. It is useful in many applications and so we single it out and give it a name. To call it kinetic energy is incidental (any name, from *alligator* to *zebra,* will do as well scientifically). We will show that the scientific concept so defined does have the meaning of energy in everyday use and so the name *kinetic energy* is appropriate. Before we can discuss these points in detail we have to introduce the concept of *work.*

Work. Work in a scientific sense is defined in the following way: when a force of F newtons is exerted on a material body of mass m kg (Fig. 5.1) and as a result the body moves in the direction of the force

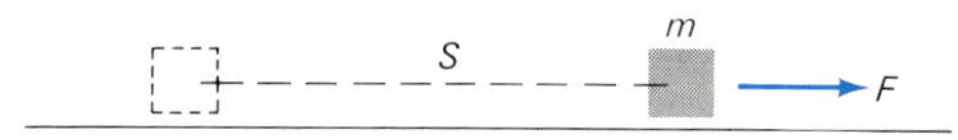

Figure 5.1 Illustration of the definition of work.

a distance S m, then we say that the force, or rather the agent that exerts the force, has done an amount of work equal to FS (the unit will be shown to be the joule also) on the body. In the English system the units of force and distance are pound and foot; the unit of work is called foot-pound (1.356 joules). Again this is a mere definition and the choice of the word *work* is incidental. On the other hand, we can show that the scientific term *work* so defined is very close to the concept of work in the economic sense, and the choice of the term is more than coincidental. Let us consider the following example.

Suppose we use our thumb to exert a force F newtons on a thumbtack to drive it a depth S m into a board (Fig. 5.2). According to our definition, we say our thumb has done an amount of work FS joules. From the economic point of view, driving nails into boards is indeed a kind of work performed by some workers such as those engaged in assembling furniture. The economic work involved in driving two nails is twice as much as that of one nail. From the scientific point of view when the thumb pushes two tacks one after another, the total work done $(FS + FS)$ is twice that of one in agreement with the economic sense. Moreover, when the thumb pushes the two tacks together, the force exerted by the thumb is twice as much $(2F)$ and so the scientific work done is $2F \times S$, which is also twice as much as that of driving one tack. Finally, a nail halfway driven in (distance $= \frac{1}{2}S$) corresponds to an economic work only half that of driving one nail all the

Figure 5.2 A force acting on a thumbtack.

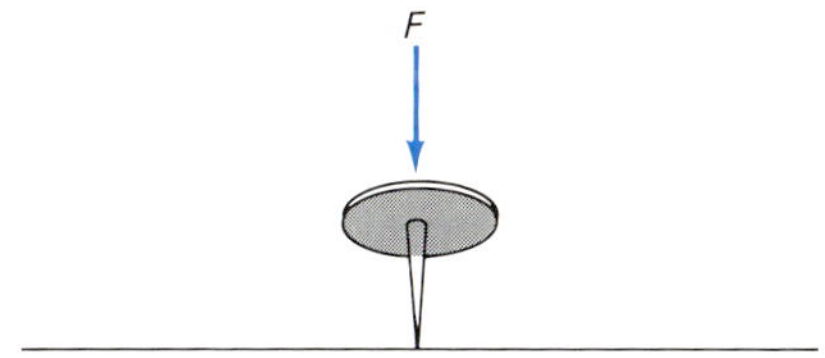

way in. The scientific work done in this case is $F(S/2)$, which is also one half of the work FS to drive a nail all the way in. Thus the quantity of work defined scientifically $F \times S$ is indeed proportional to the economic work and may be used as a measure of the latter.

In fact, in many cases of physical labor when work is done, it is usually associated with a force F exerted through a tool that advances a distance S, for example, a knife cutting through a depth S. The economic accomplishment of the action (the amount of meat cutting done by a butcher, for example) is proportional to the force F exerted as well as to the depth S cut through. Thus it is proportional to the product F and S, which equals the work, as scientifically defined. The same analysis can be applied to many other examples of working, such as carrying weight upstairs, engaging an ox to turn the grindstone of a flour mill, and so on.

The work of carrying weight upstairs in a skyscraper may be measured by pound-floor. One pound-floor is the work involved in lifting 1 lb of weight one floor up. To move 5 lb 10 floors up involves a work of $5 \times 10 = 50$ pound-floors, which is 50 times that of 1 pound-floor, and the mover should be paid 50 times as much as for 1 pound-floor. In grinding wheat in a flour mill turned by an ox, the amount of wheat ground is proportional to the distance the ox has walked and also to the force exerted by the ox. It is thus proportional to the quantity of work, defined by the product of force and distance. In all these examples the scientific quantity of work is proportional to the work in an economic sense and therefore the use of the term *work* for $F \times S$ is appropriate. Our scientific knowledge of work is thus readily applicable to these economical problems. On the other hand, there are other kinds of economic work that cannot be reduced to the scientific work here defined, for example, a doctor's visit and a concert performance. We are here concerned with only a limited kind of economic work.

In all the examples an agent is involved that does the work. It is either a human or an animal or an engine. Animals consume food and engines consume fuel. All such agents have to be paid for. Thus work is not free. People have tried hard to develop devices that would do work free but never succeeded. Food and fuel contain energy. Work is done at the expense of energy. Energy is not free; thus work is not free. Remember that work is not done just by exerting a force F on a body—the body has to move through a distance S. Just carrying a weight on the shoulder but standing still involves no work done on the weight. The weight could also be supported by a wooden pole, which does not involve energy and does not have to be paid for. The relation between work and energy is established scientifically by a theorem developed below.

Relation Between Work and Energy. Let us consider the use of a hammer of mass m to drive a nail into a board (Fig. 5.3). The hammer must be swung fast. In other words, it must have a fairly large velocity v. Therefore it is associated with a large amount of kinetic energy $\frac{1}{2}mv^2$. After the nail is driven in, the hammer comes to a stop (velocity becomes zero) and the kinetic energy is reduced to zero. This change, the stopping of the hammer, seems to take place instantly but actually it does not. It takes place over a period of time t even though the period t is very short, a tiny fraction of a second. During the period t the hammer is being decelerated. The acceleration of the hammer is $(0 - v) \div t$, or $-v/t$, with the negative sign indicating deceleration. According to Newton's second law, the force acting

Figure 5.3 Using a hammer to drive a nail.

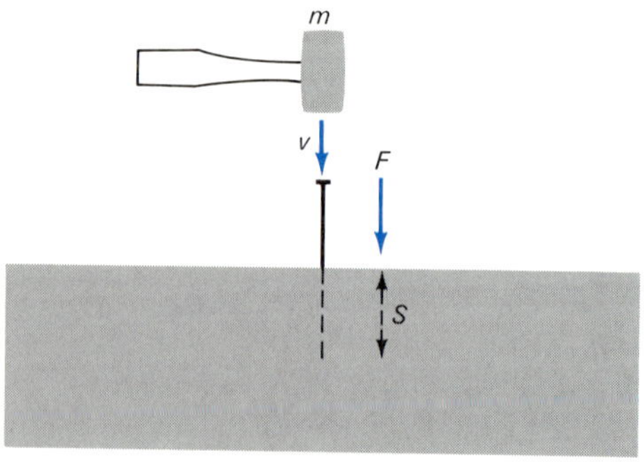

on the hammer by the reaction of the nail equals the mass of the hammer times the acceleration, i.e., $-mv/t$. According to Newton's third law the force acting on the nail (by the hammer) is equal and opposite to the reaction. The reaction equals $-mv/t$, so the force equals mv/t. We now calculate the work done by the hammer in driving the nail a depth of S. The work equals the force times the distance, i.e., $(mv/t)S$. This can be changed to $mv(S/t)$. Now S/t is the average velocity of the nail over the period t. This velocity can be calculated as follows. Because the nail is in touch with the hammer its velocity is the same as the hammer, which begins with v and ends with zero. Thus the average velocity is $\frac{1}{2}(v + 0)$,[1] or $\frac{1}{2}v$. Substituting $\frac{1}{2}v$ for S/t in the previous expression, we finally conclude that the work done in driving the nail is $mv(v/2) = \frac{1}{2}mv^2$. Therefore we establish the conclusion that in such a process *the work done by the agent equals the loss of kinetic energy of the agent.*

The same conclusion can be established in general. In all such cases it can be proved that the work done by an agent in a process equals the decrease of kinetic energy of the agent. This relation manifests itself in many practical cases, such as the use of a wrecking ball to knock down a building. The demolition of a building requires a certain amount of work—to push down a wall involves the use of a force F to move the wall over a distance S before it will fall by its own weight. Thus an agent has to do an amount of work FS. This work is done at the expense of the kinetic energy of the wrecking ball. The ball is made to swing with a velocity v and thus begets a kinetic energy $\frac{1}{2}mv^2$. After knocking down the wall, the ball comes nearly to a stop—kinetic energy is reduced and converted to work.

The upshot of the theorem is that one way to get some work done is by making use of the kinetic energy. Many economic activities involve doing work. Thus kinetic energy assumes an economic significance.

[1] Strictly speaking, this way of finding the average velocity is valid only for cases of uniform acceleration, which is assumed here.

By common sense the heavier (large m) and the faster (large v) the wrecking ball, the more wrecking job done. On the other hand, the larger m and the larger v correspond to a larger value of $\frac{1}{2}mv^2$, so the kinetic energy is larger. Now we see that the scientific definition of kinetic energy is indeed in agreement with the common-sense use of the term *energy,* that is, an ability to get a job done—the more energy, the more job done. However, this ability represented by kinetic energy, though directly proportional to the mass m, is not directly proportional to the speed v, but to the square of it. And there is the numerical factor $\frac{1}{2}$. These are not obvious from common sense. They emerge through the preceding theorem based on Newton's laws.

Because kinetic energy and work can be equated, their units must be the same. Thus in the mks unit system the unit of work is also the *joule.* A force F newtons acting on an object advancing a distance of S m is doing an amount of work FS joules. In the English system a force of F lb working over a distance of S ft is doing an amount of work FS foot-pounds.

Kinetic energy can be used to do work not only constructively but also destructively. Whether wrecking an old building is constructive or destructive is a matter of semantics. In an auto crash the kinetic energy of the car is doing the "work" of property damage and personal injury. Thus a speed twice as large will produce four times as much damage. Bullets and cannon balls do damage because of their kinetic energy, made large by a very high speed. An amount of water equal in weight to a cannon ball and moving with the same speed will be as destructive. In a landslide in Italy thousand tons of rocks poured into a water reservoir. The displaced water raced across a valley and demolished a whole village on the other side.

Where is the kinetic energy of the swinging hammer coming from? It originates from the exertion of the hand muscle, which makes it move faster and faster. We can prove in the following that the work done by the hand muscle on the hammer equals the increase of kinetic energy of the hammer. Thus the

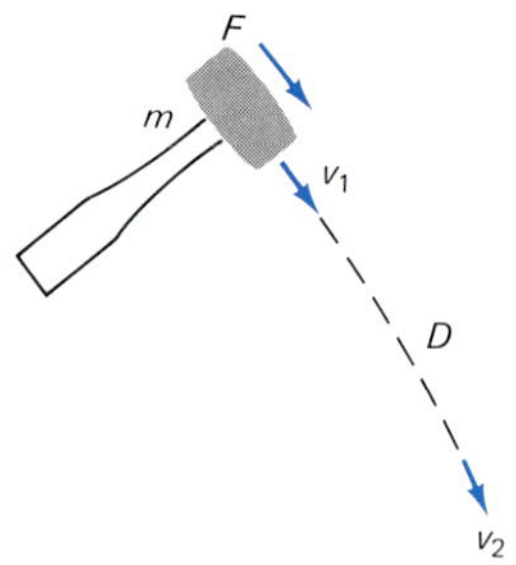

Figure 5.4 Acceleration of the hammer.

conversion between kinetic energy and work can proceed both ways.

When the hand exerts a force F on the hammer during the swing (Fig. 5.4), the hammer is accelerated with an acceleration a determined by Newton's second law:

$$F = ma$$

The work done by the hand in swinging through a distance D is the following:

$$W = FD = maD$$

Let v_1, v_2 be the velocities of the hammer at the beginning and at the end of the time period of swing T. Then

$$a = \frac{v_2 - v_1}{T}$$

The distance D equals the average velocity times the time T:

$$D = \tfrac{1}{2}(v_2 + v_1)T$$

Therefore

$$W = maD = m\left(\frac{v_2 - v_1}{T}\right)\frac{1}{2}(v_2 + v_1)T = \frac{m}{2}(v_2^2 - v_1^2)$$

or

$$W = \tfrac{1}{2}mv_2^2 - \tfrac{1}{2}mv_1^2$$

In other words, the increase of kinetic energy equals the work done. Work and energy are convertible both ways.

Source of Energy. Where is the kinetic energy of a wrecking ball coming from? It comes from the work done by the wrecking truck, whose engine swings the wrecking ball. Thus the engine does the work, which increases the kinetic energy of the wrecking ball, which in turn does the work of wrecking the building. The amount of wrecking work is the same as the amount of work done by the engine. Why not let the engine do it directly? Why use the ball? Although the engine provides all the energy for the wrecking, the energy must be delivered through an agent that exerts an extremely large force on a selected point on the building. The wrecking ball fulfills these requirements. In most engineering processes many stages of transformation of energy are necessary before the energy can be delivered to do the job desired.

If we could generate an enormously large amount of kinetic energy out of nothing, we would become rich instantly. In trying to accomplish this many schemes have been contrived, but none have succeeded. This is related to a basic principle of energy, i.e., conservation of energy, which will be developed shortly. This principle implies that energy cannot be created out of nothing. It is obtained by transformation of another form of energy. The kinetic energy of the hammer comes from the exertion of force by the muscle. The gaining of kinetic energy of the hammer is at the expense of energy of the muscle. Biological studies have shown that the energy of muscle originated from the energy of food we ingested. Muscle energy is not free; it bears the cost of the food. The kinetic energy of the wrecking ball is obtained from the engine of the wrecking truck. The energy of the engine comes from the fuel it burns.

Thus there are a variety of forms of energy. Each can be transformed into another but no energy can be created out of nothing. The other forms of energy will be discussed in Chapters 7, 11, 13, and 15. The transformation of energy will be discussed in Chapter 16.

Our biological and economic lives depend on energy. Because energy cannot be created out of nothing, it must come from known sources of energy.

Some forms of energy are limited and cannot be replenished once depleted. Thus we have a problem with the energy resources on the earth. This problem has been discussed in Chapter 3. The use of some of the energy resources, such as coal and gasoline, produces undesirable side effects. It leads to the pollution of air, water, and land, which is a major environmental problem today. This aspect will be discussed in Chapter 17. On the other hand, the current problems of resources and pollution are rooted in the nature of energy and the relation of energy to our biological and economic life. A proper understanding of the physical science of energy is necessary to provide the basis for discussing all the other problems.

5.2 Potential Energy and Conservation of Mechanical Energy

Potential Energy. We now consider another form of energy, namely, *potential energy.* Let us first consider a simple example. A particle of mass m at a height h_1 having a velocity v_1 vertically downward is accelerated by gravity (Fig. 5.5). At a later time it descends to the height h_2 with velocity increased to v_2. From what we have discussed of uniformly accelerated motion we learn that

$$v_2 = v_1 + gt$$

$$h_1 - h_2 = v_1 t + \tfrac{1}{2}gt^2$$

Figure 5.5 The free fall of a particle of mass m.

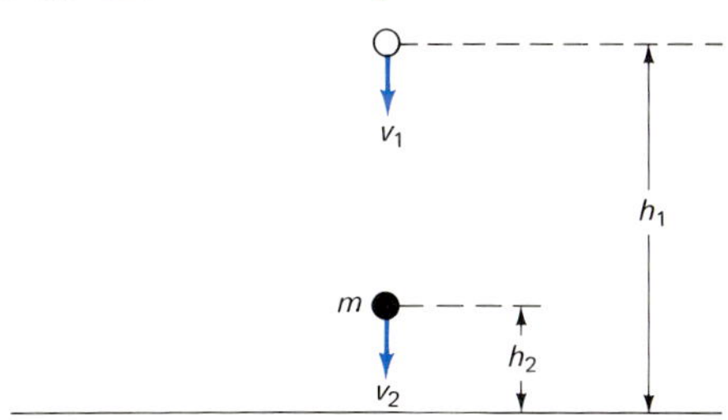

The quantity g is the gravitational acceleration. Eliminating t from these two equations we have

$$h_1 - h_2 = v_1\left(\frac{v_2 - v_1}{g}\right) + \frac{1}{2}g\left(\frac{v_2 - v_1}{g}\right)^2$$

$$= (v_2 - v_1)\frac{1}{2g}[2v_1 + (v_2 - v_1)]$$

$$= \frac{v_2^2 - v_1^2}{2g}$$

Thus we deduce the relation

$$v_2^2 = v_1^2 + 2g(h_1 - h_2)$$

This equation may be changed to the following form:

$$\tfrac{1}{2}v_2^2 + gh_2 = \tfrac{1}{2}v_1^2 + gh_1$$

Multiplying both sides with the mass m, we obtain

$$\tfrac{1}{2}mv_2^2 + mgh_2 = \tfrac{1}{2}mv_1^2 + mgh_1$$

In this equation $\tfrac{1}{2}mv_2^2$ and $\tfrac{1}{2}mv_1^2$ are the kinetic energies at the final and initial positions, respectively. Now we define a quantity called the potential energy of the mass m by the product mgh. This quantity is associated with the spatial position of the mass m, the height h being related to the position. Don't worry about why it is called potential energy for the time being; it is just a name for the amount mgh. Then, mgh_2 and mgh_1 of the preceding equation represent the potential energies at the final and initial positions, respectively. The preceding equation thus states that the sum of kinetic and potential energies, i.e., the total energy, in the final position is the same as that in the initial position. We can take another time and another final position and prove the same. Thus the total energy is always the same at any point during the course of the fall. Or simply stated, the total energy is always the same. This is the simplest manifestation of a general principle called the *principle of conservation of energy.* A number of points need to be explained here.

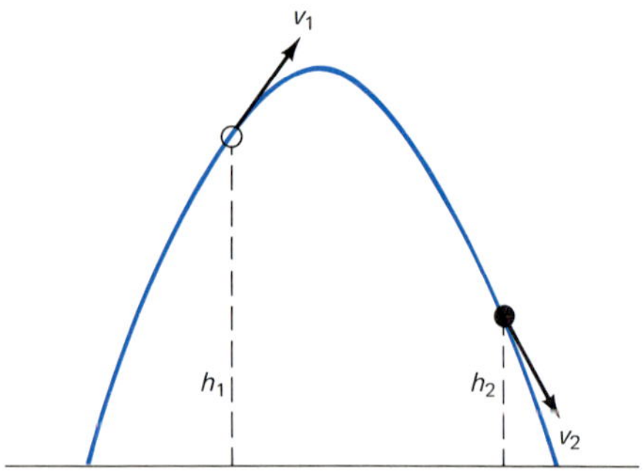

Figure 5.6 Conservation of energy of a projectile during ascent and descent in its parabolic trajectory.

Conservation of Energy. Like kinetic energy, the quantity *potential energy* is first defined by a specific mathematical expression, *mgh*. It is merely a convenient way to render the last equation in verbal form, no more and no less. On the other hand, the statement that the total energy, kinetic and potential, remains always the same in the course of motion can be proved to hold not only in the case of vertical free fall discussed here but also in all other motions under the influence of gravity, such as the vertical motion upward, the parabolic motion of a projectile (Fig. 5.6), the vibration of a pendulum (Fig. 5.7), and so on. (The general proof of the theorem requires the use of calculus and so will not be given here.) Therefore the statement of conservation of energy is one of great significance and the concept of potential energy is a very useful one. Otherwise the quantity *mgh,* given whatever name, would not mean very much.

Figure 5.7 Conservation of energy of a pendulum during the swing.

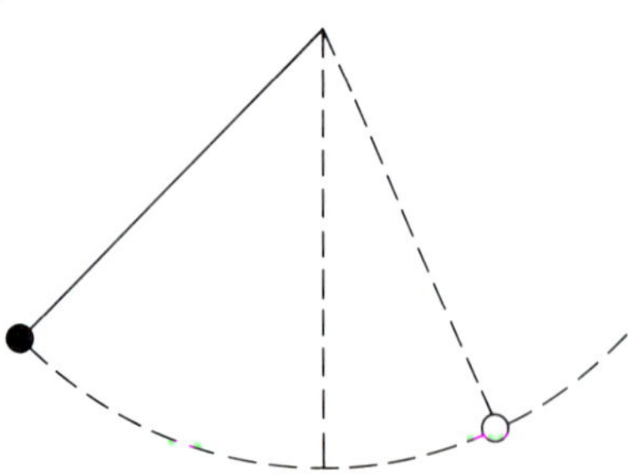

The principle implies that under the action of the gravity force, even though the kinetic energy may decrease as the height increases, it will recover its loss when the particle returns to its original height, for then the potential energy value returns to the original value and so does the kinetic energy. Therefore the ability of doing work is not lost as the particle ascends because we can always wait for the particle to return to the original height and recover the lost kinetic energy. When the particle moves to a higher level and thus appears with a smaller kinetic energy, we say that the lost kinetic energy is not really lost but has converted into another kind of latent energy that is recoverable later. The situation may be compared with the deposit of cash in a bank account. The total amount of cash is decreased but the total assets, cash plus deposit, remain the same. This latent energy is characterized by the particle's occupying a higher position. Thus we may associate this latent energy with the height *h* of the particle. The potential energy *mgh,* defined before, which increases with the height *h*, may be interpreted as a mathematical representation of this latent energy, because the sum of it and the kinetic energy is a constant and so any decrease in kinetic energy is reflected as an increase in potential energy. Interpreted this way the term *potential energy* begets meaning; it is a form of energy, albeit a latent one, similar to kinetic energy and convertible to it. Now we may say that when a mass moves under the gravity force, its kinetic energy may be changed to potential energy and vice versa with the total energy remaining the same, just as cash may be converted into a bank deposit with the total assets remaining the same. Potential energy has no other use except to be converted into kinetic energy; then it can be used to do work. A bank deposit has no other use than for conversion into cash; then it can be used for shopping.

The significance of the principle of conservation of energy is that by the action of gravity the ability to do work (energy) cannot be increased or decreased. In other words, you cannot get energy out of no energy. Later we will show that this principle holds not only for gravity force but also for many other forces, so that when the whole universe is taken together the total energy of the universe is always the same no matter what changes are taking place inside. Obviously this statement is of utmost importance from a practical, economic point of view. In this general sense many other forms of energy, which will be introduced in later chapters, are involved. The particular principle presented here may be called the *principle of conservation of mechanical energy,* which includes kinetic energy and potential energy. And the potential energy here discussed is one associated with gravity force and should be called gravitational potential energy. Other forces will give rise to other potential energies.

Units. We now consider the units of energy in more detail. In scientific work based on the cgs system of units, the unit of energy is the *erg*. A 1-g mass moving with a speed of 1 cm/sec will have an amount of kinetic energy equal to $\frac{1}{2}mv^2 = \frac{1}{2} \times 1 \times 1^2 = \frac{1}{2}$ erg. In the mks system a mass of 1 kg moving with a speed of 1 m/sec has an amount of energy of $\frac{1}{2} \times 1 \times 1^2 = \frac{1}{2}$ joule. Converting this amount to the cgs unit, $\frac{1}{2}$ joule equals $\frac{1}{2} \times 1{,}000 \times (100)^2 = \frac{1}{2} \times 10^7$ ergs. Thus

$$1 \text{ joule} = 10^7 \text{ ergs}$$

Another unit of energy commonly used in atomic and nuclear physics is called the *electron volt,* equal to 1.6×10^{-19} joules, designated by *eV*. One million times of it is called a *million electron volt* (*MeV*). These units will be explained further in Sec. 5.6. The calorie, primarily a unit of heat, is also a unit of energy; see Sec. 7.1.

$$1 \text{ cal} = 4.2 \text{ joules}$$

In electric utility the unit of energy is *kilowatt-hour* (kwh), which will be introduced in the following section.

Power. In engineering and economic applications we not only are concerned with the amount of work done or energy consumed but also with the time rate at which work is produced or energy is consumed. The amount of money we pay for the use of electricity is gauged by the total amount of energy consumed. However, as far as working out the monthly budget is concerned, it is the time rate of energy consumption, i.e., the energy consumption per month, that is important. When we compare two automobile engines to decide which one is more powerful, we are not comparing the total energy production—any engine, given a long enough time, will produce as much energy as desired. What we are comparing is the amount of energy produced in the same length of time, i.e., the time rate of energy production. The more powerful one is that which produces a larger amount of energy in the same length of time. Thus it can accelerate the car faster. The time rate of energy production or consumption is defined as the *power*. The unit is thus a joule per second. The unit joule/sec has another name, *watt* (w). In the electric power industry, a unit 1,000 times larger, called *kilowatt* (kw), is used. In the English system, the unit of power is the *horsepower,* equal to 550 foot-pounds/sec; it equals approximately $\frac{3}{4}$ kw.

A household, consuming electricity at the average power rate of P kw over a time span of T hours, will have consumed a total amount of energy equal to $(P \times 1{,}000) \times (T \times 60 \times 60)$ joules, or $3{,}600{,}000PT$ joules. This is the amount that determines the electric bill for that period. In practice the electric bill is usually figured not on the unit joule but on the unit *kilowatt-hour* (kwh), which is the energy consumed in one hour at the rate of 1 kw. In the above example the total amount of energy is PT kilowatt-hours. It is obvious that 1 kwh is equivalent to 3,600,000 joules. At the present time the cost of electric energy for household use is of the order of 3¢/kwh, with rates varying according to locality (1¢/kwh to 5¢/kwh).

Practical Uses of Energy. Electric energy will be discussed more in detail in Sec. 5.6. At this point

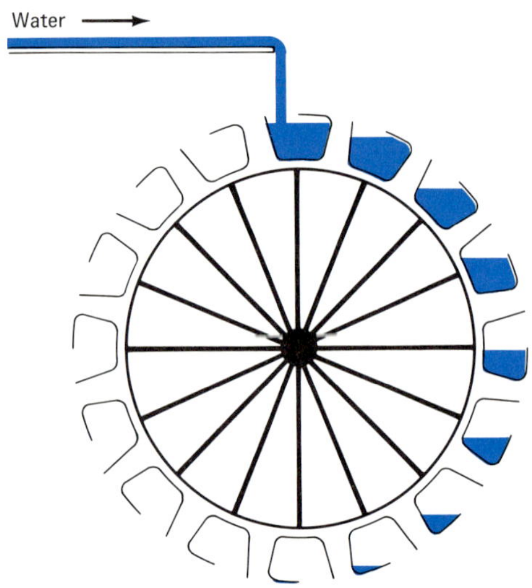

Figure 5.8 Water turns a watermill.

we have already introduced kinetic energy and potential energy and we are in a position to discuss some of the energy sources based on them that are of practical significance.

We introduced potential energy through the gravitation force of the earth. Gravitational potential energy is a source of energy of practical significance as manifested in the use of hydraulic energy. In the old-fashioned water mill, it is the potential energy of the water at a higher altitude that is converted into kinetic energy when water drops to a lower level (Fig. 5.8). And it is this kinetic energy that pushes the wheel of the water mill, which in turn pushes the grindstone. (In the absence of such an energy source the grindstone has to be turned by a human or an animal.) The work done on the grindstone to grind the grain thus comes from the potential energy of the water.

In modern times water power is used in a more sophisticated way but the principle is the same—the potential energy of water is converted into kinetic energy when water drops to a lower level. The kinetic energy is used to push a wheel (the turbine) in a way similar to that in the water mill, but this is done for the purpose of generating electric energy (Fig. 5.9), which can be used for many purposes in many locations more conveniently.

Rocks rolling downhill also gain kinetic energy. A large meteorite descending on earth from a very high "altitude" gains a tremendously large amount of kinetic energy upon reaching earth and can dig a big hole in the earth. From a cosmic point of view gravitational potential energy is important in bringing about many of the evolutionary processes of the stars and planets. We will return to this problem in Chapters 21 through 25 when we consider the origin and evolution of the physical universe.

If the gravitational force should be replaced by another kind of force, everything else remaining the same, then there would still be a potential energy and a law of conservation of energy. Thus other forces may also give rise to potential energies, as already mentioned. Electric force is one. Electrical potential energy is gauged by the voltage in the household current. Another example is the force exerted by a compressed or stretched spring. A compressed or wound-up spring (Fig. 5.10) contains energy (can do work or produce motion) and indeed is a form of energy source in many practical applications. This is the way a mechanical clock or a watch works. This is also the way energy is supplied in mechanical toys that have to be wound up before they can be used.

Let us now turn our attention to kinetic energy as a practical source of energy. It is less useful than potential energy because of practical limitations. Kinetic energy is associated with a velocity v. This means that the mass involved will be moving away and therefore cannot be confined in a casing for practical applications. On the other hand, kinetic energy can be manifested in a rapidly rotating disk. Even though the speed of every constituent part of the disk may be very large, the parts are all moving in circles and the whole thing can be confined in a casing to be used for practical purposes. Indeed, this has been considered as the energy source for an automobile (Sec. 9.5). Because it burns no fuel, this

Figure 5.9 Hydroelectric power plant. [Tennessee Valley Authority]

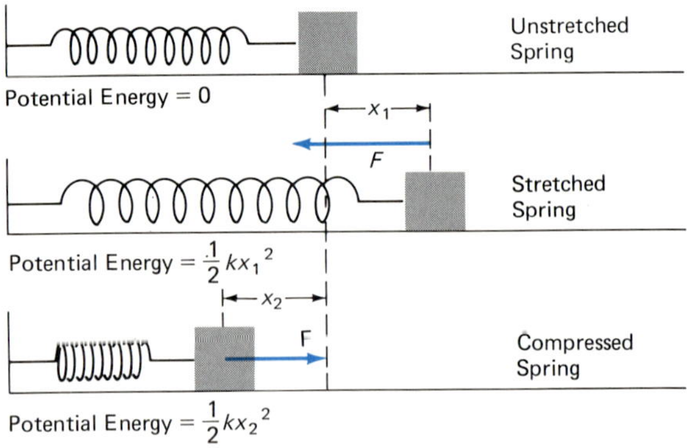

Figure 5.10 A stretched or compressed spring possesses potential energy.

car would not pollute the air and thus might be a partial solution to the air pollution problem.

Dissipative Force and Conservation of Energy. Although many forces are associated with potential energies, not all are. Friction force (from solid surface) and viscous force (from fluid media) are not associated with a potential energy. In fact, when this kind of force is acting, the total mechanical energy is not conserved. It keeps on decreasing until it is reduced to zero and everything stops. Friction and viscous forces are always present in nature. As a result, all motion on earth eventually comes to a stop if left unaided. The situation described by the principle of conservation of energy is an idealized case, minimizing the effect of friction. In all practical applications the effect of friction must be included.

Thus forces may be classified as of two kinds: (1) those, called *conservative forces,* where total mechanical energy is conserved; (2) those, called *dissipative forces,* where total mechanical energy is not conserved and tends to decrease to zero. Why do the two kinds of force work differently? Why does the principle of conservation of energy fail when dissipative forces are involved? The answer to these questions, to be discussed in detail in Chapter 7, is that the energy lost because of the action of the dissipative force is not really lost but is converted to yet another form of energy—this time it is heat energy. Including heat as a form of energy, we conclude that the grand total of all energies is still conserved even though dissipative forces are acting. Thus the principle does apply universally in a general sense. (Ultimately, friction force is reducible to molecular force and thus to the fundamental conservative electromagnetic force on the microscopic level. Therefore the principle still holds on the microscopic level.)

Energy Versus Force. We are now in a position to compare the two concepts *energy* and *force.* In conventional English usage the two terms are nearly interchangeable. In science they are poles apart. Let us first discuss an example involving the use of the lever. A force F is applied on a lever to support a weight W (Fig. 5.11). If the arm length n is longer than the other arm m, then a smaller force F can support a larger weight W, which is the principle of many weight-lifting devices. On the other hand, the work done by the force F and the energy gained by the weight W are exactly the same. When the point at F moves down a short distance r the position of W moves up a distance s. Although F is smaller than W, r is greater than s by the same proportion as can be proved by the principle of the lever and the geometrical relations of the similar triangles. Thus the product Fr is equal to the other product Ws and therefore the work put in at one end equals the energy gained at the other.

This example shows that forces of the input and output of a machine (the lever) are not the same but the energies are. It seems as if we can cheat

Figure 5.11 Conservation of energy in the action of a lever.

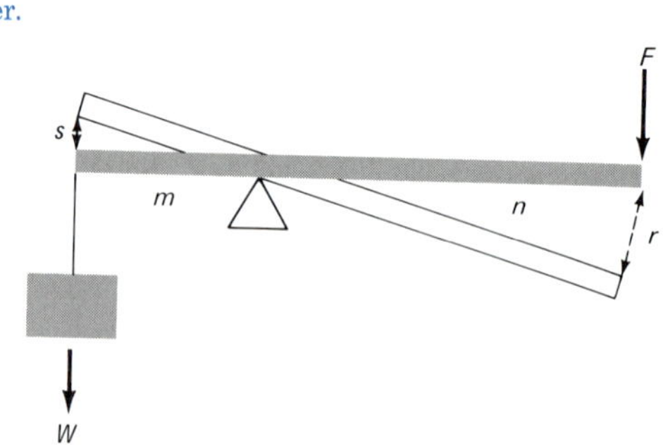

nature out of force (a small force supporting a large weight), but we cannot cheat nature out of energy (the smaller the force F, the larger the distance r and thus the same amount of work is needed to raise the weight W to a given height s). Thus energy is conserved but force is not.

Energy, being conserved, is an entity (a quantity), an *extensive* property. Force is a quality, an *intensive* property. The concept of force is basic in the dynamic theory discussed in the preceding chapter. The concept of energy is important in other ways (conservation principles, symmetry principles). Force deals with changes; energy deals with the unchanged. They deal with two complementary aspects of physical science.

5.3 Applications of the Principle of Conservation of Energy

The principle of conservation of energy is important both in theory and in practice. In practice our biological life and economic life both involve energy in a significant way. A large part of our environmental problems will be discussed in terms of the concept of energy. In theory the principle helps develop science conceptually and also helps solve theoretical problems. In this section we consider how the principle may be used to help solve problems. The conceptual development will be discussed in the next section.

Consider a *simple pendulum* consisting of a small, heavy ball hung on a string. The ball is free to oscillate. Here, the particle is constrained by the string. Yet, according to the general principle of conservation of energy, the total of kinetic energy and potential energy at any moment remains always the same as the ball swings. When it swings to the extreme left (or right), it stops momentarily at point A before it swings down (Fig. 5.12). At the moment when it stops the velocity v is zero and therefore the

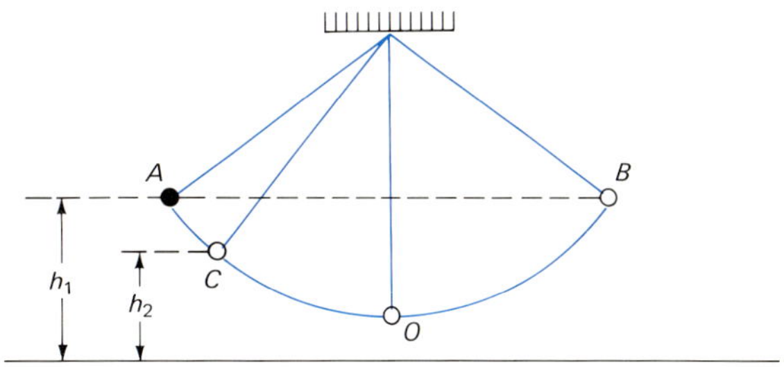

Figure 5.12 Conservation of energy of a swinging pendulum.

kinetic energy $\frac{1}{2}mv^2$ is zero. At this point all the energy is in the form of potential energy. As the ball swings down, the height of the ball h decreases and so the potential energy mgh decreases. At the same time the ball picks up speed and the kinetic energy increases. The principle of conservation of energy implies that the loss of potential energy is exactly the same as the gain of kinetic energy; in other words, the potential energy has changed to kinetic energy. On the basis of this principle we can make the following predictions:

1. The speed of the ball will be the greatest when the position of the ball is the lowest (point O in Fig. 5.12); at this point the potential energy is at its minimum, whereas the kinetic energy—and therefore the velocity—is at its maximum.

2. The ball will swing to the same height on the other side before it swings down again. The highest point B is one where the ball stops momentarily and thus has zero kinetic energy; the potential energy at B, being the total energy, thus equals that at point A. Therefore the height of the point B is the same as that of the point A.

3. The speed of the ball at any position C may be calculated from the decrease of the potential energy $mg(h_1 - h_2)$, which equals the increase of the kinetic energy $\frac{1}{2}mv^2$:

$$\tfrac{1}{2}mv^2 = mg(h_1 - h_2)$$

Thus

$$v = \sqrt{2g(h_1 - h_2)}$$

These are important and useful conclusions. They

also can be derived by using the Newton second law without the conservation principle, but not as easily or as quickly. The principle of conservation of energy, though it does not add anything new that is not in the Newtonian laws, does enable us to solve the problem more conveniently and more simply.

As long as there is no air friction, the ball will keep on swinging between two points of equal height, the energy being converted alternately between kinetic energy and potential energy. In practice, the ball swings less and less and eventually comes to a stop, because the energy is gradually wasted by air friction—in the same way that a bank deposit is reduced to zero by service charge deductions for alternate deposits and withdrawals of the same funds.

The usefulness of the principle of conservation of energy may be illustrated by the following example: Consider a ball swinging on a string. As the ball swings down from the right in Fig. 5.13, put an obstacle at point *A* to stop the string at that point so that the ball will now swing to the left with *A* as the center instead of *O*. How far will the ball swing to the left? Move point *A* down to *B*, then to *C*, and so on. How far will the ball swing to the left each time? The answer can be deduced readily from the principle of conservation of energy. The ball will swing as far as it must to reach the same height at which it started, as shown in the figure. The same conclusion cannot be quickly deduced from Newton's second law. Moreover, by the Newton law, each case will have to be dealt with separately. The conservation principle provides a general solution applicable to all cases.

Figure 5.13 Another manifestation of conservation of energy

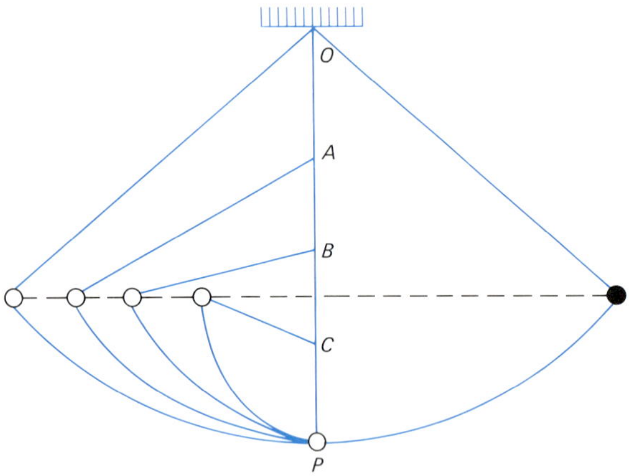

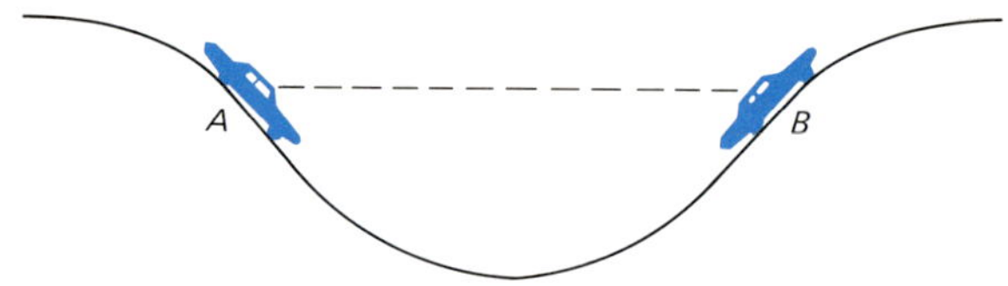

Figure 5.14 A car going down a hill and then up another.

Consider a highway dipping down to the bottom of a gorge and then climbing up the other side. A car starts at rest at point *A* (Fig. 5.14) with the engine turned off and begins coasting down the road without braking. After it reaches the bottom, the car will continue to coast up the other side. How far would it go by coasting if there were no friction of any kind? The answer may be obtained by the principle of conservation of energy: to point *B*, at the same height as it started, even though in practice the car will stop short of point *B* because of friction. This conclusion holds no matter how deep or how wide the gorge is. If the problem is to be solved by the Newton law, the solution will have to be worked out based on the shape of the gorge. If the shape is rather complicated, the mathematical method involved in obtaining the solution will be very complicated, even though the final answer will be rather simple. The use of the conservation principle leads directly to the simple final answer.

A similar result may be obtained for a bead sliding down a frictionless wire. The wire may even be twisted into a number of loops, as shown in Fig. 5.15. As long as there is no friction the bead will slide to a point at the same height as that at which it started. The same principle underlies the operation of a roller coaster. In practice the answer will be modified somewhat by the presence of the friction force.

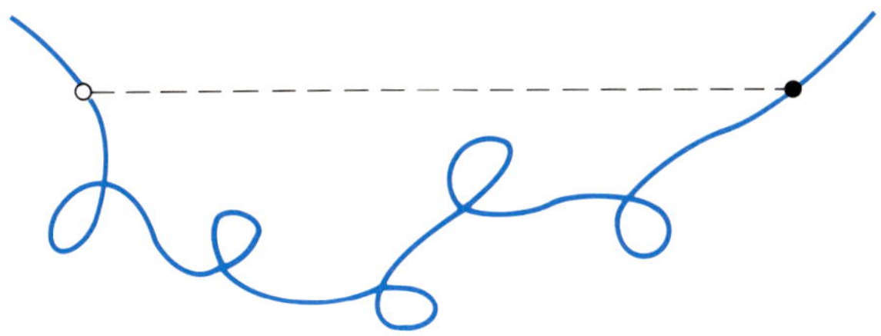

Figure 5.15 Another manifestation of conservation of energy.

The usefulness of the conservation principle as demonstrated in the preceding examples is based on the fact that the principle relates the final condition directly to the initial condition, bypassing the intermediate details. In the dynamical theory the final condition is deduced from the initial condition step by step through all the intermediate stages, none of which may be omitted. Sometimes the intermediate stages may involve complicated situations that are of no particular consequence. The dynamical theory in this case becomes cumbersome; the conservation principle is most useful in this situation for obtaining the desired results the quickest way. However, the conservation principle solves *some* problems but not *all* problems; the dynamical law solves all problems.

The dynamical law deals with the *change,* whereas the conservation principle deals with the *unchanged.* The philosophical inclinations of the two approaches are quite different. Because the conservation principle is derivable from the dynamical (Newton) law, the two are equivalent to each other and the difference in philosophical orientation is thus superficial, not intrinsic. In fact, the relation between the dynamical law and the conservation principle indicates how the two different philosophical views may be reconciled.

Figure 5.16 Conservation of energy in any trajectory in the gravitational field.

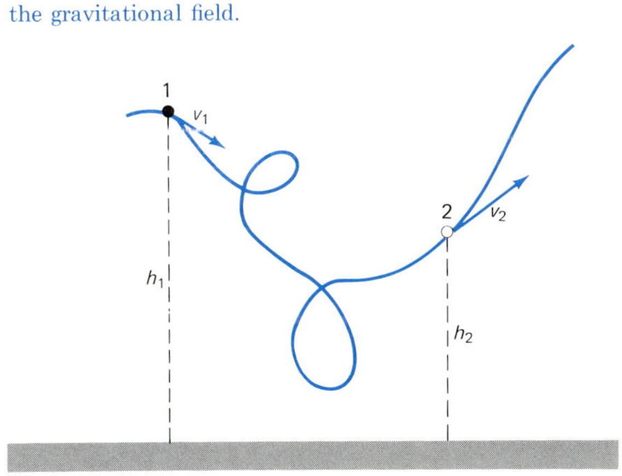

5.4 The Conceptual Significance of Energy and the Conservation Principle

We now consider the significance of the energy concept in pure science. Disregarding the everyday practical usage of the notion, the concept of energy was first introduced in mechanics in connection with the development of the theory based on Newton's dynamical laws. From Newton's second law we derived a general equation applicable for gravitational problems (Fig. 5.16):

$$\tfrac{1}{2}mv_1^2 + mgh_1 = \tfrac{1}{2}mv_2^2 + mgh_2$$

This holds true whether there is a string (like in a simple pendulum) or no string, a rail (like in a roller coaster) or no rail, as long as gravity force is the only external force applying and there is no friction force. We name $\frac{1}{2}mv^2$ kinetic energy, mgh potential energy, and the preceding equation reads that the total energy remains always the same. The energy concept and the conservation principle are logical deductions of the dynamical theory and thus do not seem to represent anything basic. As such its significance seems to be more practical than fundamental—it helps solve problems. From the fundamental point of view the dynamical theory based on laws of motion and laws of force is complete and no new additions are called for. In fact, Newton developed his theory without introducing the energy concept at all. However, the energy concept and the conservation principle turned out to be so useful and powerful that they are used beyond the Newtonian theory. For example, energy was related to heat, a phenomenon beyond mechanics, and a more general principle of conservation of energy, including heat energy, was later introduced.

Now we know that heat is reducible to mechanics of the atoms and the preceding generalization appears only natural. However, there are further applications to areas truly beyond Newtonian mechanics, where we have to conclude that the energy concept and the conservation principle are more basic, more fundamental than the Newtonian laws. Historically the Newtonian theory was developed first, and the energy concept first came to us in its special form within the framework of the Newtonian theory. Thus it appears that the Newtonian concepts were more basic.

As a matter of fact the Newtonian theory is not final and has undergone three major generalizations: the field theory, the relativity theory, and the quantum theory. Each of the three revisions leads to a more general theory applicable to a wider field. The Newtonian concepts are modified and revised, but interestingly the energy concept and the conservation principle remained intact. If conservation of energy is to be regarded as a mere consequence of the Newtonian theory, then when the latter is revised, the conservation principle is generally expected to require revision also. That it comes out intact is unusual; it appears as if a river flows higher than its source. On the other hand, if we consider the energy concept and the conservation principle to be more basic and more fundamental, then this is no surprise. It is only at the very late stage of the history of science that we realize energy to be a basic concept. In this book we consider matter and energy as two basic concepts and present physical science centered on these two concepts.

What makes the conservation principle so basic that it survived the three great revolutions in physics? This point cannot be discussed in detail here except to say that the reason is traced to the relation between conservation principles and *symmetric principles* in nature.

The first revolution of the field theory originated in electromagnetic studies that started off in a classic Newtonian fashion. Thus the ideas of force (electric and magnetic) and energy were employed from the start. Later, however, it was found necessary to break away from the Newtonian approach, and the new concept of field was introduced. Field is a property of space—empty space, which in the Newtonian framework has no property at all. In the Newtonian view all properties of nature are associated with matter. The empty space between material particles is a complete void, having no attributes and no significance. This void becomes the essence in the field theory. It is endowed with such attributes as electric field, magnetic field, and so on. In particular, it begets energy.

In our presentation, energy began as an attribute of matter ($\frac{1}{2}mv^2$ and mgh, all dependent on mass m). In the field theory it is not always tied to matter but is associated with the field as well. An electromagnetic wave is a wave of the field, the electromagnetic field, not a wave of a material medium. When a radio wave is sent to the astronauts on the moon the wave carries energy with it and the energy triggers mechanical actions remotely controlled on the earth. This energy is not associated with matter. No material substance is sent to the moon with the radio wave. It is not even transmitted by material substance, for the space between the earth and the moon is nothing but vacuum. The propagation of energy to the moon is a sheer property of empty space.

The second revolution, brought about by relativity theory, originated in the study of the electromagnetic field. It was found necessary to modify many of the previous Newtonian concepts. One important revision concerns the nature of mass and energy. The theory of relativity deems that mass and energy are not independent but are equivalent to each other—m kg of mass are equivalent to mc^2 joules of energy, where c is a constant 3×10^8 equal to the speed of light in m/sec. Denoting energy by E this may be expressed by the equation

$$E = mc^2$$

Previously we had two separate laws; one, concerning conservation of mass, originated in Lavoisier's study of chemistry. The other, concerning conservation of energy, originated in physics. Once mass and energy are equivalent one may be changed to the other and the two cannot be conserved separately. Instead there is only one conservation law governing mass and energy together. This is the second revision of the concept of energy.

The third revolution, brought about by the quantum theory, encompasses many aspects. One concerns a modification of the field theory. In the *quantum field theory* we developed a theoretical framework in which material particles such as the electron and positron (a positive electron) may be created out of energy and may be annihilated and return to energy. In nuclear physics such creation and annihilation processes have actually been observed and the theory provides a proper explanation of them. Thus the original Newtonian concept of the particle, permanent and unchanging, is completely uprooted. Matter, instead of being the ultimate, basic entity of the physical world, now becomes a mere attribute of the fields, which are now the ultimate entities. The conversion of matter to energy is a manifestation of the field. The concepts of energy and matter are even more closely related.

In spite of these revisions, for most problems in everyday life, including most of the energy and environmental problems to be discussed later, the Newtonian concepts of energy are still essentially correct, except when we are involved with electromagnetic radiation and nuclear energy. Therefore we will base most of our discussions on the Newtonian theory, considering energy to be a separate entity from matter, obeying its own conservation law.

*5.5 Equilibrium

A special topic of statics, namely, *equilibrium,* will be singled out for discussion here to illustrate the use of the energy concept in solving some of the problems.

A rigid body is in equilibrium when all forces acting on it balance out among themselves. There are three kinds of equilibrium, *stable, unstable,* and *neutral.*

A cone standing on its base is said to be in stable equilibrium (Fig. 5.17). A cone balanced precariously on its apex is said to be in unstable equilibrium. The terms *stable* and *unstable* are used in the conventional sense and thus do not need further explanation. A cone resting on its side is said to be in neutral equilibrium, neither stable nor unstable. In this case the equilibrium configuration may change (rolling on the side), but the new configuration is one identical with the old. Thus it changes but remains in equilibrium at the same time.

A bead balanced on a saddle is in unstable equilibrium (Fig. 5.18). A bead placed inside a bowl is in stable equilibrium. A bead resting on a plane is in neutral equilibrium. We may imagine the bead on the saddle as being made of two hemispheres. The pull of gravity on the left half tends to pull the bead to the left. That on the right tends to pull to the right. Because the two halves are symmetrical, the two forces are equal in magnitude and therefore balance out exactly. The bead thus remains at rest and is in equilibrium. However, when a slight disturbance is introduced, say, a gentle breeze blowing on it that causes the bead to tip slightly to the left,

Figure 5.17 Three kinds of equilibrium.

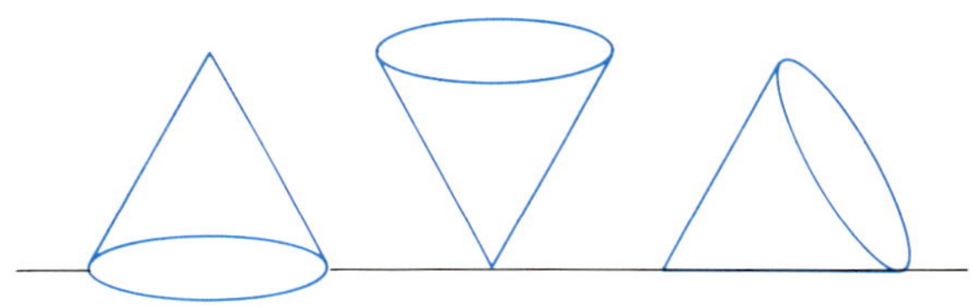

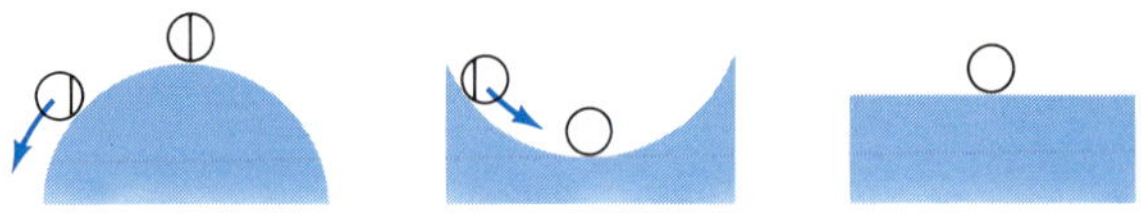

Figure 5.18 The stability of the equilibrium explained.

then the proportion of the two gravity forces pulling to the left and to the right is changed in favor of the former, shown in Fig. 5.18. This causes the bead to tip further left; eventually the bead rolls down the saddle to the left. This is a self-accelerating effect that characterizes unstable equilibrium. A slight imbalance tends to magnify the imbalance. Thus unstable equilibrium is always precarious and does not last long.

What characterizes stable equilibrium is that a slight imbalance will give rise to a change that tends to restore the imbalance. This can be seen in the case of the bead in a bowl, as illustrated in Fig. 5.18. Therefore the equilibrium is stable against small disturbances, which are always present at any time under any circumstances. Stable equilibrium is one that has a built-in mechanism to correct any slight deviations and to restore the equilibrium; thus it is long-lasting. Neutral equilibrium is one in which a slight disturbance causes a change that neither accelerates nor restores the change.

The equilibrium problem may be looked upon from the potential energy point of view (Fig. 5.19). The unstable equilibrium position is one in which the potential energy is a maximum (compared with neighboring positions). The stable equilibrium position is one in which the potential energy is a minimum. The neutral equilibrium position is one in which the potential energy is a constant. The variation of the potential energy in the neighborhood of the equilibrium position in the three cases is shown by the curves in Fig. 5.19. The potential energy of the bead is simply proportional to the height.

The natural tendency of heavy things to fall downward may be rephrased by saying that the potential energy of an object tends to decrease naturally (decrease of mgh means decrease of height h, thus corresponding to a downward movement). In all three cases of equilibrium the potential energy is locally constant (the curves are flat in the immediate neighborhood of the equilibrium position). No decrease is possible and thus no motion will take place naturally. On the other hand, if forced away from the equilibrium position by a slight disturbance, which is naturally present, the deviation accelerates itself in unstable equilibrium because the potential energy can now keep on decreasing. (It starts from a maximum.) In stable equilibrium the deviation tends to correct itself because the potential energy tends to return to minimum at the equilibrium point. In neutral equilibrium the object stays wherever it goes because potential energy is constant and no further changes are possible. Thus the phenomena of equilibrium and the associated stability characteristics can both be explained by the behavior of the potential energy of the system.

Equilibrium, implying the counterbalancing of a number of opposing forces, is a concept that finds great use in many other fields beyond physics. It appears not only in chemistry and biology but also in economics and political science. Although everybody uses the concept of equilibrium, few are careful enough to analyze the stability of equilibrium. Failing to recognize the difference between stable, unstable, and neutral equilibrium could lead to confusing and unfortunate consequences.

In studying economics the free market pricing system may be considered as a system in equilibrium

Figure 5.19 Relation between the potential energy function and the stability of the equilibrium.

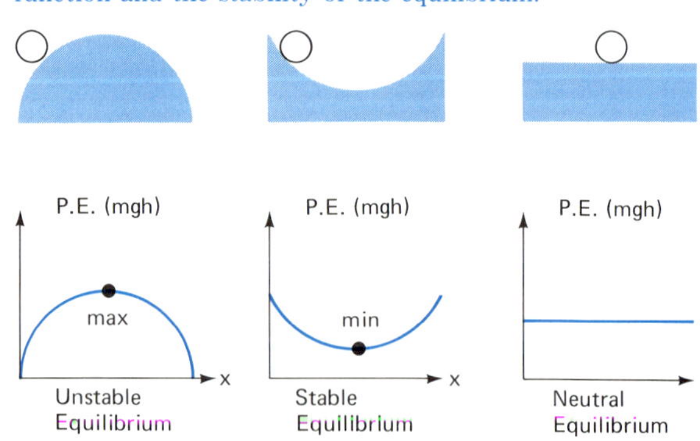

based on the balancing of supply and demand. It is easy to see that this is a stable equilibrium. A shortage of supply of eggs, because of a chicken epidemic, say, drives up the price; the high price induces more production, resulting in the increase of supply and the lowering of price. The system thus returns to equilibrium by itself, restoring automatically the deviation. The free market is such that any deviation in price is self-correcting, which is the essential feature of stable equilibrium. It does not need artificial control to bring the price down. This view has been expressed in the philosophy of laissez faire espoused by the conservative economists.

But there are other economic operations characterized by unstable equilibrium. The operation of a bank is based on the equilibrium between deposits and loans. Here a disturbance in the balance may not be self-correcting but instead may amplify itself. An accidental excessive withdrawal by the depositors lowers the reserve, undermines the confidence of the public and induces additional withdrawal, and so on, until the bank collapses under panic. A crowd of screaming, hysterical depositors gathered outside a closed bank was a typical scene of the Depression era. A system in which a disturbance is not self-correcting but amplifies itself is a system in unstable equilibrium. Because there is no self-correcting mechanism built in, the application of the laissez-faire doctrine to banking is destined to failure. The fact is so obvious that nowadays all economists, including the most conservative (such as those of the Chicago School), agree that banks must be regulated and controlled by governments. The use of government control to carry out economic policy has been advocated by the more liberal economists; the radicals want to extend government control to everything.

The doctrines of laissez faire and government control fight each other in economics as uncompromising ideologies. If the problem is analyzed from the stability point of view and the difference between the two kinds of equilibrium is recognized, the problem is taken out of philosophy and ideology and becomes a mere problem of economic mechanics. There is really no contradiction in adopting laissez faire for the egg market and applying government control to banking because the two problems are basically different. One is a case of stable equilibrium and the other a case of unstable equilibrium; the two should be treated differently. What is good for stable equilibrium is generally not good for unstable equilibrium. To apply one doctrine to all cases is naively wrong. The important thing is to determine which equilibrium is stable or unstable and then to deal with it properly. If the economists learn to distinguish the two kinds of equilibrium, much controversy may be eliminated, and economics can progress faster from an art to a science.

The economic development of recent years has brought forth a new phenomenon, the creeping inflation, or the wage–price escalation, before which the doctrines of both the conservatives and the liberals are helpless. At the heart of the problem is the emergence to dominance of the service industries, which increased in the postwar decades from 45 to 60 per cent of the total employment. Services, such as a doctor's visit, cannot be overproduced and cannot be stocked up, as chicken eggs can. Moreover, there are little substitutes to choose from. The self-restoring, stabilizing factor is absent. The pricing of the service industries, upon close analysis, is more like a system in neutral equilibrium; the price–wage escalation is more like the rolling of a cone on a plane. It differs from both stable and unstable equilibrium, and both the classic measures, laissez faire and government control (such as the price and wage control of the World War II vintage), are not helpful. A wholly new policy is called for to deal with the new situation. It was here that the confusion, inaction, and misdirection of the economic policies of recent years originated.

Turning from economics to political science, we note that the idea of equilibrium has been in politics from time immemorial. It manifested itself in the practice of balance of power. This strategy played a significant role in international politics in nineteenth-century Europe. When successful, balance of power leads to peace. Yet few politicians realize that the balance of power of the nineteenth century eventually developed into one of unstable equilibrium, which may suddenly collapse upon a slight disturbance. In fact, a shot at Serajevo ignited the holocaust of World War I.

Let us start with two-power equilibrium. When two powers have equal strength, the equilibrium results in peace. But this equilibrium is unstable. Pre-emptive war and surprise attack are logically desirable choices, and therefore war may break out without provocation. Any imbalance of power easily leads to war. Any such war is a war of total annihilation. Examples are the struggle between Athens and Sparta and that between Rome and Carthage. Such a war is usually a lock-horn, cutthroat struggle. The nuclear stalemate of two superpowers, like two scorpions in a small bottle, is dangerous because the equilibrium is unstable.

The multiple-power equilibrium game played by the British prime ministers in the nineteenth century is not a wise policy for peace in the long run. Small powers tend to gang up to strengthen their positions and eventually the multiple powers are polarized into two contending blocks. From this point on, it is a two-power game. A slight disturbance tips the balance and the lock-horn struggle is on. World War I did to Europe the same damage as the Peloponnesian War did to the Greeks and the Punic Wars did to the Romans.

Are there any other games we can play besides the two-power and multiple-power games, both of which are disastrous? Many other numbers exist between 2 and infinity. Let us try 3. The three-power game turns out to be completely different. No two powers of the three will engage themselves in a lock-horn, cutthroat war of total annihilation because both realize that the ultimate winner will be the third power standing by. There will be no pre-emptive war, no surprise attack. Once war starts between two of them, it will be over quickly before both are weakened enough to become prey of the third power. Moreover, no one power will become strong enough to be able to overrun the two others combined because before this happens the other two will gang up to "cut him down to size." Thus any disturbance leading to imbalance tends to produce an effect that will correct the imbalance and restore the equilibrium. Hence the equilibrium is stable according to our definition and the peace maintained can be long-lasting. Accidental disturbances always occur (such as the assassination of an archduke) and cannot be completely controlled and eliminated. What is needed to maintain a stable peace is an equilibrium that is stable. Unstable equilibrium is no peace; it is but a prelude to war.

In actual history there was a period in the Orient (third century A.D.), called the Period of Three Kingdoms, when the civilized world of the Orient was dominated by three contending powers. The mechanisms of self-correction on the imbalance of power were demonstrated and described most vividly in the *Romance of the Three Kingdoms*.[2] Such an international relation created a period of contention without destruction, struggle without annihilation, and peace without corruption. The heroes and stories of the period are most colorful; they occupy a prominent role in literature and drama, and are familiar to schoolchildren in the whole Oriental world.

Recent changes in relations with Peking and Moscow seem to signal the emergence of a new Period of Three Kingdoms. Considering the human propensity to war, such an arrangement is perhaps a better substitute to a nuclear Peloponnesian War that may break out any moment between the two scorpions in a small bottle.[3]

If the politicians learn to analyze the stability of equilibrium like the physical scientists do, millions

[2]Lo Kuan-Chung (ca. 1330 to ca. 1400), *Romance of the Three Kingdoms,* trans. by C. H. Brewitt-Taylor (Rutland, Vt.: C. E. Tuttle Co., 1959).

[3]Henry Kissinger obviously played the game of balance of power. Judging by his television interview of February 1, 1973, it seems that he was playing the nineteenth-century game. His emphasis is equilibrium, his basis is mutual self-interest, and his number of powers is 5. There is no critical analysis of the stability of the equilibrium. Stability requires that the number of powers be 3 and that frivolous self-interest be sacrificed to maintain a stable, lasting peace. For details on the point of view here sketched see *U.S. Congressional Record,* May 24, 1972, pp. S8328–S8331.

of lives and billions of dollars could be saved. Here is another aspect of science that is relevant to everybody's life. If by analogy, by guesswork, by whatever meager wisdom we may gain through studying science, we can find better ways to promote economical well-being and to maintain international peace, our efforts will be more than amply rewarded.

5.6 Electric Energy

Electric energy has been in use for merely a century, but it has transformed practically every aspect of our life. It has become the symbol of modern technological civilization. It brings us convenience and progress but recently it also brings us environmental pollution. The physical basis of this technological development is thus a relevant concern.

From the following discussion it will be clear that electricity itself does not give us energy. Because energy is conserved, the electric energy we use comes from the energy input at the power plant, either chemical energy from the fuel in a steam plant or gravitational energy of the water in a hydroelectric plant. In the power plant the input energy is converted into the energy of the electric current. Electricity is merely an instrument for utilization of the energy of the fuel or the water. Why not use the original energy directly? We could, but not as conveniently. Electric energy can be made available anywhere, any time, just by flipping a switch. Not too long ago a university used to hire a laborer whose sole duty was to wind all the clocks in campus buildings. By changing to electric clocks this expense was saved. On the other hand, in some applications the electric intermediary seems to be unnecessary and wasteful. Heating by burning fuel can be done directly. First changing heat energy from the fuel to electric energy and then back to heat energy in an electric heater is indeed roundabout and unnecessary. The only gain is convenience. It is definitely wasteful because a large fraction of energy is lost in the processes of transformation. Besides, the lost energy has adverse environmental effects (thermal pollution, see Chapters 2 and 17).

Electric Current. What is an electric current? Where is its energy coming from? How is its energy put to work? These are our problems here. An electric current is a current of electricity or electric charge in the same way as a water current is a current of flowing water. Thus we may use water current as an analogy, replacing water by electric charge in all comparisons. For this purpose we need only consider electric charge to be a substance, or rather an entity, that is the source of the electromagnetic force—the second of the four basic forces. Actually electric charge is an attribute of many elementary particles. In an electric current it is carried by the electron; an electric current is generally a stream of electrons. Because the electron mass is very small, electricity may be regarded as weightless. Electric charge is measured by the unit *coulomb.* The charge carried by one electron is 1.6×10^{-19} coulomb. The coulomb is the unit of electric charge used in the mks system. Other electrical units in the mks system include the ampere, volt, ohm, farad, and henry. This system of electrical units is called the practical system of units.

The water current through a tube is measured by the amount of water flowing through a given cross section of the tube in 1 sec. It makes no difference which cross section along the tube is taken for this purpose because the amount of flow through any cross section is all the same. Similarly, the amount of flow of electric charge through a wire, defined as the *electric current,* is measured by the amount of electric charges passing through any cross section of the wire in 1 sec (Fig. 5.20). The unit, coulomb/sec, is given another name, *ampere.*

Figure 5.20 The amount of electric charge flowing through a cross section in 1 second's time.

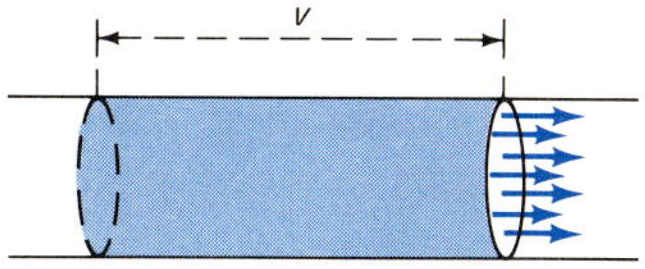

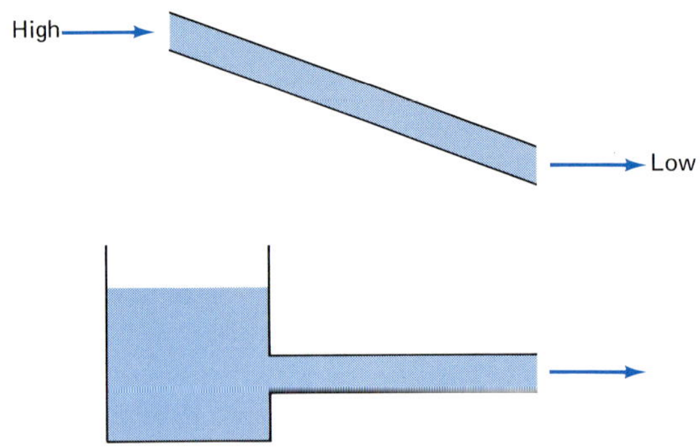

Figure 5.21 Water flows under the force of gravity.

What makes electricity flow? In a water current the flow is usually the result of the gravity force. One end of the tube must be higher than the other. Or equivalently, one end must have a higher pressure (Fig. 5.21). This is how tapwater flows into the house; the pressure comes from water in the storage tank, which is usually higher than most buildings.

One way electric current may be produced is to use a battery or a dry cell. The positive electrode, the terminal marked with a (+) sign (the central terminal of a dry cell) corresponds to the high-pressure end of the water current. Thus positive electricity flows out of it just as water flows out of the tower. After flowing through the wire, electric charge returns to the battery through the negative electrode. Electricity at the negative electrode is like tapwater that has run through the house and eventually returned to the reservoir. The battery works in such a way that when electricity at the negative electrode is brought to the positive electrode through the interior of the battery, it is like water in the reservoir pumped to a high tower ready to flow down the system again. The battery works like a pump; thus electricity keeps on flowing in the closed *circuit.*

How does the battery do the pumping? We need not be concerned with the detailed mechanism of the pumping, just as we need not be concerned with the gear mechanism of a coin-operated soft drink machine. The important thing is the energy used in pumping, just as money is the important element in the coin machine. A water pump requires energy to operate either by the arm muscle or by a motor. The work done by the arm or the motor on the pump is converted into potential energy of water, which eventually becomes the kinetic energy of water in the water current. Similarly, the battery must have an energy source. The energy is converted into potential energy of electricity on the positive electrode. When flowing to the negative electrode in the circuit, the electricity has its potential energy converted into kinetic energy or other energy that in turn does the work of the electric devices in the circuit. The energy source of the battery or the dry cell is chemical energy; there is a chemical reaction going on when electric current flows in the dry cell and the reaction releases energy to power the dry cell. When all the chemicals are used up there is no more energy left and the dry cell becomes useless. A battery can be recharged and used again, i.e., the chemicals may be regenerated by forcing a current through the battery in the reverse direction (and thus forcing the chemical reaction to go backward). In the recharging process the charging current adds energy to the battery to replenish its energy content. As long as the energy is there, how to use it is of secondary importance. The detailed mechanism of energy conversion here involved is complicated and need not be a concern for the present discussion.

Electromotive Force. Thus we may regard a battery as a pump with its chemicals supplying the energy. There are other devices, such as the generator in an electric power plant, that can provide pumping action and thus can generate electric current. Any such device is called a *seat of electromotive force* (emf). The use of the term *force* here is a little unfortunate because it is already defined in the Newtonian sense but is used here not in that sense. Here it merely means a capability of generating an electric current. Because the effects of a current are gauged by the energy output, this ability of generating a current may also be gauged by its energy output. Thus *electromotive force* may be defined quantitatively as the *energy* (not force) generated when one unit of electric charge passes through the seat of electromotive force. The unit of electromotive force so defined in the practical unit system is called a *volt* (v). A flashlight cell has an electromotive force of 1.5 v; it means that when one unit of charge (1 coulomb) passes through the device, it gains 1.5 units (joules) of energy. A larger electromotive force produces more energy on a unit charge, thus producing a larger electric effect (light, heat, motion, and so on). Thus the voltage is a measure of the strength of the electric effects produced. Using the water current analogy, we may compare voltage with the energy a pump imparts on a unit mass of water when passing through it. This energy manifests itself in the form of potential energy of water in the tower. A more powerful pump is one that can raise water to a higher tower. When running down from a higher tower, the water produces a more powerful current and can do more work; for example, when ejected from a hose, it can wash a car clean faster.

Electric Potential. As water runs down a tower in a closed circuit of circulation, the potential energy decreases gradually. At point A in Fig. 5.22, half as high as the water in the tower, this energy is half gone. At point B, one-fourth the height of the water, only one-fourth of the energy is left. From A to B, the water expended one-fourth of its potential energy. Between any two points the energy expended equals the potential energy difference between the two points. The potential energy expended in flowing downward is converted into other forms of energy and may be utilized; a little water wheel, for example, may be inserted in the course of the flow to engage a little machine to do work in the same way as a water mill does.

A similar situation exists in a circuit of electric current. As a unit charge passes through the seat of electromotive force E, it gains an amount of energy equal to E joules. As it proceeds forward in the circuit (Fig. 5.23), the energy is expended gradually and we may speak of an *electrical potential* V (analogous to the previously mentioned gravitational potential of water) at any point, such as A and B,

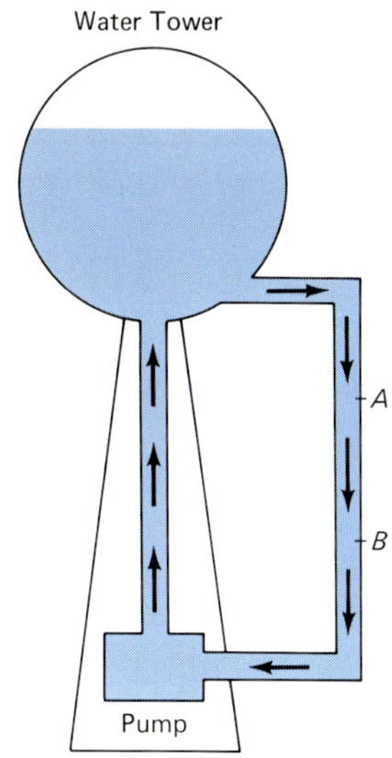

Figure 5.22 Circulation of water.

representing the amount of energy remaining at the point. Moreover, the energy expended by a unit charge between point A and point B can be calculated by the difference of this electric potential $V_A - V_B$. The energy expended may be utilized to do work, i.e., operating appliances, lighting, heating, and so on. Because the electric potential is a fraction of the electromotive force, it is measured by the same unit, the volt.

Electric potential corresponds to the elevation of water. In a water mill the elevation itself does not

Figure 5.23 Circulation of electricity.

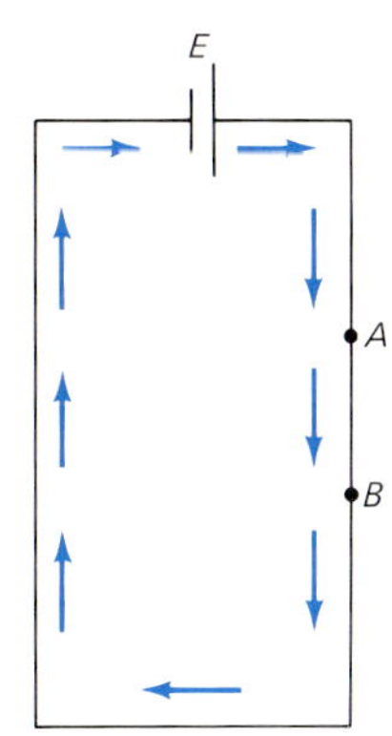

tell the total energy available, which depends also on the amount of water available. Similarly, the total electric energy is determined by the product of E and charge Q. The elevation is also important in other aspects. For example, a water mill with a large elevation drop must be more sturdily built. Analogously, the "sturdiness" of an electric appliance is usually gauged by the electric potential difference to be applied between the two terminals of the appliance.

From the unit volt is derived another unit of *energy* called the *electron volt* (eV), which is used frequently in atomic physics and nuclear physics. It is the energy gained by one electron with a charge of 1.6×10^{-19} coulomb when dropping through a potential difference of 1 v.

$$1 \text{ eV} = 1.6 \times 10^{-19} \text{ coulomb} \times 1 \text{ v}$$
$$= 1.6 \times 10^{-19} \text{ joule}$$

Larger units of 10^3, 10^6, 10^9 eV are called kiloelectron volt (KeV), million electron volt (MeV), and billion electron volt (BeV or GeV). Chemical energy released in a chemical reaction of a molecule is of the order of a few electron volts, whereas nuclear energy released in a nuclear reaction of a nucleus is of the order of a few million electron volts. Thus the unit electron volt is frequently used in atomic physics and million electron volt is used frequently in nuclear physics. Note that the electron volt is a unit of *energy,* not a unit of *voltage,* in spite of its name.

Ohm's Law. When water is forced through a thin horizontal tube from a point A to another point B, the amount of the current is proportional to the pressure difference at the two ends and is inversely proportional to the length of the tube from A to B (the Poiseuille law of fluid flow). In a similar manner, when an electric current flows along a thin wire from A to B in Fig. 5.24, the amount of the current I is proportional to the difference of electric potential at the two ends $V_A - V_B$ and inversely proportional to the length L of the wire from A to B. Therefore, according to a theorem in algebra, I is proportional to $(V_A - V_B)/L$:

$$I \sim \frac{V_A - V_B}{L}$$

It is obvious that the volume of flow is proportional to the cross-sectional area S:

$$I \sim (V_A - V_B)\frac{S}{L}$$

According to another theorem in algebra, I equals the right-hand side multiplied by a constant designated by σ:

$$I = \sigma\frac{S}{L}(V_A - V_B)$$

The constant σ is called the *conductivity* of the wire material. Introducing the *resistivity* ρ as the reciprocal of σ, we rewrite the preceding equation thus:

$$V_A - V_B = \rho\frac{L}{S}I$$

or

$$V_A - V_B = RI$$

where R, equal to $\rho L/S$, is called the *resistance* of the wire between A and B. Resistance is thus proportional to the resistivity and the length and inversely proportional to the cross-sectional area of the wire. The preceding equation is usually referred to as *Ohm's law.*

The potential difference $V_A - V_B$ equals the energy released when one unit of charge passes from A to

Figure 5.24 Illustration of Ohm's law.

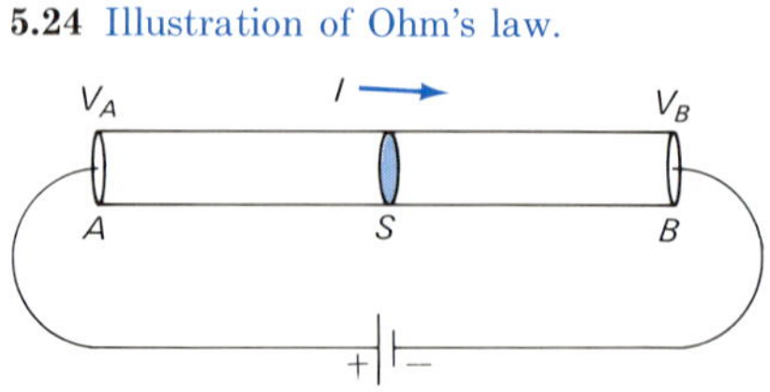

B. When a current I is established, I units (coulomb) of charge flow from A to B in 1 sec. The total energy released in 1 sec is thus $(V_A - V_B)I$ joules. The rate of energy release per unit time (second) has been defined as the power P. Therefore

$$P = (V_A - V_B)I$$

Substituting by the previous equation, we have

$$P = RI^2$$

The total energy H delivered in a period of t sec is thus

$$H = RI^2t \text{ joules}$$

A light bulb is essentially a piece of wire of high resistance enclosed in a glass bulb. The specifications marked on it, say, 110 v and 75 w, mean that when a potential difference of 110 v (that of the household electric current) is applied across the two terminals of the bulb, it will generate energy at the rate of 75 w, i.e., 75 joules/sec. This energy is converted into heat and light, which we can sense emanating from the bulb. A 150-w bulb generates energy twice as much and therefore the bulb is much brighter.

Electric Motor. How the energy is converted into heat and light is not obvious at this stage of discussion, but will be made clear after we have discussed the other forms of energy in Chapters 7 and 13. The energy we are familiar with at this point is mechanical energy. Electric energy of a current can also be converted into mechanical energy. The device that makes this conversion is called an *electric motor,* which we will discuss shortly. In our everyday life we use heat energy (house heating), light energy (lighting), and mechanical energy (operating machines and appliances). Electric energy can easily be converted into any of them and thus is the most convenient form of energy.

The principle of the motor is based on the third property of the electromagnetic force we discussed in the previous chapter, Sec. 4.4—a moving electric charge exerts a force on a magnet. An electric current, being a stream of moving charges, can thus exert a force on a compass needle and cause it to turn away from the normal north-south orientation (Fig. 5.25). Indeed, this was the way this property was first discovered by Oersted during a classroom demonstration. This is a discovery that has far-reaching consequences. It shows that electricity and magnetism are not two separate independent phenomena but are closely related. They are two facets of the same entity. It leads to the invention of the electric motor and thus makes possible the use of electric energy for operating machines and appliances.

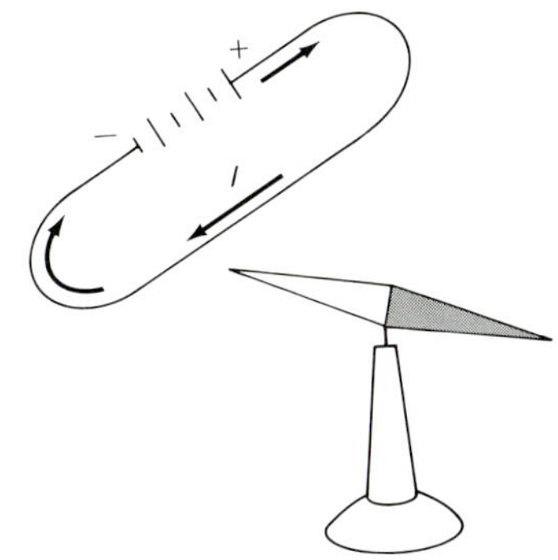

Figure 5.25 Compass needle deflected by an electric current.

A motor consists essentially of a current coil and a magnet exerting force against each other, causing one of them, usually the current-carrying coil, to rotate. The rotational motion can then be applied to operate machines and appliances such as the electric fan, electric shaver, and so on. The mechanical work performed by the motor comes from the electric energy of the current. The amount of mechanical energy delivered per unit time (power) essentially equals the electric power consumed $(V_A - V_B)I$.

Electric Generator. Batteries and dry cells are one kind of device that produces electromotive force. Their energy comes from the chemicals enclosed inside. The electric current produced has a fixed voltage and always flows in the same direction. Such a current is called a *direct current* (DC).

If electromotive forces were limited to this kind, the use of electricity would be rather limited. However, the use of electric energy is vastly expanded by the discovery of another source of electromotive force based on a completely different principle—electromagnetic induction, which is the fourth basic property of the electromagnetic force we discussed in the previous chapter (Sec. 4.4). To illustrate the principle, we consider a closed-wire loop (see Fig. 5.26) turning between the north and south poles of a magnet. As it turns, because of the relative motion of the magnet with respect to the current loop, electromagnetic induction generates an electric field and thus an electromotive force in the wire loop. As a result a current flows in the wire loop. This current can be taken out by lead wires and will do work when it flows through an appliance. The lead wires can be very long so that the site of electric energy generation and the site of electric energy utilization can be far apart. In practice the site of electric energy generation is separately located in the *power station* and the energy produced is fed to consumers by *transmission lines.*

It should be emphasized again that electricity itself does not provide energy. Energy must come from a separate source. Electricity is just the instrument by which the energy is transformed and used in a most convenient way. This point may be explained as follows. In Fig. 5.26 when the loop turns, a current is produced in the closed loop and will do work when an appliance is connected into the circuit.

Figure 5.26 Electric generator.

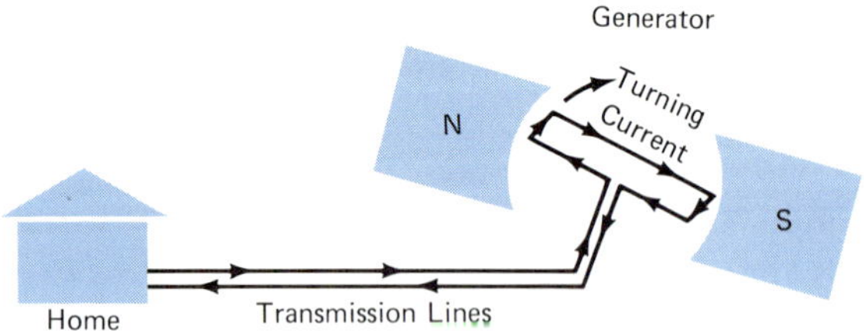

If the loop is not closed, say, by flipping a switch to make the circuit open, then no current will flow and no work will be done. The turning of the loop, like the turning of a bicycle wheel when the bicycle is turned upsidedown, does not require much effort, and in fact will proceed all by itself for a long period of time without being driven by an external agent. However, when the switch is flipped closed and the current flows through an appliance—let us say, an electric saw cutting lumber—the loop no longer turns by itself like a free bicycle wheel. It now has to be driven with effort by hand or by other power sources. The energy of the electromotive force comes from an action analogous to a water pump, which requires an engine to supply energy. The loop now requires an agent to exert a force F to turn it. When the force advances a distance d, an amount of work Fd is done by the agent, and this is the work that is converted to electric energy, which can do work in the appliance. Thus energy has to be put into the generator from an external agent before the appliance will work.

Before modern times a lumber jack used a hand saw to cut lumber. Now we may imagine the lumber jack goes to the power plant and uses his strong arm to turn the coil. As the coil turns, the electric current flows into the electric saw, which cuts the lumber. Again the energy used in cutting the lumber comes from the strong arm of the lumber jack, but this time, it is used through an intermediary—the electric power station and electric power saw system—for convenience because electric energy can be instantly switched on and off and can be used anywhere anytime. Other forms of energy are not that convenient—a fire takes half an hour to kindle; an ox cannot be used in a bedroom. Thus electricity is just an instrument to convert energy to be used in the most convenient way. The source of energy must be elsewhere.

In practice, the energy source, the agent that turns the coil, is supplied either by a *heat engine,* which is a device converting heat energy from fuels into mechanical energy (to be discussed in more detail in Chapter 16), or by hydraulic power, which originates in gravitational energy. On a hot summer day when everybody turns on his air conditioner, the load at the power plant may become so heavy that the engine cannot pull it any more; this results in a power breakdown, a blackout or brownout. When we consume electric energy we are actually consuming the energy expended at the power plant from the fuels or from the hydraulic power.

The electric current produced by electromagnetic induction in a power station is not a direct current. It is a current changing in magnitude and direction periodically, as shown in Fig. 5.27. Generally it changes direction 60 cycles/sec (in some countries the frequency is 50 cycles/sec). Such a current is called *alternating current* (AC). In most of the applications it can be used equally well as a direct current.

*5.7 General Conservation Principles

The term *conservation* refers to some quantity that is unchanging, persistent, and indestructible. At first glance everything in nature seems to be in a state of flux; nothing is constant and everlasting. The Newtonian laws of motion specifically tell us how things change. Buddhist thought exploits the concept of change thoroughly; nothing is everlasting and everything is reduced to nothing.

Figure 5.27 The magnitude and direction of an alternating current change in the course of time according to a sine function.

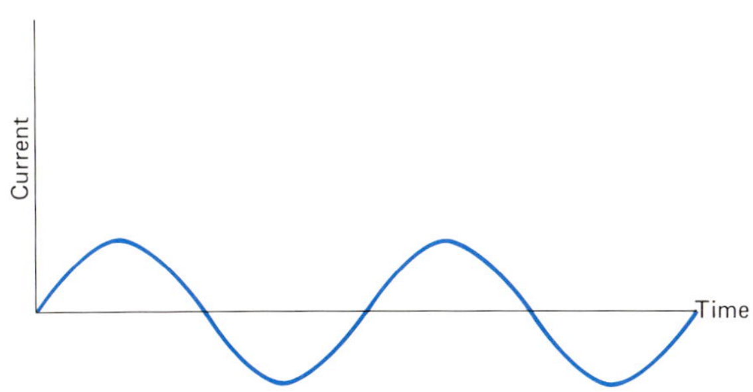

From nothingness to something is a process of *creation*. From something to nothingness is a process of *annihilation*. These ideas are made all the more familiar because of our experience in the biological world characterized by birth, growth, and death. These concepts lie at the base of our consciousness and it is only natural to use them to deal with our other experiences in the physical and material world. Monetary assets can be made to grow by good investment and can diminish by poor management. On the other hand, the concepts of conservation and transformation are also present in our everyday experience. Monetary assets can be transformed from one form to another (cash to bank deposit, for example), but the total amount remains the same.

In the Newtonian theory there is one unchanging element. The mass of an object is supposed to be fixed once and for all and will never change under any circumstances. Even though in chemical changes there are occasions in which it appears as if matter is created out of nothing or annihilated into nothing, we found out that in these cases matter is also conserved, the only thing unusual being that the form of matter before or after transformation is invisible. Moreover, in such transformations, the total mass of matter involved remains always the same. If mass is conserved, there may well be other quantities that are also conserved.

Indeed, in the modern scientific view, a few basic entities besides mass, such as energy, are conserved. Each of these quantities can assume a number of forms, and all natural phenomena, all the changes, are looked upon as transformations from one form to another. Creation and annihilation are explained this way. Not only the birth, growth, and death of living organisms but also the appearance and disappearance of material substance, including the creation and annihilation of elementary particles, are explained in terms of the transformations of the conserved entities. The basic point of view of this book is to consider physical and chemical phenomena in terms of transformations of the conserved quantities matter and energy.

Are there any other conservation principles in the

Newtonian theory? There are. Besides the principle of conservation of energy there are two more, one on *momentum* and the other on *angular momentum*.[4] As far as practical everyday life is concerned, these two do not appear as important as the one on energy and so only a brief discussion will be given in this section.

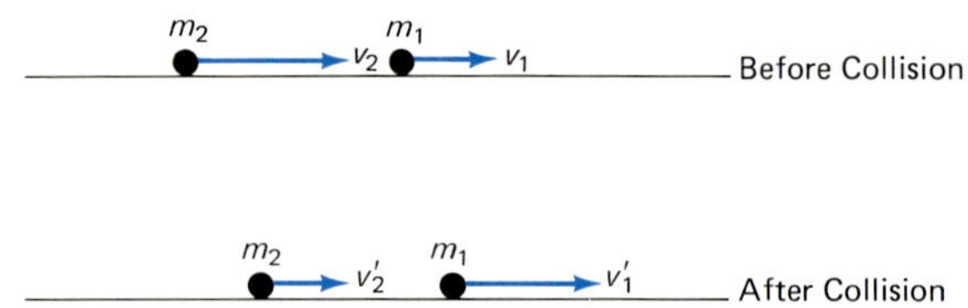

Figure 5.28 The collision process.

Principle of Conservation of Momentum. The *momentum* of a particle is defined as the product of the mass and the velocity of the particle. The law of conservation of momentum states that for an isolated system of interacting particles, the total momentum remains always the same. Thus in a collision of two particles (Fig. 5.28), the total momentum before and after collision are the same, and the principle of conservation of momentum may be expressed by the following equation:

$$m_1v_1 + m_2v_2 = m_1v_1' + m_2v_2'$$

where m_1 and m_2 are the masses of the two particles, v_1 and v_2 are their velocities before the collision, and v_1' and v_2' are their velocities after the collision.

This relation can be derived from the dynamical laws of Newton. First, we reformulate Newton's second law in terms of momentum. Remembering that acceleration is the rate of change of velocity, we can write

$$F = ma$$

$$= m\frac{v_2 - v_1}{t}$$

$$= \frac{mv_2 - mv_1}{t}$$

The quantities mv_2 and mv_1 are the momentum values after and before the collision, respectively. The equation may thus be stated by saying that the force equals the rate of change of momentum.

When two balls collide with each other, the collision process is not instant, as it seems, but actually takes a finite length of time t, albeit a short one. During this period t the two balls are in physical contact. The ball m_1 has its momentum changed from m_1v_1 to m_1v_1', and the ball m_2 has its momentum changed from mv_2 to mv_2'. Because the rate of change of momentum is equal to force, ball m_1 experiences a force $(mv_1' - mv_1)/t$ and m_2 experiences a force $(m_2v_2' - m_2v_2)/t$. Where do these two forces come from? They come from the other ball through the contact—when two objects are in contact, they usually exert forces on each other, the two forces being action and reaction (Fig. 5.29). They are equal and opposite. Therefore,

$$\frac{m_1v_1' - m_1v_1}{t} = -\frac{m_2v_2' - m_2v_2}{t}$$

Hence

$$m_1v_1' - m_1v_1 = -(m_2v_2' - m_2v_2)$$

In other words

$$m_1v_1' + m_2v_2' = m_1v_1 + m_2v_2$$

Figure 5.29 Action and reaction between two balls when they collide.

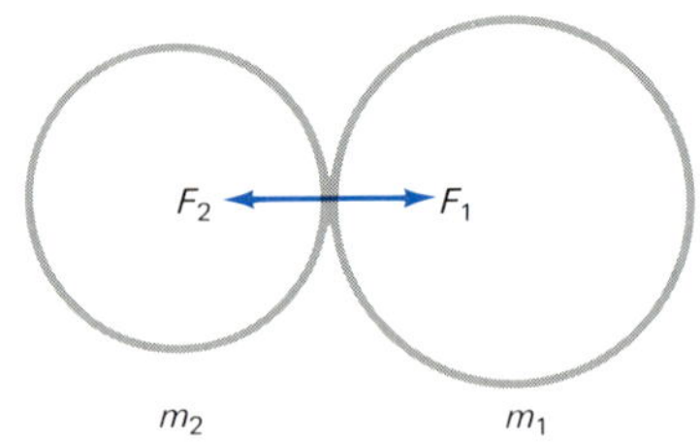

[4]In Newtonian theory the laws of motion are expressed as differential equations of the second order. The three conservation laws in Newtonian theory represent three first integrals of the differential equation. The integration constants represent the conserved quantities. The laws of motion deal with the changes; the emphasis is on the contingency. The conservation principles deal with the permanent; the emphasis is on the constant. The two views are not contradictory but are equivalent mathematically through the integration of the differential equation.

and we have proved the total momentum before and after the collision to be the same.

Physically we may visualize the situation as follows: before the balls establish contact each moves with a constant velocity until m_2 catches up with m_1 from behind. Once contact is established, action and reaction between the balls occur and their velocities begin to change. At the same time the forces cause the balls to be compressed and even flattened a little—every solid is elastic to a certain extent and under the action of a force will yield a little, resulting in an elastic deformation. As the two balls squeeze each other flat slightly, the velocity of the one in the front, m_1, increases and that of the one behind, m_2, decreases until they become equal to each other. At this instant the two balls, moving with equal velocity, do not compress each other further. Like any elastically compressed body, such as a compressed spring, they tend to rebound to recover their original shape. Thus the rebound pushes the ball in the front and makes it move even faster and pushes the ball behind and makes it move even slower until finally the contact is broken and the two balls move apart with different velocities v_1' and v_2'.

Coefficient of Restitution. The difference $v_1' - v_2'$, which is the relative velocity by which the two balls move apart, is a result of the elastic rebound and its magnitude varies proportionately to the degree of elasticity. For those material substances that have no elasticity at all, such as two pieces of putty, they stick together after collision. The rebound stage is completely absent; the two final velocities v_2' and v_1' are the same and the relative velocity of separation is zero. Collisions of this kind are called *perfectly inelastic collisions.* On the other hand, in the other extreme, the largest possible velocity of separation $v_1' - v_2'$ is one equal to $v_2 - v_1$, the velocity by which the two balls come together (proved later). In this case the collision is called *perfectly elastic collision.* Most practical cases are in between, with the velocity of separation less than the velocity of approach. Moreover, the ratio

$$\frac{\text{Velocity of separation}}{\text{Velocity of approach}} = \frac{v_1' - v_2'}{v_2 - v_1} \equiv e$$

is a constant, always the same for a given pair of balls colliding with any velocities v_1 and v_2. This constant, defined as the *coefficient of restitution e,* equals zero for perfectly inelastic collisions, equals unity for perfectly elastic collisions, and falls between zero and unity for other intermediate cases. From the condition of momentum conservation and the value of e (determined by the materials of the colliding bodies and listed in reference works), we have two equations, which are sufficient to determine the two unknown quantities v_1' and v_2'. The velocities after collision can thus be predicted, and the dynamical problem can thus be solved.

Energy Not Necessarily Conserved in Collision. Let us consider what happens to energy after a collision. Before the collision the total kinetic energy is

$$E_b = \tfrac{1}{2}m_1v_1^2 + \tfrac{1}{2}m_2v_2^2$$

After the collision the total kinetic energy is

$$E_a = \tfrac{1}{2}m_1v_1'^2 + \tfrac{1}{2}m_2v_2'^2$$

Are they the same? Sometimes, but not always! We can show that only for perfectly elastic collisions E_a equals E_b (mathematical proof omitted here). The energy after an inelastic collision is less than that before. Where does the energy go? What about the principle of conservation of energy?

During the first phase of the collision process, the two balls compress together, resulting in elastic deformation of the balls (slightly flattened). It takes work to produce the deformation, just as it takes work to knock down a wall in the wrecking job. The work is done at the cost of reduction of kinetic energy. In the perfectly elastic collision, the work done in flattening the balls is recovered and changed back into kinetic energy in the second phase of rebound of the collision process, just as a compressed spring snaps back. Thus no energy is lost. In all other cases the recovery is not complete, some of the work done in the compression phase is not changed back into kinetic energy; this results in an energy loss. This part of the work produces permanent damage to the balls (when two cars collide, both are damaged). If this part of energy is included, the energy balance sheet still checks out. Thus energy is conserved from a more general point of view, even though kinetic energy alone is not conserved in inelastic collisions. This discussion also shows why the velocity of separation is at best equal to velocity of approach because otherwise the energy after collision would be greater than that before collision, in violation of the principle of conservation of energy.

Thus one should be careful in applying the principle of conservation of energy in solving problems, particularly when collision is involved. The following example of the *ballistic pendulum* illustrates the point very clearly.

Ballistic Pendulum. A wooden block of M kg is hung by a string of length l m acting as a simple pendulum (Fig. 5.30). A bullet of velocity v is fired into the block. As a result the block swings to a maximum height h above the original position before it swings back. What is the speed of the bullet? This is actually the way the bullet speed is measured, which, being very fast, cannot easily be measured otherwise. Thus the name *ballistic pendulum*.

Figure 5.30 The ballistic pendulum.

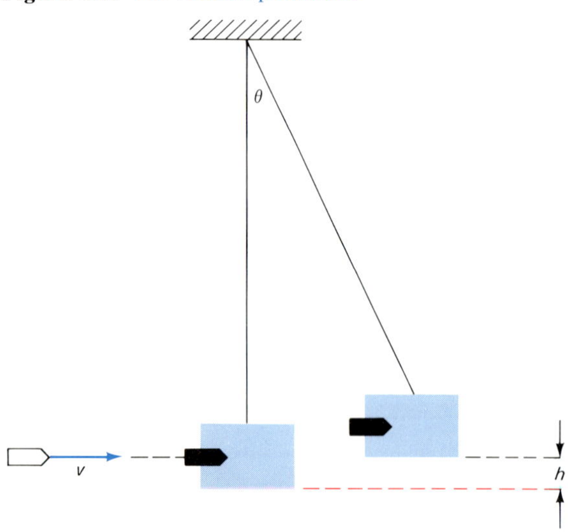

First of all we recognize that the process consists of two steps governed by different principles. The two have to be treated separately and differently. The first step is the embedding of the bullet into the block. This being a perfectly inelastic collision process, the principle of conservation of momentum can be applied. The total momentum before the collision is mv of the bullet, that of the block being zero. The momentum after the collision is $(M + m)u$, where u is the velocity of the block with the bullet embedded right after the collision. The principle states that

$$mv = (M + m)u$$

and therefore

$$u = \frac{m}{M + m} v$$

The solution of the first part of the problem provides the initial condition for solving the second part of the problem. The bullet-studded block with an initial velocity u now swings as a pendulum. Because the motion is one under the gravity force, which is conservative, the principle of conservation of energy applies. At the highest point of the swing the initial kinetic energy is wholly converted into potential energy. Thus

$$\tfrac{1}{2}(M + m)u^2 = (M + m)gh$$

Therefore

$$u = \sqrt{2gh}$$

Combining the two equations, we get

$$v = \frac{M + m}{m} u$$

$$= \frac{M + m}{m} \sqrt{2gh}$$

The velocity of the bullet thus can be determined from a measurement of the height h of the swing.

One should keep in mind that in the first part of the problem, the principle of energy conservation is not applicable. The kinetic energy before and that after the collision are not the same because the collision is perfectly inelastic, in which we know kinetic energy is not conserved.

Another Application. When a ball drops to the ground and bounces back, the rebound may be regarded as a collision process between the ball and another "ball," a gigantic one that is the earth. Because the earth is so massive, its state of motion is not changed much by the collision of a ball of ordinary size, and its velocities before and after collision may both be regarded as zero. Therefore the coefficient of restitution is simply the ratio of the velocities of the ball right after the collision (before leaving the ground), v_2, and right before the collision (before hitting the ground), v_1:

$$e = \frac{v_2}{v_1}$$

The velocity v_1 is the result of acceleration in the free fall from an original height h_1 (Fig. 5.31). Thus, according to the principle of conservation of energy:

$$\tfrac{1}{2}mv_1^2 = mgh_1$$

or

$$v_1 = \sqrt{2gh_1}$$

After the collision the ball rises with a velocity of v_2 and will reach a height h_2 determined by a similar relation

Figure 5.31 Successive bounces of a ball.

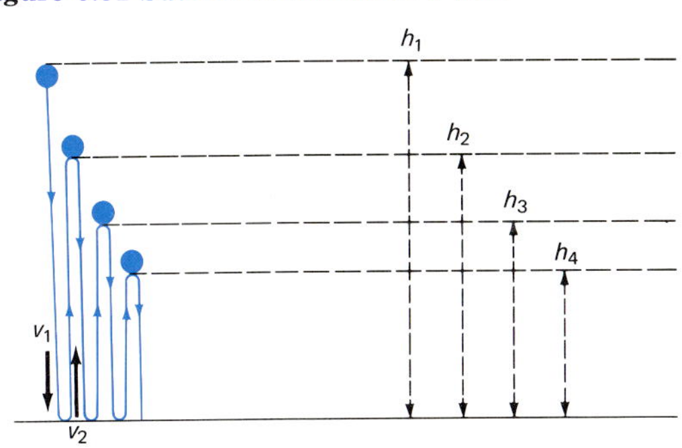

$$\tfrac{1}{2}mv_2^2 = mgh_2$$

Thus

$$v_2 = \sqrt{2gh_2}$$

When the coefficient of restitution e equals unity, v_2 will be the same as v_1 and h_2 the same as h_1. The ball will rebound to the same height and then fall and rebound again and again without end. The total energy will remain always the same; this is expected because mechanical energy is conserved in perfectly elastic collisions. On the other hand, in inelastic collisions, e is less than unity and v_2 is less than v_1. Therefore h_2 is less than h_1. The rebound is to reach a height short of the original, indicating that energy is lost during the collision. After bouncing several times, the rebound height becomes so small that the ball practically comes to a stop, and the total energy is reduced to zero as expected. Incidentally, the ratio $\sqrt{h_2/h_1}$ offers a convenient way to determine the coefficient of restitution.

Remarks. The discussion of the collision problem here also serves the purpose of illustrating the methodology of Newtonian theory. A seemingly instant event (collision) is analyzed into a sequence of events covering a very short period of time and the dynamical laws are applied methodically. The macroscopic phenomenon of collision is thus explained in terms of the space–time changes taking place on a microscopic level. The ballistic pendulum problem also illustrates the analytic procedure by which a complex process is analyzed into different phases, each dealt with by a different principle. The solution is obtained by matching the final condition of one phase with the initial condition of the next. The central idea is to reduce all natural phenomena into space–time changes, which can then be dealt with by mechanical laws. This is the basic strategy of the Newtonian theory for solving any scientific problems.

In atomic and nuclear physics, the collision between molecules, atoms, and nuclear particles is generally perfectly elastic. Both principles of energy and momentum conservation apply. The theory so developed lies at the base of a large part of atomic and nuclear physics. Epoch-making discoveries are made based on such a simple application of the Newtonian theory (artificial transmutation, Compton scattering and the discovery of the neutron). Some of the applications are demonstrated by home work problems at the end of the chapter. There are also inelastic collisions in atomic and nuclear physics. (The atoms may be exited or ionized after collision, thus consuming a part of the original kinetic energy.)

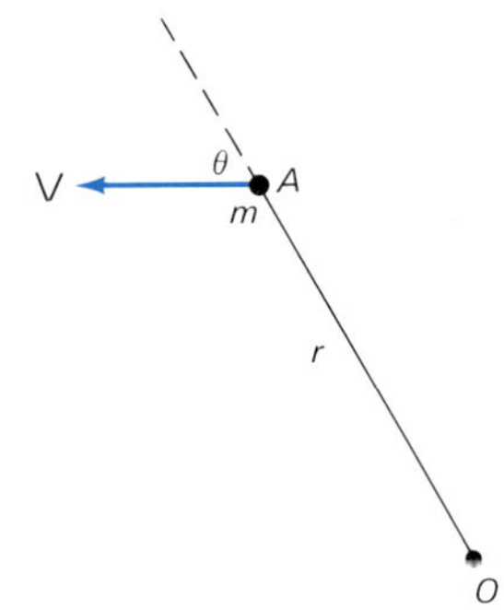

Figure 5.32 Defining angular momentum.

Principle of Conservation of Angular Momentum. We now briefly mention the principle of conservation of angular momentum. The *angular momentum* of a particle of mass m and velocity v at a point A in Fig. 5.32 with respect to a fixed reference point O is defined by the product mrv, where r is the distance $\overline{OA}$ provided that the velocity is perpendicular to $\overline{OA}$. (If not perpendicular and the two make an angle of θ, then angular momentum is defined by $mvr \sin\theta$.)

For an isolated system of interacting particles it can be proved that the total angular momentum of the system is always the same. This is the principle of conservation of angular momentum. A simple example is provided by the planetary motion. It can be proved that the angular momentum of a planet with respect to the sun is always the same. The planet moves slower at the aphelion but faster at the perihelion but the product mvr is the same. To mention another example: as an ice skater spins, the speed of spin increases (v larger) when the arms are lowered (r smaller) and decreases (v smaller) when the arms are stretched out (r increases). The product vr and thus the angular momentum remain the same.

Problems

5.1 The following is a list of good driving tips that will save gasoline. Explain why.

a. Drive at moderate speed; high speed penalizes gasoline mileage.

b. Avoid "jack rabbit" starts; always accelerate gradually.

c. Avoid stop-and-go driving; maintain a steady speed.

d. When entering an expressway, take full advantage of the acceleration lane to reach cruising speed gradually rather than having to push the accelerator to the floor board.

e. When going uphill, build up speed ahead of time rather than accelerating hard on the hill.

5.2 What is the kinetic energy of a median-weight automobile of 1,800 kg cruising at a speed of 100 km/hr (62.5 mph) in the units of joule, erg, and kwh (kilowatt-hour)? Assume 3¢/kwh (utility rate). How much is the energy worth? This amount is wasted when the car is braked to stop.

5.3 In the preceding problem, the car moves from a plain onto a hill. With the engine turned off how many meters in elevation can the car climb on the hill? Ignore all frictions.

5.4 Carry a weight of 50 kg (110 lb) from the ground floor to the sixth floor at a height of 18 m. How much work is done in joules and in kwh? Assume 3¢/kwh. How much is the job worth? Do you want to take the job? If the lifting is done by an elevator, how much is the energy cost for this service? What do you conclude from this comparison?

5.5 What is the speed of the weight in the preceding problem when it reaches the ground if it is dropped from the sixth floor?

5.6 A meal has an energy content of about 1,000 Kcal (1 Kcal = 1,000 cal). If the food is burned and the heat energy produced is sold at the utility rate of 3¢/kwh, which is more than its worth, how much can one get? How does it compare with the price of the meal?

5.7 Human muscle energy eventually comes from energy of the food. In the preceding problem if one half of the food energy is converted into muscle energy (the other half for physiological "housekeeping" of the body), how much is the cost of muscle energy per kwh? Assume the meal cost \$2.50. What do you conclude from a historical, economic, and humanistic point of view?

5.8 A light bulb of 100 w burns for 3 hours. How much is the energy cost, based on the utility rate of 3¢/kwh? The household current voltage is 110 v. What is the current in the light bulb? What is the resistance of the bulb?

5.9 A marble is dropped from a height of 10 ft to the floor. After rebound the third time, it rises as high as 1.25 ft. Find the coefficient of restitution.

5.10 A hole is drilled through the earth (assuming the earth is solid all way through) vertically downward to reach the other side of the earth. A ball is dropped into the hole. Describe the motion of the ball as time goes on. Will it shoot out on the other side? By what speed? What will happen next? (Neglect air friction.)

5.11 If the moon (1/81 of the earth mass) should crash into the earth with a speed of 24,600 mph, what speed would the earth gain?

5.12 In the preceding problem compare the kinetic energy the earth would gain with the kinetic energy of the moon just before the crash. Are they the same? Why?

5.13 A bomb at rest explodes into two pieces, one being 1,000 times heavier than the other. Which piece is more dangerous? Why? (*Hint:* The firing of a pistol may be regarded as such an explosion!)

5.14 A subatomic particle called the pion at rest can spontaneously change itself into another particle called the muon, which appears with a definite amount of kinetic energy. Show that there must be another particle accompanying the muon. (This particle is called the neutrino.)

5.15 The muon at rest can spontaneously change itself into an electron with a variety of energies. Show that there must be at least two other particles accompanying the electron. Actually two neutrinos are emitted with the electron.

6

Physical Transformations of Matter

*6.1 Macroscopic Properties of Matter

In the last two chapters we have developed the basic principles in physical science centered on two basic concepts—matter and energy. Now we apply these principles to study the properties of matter in bulk. Matter in bulk can assume three forms: solid, liquid, and gas. Their properties comprise the *macroscopic* properties of matter distinct from the *microscopic* properties originating in molecular, atomic, and nuclear structures. In studying the macroscopic properties of matter, we come to know one form of energy—heat. The nature of heat and its relation to the properties of matter will be the subject of the next chapter. The subject material in these two chapters belongs traditionally to classical physics. In Chapters 10 and 11 we deal with molecular structure of matter and the energy associated with it; the material in these chapters belongs traditionally to chemistry. In Chapters 12 and 13 we discuss atomic structure of matter and the energy associated with it; this part is usually called atomic physics. Finally, in Chapters 14 and 15 we discuss nuclear structure of matter and the energy associated with it; this part is usually called nuclear physics.

The three forms of matter—solid, liquid, and gas—are called the three *phases* of a substance. A substance can change from one phase to another; such a change is called a *phase transition*. In a given phase a substance can exist in a variety of *states*. For example, a gas may be compressed to a smaller volume or allowed to expand to a larger volume. These are different states. A change of temperature or pressure also means a change of state. A state of a gas is thus characterized by its volume, pressure, and temperature. Change of state and change of phase are typical physical transformations of matter and are among the important macroscopic properties of matter with which we are concerned here.

Many properties of matter usually change with the phase and the state. They are studied and specified for a given phase and a given state. These properties can be classified into three general categories: mechanical, electromagnetic, and thermal. Acoustic properties are in reality mechanical and optical properties are in reality electromagnetic; they do not represent new categories. All these properties represent the response of the material system to external influences of a mechanical, electromagnetic, and thermal nature—quantitative relations between the cause and the effect. These responses may also be regarded as a kind of transformation of the material system. Thus all properties of matter may be related to transformations of matter. In this section we give an overall view of macroscopic properties of matter in general. Each of the specific properties will be discussed in later sections of this and the following chapters.

Each cause–effect relation usually involves two kinds of physical quantities, one *extensive,* representing the effect, and the other *intensive,* representing the cause. An extensive quantity will be reduced by one half when the substance is reduced by one half (for example, the mass). An intensive quantity does not change when the substance is reduced by one half (for example, the temperature). The former represents a quantity; the latter represents a quality. Thus in the mechanical response of a gas the intensive and the extensive quantities are, respectively, the pressure (generally called the *stress*) and the change of volume (generally called the *strain*). In the electromagnetic response of a material substance the two are, respectively, the electric field (or magnetic field) and the electric (or magnetic) *polarization.* In the thermal response the two are, respectively, temperature and quantity of heat.

The relation between the two quantities is different from one substance to another and thus represents a material-specific property. Most times the relation is one of direct proportionality and may be specified by a proportionality constant, which is a *material-specific constant.* Thus the relation between the stress and the strain is represented by a linear equation specified by a proportionality constant called the *elastic modulus* of the substance; those between the fields and the polarizations are represented by linear equations specified by constants called the *dielectric constant* and the *magnetic permeability;* and that between the temperature and the quantity of heat is represented by a linear equation specified by a constant called the *specific heat.* These material constants thus summarize the macroscopic properties of matter. The empirical study of the macroscopic properties of matter is reduced to the experimental determination of these material constants.

From these basic properties other properties of matter and other material constants may be deduced. Because sound is essentially a mechanical phenomenon, acoustic properties may be reduced to mechanical properties. For example, the speed of sound v may be related to and thus calculated from the pressure P and the density ρ of the air ($v = \sqrt{1.4P/\rho}$), both being mechanical quantities. Light is an electromagnetic phenomenon and the index of refraction n, a material constant, may be related to the dielectric constant K and permeability μ ($n = \sqrt{K\mu}$). Complex phenomena may be reduced to combinations of simple ones and other constants may be related to the basic constants.

Besides the cause–effect relations in each of the three categories, there are phenomena involving a cause in one category and an effect in another. For example, the phenomenon of *thermal expansion* involves a thermal cause (heating) leading to a mechanical effect (expansion). The *piezoelectric* effect involves a mechanical cause and an electric effect. The thermoelectric effect involves a thermal cause and an electric effect. Additional material constants are introduced in connection with these phenomena. All these properties may be related in a diagram shown in Fig. 6.1.

The preceding description of properties of matter covers only the static properties. In some of the dynamic processes the property of the material system may also be represented by proper material constants. The *coefficient of viscosity, electric conductivity,* and *heat conductivity* tell us how a material medium gives rise to a mechanical, an electric, or a heat current in response to the respective exter-

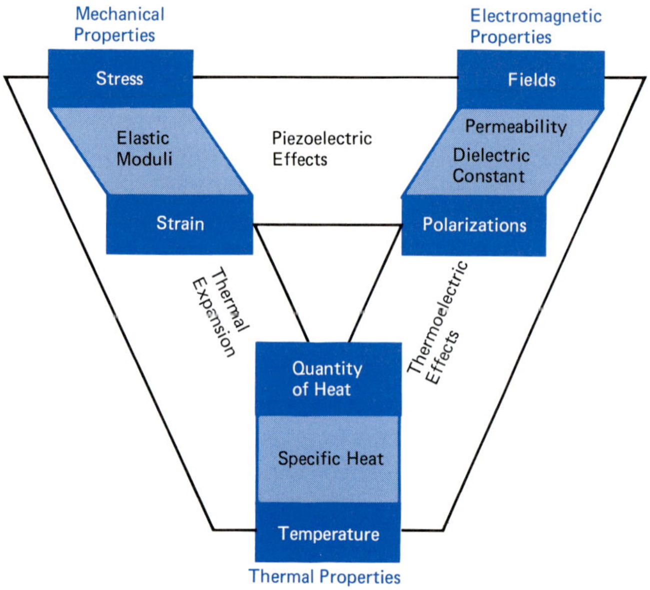

Figure 6.1 Diagram by which macroscopic properties of matter are organized.

nal influences. They represent the dynamical properties of matter. Here again we can distinguish an extensive quantity representing the effect, which is a *current density,* and an intensive quantity representing the cause, which is a *gradient* (rate of change) of a mechanical, electric, or thermal quantity. Again the two are usually related by linear equations; they are proportional to each other and the preceding coefficients are just the proportionality constants. (Thus the coefficient of viscosity relates the current density of momentum transfer to the velocity gradient; the electric conductivity relates the electric current density to the electric potential gradient; and the heat conductivity relates the heat current density to the temperature gradient.)

So there are a large number of constants representing a variety of properties of matter. These constants change to different values when there is a change of state or change of phase. They differ from one substance to another. Before the advent of atomic theory the study of properties of matter consisted largely of the determination of these constants. Voluminous data have been accumulated. The following questions arise: "Are these data the ultimate we can learn?" "Are there any relations between them that can simplify them further, at least conceptually?" "Are they explainable in terms of a rational scheme?"

The Microscopic Theory. The answer is found in the atomic theory, according to which these properties are not fundamental and basic but can be analyzed further in terms of the atomic structure of matter. The numbers of atomic species and their constituent parts are much smaller than the number of material substances. The properties of elementary particles and the forces between them are specified by a smaller number of *universal constants,* such as the charge and the mass of the electron, the gravitation constant, which are the same for all states, all phases, and all material substances. When the macroscopic properties are explained in terms of the atomic structure, the material constants can be related to and thus calculated from the universal constants. This way macroscopic properties of matter are greatly simplified.

For example, the fact that aluminum has a density of 2.7 g/cc was once thought to be a property of matter that does not need any further explanation. But from the atomic theory point of view the density is determined by the interatomic distances of aluminum, which in turn are determined by the interatomic forces in the aluminum crystal. Once we have sufficient information on the forces, we are in a position to figure out the (equilibrium) distances among the atoms forming the crystal lattice. Then we can calculate how many atoms there will be in a unit volume. Multiplying this number by the mass of one atom, we get the density of the material.

All other macroscopic properties may be treated in the same way. In fact, a large part of modern physical science is devoted to this line of investigation. If the theory of microscopic structure were complete, all macroscopic properties would be predictable from basic principles and properties of the elementary particles. No laboratories and no experiments would be needed any more to find out the fact of nature; the computers would turn out all needed information. The development of the microscopic theory at the present time, though largely complete in principle, is far from complete in details. We still have to perform experiments to learn the facts of nature; this is still the easiest way. The reasons that we do not do it by computer are that the calculation is too complicated and the treatment is still not perfect, that computer time is much too expensive, and that there are not enough computers around. (There was an estimate that even if all the matter in the whole universe were used to make computers, there would still not be enough computing facilities to do the calculations required.) On the other hand, one can be sure that nothing drastically new will come out of shaking test tubes. The macroscopic properties of matter are in principle understood in terms of the known basic principles and the properties of the elementary particles.

Change of state is a typical macroscopic property and will be studied in detail as an example. It is also a topic relevant to later discussions on engines and energy problems. The heat-related properties will be discussed in the next chapter. Others will be omitted.

6.2 Change of State

Consider a material substance of a fixed amount in a given phase and state. The state of a gas is specified by three variables—temperature t, pressure P, and volume V. These variables are called *state variables*. Once they are determined, the state of the gas is completely fixed.

The state of a substance is changed when a state variable is changed. In this section we consider properties of matter related to change of state. Consider a fixed amount of gas, say, 1 g of air, placed in a cylinder covered with a piston (Fig. 6.2). When a weight is placed on the piston, thus exerting a force on it, the piston will go down and stop at a certain position; in this position the gas assumes a definite volume. When the weight is increased, the volume is reduced and there is a definite cause–effect relation. The effect may be specified by the change of volume V. The cause may be specified by the pressure P, defined as the force applied on the gas per unit area of the piston. (When we say the tire pressure in 28 lb, we mean the force per unit area, i.e., per square inch of the tire, is 28 lb above the atmospheric pressure.) According to Newton's third law, the gas will exert an equal but opposite pressure on the piston. Moreover, the gas exerts the same pressure on the bottom and on the sides of the cylinder, this being the characteristic of a fluid. The cause–effect relation between P and V de-

Figure 6.2 Relation between the pressure and volume of a gas, and the molecular explanation of gas pressure.

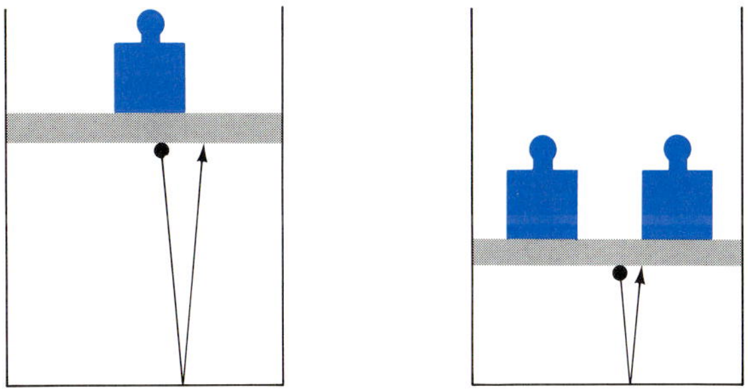

pends on the temperature t of the gas. How are the three related? This is the problem we want to investigate.

What has been said of a gas in the preceding remains true for a liquid and a solid. However, the change of the volume is much smaller than that of a gas, so small that it is usually neglected; yet there is a slight change when pressure is changed. Again the relation between the pressure P and volume V depends on the temperature t of the substance.

The three state variables P, V, t of a given amount of gas are related in such a way that two of them are independent, whereas the third is dependent on the other two.[1] For the 1 g of air in the cylinder we can change the state of the gas by changing the pressure in any way we wish (by applying weight). Once the pressure is fixed, we still can change the state of the gas by changing the temperature in any way we wish (by heating). But once the pressure and the temperature are fixed, the volume becomes fixed automatically and cannot be changed any more. This is an experimental fact. The state of the gas is thus determined by only two independent variables P and t. The third variable, the volume, is dependent on the pressure and the temperature and thus is a function of them:

$$V = f(P,t)$$

This equation may be considered as an algebraic equation for the variable t and may be solved for t. The other variables V and P are thus regarded as parameters and will appear in the solution of t. Thus

$$t = g(P,V)$$

meaning that t is a function of P and V. Similarly, P may be expressed as a function of t and V. The three equations are equivalent to one another. Any two of the three variables P, V, and t may be considered as the independent variables. The relation of the three variables expressed in any one of the three forms is called the *equation of state*. The equation of state represents an important property concerning the change of state and is the central problem of this section. Each substance has its own equation of state.

[1] The dependence on the amount of gas will be introduced later in the section. Right now we limit our discussion to a fixed amount of gas.

In the macroscopic study the equation of state, summarizing all properties concerning change of states, is considered a fact of nature not to be explained further. However, many times the form of the equation of state reveals significant information on the microscopic structure. The study of the equation of state thus helps establish the microscopic theory. In the microscopic theory our purpose is to derive theoretically the equation of state from the molecular properties. (The equation of state is important not only in change of state but also in other properties of matter. Together with the specific heat at constant volume, the two pieces of information form a basis from which many other macroscopic properties may be predicted by the laws of thermodynamics. Thermodynamics is a macroscopic theory establishing the relations between the macroscopic properties. Many properties may thus be related to and calculated from the equation of state together with the specific heat even without reference to the microscopic theory.)

Boyle's Law. We limit ourselves here to the problem of determining the equation of state of the gases by experiment. Usually a single experiment can tell us the relation between two variables. When three variables P, V, and t are involved we first fix one of them, say t, at a constant value and determine the relation between the other two, say, P and V, which may be expressed by a curve. We then repeat with the first variable t fixed at another value and determine the curve relating P and V again. This is repeated until a large number of values of t are covered. A number of P–V curves are thus obtained that may be pieced together to form a surface as shown in Fig. 6.3. A surface may be represented by an equation in the form of $P = f(V,t)$. This equation will represent the equation of state we want to obtain.

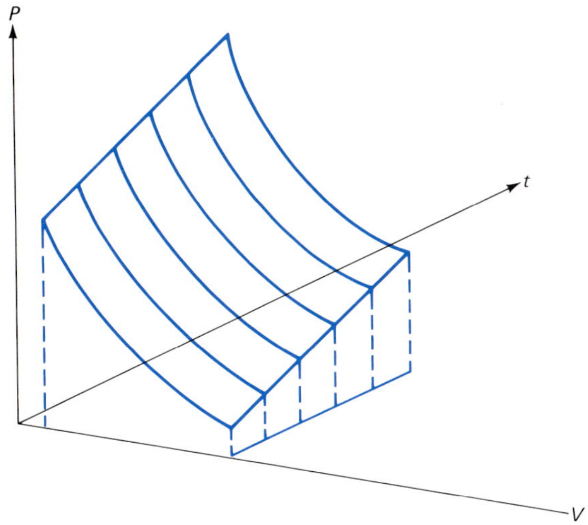

Figure 6.3 The surface representing the equation of state of a gas.

To determine the relation between pressure and volume at a fixed temperature, the cylinder of Fig. 6.2 may be immersed in a bath of water kept at a constant temperature to assure constancy of temperature of the gas inside the cylinder. We add weights one at a time and observe the change of volume. For each value of P we have a corresponding value of V; the two numbers (P,V) may be regarded as the coordinates of a point and may be plotted on a graph. Having obtained a series of points, we plot a curve passing through the points as shown in Fig. 6.4. For most gases this curve turns out to be *nearly* a *hyperbola* representable by the equation

$$PV = \text{constant}$$

This relation is known as *Boyle's law*.[2]

[2] Boyle's law and many other laws in this and the following chapters form another category of natural laws that deal with properties of matter at various levels of aggregation. They were first discovered by experiment and therefore they are *empirical laws* of nature—the starting point of scientific investigation. Now, most of them have been explained by the dynamical theories based on the few basic dynamical laws and force laws. From the theoretical point of view these laws may now be regarded as theorems in the dynamical theory. The ultimate purpose of scientific investigation is to incorporate all empirical laws into the dynamical theory.

Kinetic Theory of Gases. Boyle's law specifies a macroscopic property. Once we have a theory of atomic structure of matter we expect an explanation of any macroscopic property in terms of the microscopic theory. Indeed Boyle's law can be explained on a molecular basis. All material substances are made of molecules that attract one another. When the attractive forces are sufficiently strong, the molecules stick together in a fixed spatial relation, forming a body with a fixed volume and fixed shape. Such a body is a solid. If the forces are not sufficiently strong, the body formed has a fixed volume but not a fixed shape and becomes a liquid. When the forces are weak the molecules do not associate together in a definite spatial relation, and the body does not even have a definite volume, far less a definite shape; it becomes a gas. A gas is regarded as a large collection of sparsely populated molecules, each moving nearly independently of the others with a high speed, of the order of many hundred meters per second. That bulk substance is made of discrete particles is an old idea traceable back to Greek atomic theory. It was introduced in modern chemistry in the early nineteenth century. The element that is new here (introduced in the mid-nineteenth century) is that the molecules are moving at high

Figure 6.4 Experimental determination of the P–V relation.

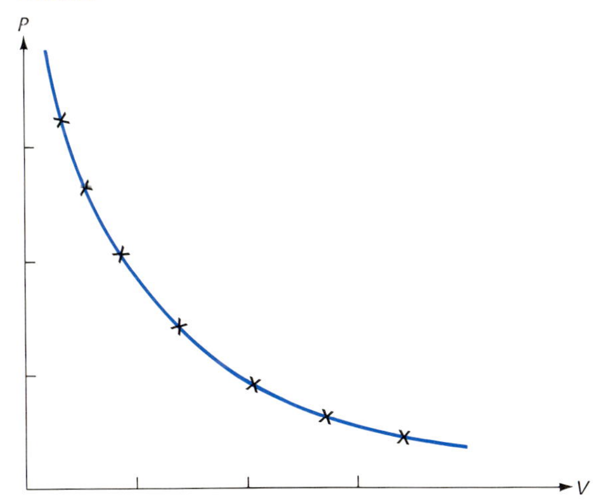

speeds. The theory is therefore called the *kinetic theory of gases*.

Thus a mechanical picture is presented of a gas. Considered as a collection of particles endowed with velocities (initial conditions) interacting with one another and with the container wall by forces, a gas becomes a mechanical system amenable to treatment by Newtonian theory. Because the molecules are sparsely populated and move with high speeds, the forces between the molecules and the gravity force of the earth on the molecules have negligible effects and may be neglected in the first approximation. Thus the molecules may be regarded as free particles, i.e., they are under the action of no force, and Newton's first law applies. They will be moving along straight lines with the same speeds until they hit another molecule or the wall of the container. What happens at such a collision is determined by Newton's second law (Sec. 5.7). The molecules bounce back from the collision and move on. Thus the molecules fly about in the container and fill up all the available space; this explains the fact that a gas assumes the volume and the shape of the container. In fact, it is for this purpose that the kinetic element of the theory was introduced.

A more significant conclusion may be derived. When a molecule collides on a wall, the two parties interact by force and this interaction has two aspects: action and reaction. One is the force the wall exerts on the molecule, which pushes the molecule backward in rebound. The other is the reaction, the force the molecule exerts on the wall. This force is impulsive. Yet when a large number of molecules are colliding on the wall from time to time all the impulsive forces merge together to appear as a continuously sustained force. This provides us with an explanation of the *pressure* of a gas, a force the gas exerts on the container wall, a basic property of a gas.

By the kinetic theory of gases, we can learn more about the pressure P of a gas, i.e., its relation with the volume V. Consider the gas in a cylinder in Fig. 6.2. We ask, "What will happen to the pressure when the volume is compressed to one half of the original?" The molecules exert force on the piston just the same, but the collisions are more frequent. The volume is now smaller by one half. It now takes less time for a molecule that has bounced back from the piston to travel to the bottom of the cylinder and then return to hit the piston the second time. In fact, the time of the round trip is reduced by one half (assuming no collision between molecules). This can be seen in the simple case shown in Fig. 6.2 and indeed, in the general case, can be proved rigorously by elaborate calculations. Thus the number of collisions is doubled and the pressure is expected to double. Therefore the inverse proportion relationship between P and V is explained by the kinetic theory of gases.

Charles' Law. We now consider temperature dependence. Instead of plotting many P–V curves at various temperatures, we proceed by a different method, which leads to the same result. We now keep the pressure constant by maintaining a fixed weight on the piston and observe the correlation between the temperature (in degrees centigrade) and volume. The increase of temperature may be achieved by heating the bath of water. As the temperature t increases, the piston moves up and the volume V increases accordingly. The corresponding values V and t may be plotted as the coordinates of a point and the experiment yields a series of points that may be joined by a curve. This curve for most gases turns out to be *nearly* a straight line, as shown in Fig. 6.5. A straight line may be represented by the following standard equation in analytic geometry:

$$V = mt + V_0$$

where m represents the *slope* of the line and V_0 represents the intersection of the line with the ordinate. The equation may be changed to another form as follows:

$$V = V_0(1 + at)$$

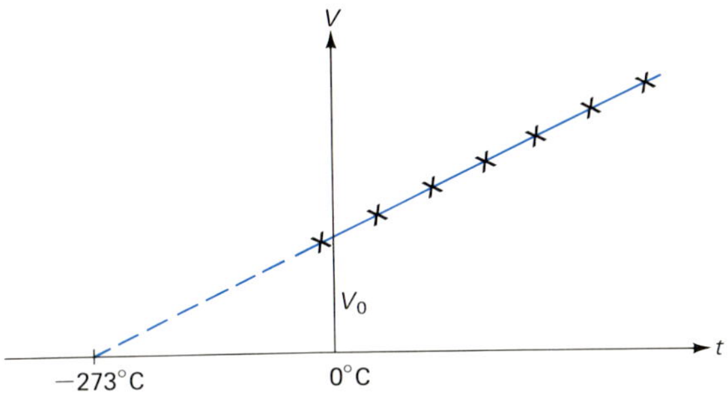

Figure 6.5 Experimental determination of the V–t relation.

in which a new constant a is introduced that is related to m by $m = V_o a$. When $t = 0$ we have $V = V_o$. This means that V_o is the volume at 0°C. The value of the constant a is found to be the same for *all* gases, equal to 1/273. This relation is known as *Charles' law.*

The preceding equation may be changed to the following:

$$V = V_o\left(1 + \frac{t}{273}\right)$$

$$= \frac{V_o}{273}(t + 273)$$

The zero point of any temperature scale is arbitrary and we may use any point as the starting point of the scale. Suppose we choose −273°C as the starting point and use the same scale division of the centigrade scale. The new temperature scale so constructed is called the *absolute temperature scale.* Its unit is designated by °K. Denoting the temperature value in this scale by T, we have

$$T = t + 273$$

and the previous equation may be written as

$$V = \left(\frac{V_o}{273}\right)T$$

This equation shows that V and T are *directly* proportional to each other. If Charles' law should be exactly correct (actually it is only approximately correct), the volume of the gas would be zero when T is zero, i.e., $t = -273$°C. Because the number 273 is the same for all gases, the preceding discussion implies that *all* gases would be reduced to zero volume at −273°C. They could not be reduced any further and so this temperature seems to be at the rock bottom of the temperature scale, there being no other temperature below it. This point is thus the natural starting point of the temperature scale and this is exactly what we chose in the new scale T. Because this point is determined by a common property of all gases it is not an arbitrary point related to a specific material such as the freezing point of water. It refers to a basic property of the temperature scale itself and thus has an absolute meaning. Because there is no temperature below this point it is called *absolute zero,* and the scale is called the *absolute scale.* This discussion is based on Charles' law. From the experimental point of view Charles' law is only approximately correct. So is the preceding discussion. Nevertheless the existence of absolute zero and the absolute scale can be established by more rigorous reasoning, which we will not undertake here.

Equation of State of an Ideal Gas. Boyle's law and Charles' law are two special cases of the equation of state. The two together are sufficient to determine the equation of state. Boyle's law states that when T is kept constant, V is inversely proportional to P:

$$V \sim \frac{1}{P}$$

Charles' law states that when P is kept constant, V is proportional to T:

$$V \sim T$$

When both P and T vary, we conclude by algebra that V is proportional to the product of $1/P$ and T:

$$V \sim \frac{T}{P},$$

or

$$V = \text{constant} \times \frac{T}{P}$$

or

$$PV = \text{constant} \times T$$

So far we consider a fixed amount of gas, say, 1 g of air. If we consider 2 g of air, the equation of state would have the same form as the preceding but the constant is expected to be different. Indeed, if P and T are the same, V of 2 g of air is expected to be two times larger than that of 1 g of air, volume being an extensive quantity. Thus the constant in the preceding equation for 2 g of air is expected to be twice as large as that in 1 g of air. In other words, the constant has a value proportional to the mass of the gas.

Let us introduce a new unit of mass called the *mole*. The mole is a unit familiar in chemistry defined as the amount of substance whose mass value in gram unit equals the *molecular weight* of the substance. The exact definition of molecular weight will be introduced in Chapter 10. Roughly speaking, it is the relative weight of the molecule referring to the hydrogen atom as unity, e.g., 2 for the hydrogen molecule H_2 and 18 for the water molecule. Thus 1 mole of hydrogen molecule is 2 g, 1 mole of water is 18 g, and so on. It is a peculiar unit of mass in the sense that the size of the unit is not fixed once and for all but varies from one substance to another. But it is useful in special situations. The reason is that 1 mole of any substance has the same number of molecules; this number is called the Avogadro number, N_0, which is a universal constant. Indicating the number of moles of a gas by n, we write the preceding equation as follows:

$$PV = nRT$$

where R is another constant. The constant R turns out to be the same for *all* gases. (If we use gram as the mass unit the constant appearing in the preceding equation will *not* be the same for all gases; this is the reason of using the unit mole.) The constant R is thus a *universal constant,* the numerical value of which is 8.317 joules/mole/°K (which is about 2 cal/mole/°K in terms of the unit calorie, to be introduced later). It is called the *universal gas constant.*

Properties of matter are expected to be different from one substance to another. For the gases the values of a and R are all the same, which is quite unusual. The appearance of a universal constant (R in the present case) usually indicates the existence of a universal law. The equation of state thus reveals something more fundamental. It seems to refer to a property of the gas that is determined by a structural feature common to all gases rather than by a specific property peculiar to one gas. Indeed, this is the starting point of the kinetic theory of gases, according to which the property of a gas is determined by the *swarm* of molecules—what each molecule is made of matters little.

Boyle's law and Charles' law, and thus the preceding equation of state, are only approximately correct. More precise experiments show that all gases deviate from the preceding equation to some extent. However, the equation is simple and revealing; it is useful and important in spite of its approximateness. We may imagine a hypothetical gas obeying this equation *exactly* and call it the *ideal gas.* The equation is thus called the *equation of state of the ideal gas,* to distinguish from those for *real gases.*

Boyle's law refers to a particular kind of change of state in which the temperature is kept constant. Such a change is called an *isothermal process.* In general, change of state may involve the change of all three variables P, V, and T. Another special case, the *adiabatic process,* will be introduced here to prepare our study of heat engines later. This is a process in which the gas is insulated from the surroundings so that no heat can enter or leave the gas.

When pressure and volume change, the temperature of the gas in this case may rise (upon compression) or fall (upon expansion). The pressure–volume relation in this process is different from Boyle's law and is found to be given by the following equation:

$$PV^{\gamma} = \text{constant} \qquad \text{(adiabatic)}$$

where γ is a constant (see Sec. 7.2). For air the value of γ is 1.40. Other gases may have a different value. This relation may be derived theoretically from the equation of state together with the condition of heat isolation. We shall not go into this mathematical derivation.

From the preceding equation and the equation of state we can prove another relation pertaining to the adiabatic process that will be used in later discussion:

$$TV^{\gamma-1} = \text{constant} \qquad \text{(adiabatic)}$$

Van Der Waals' Equation. For the actual gases we desire more exact equations of state than the previous one, and many forms have been proposed. They are called *equations of state of the real gases.* They are expected to be more complicated than that of the ideal gas so that they will fit the experimental results better. They should also reduce to the latter as a first approximation. It happens that no single equation can be found to fit all gases. This suggests that the modification of the equation of state of the ideal gas depends on the specific properties of the molecules themselves, which are different from one gas to another. The study of the equation of state of the real gases thus throws light on the properties of the molecules themselves in addition to the gross structure (the swarm) of a gas.

The most important and the simplest of the equations of state of the real gases is *van der Waals' equation,* which has the following form for 1 mole of gas ($n = 1$):

$$(P + a/V^2)(V - b) = RT$$

where a and b are two constants. Although the same form of the equation applies to all gases, the values of a and b (see Table 6.1) are different from one gas to another. They are *material constants* representing specific properties of a particular gas. The equation modifies the ideal gas equation by adding a small term a/V^2 to the pressure P and subtracting a small term b from V. When a and b are negligible compared with P and V the equation reduces to the ideal gas equation. This equation, though simple, is able to account for a large part of the deviation of the real gases from the ideal gas equation.

Because only quantities of the same kind may be added together, the term a/V^2 must be in the nature of a pressure and the term b must be in the nature of a volume. The equation thus suggests an interpretation in terms of the molecular theory. According to the kinetic theory of gases, a gas is essentially an empty space with a swarm of molecules flying

Table 6.1 Van der Waals' Constants a and b for 1 cc of Gas

Substance	a	b
Helium	6.8×10^{-5} atmosphere	1.06×10^{-3} cc
Argon	$268. \times 10^{-5}$	1.43×10^{-3}
Oxygen	$273. \times 10^{-5}$	1.43×10^{-3}
Nitrogen	$272. \times 10^{-5}$	1.73×10^{-3}
Hydrogen	48.7×10^{-5}	1.18×10^{-3}
Carbon dioxide	$717. \times 10^{-5}$	1.91×10^{-3}
Ammonia	$833. \times 10^{-5}$	1.66×10^{-3}

about inside. When a gas is compressed the volume is reduced at the expense of the empty space between the molecules. The volume occupied by the molecules themselves cannot be reduced. It is thus natural to interpret b as the total physical volume of the molecules themselves so that $(V - b)$ is the volume of the empty space between molecules that is reducible under pressure and therefore should enter into relation with pressure in the equation of state in the way previously discussed. The additional pressure a/V^2 may be interpreted as originating from the mutual attraction among the molecules. The molecules are assumed to exert weak attractive forces (called *intermolecular force* or *van der Waals' force*) on one another tending to pull themselves together; the effect on the volume is the same as if an additional pressure is applied from outside. The idea of intermolecular force is a useful one; we may use it to explain the phenomena of adsorption, surface tension, and so on. The origin of the intermolecular force is a problem of the microscopic theory and is well accounted for by the present theory of atomic structure. The constant a here gives us experimental information on the intermolecular forces. On the other hand, the constant b gives us information on the physical volumes of the molecules themselves. In fact, historically the first determination of the molecular size, weight, and number involved the use of the constant b (Loschmidt's method).

The relation between P, V, and T given by van der Waals' equation are shown by a series of curves relating P to V at a number of temperatures T_1, T_2, . . . in Fig. 6.6. When V is large the equation approaches closely the equation of an ideal gas for which the curves assume the shape of a hyperbola. This happens in the right-hand side of the diagram. When V is small, the situation is quite different. In fact, in the diagram it is shown that in the lower left-hand side, a wiggle appears in the curves.

The corresponding experimental results are shown in solid curves in Fig. 6.7. The curves agree closely with the theoretical curves except that the wiggle is replaced by a horizontal line. The horizontal line means that in this region the volume is changed from V_2 to V_1, with pressure remaining always the same. What actually happens is this: a gas of temperature T_b when compressed to volume V_2 will start to condense into a liquid. As condensation proceeds, the volume becomes smaller until the whole gas of volume V_2 is converted into a liquid of volume V_1. In

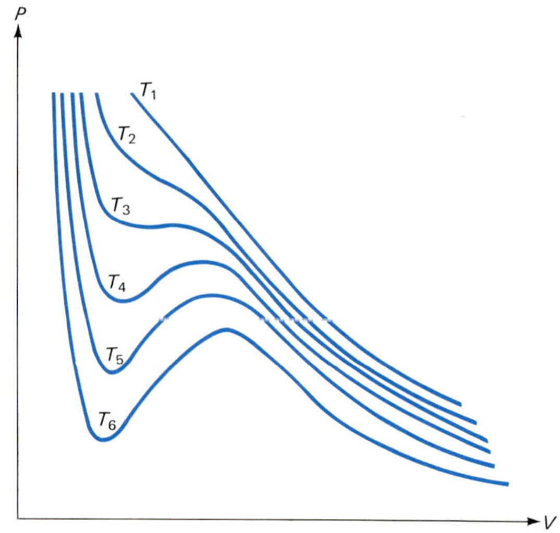

Figure 6.6 Curves representing P–V relations for a number of temperatures according to the van der Waals equation of state.

Figure 6.7 Experimental curves representing P–V relations for a number of temperatures.

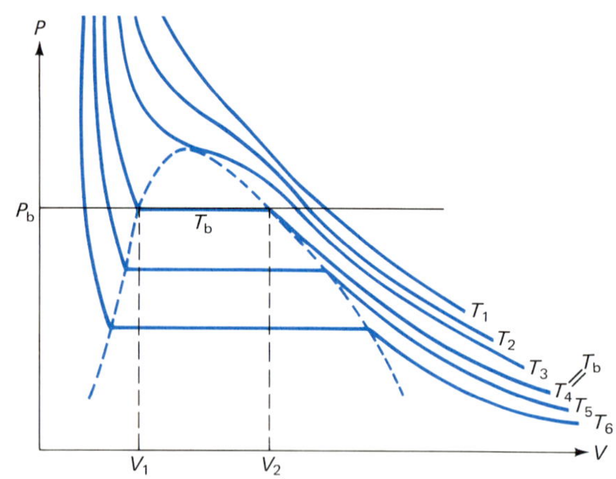

the condensation process the pressure remains always the same P_b. Thus the horizontal line represents the process of condensation. We will return to the problem of condensation more in detail in Sec. 7.3. After the point V_1 a further reduction of the volume (now a liquid) calls for a tremendously large increase in pressure as shown by the steep curves to the left of V_1. This is expected of a liquid which is very hard to compress.

Thus the molecular theory underlying van der Waals' equation not only explains the behavior of a gas but also that of a liquid as well. The equation also includes the phase transition from gas to liquid. That the transition from gas to liquid (condensation) does not follow the wiggle of van der Waals' equation, but follows a horizontal line can also be explained by a sophisticated argument (the Maxwell rule) that we will not go into here.

Problems

6.1 Consider an automobile engine of a compression ratio 10. If the gas were compressed isothermally to one tenth its original volume, what would be the pressure of the compressed gas on the piston?

6.2 In the preceding problem, the gas in the cylinder is compressed adiabatically in actual operation. What is the pressure in this case?

6.3 In problem 6.2 the compressed air–gasoline-vapor mixture is ignited and the temperature changes from 700°C to 2,200°C, say. What is the pressure after the heating?

6.4 Make use of Boyle's law to devise a pressure gauge.

6.5 Charles' law implies one point that is not satisfactory—at absolute zero the volume of a material substance becomes zero! How can this shortcoming be removed?

7 Heat Energy

7.1 What Is Heat?

Heat is a physical phenomenon apparently unrelated to mechanics. Yet after a lengthy investigation it was found that heat is not a separate and independent phenomenon but a manifestation of mechanics on the molecular level. Thus it is reducible to and explainable by mechanics.

The concept of heat originates from the hot and cold sensation. The degree of hot or cold is gauged by the quantity *temperature,* which may be measured by a thermometer. The thermometer may be graduated by the Farenheit scale, which is used in weather reports and for commercial purposes, or by the centigrade scale, which is used in most scientific work. In the centigrade scale the freezing point of water is taken as the zero point of the scale and the boiling point of water is taken as 100°C to determine the scale division.

A basic property of heat is that when two objects of different temperatures are brought into contact, they will interact in such a way that the hot one cools, the cold one warms, and eventually the two temperatures become the same somewhere between the two original temperatures. From the very beginning scientists tried to understand heat by assuming the existence of an entity called *heat* that could be kept inside a body and also transferred from one body to another. The more heat quantity transferred into a body (e.g., by heating), the higher the temperature. When two objects are in contact, heat flows in such a way as to equalize the temperatures, just as water flows from one container to another to equalize water level. Thus in any thermal phenomenon there are two quantities involved: temperature, which is qualitative and intensive, and heat, which is quantitative and extensive.

Theory of Heat. What is this heat entity? In the eighteenth century, following the success of Newton's materialistic theory, all natural phenomena were looked upon from a materialistic point of view. Heat was no exception and the heat entity was considered as a fluidlike substance. Because a hot object is not any heavier than it is when cold, the heat substance must be one of no mass or of negligible mass. It must then be a kind of substance very different from the Newtonian kind of material substance characterized by mass. At that time it was thought that many different kinds of substance may exist. (Now we know that there is only one kind—the Newtonian kind, characterized by mass.) This heat substance was given the name *caloric*.

The caloric theory was found to be untenable later and was replaced by another theory—the mechanical theory of heat, according to which heat is a form of *energy*, not a *substance*. The history of contention between the two theories is illustrative of the development of scientific ideas and is worth mentioning here. The mechanical theory originated from the observation that when mechanical work is done against friction force, heat is generated. Rumford (an American Tory during the Revolution) noticed that in the process of drilling a cannon barrel, a large amount of mechanical work is spent and a large amount of heat is generated. This phenomenon, nevertheless, can be accounted for by the caloric theory. According to this theory the heat substance caloric is absorbed inside the material substance, in this case the cannon barrel. In the drilling process iron filings are chipped off from the barrel and in the process, as the theory goes, the caloric contained inside the material body becomes freed, manifesting itself in the form of heat.

To counter the caloric theory, Rumford performed an experiment in which he used a blunt drill on the cannon barrel. This time the drilling action did not chip off metal filings, but produced heat just the same. In fact, as long as drilling continues and mechanical work is done by the force acting on the drill against the friction, heat is produced. Heat seems inexhaustible as long as work is done. This counters the idea that heat is a substance and is exhaustible. Rumford was able to figure out in this experiment a *mechanical equivalent of heat*—a definite amount of mechanical work spent that would generate a unit amount of heat. If heat can be generated indefinitely by the same rate when work is expended, it seems more natural to suppose that heat is equivalent to energy and is converted from energy by a definite exchange ratio. This is the essence of the mechanical theory of heat.

However, the Rumford experiment did not do away with the caloric theory. The result still can be explained by the caloric theory, even if a little far-fetched. One may still think of heat as a substance absorbed in material bodies, just as water is absorbed in a sponge. By pressing the sponge, water may be squeezed out. This might well be what the blunt drill does, as the caloric theory goes. Whereas the water in a sponge is exhaustible one might still think that the caloric content is so large that the blunt drill could only scratch the surface, not exhausting it all in our lifetime.

Later, another experiment was carried out by Davy that dealt a serious blow to the caloric theory. He rubbed two pieces of ice against each other and found as a result that the ice melted and changed into water. According to the caloric theory, in the rubbing action caloric is squeezed out of ice and as a result the ice would be colder, being deprived of its heat content. But the ice changed into water, and water has more heat content than ice because we have to add heat to ice to change it to water. Therefore the theory leads to a contradiction.

Nevertheless, the caloric theory dies hard, because the mechanical theory itself is not wholly satisfactory. If heat is indeed a form of energy and is convertible from energy by a definite exchange rate, then heat should be able to change back into energy by the same rate. Whereas work can easily be converted into heat, the reverse is not always possible. By a complicated process such as that involved in a steam engine, work can be produced at the expense of heat, but not at the same exchange ratio. In such a conversion some heat is always wasted; the conversion can never be 100 per cent. Thus heat and work do not seem to be equivalent.

So neither of the two theories can establish a clear superiority over the other. Both have difficulties and thus are provisional. In fact, in the first half of the nineteenth century, the two theories stood side by side and scientists used either or both theories for one problem or another.

In 1824 the caloric theory received a significant boost by the work of Carnot. He studied the efficiency of heat engines and derived an important theorem on it based on the caloric theory. This theorem, bearing his name, tells us how much work can be produced from a certain amount of heat; the result is dependent on the temperatures at which the engine operates. The conclusion of this theorem is still correct today even though the caloric theory, by which Carnot derived the theorem, is now outdated. In other words, the Carnot theorem had to be reproved later, using the mechanical theory. This theorem answers the question the early mechanical theory could not answer, thus increasing the credibility of the caloric theory at that time. In this work Carnot considered heat as a substance going through a heat engine in the same way that water goes through a watermill. The amount of water coming out of the watermill is exactly the same as the amount entering the mill. The work done by the mill comes from the potential energy of the water. Carnot imagined that the same amount of caloric entering and leaving the heat engine and the "potential energy" of the caloric, gauged by temperature, was converted into mechanical work. This view has been proved wrong, but his conclusion and argument, by a happy accident, happened to be correct.

Starting in 1840, Joule performed a series of experiments to determine more accurately the mechanical equivalent of heat in the conversion of work to heat and found that in all cases, whether the energy is mechanical or electric, for a certain amount of work spent there is always the same amount of heat generated. To be exact, 1 *calorie* of heat, defined as the amount necessary to raise 1 g of water by 1°C, is equivalent to 4.2 joules of energy, this number being the mechanical equivalent of heat:

$$1 \text{ cal} = 4.2 \text{ joules}$$

The series of experiments proved without doubt that heat is generated from work by a definite conversion ratio. Thus it is not a substance. The unit of heat measurement, the calorie, thus may also be regarded as an energy unit, according to the preceding conversion ratio.

Hence in the middle of the nineteenth century, scientists were faced with the dilemma of having two theories, each one with definite merits. The dilemma was resolved by Kelvin and Clausius in 1850, who realized that there are two basic, independent principles governing the behavior of heat. The two previous theories had each been concerned with one of the two basic principles. Each was correct to a certain extent but not complete without the other. The complete theory is based on both principles. Because the two principles are independent, there is no ground for contradiction. The two principles are called *the first law and the second law of thermodynamics* and the branch of science that follows is called *thermodynamics.* It represents a synthesis of two partially correct but incomplete theories. Their merits are preserved but the contradiction is removed. The first law confirms the mechanical nature of heat. But a separate and independent second law was introduced to govern the conversion of heat back to work. (See Chapter 16 for the details.) On this basis the Carnot theorem was rederived, based on the mechanical theory of heat.

The development of the theory of heat is a good example to illustrate the progress of science by the process of *synthesis.* Special, incomplete theories are developed first. They may contradict each other. Then a synthesis takes place, incorporating the merits of the special theories into a general, unified theory and eliminating their contradictions. The process of synthesis points the way in which the other less completely developed sciences are to be developed. In many of them we still have several special, incomplete, and contradictory theories contending with one another. As a result, one gains the impression that this is the way theories are and always will be. Far from it; this situation is a passing phase, a kind of growing pain. Eventually the contradictory theories must be unified by a synthesis.

Heat as Energy of Molecular Motion. How do we visualize heat as energy? With the advent of the molecular theory, heat of a body is regarded as the kinetic energy of the random motion of the molecules the body is made of. Take a gas as an example. A volume of gas is visualized as essentially an empty space populated sparsely by a swarm of small particles, that is, molecules, each moving with a high speed v in a random direction (Fig. 7.1). Each molecule has a mass m. It thus has a kinetic energy $\frac{1}{2}mv^2$. In the current view the total kinetic energy of all molecules represents the total heat content of the gas. When heat is added to a gas by a burner and its temperature is raised, the total kinetic energy of the molecules is increased and each molecule moves, on the average, faster than before. Energy is added to the gas in the form of heat. When a hot gas and a cold gas of the same kind are put into contact and heat "flows" from one to the other, what happens is that molecules of the two gases collide with one another through the surface of contact, so that after numerous collisions the slow-moving ones in the cool gas are speeded up and the fast-moving ones in the hot gas are slowed down. The equalization of temperature represents the equalization of average molecular speed or average kinetic energy through collisions. The transfer of heat thus represents a transfer of kinetic energy from one gas to the other. The molecular motion here discussed, though not directly visible, produces a hot or cold sensation on contact and is recognized phenomenologically as heat.

Figure 7.1 Molecules and their motion in a gas.

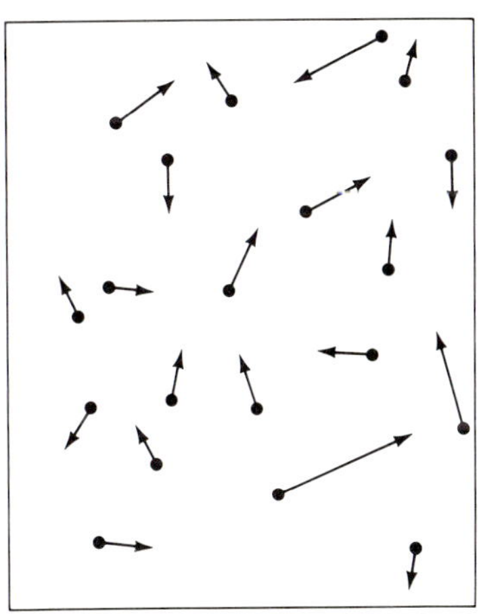

When a pendulum swings in the air, its amplitude gradually decreases and eventually the pendulum bob comes to a stop. The energy (kinetic and potential together) of the pendulum gradually decreases to zero and we say that this is due to the friction (or viscous) force of the air, which is not conservative but dissipative. This can now be explained by saying that the pendulum, as it swings in the air, collides with the air molecules and thus loses its kinetic energy to the air molecules gradually. If we consider the energy of the pendulum only, it decreases in time. But if we take the pendulum and the air molecules together, we find that the energy lost by the pendulum is gained by the air molecules and manifests itself as the heat energy of the air (some of the heat also goes into the pendulum), and the total energy remains unchanged.

Thus with the help of the mechanical theory of heat, we find that the law of conservation of energy remains valid even in the case of dissipative force if in addition to the kinetic and potential energies we include heat energy in the total sum. This discussion also brings out the point that the mechanical theory of heat reduces heat to a mechanical phenomenon and that the theoretical study of heat can be carried out by applying the well-known mechanical laws of Newton. A gas now appears as a Newtonian system of a large number of particles and any problem of it can be worked out by the Newtonian theory in principle, even though the detailed mathematical solution may be quite complicated. Nevertheless, we can cut through the mathematical complexity by using a statistical method. Many useful conclusions may thus be derived.

Such a theory has been developed (the kinetic theory of gases). The theory enables us to calculate the average speed of molecules in a gas. For example, the average speed of air molecules under the standard conditions (1 atmosphere pressure and 0°C temperature) is 5×10^2 m/sec. As mentioned before, by a rather involved argument encompassing the theory of viscosity, the kinetic theory of gases was able to tell us for the first time, in the middle of the nineteenth century, the size and the number and therefore the absolute weight of each kind of molecules (Loschmidt's method). Earlier in chemistry we learned the relative weights of atoms and molecules and defined them as *atomic weights* and *molecular weights*. Thus the atomic weight of hydrogen is 1, the molecular weight of nitrogen is 28, and so on. What the unit atomic weight corresponds to in absolute weight was not known in chemistry but could be determined by the kinetic theory of gases and later by many other more accurate methods. The result is that one unit of atomic weight (the weight of one hydrogen atom) equals 1.66×10^{-27} kg. From this and the weight-to-volume ratio, which is the density, we can figure out the size of the radius of molecules, which turns out to be of the order of 10^{-10} m.

What about a solid? The molecules of a solid are bound together by forces between molecules; they form a regular lattice pattern (Fig. 7.2) so that the solid assumes a definite shape of itself without the help of a container (as in a liquid or a gas). This does not mean that each of the molecules is standing still in the lattice structure. In fact, each one can move back and forth a little bit about its assigned lattice position in a vibrational motion. The energy of this vibrational motion, kinetic and potential, represents the heat energy content of the solid.

Figure 7.2 Molecules and their motion in a solid.

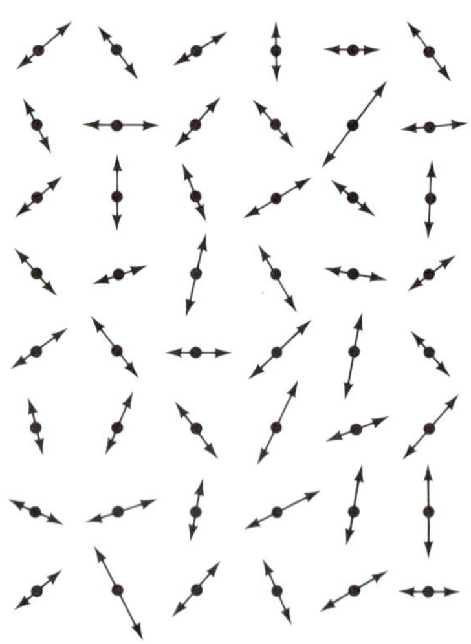

Why the conversion of heat to work is different from that of work to heat is an important topic, to be discussed in more detail in Chapter 16. Nevertheless, a brief qualitative discussion may be given here. The macroscopic mechanical energy corresponds to the *ordered* motion of the molecules of a substance. When we swing a hammer with a velocity v, every molecule of the hammer moves with the same velocity v in the same direction. The kinetic energies of all molecules add together to become the kinetic energy of the hammer, which can do work because the hammer moves in one direction. If all molecules with the same speed v now move in random direction, as in a gas, the kinetic energies of the molecules add together to become the *heat* energy of the body, which cannot do work readily—a gas cannot hit like a hammer because of the randomized directions of the molecules. Ordered motion can naturally and spontaneously change into random motion but the reverse is not true, by the law of probability. Therefore work can be readily converted completely into heat, but the reverse is not true.

Thus a new principle, the *probability principle*, is introduced into physical science. This gamblers' principle is a mathematical truism independent of physical laws. It is thus an independent principle in addition to the dynamical laws and laws of forces. The theory that follows is thus beyond the dynamical theory and is called the *statistical theory*. It encompasses the *kinetic theory of gases, thermodynamics*, and *statistical mechanics*. Statistical theory is an extension of dynamical theory. In fact, the probability principle can be incorporated into any dynamical theory to develop a statistical theory. The theoretical structure of physical science is now complete. It consists of only two theories: the dynamical theory and the statistical theory. The former is based on the three dynamical laws of Newton (modified by the field theory, the relativity theory, and the quantum theory), and four basic laws of forces (possibly two more). The latter is an extension of the former by adding the probability principle. Except for high-energy physics and cosmology the preceding eight basic laws in the two theories account for all happenings in the physical world.

Calorimetry. Having explained the nature of heat, we now consider the measurement of the quantity of heat. Because heat is a form of energy, its unit should be the same as that of energy. Historically, heat had been measured as a quantity, without knowing its identity with energy, and the unit calorie (cal) had been in use. Later when the equivalence of heat with energy was established, the units of heat and energy became interchangeable. As mentioned already, the mechanical equivalent of heat is 4.2 joules for 1 calorie.

The measurement of the heat quantity is not as simple as that of temperature, which can be measured by a thermometer, or as that of weight, which can be measured by a scale. The amount of heat is measured by transferring the heat to a definite amount of water in a container and then measuring the rise of temperature of the water. If the water temperature is raised by T°C and the mass of water is m g, the total amount of heat transferred to water is $m \times T \times 1$ or mT cal. Any amount of heat can be measured in this way. Of course the container of water, raised in temperature together with the water, also takes up some amount of heat and this effect has to be included in the measuring process. The apparatus used for this purpose is called a *calorimeter* (Fig. 7.3). Examples of its use will be introduced later.

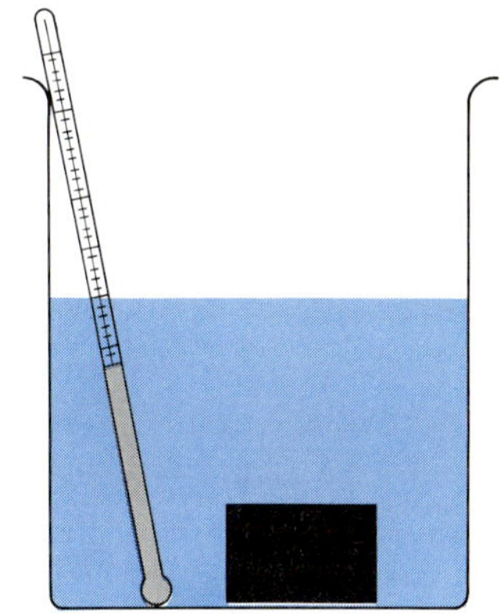

Figure 7.3 Calorimeter.

Heat Transfer. We have mentioned that heat flows from one material body to another in a way to equalize the temperatures of the two bodies. This phenomenon, called *transfer of heat,* will now be discussed further. Heat transfer is involved in many industrial and environmental problems. The heat coming onto the earth's surface and that going away from it strike a delicate balance on which hinges the world's climate. Environmental changes tipping the balance of the heat transfer may have catastrophic consequences.

There are three ways heat can be transferred: *conduction, convection,* and *radiation.* Conduction is the way in which heat is transferred by a material medium without the material medium itself being involved in motion. The flow of heat in a solid medium is exclusively conduction. On the other hand, flow of heat through a liquid or a gas may involve the motion of the medium itself. When this takes place, such as when water is heated in a kettle, the heat transfer is called convection. In addition, heat may be transferred even without a material medium in between, that is, through a vacuum, which occurs with the flow of heat from the sun to the earth. Such a transfer is called radiation.

The law of heat conduction is relatively simple. Consider the heat transfer through a glass windowpane from indoors to ourdoors. This transfer results in a loss of heat of the house, which has to be replenished by heating in wintertime. The *law of heat conduction* states that the heat loss Q is proportional first to the temperature difference between the inside and outside, $T_i - T_o$, which is understandable (the colder the outside, the more heat loss and the higher heating cost). Next it is proportional to the area A of the windowpane, which is obvious (the more glass window areas, the greater the heating cost). In addition, it is inversely proportional to the thickness d of the pane, which is reasonable (the thinner the pane, the greater the heat loss). Finally, it is proportional to the length of time t of the heat flow, which is obvious. Thus mathematically:

$$Q \sim \frac{A(T_i - T_o)t}{d}$$

or equivalently:

$$Q = K\frac{A(T_i - T_o)t}{d}$$

where K is a constant called the *heat conductivity*. K is a material-specific constant, differing from one substance to another. Materials with a large value of K are called *heat conductors,* whereas materials with a small value of K are called *heat insulators.* Glass is an insulator; otherwise the heat loss and heating cost would be intolerable. The wall material is also an insulator to keep heat loss low. Because the wall is much thicker (d large) than glass, the heat loss through the wall is not as large as that through the glass window. (See Problems 7.1 and 7.2 for energy-saving measures.)

The interior of the earth is quite hot and therefore a process of heat conduction is going on transferring earth heat from the interior to the surface. The rate of heat flow can be calculated by the preceding equation. This is a point related to the problem of the origin and evolution of the earth (Chapters 23 and 24). There is a similar problem with the moon, and one of the Apollo experiments involves the measurement of such a heat flow. The earth heat represents an energy source, i.e., geothermal energy, which is being considered for utilization in the current energy crisis.

The law of convection is much more complicated because it involves the motion of a liquid or a gas, which is itself a complicated mechanical problem. Transfer of heat through the ocean and the atmosphere is largely due to convection (ocean currents and air currents). In studying the deep interior of the earth we find some convection also takes place (Chapter 24).

The *law of radiation heat transfer* is rather simple. A body of surface area A with a temperature T°K (in the absolute scale) placed in an environment of a lower temperature T_o°K will lose heat energy Q at a rate Q/t proportional to A and $(T^4 - T_o^4)$. Thus mathematically:

$$Q \sim A(T^4 - T_o^4)t$$

or equivalently:

$$Q = SA(T^4 - T_o^4)t$$

where S is a constant differing from one material to another. Black bodies have large values of S, whereas white bodies have small values of S. The material with the largest value of S is called a *black body,* an important concept in the theory of radiation. The value of S for the black body is a universal constant called the *Stefan constant.*

The preceding equation is derived from the *Stefan–Boltzmann law* of thermal radiation (Sec. 13.2), which says that a hot object of a temperature T°K emits heat energy by the rate of SAT^4 cal/sec. At the same time, the surroundings of temperature T_o also emit heat energy, and some of it is intercepted by the body. The rate of heat interception is SAT_o^4. (Why? See Problem 7.3.) The net energy loss of the hot body is the difference of the two, given in the preceding equation.

The way heat from the sun reaches earth is by radiation. The space in between is a vacuum. Earth itself, being kept at a temperature higher than the outer space surrounding it, also loses heat by radiation. Many factors affect the rates by which heat is received and radiated away by the earth's surface—land utilization, cloud formations, atmospheric composition all have an effect. The related environmental problems have already been discussed.

7.2 Thermal Properties of Matter

Specific Heat. A basic thermal property of matter is the *specific heat.* It is a material-specific property defined as the amount of heat, in the unit of calorie, necessary to raise the temperature of 1 g of the substance by 1°C. It tells us how hard it is to heat up this substance. According to our definition of calorie, it takes 1 cal to raise 1 g of water by 1°C. Thus the specific heat of water is 1 cal/g/°C. It turns out that water is the substance with the largest specific heat. It is the most difficult to heat up. What this means is that the same amount of heat can raise other substances to a higher temperature than water of the same amount (measured by mass). This fact is an important factor in meterology. Under the influence of the daytime sunshine, the temperature of the ocean is raised less than that of the continent. Similarly, in the evening it drops less than that of the continent. Thus the climate of the continent is marked by drastic day-night temperature changes, whereas that of the ocean is not.

The concept of specific heat applies not only to solids and liquids but to gases as well. However, there is one significant difference. Solids and liquids have a definite volume but gases do not. The volume of a gas depends on the pressure and temperature of the gas. As volume varies, the amount of heat necessary to raise the temperature of a unit mass of gas by 1°C also varies. Thus the specific heat value varies with respect to the condition on volume. Usually we consider two kinds of specific heat under two specific conditions. One is called the *specific heat under constant volume* C_v, which is defined as the specific heat when the volume of the gas is kept constant. The other is called the *specific heat under constant pressure* C_p, defined as that when the pressure of the gas is kept constant. When pressure is constant, the volume cannot be the same and undergoes expansion as heat is added. In expansion the gas does work. More energy is required for this purpose. Thus C_p is usually greater than C_v. The ratio of the two specific heats, C_p/C_v, is a quantity that is important in several ways and is usually designated by the Greek letter γ. The value of γ of air is 1.4. This ratio appears in the formula of the speed of sound (Sec. 4.10). It also appears in the equations for adiabatic processes (Sec. 6.2).

The thermal condition of a material body is specified by the intensive quantity temperature and the extensive quantity heat. The two are related—the more heat content, the higher temperature. The relation is generally one of direct proportion, and the proportionality constant, called *heat capacity,* is just the specific heat multiplied by the mass.

Specific heat may be measured by a calorimeter. Suppose we want to measure the specific heat of aluminum. We take a sample of aluminum of m_a g and heat it to a temperature of T_a°C (say, boil it in water; then $T_a = 100$°C). The calorimeter consists of a copper cup of mass M_c g filled with water of M_w g, both at an initial temperature of T_i°C. After the hot aluminum sample is dropped in the water, the temperature of water rises and eventually all three reach a final temperature T_f°C. Let the specific heat of aluminum, copper, and water be C_a, C_c, and C_w ($C_w = 1$ cal/°C/g). The heat released by the hot aluminum sample is $M_aC_a(T_a - T_f)$. The heat taken up by the calorimeter (cup and water) is $M_cC_c(T_f - T_i) + M_wC_w \times (T_f - T_i)$. The two must be equal; thus

$$M_aC_a(T_a - T_f) = M_cC_c(T_f - T_i) + M_wC_w(T_f - T_i)$$

Knowing the values of M_a, M_c, M_w, T_a, T_f, T_i, C_c, and C_w, we can calculate the value of C_a. (See Problem 7.4.)

The specific heat values in the unit of cal/°C/g for a number of substances determined experimentally are listed in Table 7.1.

Specific heat may be defined by using other units of mass. With the gram as the unit of mass the specific heat values of various materials are not correlated. If we use the mole as the unit of mass, the corresponding specific heat, called the *molar specific*

Table 7.1 Specific Heat

Aluminum	0.217 cal/°C/g
Copper	0.093
Glass	0.199
Ice	0.55
Mercury	0.033
Silver	0.056
Water	1.000

heat, exhibits startling regularity among various substances. Many gases have the same molar specific heat under constant volume 3 cal/°C/mole; others 5 cal/°C/mole, and so on. The molar specific heats of most solids are nearly the same, equal to 6 cal/°C/mole. These regularities suggest that there is something in common among all gases and among all solids in their structure; we may expect the microscopic theory to explain these regularities. Indeed, we can explain them by the kinetic theory of gases.

Kinetic Theory of Gases. We have already used the kinetic theory of gases to give a qualitative explanation of Boyle's law, i.e., the product of P and V is a constant. The idea can be developed further and applied to explain the behavior of specific heat. Assume a molecule of mass m and velocity v moving vertically in a cylinder (Fig. 7.4), colliding with the piston and the bottom alternatively. After each collision on the piston, the velocity changes from v to $-v$ and the momentum changes from mv to $-mv$. Thus

$$\text{Change of momentum per collision} = mv - (-mv) = 2\,mv$$

After each collision the molecule travels a distance d downward and then another distance d upward before the next collision. The time between two collisions is $2d/v$. The frequency of collisions is the reciprocal of this time; thus

$$\text{Number of collisions per sec} = \frac{v}{2d}$$

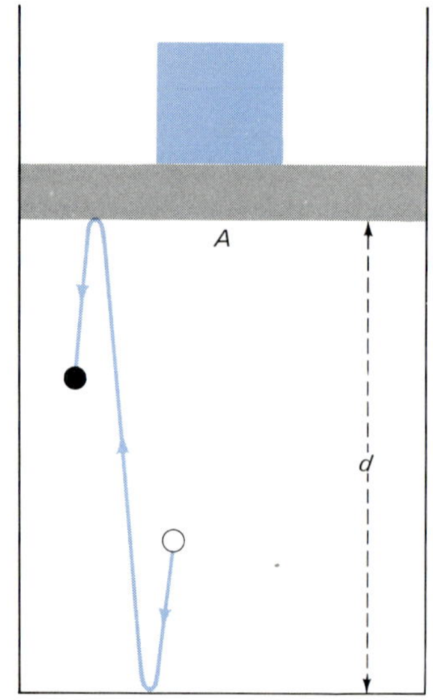

Figure 7.4 Pressure of a gas explained as due to molecular collisions on the wall of the container.

Multiplying the preceding two equations, we get

$$\text{Total change of momentum per second caused by one molecule} = 2mv \times \frac{v}{2d} = \frac{mv^2}{d}$$

Now, Newton's second law may be stated by saying that force equals the rate of change of momentum because

$$F = ma = m\,\frac{v_2 - v_1}{t} = \frac{mv_2 - mv_1}{t} = \frac{\text{change of momentum}}{\text{time}}$$

The previous equation may thus be interpreted by saying that one molecule, on the average, exerts a force of mv^2/d on the piston. It is true that this force is not continuous and even; it is impulsive—during the short period of the collision process (see previous discussion of collisions in Sec. 5.7) a large force is acting on the piston, but between collisions there is none. On the other hand, many molecules are colliding on the piston at different times and all the impulsive forces merge together to form a continuous, even force. The preceding force mv^2/d may be regarded as the average force per molecule. Let us find out the total force all molecules of the gas exert on the piston. All molecules move with different velocities. We simplify by considering them moving with the same velocity v—a suitable average of the actual velocities. Let N be the total number of molecules in the gas. If all of them were moving in the vertical direction, then the total force would be N times mv^2/d, that is, Nmv^2/d. This could not be the case—molecules are moving in all random directions and the actual force should be smaller than this because of the dispersed directions. How much smaller? We can argue that the force is smaller by one third. Molecules can move up and down, left and right, and in and out; thus the fraction up and down would be one third. Here we are approximating the random directions by three perpendicular directions only, which is a very rough approximation. More detailed mathematical calculation shows that this rough approximation turns out to be exactly correct. Thus

$$\text{Force all molecules exert on piston} = \tfrac{1}{3}Nm\left(\frac{v^2}{d}\right)$$

Now the total force equals pressure P multiplied by the area of piston A,

$$PA = \tfrac{1}{3}Nm\left(\frac{v^2}{d}\right)$$

Multiplying d on both sides, we have

$$PAd = \tfrac{1}{3}Nmv^2$$

But Ad is the volume V of the gas. Thus we arrive at the final result:

$$PV = \tfrac{1}{3}Nmv^2$$

Once again we have derived Boyle's law. More important, we have derived the macroscopic constant of Boyle's law in terms of the microscopic molecular quantities N, m, v. This provides an opening wedge to learn more about the gas on the molecular level. Again the theory illustrates the workings of the Newtonian theory. A gas is looked upon as a mechanical system—a system of particles—and is thus analyzed the Newtonian way. The mathematical complexity of a large number of molecules is overcome by an averaging procedure. As a result we have derived the pressure in terms of the mechanical quantities N, m, and v of the molecules.

Application to Specific Heat. We now apply this result to the study of specific heat. The kinetic energy of a molecule being $\frac{1}{2}mv^2$, the total kinetic energy of all molecules of the gas is $N(\frac{1}{2}mv^2)$. This can be related to PV by the preceding equation; thus

$$\text{Total kinetic energy of all molecules} = N(\tfrac{1}{2}mv^2) = \tfrac{3}{2}PV$$

Making use of the empirical law $PV = nRT$, we get

$$\text{Total kinetic energy of all molecules} = \tfrac{3}{2}nRT$$

Before we proceed further, we derive another relation from

$$N(\tfrac{1}{2}mv^2) = \tfrac{3}{2}nRT,$$

that is,

$$\tfrac{1}{2}mv^2 = \tfrac{3}{2}\left(\frac{n}{N}\right)RT = \tfrac{3}{2}\left(\frac{R}{N_o}\right)T$$

where N_o is the Avogadro number. This equation states that the average kinetic energy of a molecule is proportional to the absolute temperature T. This gives us the mechanical meaning of temperature—T is a measure of the average molecular energy. Moreover, the average is the same for all molecular species (different m) at the same temperature T because R and N_o are universal constants. These are significant conclusions in the kinetic theory of gases and are useful in many applications. The constant R/N_o is called the *Boltzmann Constant* k. Thus the average kinetic energy of a molecule is $\frac{3}{2}kT$ and this is the same for all kinds of molecules.

When the temperature is increased by 1°C from T°K to $(T+1)$°K the corresponding increase of total kinetic energy of all molecules equals

$$\tfrac{3}{2}nR(T+1) - \tfrac{3}{2}nRT = \tfrac{3}{2}nR$$

As the temperature increase is affected by heating, it is the heat energy that is converted to kinetic energy of molecules. The molar specific heat, i.e., the heat necessary to increase the temperature by 1°C per mole of material, according to the kinetic theory, is thus $\tfrac{3}{2}(nR)$ divided by n converted into calorie unit:

$$\text{Molar specific heat} = \tfrac{3}{2}nR \div n = \tfrac{3}{2}R = 3 \text{ cal/mole/°C}$$

Because R is a universal constant, the molar specific heat would be the same, 3 cal/mole/°C, for all gases. Thus the kinetic theory leads to an important prediction on the thermal properties of gases, which can be tested empirically. Experimentally, the molar specific heats of many gases (monatomic molecules) are indeed the same and have a value very close to 3 cal/mole/°C, lending strong support to the theory. [There are exceptions, though. Gases of diatomic and triatomic molecules have molar specific heat equal to 5 and 6 cal/mole/°C respectively, instead of 3. But this discrepancy can be explained satisfactorily by the additional energy needed for the rotational motion of the diatomic or triatomic molecule; thus more heat is required (5 or 6 versus 3) to raise the temperature to the same level.]

The molar specific heat discussed here is the one under constant volume—the heat energy added is used exclusively for increasing the kinetic energy of the molecules and this can be the case only when the volume is kept constant. When we determine the specific heat under constant pressure, the volume is not constant and thus the gas does work in expansion, requiring additional energy. It can be shown that molar C_p is greater than molar C_v by R, about 2 cal/mole/°C (Problem 7.11). Thus C_p equals 5 cal/mole/°C for monatomic molecules ($\gamma = \frac{5}{3}$) and equals 7 cal/mole/°C for diatomic molecules, including air ($\gamma = \frac{7}{5} = 1.4$).

Another conclusion of the kinetic theory is that when $T = 0$, the kinetic energy of the molecules is reduced to zero and all molecular motion ceases. The energy cannot be reduced further and thus the temperature has a lowest point. This corresponds to the absolute zero of temperature.

Conductivity. Just as in the mechanical and electrical properties, there is a dynamical aspect in the thermal properties of matter, i.e., heat conduction. The law of heat conduction has already been introduced in Sec. 7.1 as a law of heat transfer. It is analogous to electric conduction. We now look at it from the viewpoint of properties of matter. The extensive quantity here involved is the *heat current density* Q/tA cal/sec/cm^2, where Q is the quantity of heat passed through the cross-sectional area A cm^2 in a time of t sec (Fig. 7.5). The intensive quantity is the *temperature gradient* $(T_2 - T_1)/(x_2 - x_1)$, where T_2 and T_1 are the temperatures of the conductor at positions x_2 and x_1, the gradient representing the spatial rate of change of temperature. It is the temperature gradient that *causes* the heat current to flow. The cause–effect relation between these two quantities is established by the law of heat conduction, which asserts that the two quantities are proportional to each other:

$$\frac{Q}{At} = K\left(\frac{T_2 - T_1}{x_2 - x_1}\right)$$

The values of the thermal conductivity K for a number of substances are listed in Table 7.2.

The thermal conductivities of substances vary widely. It so happens that all thermal conductors

Figure 7.5 Conduction of heat through a rod.

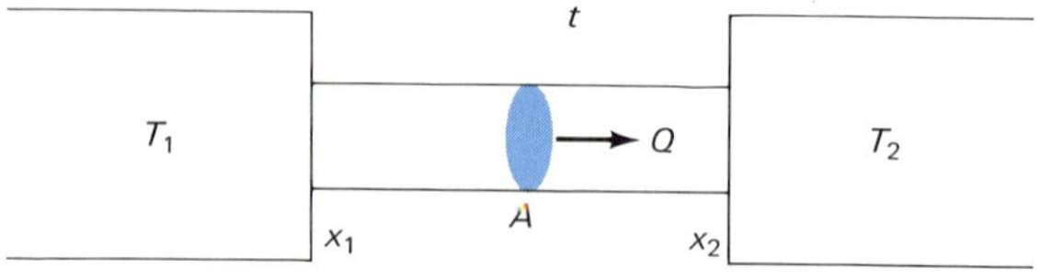

Table 7.2 Thermal Conductivity

Thermal Conductors:	
Aluminum	0.49 cal/sec/cm/°C
Copper	0.92
Lead	0.083
Mercury	0.020
Silver	0.97
Steel	0.12
Thermal Insulators:	
Insulating brick	0.00035
Concrete	0.002
Glass	0.002
Air	0.000057
Hydrogen	0.00033

are electrical conductors, all thermal insulators are electrical insulators, and vice versa. This is a very important relation. It implies that electrical conduction and thermal conduction are not mutually independent but are related in a definite manner. When we compare the thermal and the electrical conductivities quantitatively, we find that they are actually proportional to each other; this relation is known as the *Wiedemann and Franz law*. Why the two conductivities are related so closely is a question that can be answered in the microscopic theory. According to the atomic theory, conductors are those materials in which some electrons of the atoms can freely move about inside the solid. These are called *free electrons*. Insulators are those that do not have free electrons. Free electrons are the carriers of heat in heat conduction and also the carriers of electricity in electric conduction. It is the bodily movement of the free electrons that carries the heat (kinetic energy) and electricity with them and transports them through the conductor. The conduction problem is thus reduced to one of motion of a set of particles and thus may be dealt with by the Newtonian theory. Because the mechanisms by which they transport heat and electricity are very similar, the two conductivities are proportional to each other. In particular, this explains why thermal conductors are also electric conductors.

Whereas in a solid, free electrons are the carriers for the *transport phenomena,* in a gas, which has no free electrons, the molecules themselves become the carriers for the transport phenomena. There is a similar relationship between two transport phenomena in gases: heat conductivity and viscosity. Viscosity refers to the viscous force a gas exerts on an object moving through it. This force is what causes a swinging pendulum to slow down and eventually stop. From a mechanical point of view, the viscosity phenomenon actually is one in which the "velocity" is transported away from the moving body, causing it to slow down. The gas molecules themselves are the carrier. In heat conduction through a gas (separate from convection), it is also the molecules themselves acting as the heat carrier. Again in both cases the problem is reduced to a mechanical problem of the motion of the particles (molecules) acting as the carrier. Again, based on the mechanical laws, theories of viscosity and heat conductivity are developed (in the kinetic theory of gases) and again the two constants are found proportional to each other.

Low-Temperature Phenomena. At low temperatures some of the regularities of the specific heats break down. In fact, even at room temperature the molar specific heat of a diamond is much smaller than the value of 6 cal/°C/mole expected from the general rule. Such deviations may be explained by the quantum theory. In fact, evidence from the specific heats played an important part in the early development of the quantum theory.

At very low temperatures near −273°C (absolute zero), a number of strange phenomena occur that defied explanation for half a century. One is the phenomenon of *superconductivity.* When the temperature is reduced to a few degrees above absolute zero, the resistance of a conductor suddenly decreases by several million times. It practically vanishes. Kamerlingh Onnes was able to set up a current in a superconducting ring in Holland and took it to London for demonstration in a lecture. Without any electromotive force the current was still flowing when he arrived in London.

The thermal conductivity also decreases drastically together with the electric conductivity. The temperature at which a conductor becomes *superconducting* is called the *transition temperature,* which differs from one substance to another. The application of a magnetic field may change the transition temperature. There are other relevant factors such as the atomic weight. The phenomena have been investigated extensively on an experimental basis since the beginning of the century.

In the 1930's, after the completion of the development of quantum mechanics, Dirac made the famous statement that this theory solves most problems in physics and all problems in chemistry. It was believed in the physics community then that there were only two problems remaining: the nucleus of the atom and superconductivity. Only recently the theory of superconductivity made significant progress (Nobel prize of 1972). It is now believed that superconductivity is a quantum phenomenon concerning the collective behavior of a large number of electrons in a crystal lattice.

Near absolute zero most liquids freeze and become solid. Helium is an outstanding exception; at the pressure of 1 atmosphere it liquefies at 4.2° above absolute zero. When the temperature is lowered below the so-called λ point of 2.186° above absolute zero, the properties of the liquid helium suddenly change. The new properties are the liquid counterpart of superconductivity and are called *superfluidity.* It is not surprising that one of the properties that characterizes superfluidity is the abnormally low viscosity analogous to the loss of resistance of a solid in superconductivity. Other properties include the *fountain effect,* the *second sound,* and so on. It is now believed that superfluidity is a quantum phenomenon of the liquid as a whole, a macroscopic quantum phenomenon capable of direct observation. Most quantum phenomena are microscopic and are observable only indirectly.

Superconductivity is a freak of nature. It was merely a puzzle to physicists but recently has begun to figure in practical applications. Superconducting power transmission lines are under consideration. Environmentalists would like to see the conventional power transmission lines, which are an eyesore to nature lovers, be replaced by superconducting lines that are buried underground. Superconducting transmission lines have the major advantage of saving energy because their resistance is much less than that of the normal transmission lines, which consume 12 per cent of all electric energy.

7.3 Change of Phase

Change of state involves a *continuous* change of the properties of a substance, i.e., pressure, volume, and temperature. When carried to extreme, such a change may lead to a *discontinuous* change of the properties of the substance such as that which happens when a solid melts and becomes a liquid. Such a change is called a *change of phase.* There are three generally known phases of a substance: gas, liquid, and solid (Fig. 7.6). The solid of some substances may exist in a number of crystal forms; each one may be considered a separate phase. Metallurgical processes, such as tempering, usually involve the change from one solid phase to another. Thus a *phase* refers to a definite *state of aggregation* of the molecules of a substance.

The most outstanding feature concerning the state of aggregation is that the phases are discontinuous and finite. In contrast, *states* are continuous and infinite. The solid phases are discontinuous and finite because atoms are discontinuous; there are only a finite and discrete number of ways the atoms can be arranged in regular pattern to form a stable configuration. The existence of a gaseous phase in which the molecules are completely separate and unrelated is expected and understandable. On the other hand, the existence of a liquid phase (and the nonexistence of other phases) between solid and gas is not so easy to understand. In fact, liquid is the least understood state of aggregation compared with solid and gas.

From the macroscopic point of view, we take the

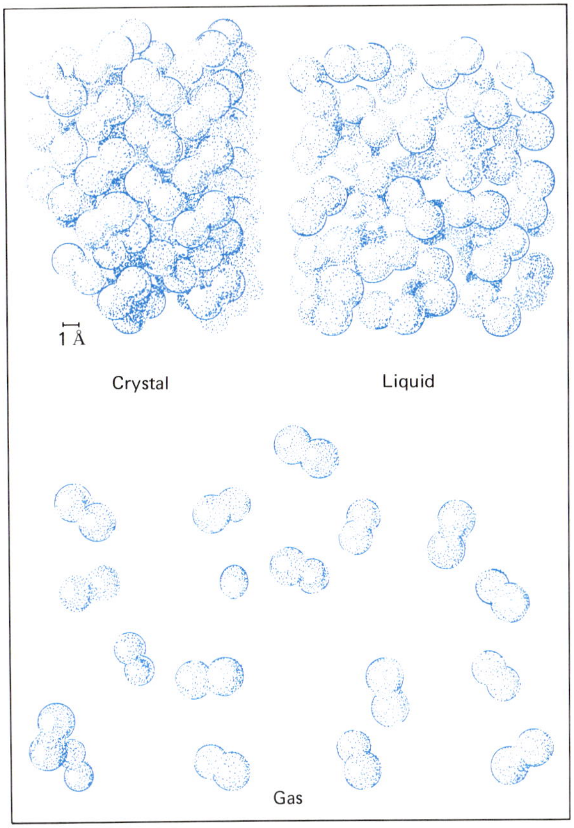

Figure 7.6 Solid, liquid, and gaseous iodine, showing diatomic molecule I_2. [From *College Chemistry,* Third Edition, By Linus Pauling. W. H. Freeman Company. Copyright © 1964.]

three phases of gas, liquid, and solid for granted and study the properties related to *phase transition,* i.e., the change of phase from one to another. We learn the facts by experimentation. The results may be analyzed by a macroscopic theory (which is largely based on thermodynamics). From the microscopic point of view all the results are to be explained in terms of the molecular properties (by the kinetic theory of gases and statistical mechanics).

The first important fact is that change of phase takes place at a sharply defined temperature. Thus under the air pressure of 1 atmosphere ice melts and changes to water at 0°C and water boils and changes to steam at 100°C. Under the same pressure below 0°C water can exist only in the form of ice; above 0°C and below 100°C water can exist only in liquid form; above 100°C water can exist only in gaseous form. The temperature at which a solid changes into liquid is called the *melting point* and that at which a liquid changes into gas is called the *boiling point.* These temperatures are different from one substance to another. Their values for a number of substances are listed in Table 7.3.

Besides melting and boiling, there is a third process, called *sublimation,* by which a solid changes directly into a gas. Dry ice (solid carbon dioxide) under normal air pressure changes into gas directly without changing into liquid first; that is why it is dry. Other substances show sublimation under a low pressure. A *sublimation point* may be defined analogously.

Table 7.3 Properties Related to Phase Transition

Substance	Melting Point	Heat of Fusion	Boiling Point	Heat of Vaporization
Mercury	−39°C	2.82 cal/g	357°C	65 cal/g
Nitrogen	−210	6.09	−196	48
Oxygen	−219	3.30	−183	51
Water	0	79.7	100	539
Ethyl alcohol	−114	24.9	78	204
Platinum	1775	27.2		

The melting point and boiling point are usually referred to under the atmospheric pressure. It is found that these two points change with respect to pressure. Thus the melting point of ice is lowered and the boiling point of water is increased when pressure is increased. These properties are familiar in everyday experience. When a truck runs over an icy road surface, the ice may melt. The pressure of the truck on ice lowers its melting point. If the new melting point is below the temperature of the ice, melting ensues. The pressure cooker is based on the fact that boiling point is increased with pressure. With water boiling at a temperature higher than 100°C, the food may be cooked faster than normally boiling at 100°C. In high altitudes the atmospheric pressure is considerably lower than that at sea level and so water boils at a temperature less than 100°C. It becomes more difficult to cook food, an experience well known to mountain climbers.

The dependence of the melting point and the boiling point of water on pressure is represented by the two curves in Fig. 7.7. At very low pressures ice may change directly to vapor at the sublimation point. The pressure dependence of the sublimation point may be represented by another curve in Fig. 7.7. The horizontal line drawn at 1 atmospheric pressure in Fig. 7.7 shows that water exists in the form of ice below 0°C, in the form of liquid between 0°C and 100°C, and in the form of steam above 100°C. At the melting point ice and water coexist; at the boiling point water and steam coexist. These apply to other points on the melting point curve and boiling point curve. The intersection of the two curves is thus a point representing a temperature and a pressure at which ice, water, and steam may coexist. Such a point is called the *triple point.*

Figure 7.7 Pressure dependence of the melting, boiling, and sublimation points.

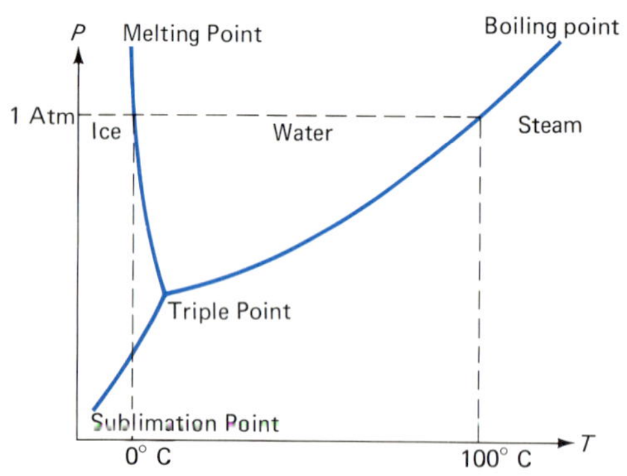

Because the sublimation point curve represents points at which ice and steam coexist, the triple point must lie on the sublimation curve. In other words, the three curves intersect at one point, the triple point. The three curves thus divide the P–T space in Fig. 7.7 into three regions corresponding to ice, water, and steam, respectively. For water the temperature and pressure of the triple point are 0.0075°C and 4.58 mm Hg, respectively.

It should be added that when water is heated to boil, its temperature cannot be raised above the boiling point. When more heat is added to boiling water, no temperature increase will result. Instead, more water is converted into steam and the added heat is used up for this purpose. After all the water at the boiling point temperature is converted into steam at the same temperature, additional heat put into the steam will then raise the temperature of the steam.

This leads to another important fact in phase transition: to convert water into steam requires a certain amount of heat. The same happens in converting ice into water. The amount of heat necessary to convert one unit mass of solid at the melting point to liquid of the same temperature is defined as the *heat of fusion.* Its value for water is 80 cal/g. The amount of heat necessary to convert one unit mass of liquid at boiling point temperature to gas at the same temperature is defined as the *heat of vaporization.* Its value for water is 540 cal/g. These heats involved in phase transition are called *latent heats.* Their values for a number of substances are listed in Table 7.3.

When an ice cube melts in a glass, it absorbs heat from the water. As a result water cools down. On a global scale the polar icecap in the ocean works the same way as an ice cube in a glass of water. Thus the latent heat of water plays an important role in world climate.

Under a constant pressure—say, the atmospheric pressure—the continuous heating of a substance originally in the solid phase will thus result in the following sequence of events: first the heat is absorbed to raise the temperature of the solid by a rate determined by the specific heat. Once the melting point is reached, the temperature will not increase any further until the melting is completed. Heat is now absorbed to melt the solid by a rate determined by the heat of fusion. While melting proceeds, the two coexisting phases, solid and liquid, have the same temperature equal to the melting point. Thus the process of melting is one taking place under a constant pressure P and a constant temperature T, the relation between P and T being given by Fig. 7.7. After the completion of melting, we have the substance in the liquid phase at the melting point temperature. Upon continued heating, the heat is absorbed to raise the temperature of the liquid until the boiling point is reached. The process of vaporization then takes place in a way analogous to melting and so need not be described here. Both processes are reversible and may be carried out in the reverse direction, corresponding to condensation and freezing. The same amount of latent heat will be released in the reverse process. Thus the condensation of 1 g of steam to water is accompanied by the release of 540 cal of heat and the freezing of 1 g of water to ice is accompanied by the release of 80 cal of heat.

During the change of phase the pressure and temperature are kept constant but the volume is not constant. When water changes into ice, it expands a little. When water changes into steam, it expands about a thousandfold. Because change of state eventually leads to change of phase, the properties of phase transition are expected to be related to the equation of state. This point has already been touched upon in Sec. 6.2 in connection with van der Waals' equation. It will now be considered from a general point of view.

Let us consider the change of state of a fixed amount of gas, varying its pressure and volume while maintaining its temperature at a constant value T_1. The P–V curve obtained is an *isothermal curve*. It is found that when the volume is large, the P–V relation obeys Boyle's law closely. As the volume is compressed and becomes smaller, the pressure increases (see lower right part of Fig. 7.8). When the pressure has increased to a certain value P_1, the gas begins to condense into liquid. Since a phase transition takes place at T_1, the boiling point under pressure P_1 is thus T_1. Hence P_1 and T_1 are related by the boiling point curve of Fig. 7.7. In the process of condensation both P_1 and T_1 remain constant, as mentioned before. The P–V curve in Fig. 7.8 corresponding to condensation is thus a horizontal straight line indicating the same pressure P_1 being maintained throughout the phase transition. It also indicates that the volume is changed from that of the gas V_2 to that of the liquid V_1. At the point (P_1, V_1) condensation is complete. Any further compression will change the volume of the liquid. Because a liquid is difficult to compress, the P–V curve goes up steeply. The solid broken curve in Fig. 7.8 thus represents the P–V relation at a constant temperature for both the gas and the liquid, including the process of phase transition in between.

Changing the temperature from T_1 to other values T_2, T_3, . . . one at a time, we determine in the same way a series of isothermal curves representing the P–V relation. The series of curves may be put together to form a surface that represents the relation between P, V, and T. This surface embodies the equations of state of the gas and the liquid, including the phase transition in between. The surface so constructed is shown in Fig. 7.9.

Figure 7.8 Condensation of a gas.

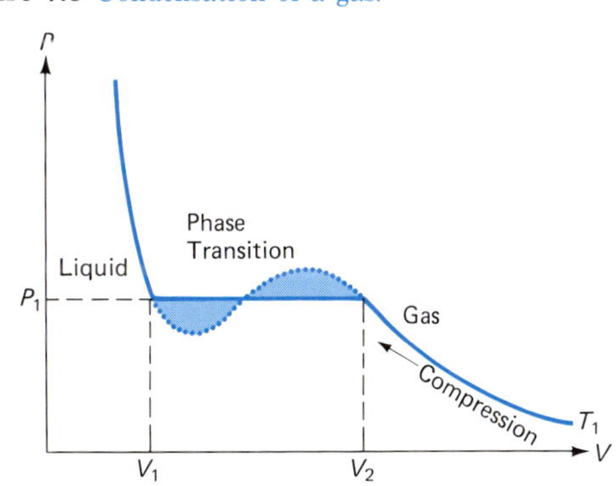

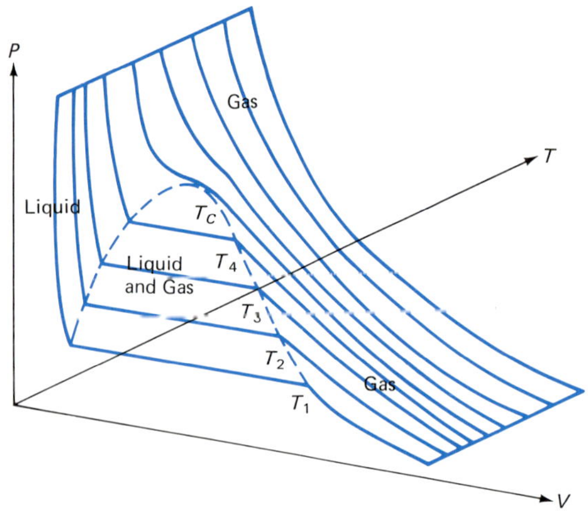

Figure 7.9 The surface representing the experimental equation of state including both the gaseous and the liquid phases.

The surface is not "smooth"; there is a "broken area" in the middle that corresponds to the region of phase transition. An outstanding feature brought out by the surface is that the phenomenon of phase transition is confined to temperatures below a certain value T_c, called the *critical temperature*. Above the critical temperature the gas never liquefies, no matter how large a pressure is applied. Critical temperature differs from one gas to another. That of hydrogen is $-240°$C, which is very low. Thus hydrogen is very difficult to liquefy. The isothermal curve corresponding to T_c is marked out in Fig. 7.9.

Can the equation of state corresponding to this surface be put down in a mathematical formula? We have already compared the experimental results with van der Waals' equation. Van der Waals' equation represents a continuous function and therefore gives rise to a continuous curve (dotted curve in Fig. 7.8). The experimental curve is a broken curve and therefore can never be expected to be represented by a single continuous equation. The wiggle of van der Waals' equation represents an unstable situation. Maxwell was able to argue by thermodynamic reasoning that the actual curve must be a straight line drawn in such a way that the two shaded areas in Fig. 7.8 are the same. Thus by van der Waals' equation and the Maxwell rule, we are able to get a theoretical equation of state that fits the experimental facts. This is as far as the van der Waals theory can go (based on the kinetic theory of gases and thermodynamics). This theory, however, is not the ultimate on the subject.

The more complete theory is to be found in the microscopic theory based on statistical mechanics. The problem of condensation is one of the most difficult in statistical mechanics in modern times. The difficulty lies in the fact that condensation involves a discontinuous change, whereas everything that goes into the theory is continuous. How can one get discontinuity out of continuity? This is the crux of the matter and is a formidable problem indeed. The atomic theory is committed to explain *all* macroscopic properties of matter and the phenomenon of condensation can be no exception. Laymen may be surprised to learn that scientists spend many man-years to find a solution to such a "trivial" problem as why steam condenses. The objective is to explain the macroscopic properties of condensation in terms of the properties of the molecules exclusively. In spite of the familiarity of water in our everyday life, this theoretical problem, urged on by the desire of reducing all natural phenomena to a few basic entities and a few basic laws, turns out to be unusually complicated. Yet significant progress has been made.

Problems

7.1 If houses were completely insulated, show that there would be no need of heating in the winter and no need of air conditioning in the summer and a sizable fraction of the total energy consumption of the society could be eliminated. In practice this is impossible. Why?

7.2 Even though complete insulation of a house is impossible, a better-built and equipped house could save heating and air-conditioning costs. Georgia Power Company researchers learned that by investing $1,000 more on a $74,500 house, the energy saving could be as much as 17 per cent. The following measures are taken. Explain why each of the following will save energy: (a) Install an automatically controlled attic ventilation fan ($125) that draws out attic air when sufficiently hot. (b) Use a light-colored roofing material. (c) Use a microwave oven in the kitchen for quick cooking. (d) Use a heat pump for both heating and air conditioning (costs $550 more than conventional facilities). (e) Insure tight construction around doors and windows. (f) Use double-paned glass with an air layer in between for windows (or add storm windows for $500 extra). (g) Install full insulation on all ceilings, floors, and walls ($250). (h) Install wall heaters in bathrooms.

7.3 Derive the law of cooling, $Q = SA(T^4 - T_o^4)t$, from the Stefan–Boltzmann law.

7.4 The example given in the text for the determination of the specific heat of aluminum requires a knowledge of the specific heat of copper, of which the calorimeter is made. How can the specific heat of copper be determined in the first place?

7.5 Work out a method to use the calorimeter to determine the heat of fusion of water.

7.6 Oxygen is 16 times heavier than hydrogen. In a mixture of the two gases, which molecule moves faster on the average? By what ratio?

7.7 Brownian movement: A pollen particle suspended in water may be considered as a "molecule" from the kinetic theory point of view. What is the average "molecular" velocity? A pollen, being a cell, has a size of about 1 μ (micron, 10^{-6} m) and a density about the same as water. This motion can actually be observed under a microscope, first discovered by Brown.

7.8 The two uranium isotopes ^{238}U and ^{235}U may be separated by the gaseous diffusion method. Explain how this is done, based on the kinetic theory of gases.

7.9 Knowing the rate of geothermal heat flow from the interior of the earth to the surface (0.02 per cent of the solar energy input to the earth surface), find the temperature gradient in the earth's crust. Pick the conductivity in the table closest to crust rock.

7.10 In problem 7.9, from the temperature gradient obtained find the temperature of the earth 15,000 ft underground. Is it hot enough that the heat energy can be utilized commercially? How?

7.11 Show that the difference between the molar specific heats C_p and C_v equals R, the universal gas constant.

7.12 How many calories of heat are needed to raise 1 g of water from $-50°C$ to $150°C$ under atmospheric pressure? Specific heats of ice, water, and steam can be found in the text. Remember that water is a triatomic molecule.

8

The Energy Industry

8.1 Energy and Society

As a living organism, man needs energy to sustain all his biological functions. Muscle action requires energy. Locomotion requires energy. All physiological functions require energy. This energy need is supplied by food taken from the animal and vegetable kingdoms. The ultimate source of this energy is the solar energy captured in photosynthesis by green plants. In the long prehistoric era, food was the only energy need of man. Food was also the only material need at that time. Its collection and consumption accounted for most human activities.

As civilization and technology advanced, the material requirements became more complex with the addition of clothing and shelter. The energy needs also multiplied, largely in association with the expanded material needs. The energy of muscular exertion was augmented by the use of power machines, and locomotion was facilitated by transportation vehicles. Both devices work more effectively than muscles and consume larger amounts of energy. The energy used in machines comes from a different source, the variety of which has been discussed in Chapter 3.

In the early times, muscle energy was employed in the collection and production of food, in making clothing, building shelters, and so on. In an industrialized society, the same purposes are carried out by machine energy. Thus with the rise of industrialization, there also arose an energy industry to supply energy for manufacturing and construction. This industry is concerned with the conversion of energy from that stored in a source (fuel) to mechanical energy to be used for various purposes. The energy need of a transportation vehicle is fulfilled by the installation of an engine that derives its energy from a fuel; the same technical problem of energy conversion is involved.

The inception of human civilization was characterized by the use of fire, which essentially is an energy source. The use of this new energy source marked the development of a number of civilizing activities. Fire was used for cooking, illumination, and heating. All these uses have been carried down to the present day, but with many sophisticated modifications. Today these functions are carried out with the help of the utility industry. Essentially, the utility industry provides us a "fire" derived from a fuel. The energy we use may come directly from the fuel—whether coal, oil, or natural gas, such as in house heating—or may come indirectly with electricity as the intermediary. Energy can be transformed; in the process the quantity is conserved. To use energy in a different form does not essentially reduce its quantity even though a fraction is lost for various reasons. Many times convenience dictates the form of energy to be used, and electricity is generally the most convenient form.

8.2 The Steam Engine

In this chapter we are concerned with the problem of conversion of heat energy generated by the burning of a fuel to mechanical energy by a *heat engine*. The earliest heat engine was the steam engine associated with the name of James Watt. The story goes that Watt was inspired to invent the engine by observing the power of steam that pushed open the lid of a kettle filled with boiling water. This story is misleading because it misses the most important point of a heat engine. The power of the engine comes from the heat, not from the steam, which is just an instrument for converting heat to mechanical energy.

It is true that steam from a water boiler is used in a cylinder to push the piston in a steam engine, thus generating mechanical energy. But this pushing cannot go on indefinitely, for the water will soon be boiled dry. What happens in a steam engine is that the water is used repeatedly without loss; after each expansion, the steam is compressed and changed back to water to be used over and over again. This is characteristic of all heat engines; all have a *working substance* (water in the steam engine) that is expanded and compressed repeatedly and used repeatedly.[1]

A problem arises. Compression requires mechanical energy. If the compression is carried out at the same temperature as that in expansion, the energy for compression is exactly the same as that produced in the expansion, because the two processes are the exact reverse of each other. Thus the gain of energy in expansion and loss in compression would have canceled each other out exactly and no useful energy would be derived as the net result of a cycle of operation.

The problem is solved by letting the steam expand at a high temperature—at the temperature of the *boiler,* where hot steam comes out—and by letting the steam be compressed at a lower temperature—at the temperature of the *condenser,* where steam is cooled and condensed into water (Fig. 8.1). More energy is produced at the high-temperature expansion (pushing the piston out and doing mechanical work) than is consumed in compression at low temperature. Therefore a net gain of mechanical energy is obtained in one cycle of operation. Here we encounter another common feature of all heat engines—in addition to a boiler at a high temperature T_1 there must be a condenser cooled by a coolant at a low temperature T_2. During the compression heat is released and is flushed away by the coolant. The working substance is changed in state and phase repeatedly between these two temperatures. Two adiabatic processes are needed to change from T_1 to T_2 and from T_2 to T_1 as shown in Fig. 8.1 to complete the cycle (called the Rankine cycle).

Thus the hot steam, after expansion in the cylinder, must be directed to the condenser to be cooled and compressed into water that will be used in the boiler again. The circulation of steam is controlled by a system of valves, and the compression is done

[1] A rare exception is the Stanley Steamer, an early American automobile powered by steam (Sec. 9.2). The spent steam is thrown away, not compressed and reused. The engine is thus not a heat engine under our definition and our theory of the heat engine (Carnot theorem) does not apply without proper modification.

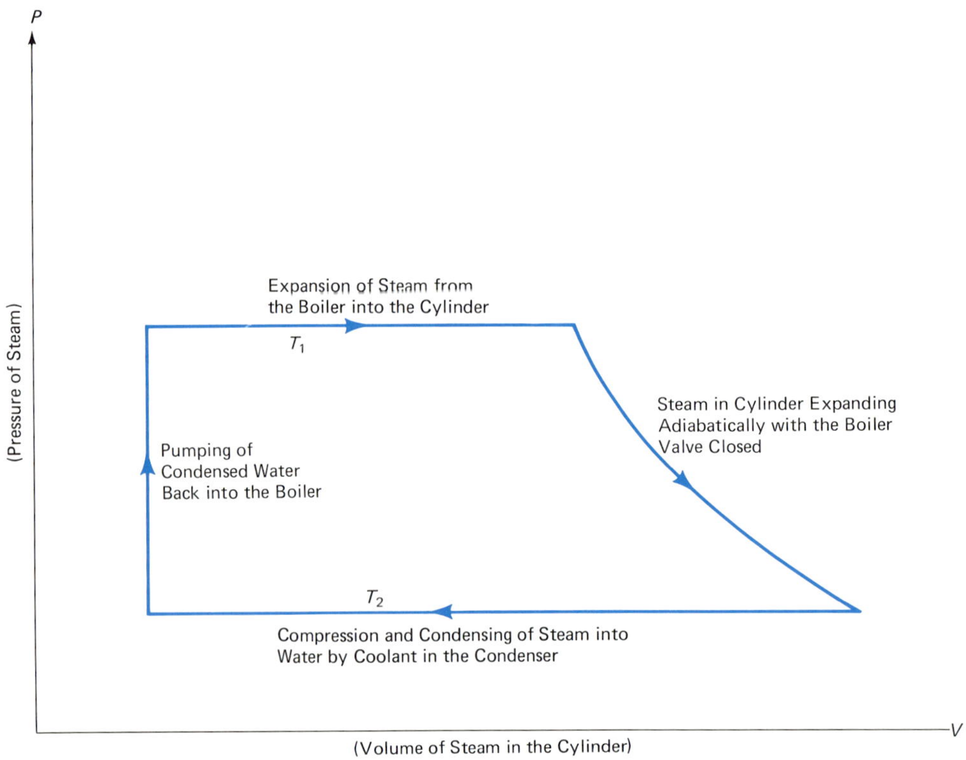

Figure 8.1 Change of state of the working substance steam in one cycle of operation of the steam engine (the Rankine cycle).

by the return stroke of the piston. In early times the valves were opened and closed by a boy, an action seemingly requiring intelligent decision. Actually, the opening and closing of the valves are definitely related to the position of the piston, and a mechanical robot can do the job as well. In fact, in all later steam engines the valves are controlled by rods connected to the flywheel in such a way that the valves will be opened and closed at the desired moment. No more humans are employed. (It was a lazy valve boy who first contrived the automatic mechanism.)

Efficiency of the Steam Engine. When students first study steam engines, this robot action catches their fancy. Yet the tricky mechanism is not the most important point of consideration. The crucial problem of a steam engine is its efficiency, i.e., the ratio of the amount of energy produced W to the amount of heat (in energy units) expended Q. It is this ratio that determines the economic performance (proceeds against costs) of the engine. The efficiency, although dependent on how well built the engine is, cannot exceed a limit set by the two temperatures T_1 and T_2; this limit is established by the *Carnot Theorem* (Chapter 16), according to which the efficiency E is at most $(T_1 - T_2)/T_1$, where T_1 and T_2 are temperatures of the boiler and condenser measured in the absolute temperature scale:

$$E \leq \frac{T_1 - T_2}{T_1} = 1 - \frac{T_2}{T_1}$$

Accordingly, in order to increase the efficiency to obtain higher performance it is necessary to raise the temperature of the boiler T_1, or to lower the temperature of the condenser T_2. Because the cooling is usually done by water taken from earth's surface, T_2 is usually the temperature of the earth's surface and cannot be lowered further without additional cost. Therefore the usual way of improving efficiency is to raise the temperature of the boiler T_1. This point has an important bearing on both the operation of the internal combustion engine and its pollution problems. The very existence of the cooling system (condenser) is the origin of another environmental problem—thermal pollution. The part of the heat Q that is not converted into mechanical energy W, i.e., $Q-W$, becomes the wasted heat energy released into the environment through the cooling system. This part cannot be reduced to zero because T_2 cannot be reduced to absolute zero; it cannot be lower than the temperature of the environment. In a steam engine operated at 30 per cent efficiency (Problem 8.1) only 30 per cent of the heat energy generated from the fuel is converted into mechanical energy. The remaining 70 per cent not only is wasted but also causes thermal pollution of the environment.

*8.3 The Internal Combustion Engine

Steam engines once were used to power railroad locomotives. With the gradual substitution of oil for coal as fuel, steam locomotives have dropped out of sight and have been replaced by diesel locomotives powered by internal combustion engines. With the increased use of automobiles, internal combustion engines have become the dominant power plant of the present time. Besides the fuel cost considerations, the internal combustion engine has the advantages of fast action and light weight over the steam engine and others, even though its efficiency is below maximum. These advantages result from the very nature of the operation of the internal combustion engine. Because fast action and light weight are exactly what are required in an automobile, the internal combustion engine was adopted in car manufacturing from the beginning. This also brings out the difficulty of finding a substitute engine for the solution of the pollution problem, which is largely associated with use of the internal combustion engine. An understanding of the basic principles of this engine is thus basic to an understanding of the related environmental problem.

In the internal combustion engine the working substance is air taken from the atmosphere, a point not generally realized. Gasoline is used as the fuel to supply heat energy to be converted into mechanical energy by cyclic changes of the state of the working substance air. Although there are no such things as the boiler and the condenser, this engine, like all others, does work between two temperatures, that of the heated air in the cylinder and that of the exhaust gas returning to the atmosphere. These temperatures determine the efficiency of the engine in the same way as the boiler and condenser temperatures do in a steam engine.

We now describe the working of the internal combustion engine in the automobile (the *Otto cycle*). The main body of the engine is a cylinder fitted with a piston that is connected with the crankshaft. The operation cycle consists of four strokes corresponding to a total of two turns of the shaft. First, a mixture of air and gasoline vapor (Fig. 8.2*a*) is sucked into the cylinder by the downward movement of the piston through the opening of the inlet valve (intake stroke). The mixing ratio for air and gasoline should be about 15 to 1 for the most efficient burning of gasoline. Next (Fig. 8.2*b*), on the return of the piston and the closing of both inlet and exhaust valves, the mixture is compressed adiabatically from a volume V_C to V_D, shown in Fig. 8.3. The volume of the mixture is reduced by a ratio V_C/V_D, called the *compression ratio* ρ. Modern high-performance engines have a compression ratio of about 10 to 1, doubling the ratio prevalent in the 1920's. Temperature and pressure typically are increased in the proc-

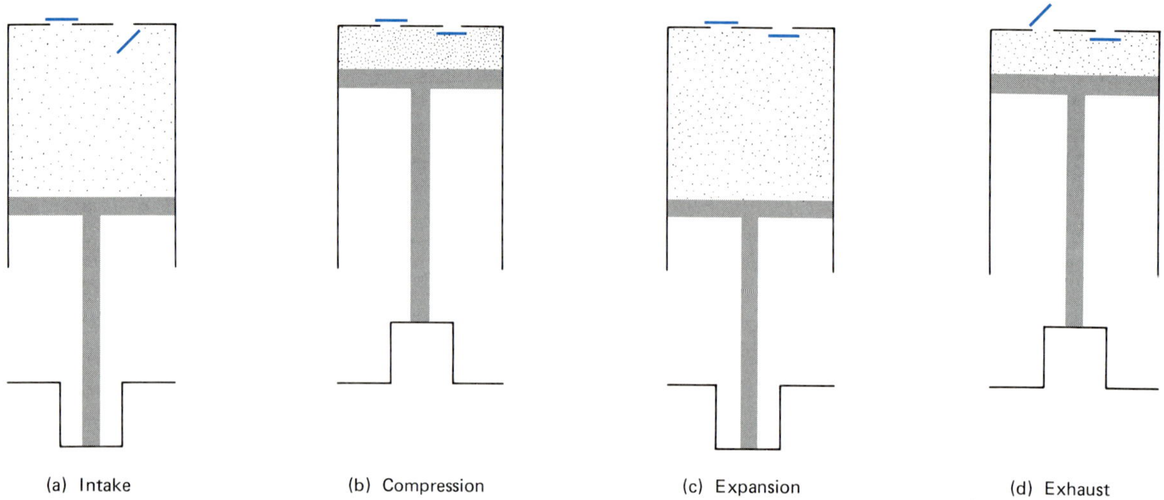

Figure 8.2 The four strokes of the internal combustion engine.

ess to 600°C and 5 atmospheres in a low-compression-ratio engine. At this point an electric spark ignites the gasoline vapor. The vapor burns quickly, instantly raising the temperature and pressure of the gas to about 2,000°C and 15 atmospheres in the same volume. The high-pressure gas then expands adiabatically, pushing down the piston and producing mechanical energy to do work (power stroke or working stroke). Temperature is lowered in the process. Finally, on the return of the piston the exhaust valve is opened, and the exhaust gas is expelled from the cylinder (exhaust stroke) and released into the atmosphere. The volume of the air released is the same as that taken in (equal to the volume of cylinder), and the pressure and temperature are equalized with those of the air by mixing. The working substance coming from the atmosphere and returning to it is itself not altered. The cycle is completed and may be repeated again, even though the air used in the next cycle is not necessarily the same body of air used in the previous cycle. In the cyclic changes of state of the working substance, heat from burning gasoline is converted into mechanical energy that is transmitted from the crankshaft to the propeller shaft to turn the wheels of the automobile. The pressure–volume changes of the working substance in the Otto cycle are shown in Fig. 8.3.

Efficiency of the Internal Combustion Engine. The theoretical efficiency of the engine, by definition,

Figure 8.3 Change of state of the working substance air in one cycle of operation of the internal combustion engine (the Otto cycle).

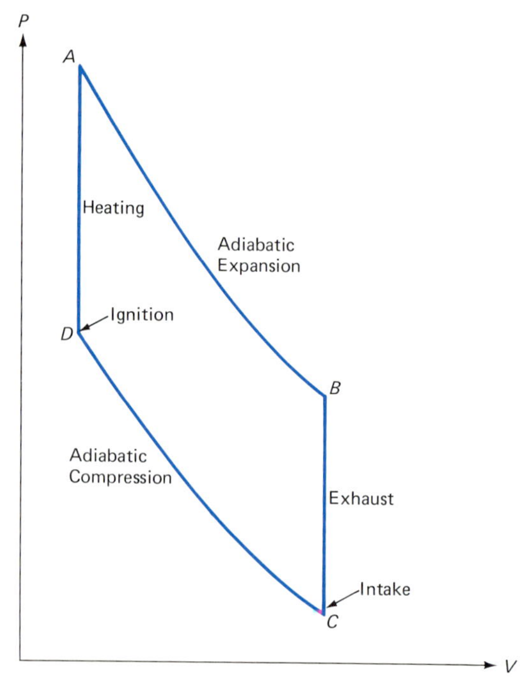

is given by the ratio $(Q_1 - Q_2)/Q_1$, where Q_1 is the total heat produced by burning gasoline (cost) and Q_2 is the heat of exhaust gas thrown away into the atmosphere, assuming the difference $Q_1 - Q_2$ is wholly converted into mechanical energy (net gain) without other losses. Let the temperatures and volumes of the gas at points A, B, C, D in Fig. 8.3 be designated by T_A, T_B, T_C, T_D (in absolute temperature scale) and V_A, V_B, V_C, V_D, respectively. Designate the specific heat of air at constant volume by C_V. The heat generated in burning the fuel causes the gas to be heated from T_D to T_A. Thus

$$Q_1 = C_V(T_A - T_D)$$

Similarly

$$Q_2 = C_V(T_B - T_C)$$

Therefore the maximum efficiency E is given by

$$E = \frac{Q_1 - Q_2}{Q_1} = 1 - \frac{T_B - T_C}{T_A - T_D}$$

By making use of the law of adiabatic expansion in Sec. 6.2, we can prove (mathematical derivation omitted here) that the efficiency is related to the compression ratio ρ or the temperatures as follows:

$$E = 1 - \frac{T_B - T_C}{T_A - T_D} = 1 - (1/\rho)^{\gamma-1}$$

$$= 1 - \frac{T_B}{T_A} = 1 - \frac{T_C}{T_D}$$

The value of γ for air is about 1.4. Knowing the compression ratio, the maximum efficiency E can be calculated. The equation shows that the higher the compression ratio ρ, the higher the efficiency E. Thus to increase the efficiency (to achieve high performance), the tendency is to increase the compression ratio.

Comparing the internal combustion engine with most other heat engines that maintain a fixed boiler and condenser temperature, say, a steam engine, we learn from the last equation that the maximum possible efficiency of an internal combustion engine is the same as that of most other heat engines operating with boiler and condenser temperatures equal to T_D and T_C (or T_A and T_B).[2] Because T_C is the temperature of the atmosphere, the efficiency is equivalent to a steam engine with boiler temperature equal to T_D, about 600°C. Yet in burning the gasoline, the air temperature reaches 2,000°C at T_A. With the temperature this high, we should get a higher efficiency. Obviously some efficiency is lost in the internal combustion engine. Two questions arise: "How does the efficiency become reduced?" and "Why operate the engine in the less efficient Otto cycle?"

Question one. The answer to the first question is this: In the internal combustion engine, the "boiler temperature" is not a constant quantity, varying from T_D up to T_A; the "condenser temperature" is also not a constant, varying from T_D down to T_C, the temperature of the atmosphere. Thus the average condenser temperature is higher than that of the environment, and therefore the efficiency is lower than that achieved when the condenser temperature equals that of the environment. The efficiency is determined by the difference between the two temperatures; the higher the boiler temperature or the lower the condenser temperature, the better. The temperature of the environment is the lowest that can be obtained without going into extra expense. The internal combustion engine does not fully use the environment as a condenser, and so efficiency is compromised. The relationship between the Otto cycle and the cycle of most other heat engines composed of two isothermal and two adiabatic processes (the Carnot cycle) may be brought out by imagining the Otto cycle as being equivalent to a series of Carnot cycles as shown in Fig. 8.4. Each of the small Carnot cycles has an efficiency determined by its boiler and condenser temperatures, which are varying from one Carnot cycle to another. Yet the efficiencies are all the same because all equal a value determined by the compression ratio, which is the same. The first small Carnot cycle with the boiler

[2]Or operating with the same compression ratio; see Sec. 16.3.

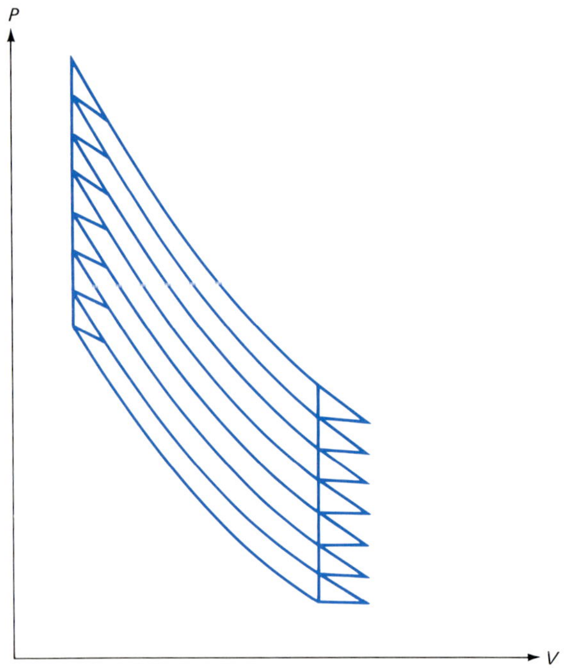

Figure 8.4 The Otto cycle replaced by a series of Carnot cycles.

temperature at T_A (2,000°C) has a condenser temperature at T_B, much higher than that of the environment T_C. Thus the efficiency is lowered.

Question two. Why not operate the engine in a Carnot cycle—using the highest temperature obtainable T_A as the boiler temperature and the environmental temperature T_C as the condenser temperature—to increase the efficiency? It is true we lose some efficiency in the Otto cycle of the engine, but efficiency is not the only consideration. When we need a more powerful engine, we do not just need an engine that will give us more energy; any small engine operating for an extremely long time will give us a very large amount of energy. What we need is an engine that will give us more energy in the same amount of time. Indeed, *power* rating is given by energy production per unit time, and its unit, the watt, is 1 *joule/sec.* The internal combustion engine works fast and thus gives us more energy per unit time. Its fuel burns internally and heating is instant. In the steam engine, for example, heat is transferred to the working substance from an external source (furnace) by an isothermal process that is slow. Although more energy is produced per cycle, it takes a much longer time to complete one cycle and thus the energy production per second, i.e., the power rating, is small. To produce the same power on a steam engine, a much more bulky and massive machine would have to be used. In automobiles we need a light-weight machine, and therefore the internal combustion engine is preferred.

Another advantage of the Otto cycle over the Carnot cycle concerns the ratio of average pressure to maximum pressure in the cycle of operation. We shall not go into the technical details on this point except to say that for comparable performance, a much higher maximum pressure is encountered in the Carnot cycle than in the Otto cycle, so the engine must be more sturdily built to withstand this pressure. This adds to the bulkiness of the engine.

Thus dollar for dollar and pound for pound, the internal combustion engine is hard to beat as an auto engine. Otherwise the auto exhaust pollution problem could be solved more easily by changing to another type of engine.

Internal Combustion Engine and Air Pollution. The use of the high-performance engine is related to many pollution problems. When the compression ratio was increased to raise efficiency, it was found necessary to raise the gasoline–air mixing ratio from 1:15 to 1:13 or 1:12. Because we use more gasoline for the same amount of oxygen in the air, not all of it will be burned. Thus more unburned hydrocarbons will be left in the exhaust, and we have an increase in the amount of a major auto pollutant—the unburned hydrocarbons. Furthermore, a high-performance engine requires gasoline of a higher octane rating, which contains a higher amount of aromatics that are incompletely burned, further adding to the hydrocarbon pollution. Moreover, at a higher temperature there is greater production of nitrogen oxides, another major auto pollutant. Finally, "engine knock" is more pronounced at a high compression ratio. An antiknock additive, tetraethyl lead, is thus required, introducing a new auto pollutant, lead. (Lead is removed from the engine by further additives, ethylene dibromide and dichloride.)

Engines other than the steam engine and internal combustion engine work generally on the same principle, so they will not be discussed further. All of them convert heat to mechanical energy. All such conversions are governed by the two basic laws of thermodynamics: the total amount of energy remains unchanged in the conversion, but only a fraction of the heat can be converted into mechanical energy.

8.4 The Electric Power Industry

Electric energy carried by a current may be readily converted into mechanical energy by the use of an electric motor. Thus whenever mechanical energy is needed in manufacturing, construction, and transportation, it may be supplied by electricity. This energy is also readily converted to heat and light and may fulfill the needs of heating, lighting, cooking, and so on. In fact, practically all energy needs may be met by electricity. Electric energy is now widely used because of its convenience. It is instantly available by the flick of a switch, whereas other fuels and energy sources take time to start. It can be used anywhere (as long as the current is there) without the effort and cost of transportation of the energy source, aside from a small amount of transportation cost incurred in the transmission lines from a power station far away.

Yet electric energy is not a primary energy source. The electric energy we tap from the household circuit comes from electric power plants where electric energy is generated at the expense of mechanical energy. The latter is produced by a heat engine by converting heat energy derived from burning a fuel. Thus the consumption of electricity is ultimately the same as the consumption of fuel. Essentially little energy is lost by conversion from mechanical to electrical energy because of the law of conservation of energy. On the other hand, a substantial amount is lost by conversion from heat to mechanical energy. However, the versatility makes electricity desirable and sometimes indispensable.

As to the environmental effect, the use of energy in electrical form in houses and plants reduces air pollution from coal and oil in the immediate neighborhood, but does not eliminate the pollution problem entirely. It merely changes the point of origin of the pollutants from the houses and factories to the power plant site. If the power plant can be located far from population centers with adequate dispersal provisions, the environment problem of energy utilization may be reduced, if not eliminated.

Electrical energy plays an important role in the use of some energy sources that are not suitable for use in individual houses and plants because of geographical and technical limitations. Water power and nuclear energy belong to this category. Energy from these sources may be tapped in a centralized power station where the energy is transformed into electrical energy to be distributed to individual houses and plants. After the depletion of fossil fuels, most future energy sources will have to be used in this way.

Energy Consumption Statistics.[3] In 1967 the energy consumption in the United States averaged 10 kw per person. This amount is divided among industry (3.55 kw), transportation (2.03 kw), household (1.97 kw), commercial (0.83 kw), and others (1.67 kw). Fuels and other energy sources are used to generate heat energy, mechanical energy, electrical energy, and nonenergy products. Heat energy is used for space heating (1.51 kw, mostly for house-

[3] Alvin M. Weinberg, "The National Energy Problem," invited paper, American Physical Society, New Orleans Meeting, November 23, 1970.

hold) and other functions such as smelting and cooking (3.33 kw mostly for industry). Mechanical energy is used for motive purposes (2 kw exclusively for transportation). Electrical energy is used for various purposes (2.1 kw, one half for industry and one quarter for household). The nonenergy uses of fuel (1.11 kw, one third for industry) include such outlets as the petrochemical industry, in which petroleum is the raw material for producing a variety of chemical products.

The amount of consumption is increasing year after year. Generation of electricity, accounting for only 21 per cent of the total energy consumption, increases by 8 per cent per year. The total energy demand doubled in the past 20 years and probably will double in less than 15 years hence.

Problems

8.1 A steam engine is operated with a superheated steam of 180°C and with a coolant of a temperature 15°C. What is the maximum efficiency of the engine?

8.2 What is the maximum efficiency of the internal combustion engine with the data given in the text? Is it better or worse than a steam engine? Why?

8.3 Where is the condenser of the steamboat engine? Is the radiator of an automobile a condenser?

8.4 All energy consumed by man eventually becomes low-temperature heat energy dumped into the environment. If total energy consumption doubles every 16 years, as it is now, how many years would it take to melt the polar ice cap to such an extent that the ocean water level would be raised 50 ft, inundating all coastal cities?

9

The Future Automobile

The introduction of the automobile has drastically changed modern society. Generally considered a blessing, it is now arousing second thoughts, for it is the main source of air pollution, and contributes greatly to noise and congestion in modern city life. Once during a student demonstration, an automobile was physically buried as the villain responsible for many modern woes. Naturally, the view has been advanced that use of the automobile should be reduced or even eliminated. Whether it is possible to turn back the clock we shall not argue here. The social ramifications of use of the automobile will be discussed at the end of this chapter to provide a broader perspective on the problem. From the point of view of physical science, there are many possibilities for conserving fuel resources and improving performance of the automobile with respect to the energy and pollution problems. These will be discussed first.

The origin of the pollution problem is the use of gasoline as fuel in the internal combustion engine. The engine converts heat energy produced by burning gasoline into mechanical energy to propel the vehicle. Gasoline is largely composed of hydrocarbons. In the combustion process gasoline vapor is mixed with air and ignited. Oxygen in the air combines with carbon and hydrogen to form carbon dioxide (CO_2) and water (H_2O). In practice, however, the combustion is usually not complete. Some unburned hydrocarbons are released into the air from the engine exhaust, forming a major component of the pollutants. Also, some carbon is burned incompletely and forms a poisonous gas, carbon monoxide (CO), another pollutant detrimental to health. Moreover, some of the nitrogen in the air inside the cylinder under high temperature during the combustion process combines with oxygen and forms a small amount of oxides, which also contributes to pollution. Finally, some gasoline additives, such as lead, become pollutants after combustion. Unburned hydrocarbons and the oxides of nitrogen together undergo a set of photochemical reactions in the atmosphere, leading to the formation of smog, such as that first seen in Los Angeles.

Even if we commit ourselves to the internal combustion engine, there is some room for improvement to reduce air pollution. During deceleration, combustion is likely to be incomplete because of the intake of the exhaust gas into the combustion chamber. Thus the engine may be redesigned to increase the intake during deceleration to assure more complete combustion. The unburned hydrocarbons in the exhaust gas may be eliminated by *exhaust burners,* of which there are two types: *direct flame afterburners* and *catalytic converters*. Current efforts to reduce auto emission are largely based on the exhaust burner approach, because it is the least troublesome for the auto industry; it simply adds a new unit to the engine without having to change the existing engine design and production facilities. Catalytic converters were required on all 1975 model cars.

Although technological advances along this line are possible, a more basic solution is to develop a different kind of engine using a different energy source to eliminate the pollutants of the internal combustion engine. After all, petroleum is already in short supply and will be depleted before long. Many alternatives are possible, thanks to the versatility of science and technology.

A vehicle needs an engine to convert the energy stored in fuels into mechanical energy. There are a variety of energy sources: petroleum, coal, natural gas, hydraulic power, wind power, solar energy, and so on. There are a variety of engines: the internal combustion engine, steam engine, electric motor, turbine, and so on. Gasoline and the internal combustion engine are used in the automobile because of technical advantages. The energy and pollution problems, however, render them undesirable. We must look into alternatives for the future automobile, for the sake of eliminating pollution and solving the problem of exhausting petroleum reserves. The use of other energy sources will be discussed in Chapter 18. This chapter will be concerned with the use of other engines.

9.1 The Wankel Rotary Engine

In connection with the pollution problem, we should note the introduction of the Wankel rotary engine[1] (Fig. 9.1). This engine made its appearance in the American market in 1971 in the Japanese Mazda, which was already four years old in Japan. Some think this is the auto engine of the (near) future, even though it is basically a four-"stroke" internal combustion engine. It was originated by the German engineer Wankel in the 1950's and developed by many others, including the pioneering Mazda maker in Japan. General Motors is now entering the field; the rotary engine will be used in the Chevrolet Vega soon. At the present, a total of 26 companies are engaged in developing the engine.

The conventional automobile engine is the *reciprocating* type, with the piston moving back and forth in strokes. It is well known that a back-and-forth movement involves more complication and vibration, whereas a rotary movement is simpler and smoother. To reverse the piston in a back-and-forth movement requires acceleration and thus requires force that changes rapidly. The result is a large amount of vibration and high levels of wear and tear. These are eliminated in a circular movement. Electric shavers can be made with reciprocating action as well as with rotary action; the latter is generally quieter and smoother. The same may happen in auto engines. The idea is quite obvious and has been pursued for a long time.

Early development of the Wankel rotary engine was discouraged because of pollution considerations. Although it produces 20 to 60 per cent less nitrogen

[1] Numerous newspaper and magazine articles have appeared on the use of Wankel rotary engines in automobiles, e.g., George Alexander, *New York Times Magazine,* October 3, 1971, p. 18. Technical data are available from Mazda Motors of America; 3040 East Ana St.; Compton, California 90221.

Figure 9.1 Heart of Mazda rotary engine; oval-shaped combustion chamber with shield-shaped rotor acting as the piston. [Mazda Motors of America]

oxides, it spews twice as much unburned hydrocarbons and a little more carbon monoxide. It runs on low-octane, unleaded gas and thus is free from lead pollution. Now it appears that its pollution is easier to clean up with exhaust burners than the pollution of piston engines. The Mazda engine uses two separate sets of spark plugs. Shortly after the firing of one, the other fires, burning up a large part of the residual fuel and increasing power output by 10 per cent. Nevertheless, the major pollution control device is a thermal reactor (Fig. 9.2) in which the exhaust gas is held for a short time and, with the help of injected fresh air, the pollutants, unburned hydrocarbons and carbon monoxide, are oxidized to form CO_2 and H_2O (Fig. 9.3). This is done without the use of catalysts. The amount of emissions of the early Mazdas since 1971 is said to be far less than the minimum required by the then applied pollution control standards, though not up to the 1975 standards set by the Clean Air Act passed by Congress in 1970. This Act set the requirements for the 1975 and later models as follows: for each mile driven, the emission of carbon monoxide should be less than 3.4 g, that of hydrocarbons less than 0.41 g, and that of nitrogen oxides (1976 and later), less than 0.4 g. These standards correspond to a 90 per cent reduction from the permissible emissions in 1970. Mazda claims that a refined version of the thermal reactor has been found to meet the 1975 standards of hydrocarbon and carbon monoxide (one half of the permissible level) set by EPA. In addition, the combustion chamber is designed to minimize the formation of nitrogen oxides and no other external control device is required to meet the 1975 standard of nitrogen oxides (one third of the permissible level). One advantage of the rotary engine with respect to pollution control is its small size and light weight, so that the auto maker can afford to install a more sophisticated and bulkier exhaust burner.

The 1975 standards specified in the 1970 Act have been postponed to 1977 and the actual 1975 standards are, for each mile driven, 15 g of CO, 1.5 g of HC, and 3.1 g of NO. On March 5, 1975, the Environmental Protection Agency postponed the 1977 standards further to 1978 and requested congressional action to postpone the standards to 1982. The reason given was that the catalytic converters used in 1975 cars turned out, unfortunately and unexpectedly, to aggravate sulfur pollution.

The successful development of the rotary engine currently used in Mazda cars is based on a number of factors common to any engineering success: better basic principles employed, ingenious engineering design, persistent efforts to overcome technical difficulties, and adequate attention to social needs. From the viewpoint of basic principles, the rotary engine

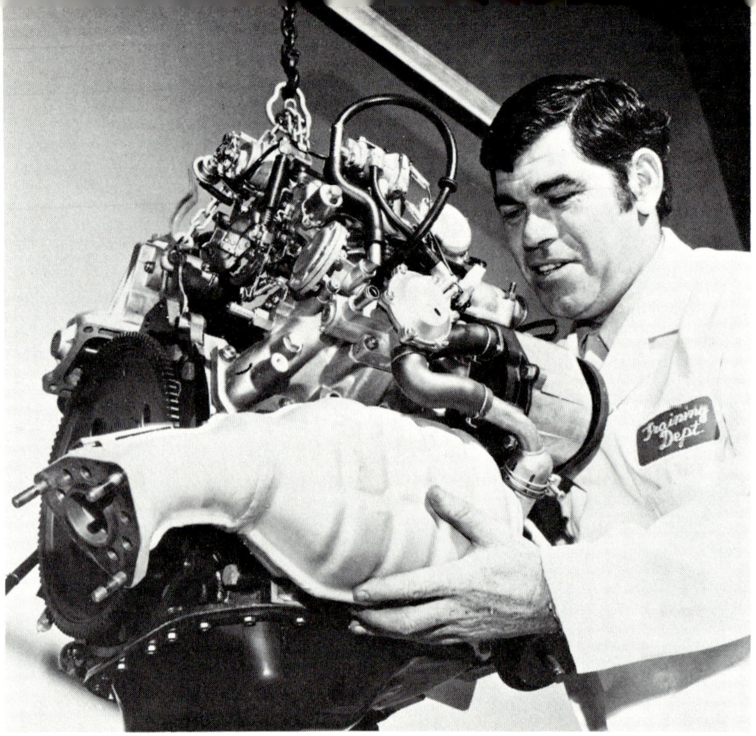

Figure 9.2 The thermal reactor, shown attached in its normal position on the lower portion of the rotary engine, is Mazda's emission control device. It has no moving parts to wear out and no special chemicals to require replacement. [Mazda Motors of America]

is superior to the reciprocal type because there are fewer vibrations and fewer moving parts. Yet, it is hard to imagine how the reciprocal piston action could be changed to rotary action to carry out the four strokes required in an internal combustion engine. The problem is solved by an ingenious design of the combustion chamber (replacing the cylinder) and the rotor (replacing the piston). Yet a sound basic design will not produce a useful machine unless many of the detailed technical problems are worked out. One problem involved in the rotary engine was the development of an airtight seal (corresponding

Figure 9.3 How the thermal reactor works. Raw exhaust from the rotary engine is injected with fresh air and enters the thermal reactor, where hydrocarbons and carbon monoxide are oxidized to form harmless carbon dioxide and water. [Mazda Motors of America]

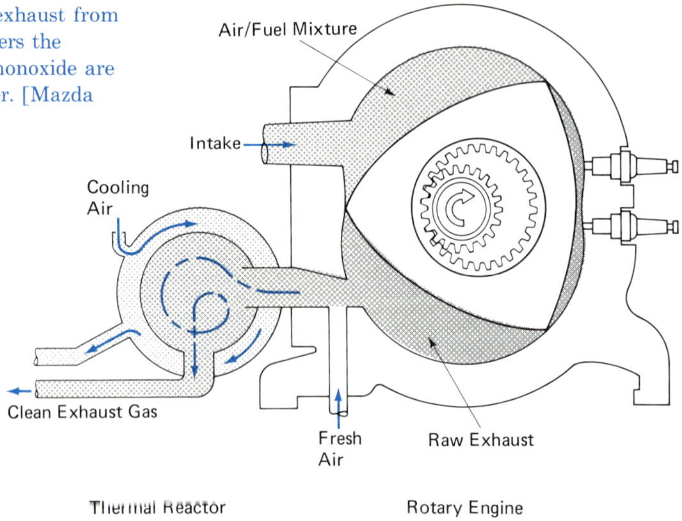

to the piston ring) to be installed between the rotor apexes and the inner wall of the combustion chamber. Many millions of dollars and many years of effort were expended just to solve this seemingly minor problem. Finally, social needs helped; pollution considerations provided the stimulus to develop the new engine.

The combustion chamber (like an oval-shaped pill box) and the rotor are shown in Fig. 9.4a (also Fig. 9.1). The shield-shaped rotor (shaded area) is rotated eccentrically by the gear mechanism shown. The gear ratio is 1.5 to 1, so that after the outer movable gear on the rotor has advanced one turn, the stationary gear has advanced one and one-half turns.

Figure 9.4 The four "strokes" of the Mazda rotary engine.

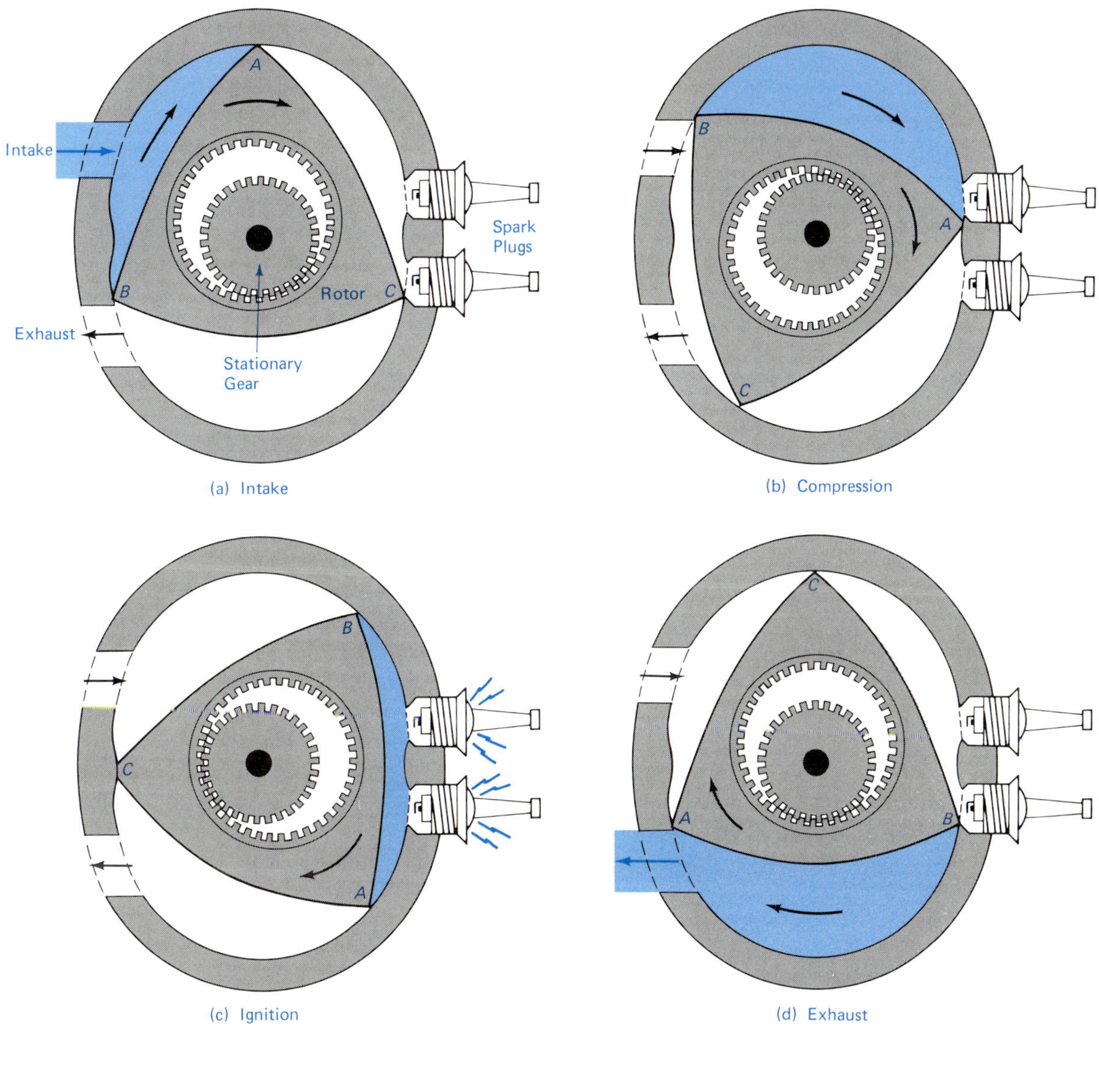

The rotor thus rotates one and one-half turns (which corresponds to a combination of one turn of "revolution" and one-half turn of "spin" of the rotor). The gear contact point has moved from the bottom (Fig. 9.4a) to the top position, and the point *A* has moved to the bottom position. Thus the rotor works symmetrically in the upper and lower halves of the combustion chamber. As the three points *A, B, C* are symmetrical to one another, the gear ratio also assures that the three points move in an exactly symmetrical manner. The inner wall of the combustion chamber is shaped to fit the trajectory of *A*, which is called an *epitrochoid*. It fits *B* and *C* equally well. Thus the tight fit of the three apexes with the inner wall of the chamber divides the space between the inner wall and the rotor into three airtight compartments that can be used for separate purposes. To ensure an airtight fit, a seal is needed between each apex of the rotor and the inner wall. It is here that the previously mentioned seal problem appears. A seal used earlier breaks down often. The breakdown damages the combustion chamber and shortens engine life. Mazda claimed to have developed a better seal, overcoming the engine-life problem.

In the position of Fig. 9.4a, a mixture of air and gasoline comes into the upper left compartment through the intake port. No valves and rods are involved. As the rotor turns clockwise, the upper-left compartment increases in size and the mixture is continuously taken in. This corresponds to the intake stroke of the reciprocating engine. After the apex *B* has passed the intake port, the gas-filled compartment is cut off from the gas supply and the rotation of the rotor pushes the gas in this compartment in the clockwise direction (Fig. 9.4b). In the process the volume decreases and the mixture is compressed. This corresponds to the compression stroke. At the compressed state the gas mixture is ignited by a spark plug (two in the Mazda engine) and expands forcefully (Fig. 9.4c). This corresponds to the ignition stroke. The expanding air pushes the rotor and performs mechanical work. The expanded gas keeps on moving clockwise and eventually becomes connected with the exhaust port (Fig. 9.4d). The continued rotation of the rotor pushes the exhaust gas out through the port; again, no valves or rods are involved. This corresponds to the exhaust stroke. The cycle is completed and may be repeated over and over again. The same cycle is repeated in the other two compartments with a phase lag of one third of a revolution one after another. Mazda also pioneered in producing a two-rotor engine with staggered phases that is equivalent to a six-cylinder engine.

The Wankel engine has other advantages: a compact size, a high power-to-weight ratio, a small number of parts, and fewer moving parts. A 185-hp Wankel weighs 237 lb, occupies 5 ft^3, has a total of 633 parts, of which 154 are movable. On the other hand, a 195-hp American V-8 weighs 600 lb, occupies 15 ft^3, and has a total of 1,029 parts, of which 388 are movable (Fig. 9.5). The compact size makes it possible to provide more room under the hood to accommodate the additional emission control device (the exhaust burner). American car makers tried first to solve the emission problem inside the internal combustion engine by design changes. They gradually realized that they would be forced to solve the problem outside the engine by adding an exhaust burner. From this point of view the drawbacks of the Wankel engine—namely, the high temperature of the exhaust gas and the high levels of carbon monoxide and hydrocarbons—actually become advantages for the operation of an exhaust burner. The high temperature speeds up chemical reaction, and the high level of combustible substances guarantees a continuous and effective combustion process in the exhaust burner to clean up the unburned hydrocarbons and carbon monoxide. The fewer parts and the fewer moving parts involved, the quieter the operation and the less likelihood of breakdown. These are superior engineering features of the rotary engine, disregarding pollution considerations. In fact, these are Mazda's selling points, presently competing with other models in the auto market. The pollution consideration is not an effective market-place selling point.

The early Wankel was poor in gasoline economy.

The parts in a Mazda rotary engine.

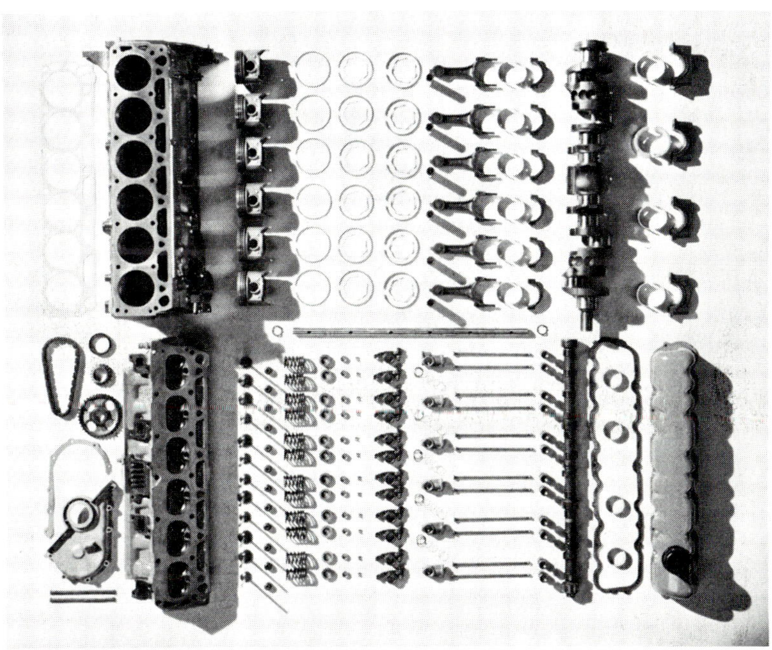

The parts in a standard piston engine.

Figure 9.5 Parts in a Mazda rotary engine compared with parts in a standard piston engine. [Mazda Motors of America]

It ran only about half the miles per gallon of a comparable conventional car. A later model, installed in a Mustang body, is getting 15 to 17 miles per gallon, comparable to conventional cars. Some in the auto industry have predicted that by 1980 the predominant engine in American cars will be rotary, not only because of the necessity for pollution control, but also because of economic considerations. Some have estimated that it can be produced at one half the cost per horsepower of a conventional engine.

9.2 The Steam Engine Car

Despite the innovation, the rotary engine is still an internal combustion engine. Looking into the future, we may list among the possible substitutes for the internal combustion engine the steam engine, the electric motor, and the turbine. The steam engine is the leading candidate. In fact, steam automobiles were on American streets early in the twentieth century. Despite rumors to the contrary, the steamer had pickup and high speed. Yet it required frequent stops for replenishment of water, and of course water freezes in winter. Factors such as these led to its demise.

Now the pollution problem has made the internal combustion engine undesirable. The steamer may have a chance to be resurrected.[2] The steamer is far superior to the internal combustion engine from the pollution point of view. It is an *external combustion engine*. The fuel burns separately from the working substance. In the internal combustion engine, burning takes place in a series of explosions. The *flame front* propagating outward from the spark plug is partially quenched on the cylinder wall because of lower temperature, so the combustion is generally not complete. The high temperature and the fast cooling are exactly the conditions for the production of nitrogen oxides. In a steamer, fuel burns in a separate furnace under better-controlled conditions: more abundant oxygen, lower pressure, and lower temperature. Hydrocarbons burn more completely, and fewer nitrogen oxides are produced. The pollutants produced in a steamer are only 1 per cent of those of the internal combustion engine.

Starting the steam is a minor problem. Some antique steamers require half an hour to build a fire of sufficient intensity. Modern ones can build up a head of steam in 15 sec. Although this may not be good enough for sport cars, it is good enough for garbage trucks, for example. In fact, two proposals have been advanced for building steam garbage trucks for Los Angeles County. A steam engine has maximum torque at the outset, and thus has remarkable acceleration. No gearshift is needed to increase power; in fact, a gear may be needed to reduce it. Other advantages are simple construction and silent operation. The internal combustion engine is noisy because it works by a series of explosions.

The major mechanical disadvantage of the steam engine is its low power-to-weight ratio, a high ratio being exactly the advantage of the internal combustion engine. A recent experimental steamer using a 1969 Chevelle body has an engine weighing 125 lb, more than the original internal combustion engine, but with a horsepower of less than the original one. Its condenser, twice as large as the radiator of the original engine, is still not sufficient to condense the steam. The large size of the condenser is one of the major problems. The early Stanley Steamer (Fig. 9.6) solved the condenser problem simply by ejecting the "dead steam" into the air, so no condenser was needed. The working substance (water) was used only once, in a single cycle, and then was discarded, instead of being used repeatedly as in other engines. So the Stanley had to stop frequently for addition of water (1 gal for 1 to 2 miles). In wintertime the discarded steam produced a blinding fog that enveloped any vehicle following the steamer in traffic. For this reason steamers were banned by the

[2] Clifford B. Hicks, *The World Book Encyclopedia,* 1971 Science Year (Chicago: Field Enterprises Educational Corporation, 1971), p. 156.

Figure 9.6 The 1907 Stanley Steamer, a steam-powered automobile.

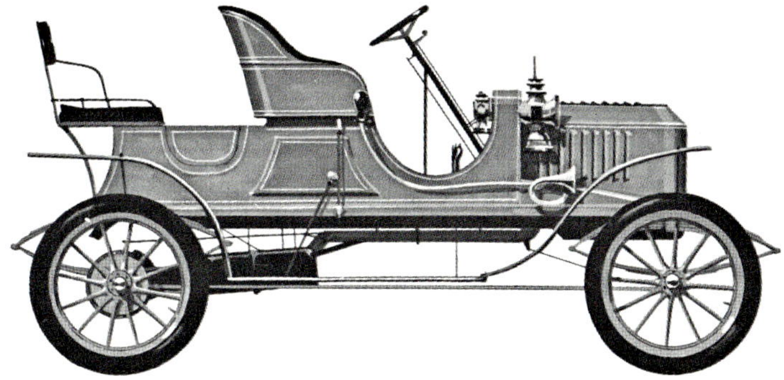

cities of Boston and Chicago. The other problem, the freezing of water during winter, remains a problem today. Besides finding a suitable antifreeze, an alternative is to use a different fluid, such as freon, as the working substance. It seems that in spite of the pollution advantage, the steam car cannot be built on an economically competitive basis. On the other hand, if the extra cost for cleaner air is paid for by the public through subsidy of steam engine cars financed by taxing the internal combustion engine cars, then the steamer may have a chance of being resurrected. This is one way the auto pollution problem may be solved (at a cost, of course).

The Stanley Steamers were handcrafted, not mass-produced. No two of them were exactly the same. Delivery was never on time, and they were very expensive. Naturally such a product cannot survive in the modern age.

9.3 The Electric Car

Next to the steam engine, other alternatives to the internal combustion engine currently under serious consideration are the electric motor and gas turbine. An electric automobile uses electrical energy stored in batteries. The fact that an electric car can work is obvious in principle and has been demonstrated by the Apollo 15 moon buggy shown in Fig. 23.10 (but at what a cost!). Its development is hindered by the lack of a battery that is sufficiently small and that can generate enough energy for uninterrupted operation for a reasonable length of time. Once the battery is used up, the car would have to stop for a recharge or a change of batteries at a "battery station." New sodium–sulfur and lithium–chlorine batteries are being developed. It is unlikely that a battery with an energy reserve comparable to that of the present car will ever be developed. Yet for commuters the daily driving distance is relatively short and the limitations of the electric car would not be serious. Batteries could be recharged overnight at home. And it is commuter traffic that mainly causes metropolitan air pollution. Because commuters are likely to have two cars in the family, one could be an electric car used exclusively for commuting or running daily errands and the other could be a conventional car used for long distance vacation trips. Because one half of auto consumption of gasoline occurs in trips 3 miles or less that can be served by the electric car, more than one half of the gasoline used now could be saved by using the electric car.

Although an electric car has certain limitations, there are a number of obvious advantages. It can start and stop instantly and is easy to operate. The machinery is simple. Most impotant, it produces no pollutants, as fuel-burning engines do. However, the elimination of pollution is not with strings. The energy in the battery has to come from some power source. Either the battery is recharged overnight in the owner's garage or exchanged at the battery station. In both cases the energy ultimately originates

Figure 9.7 The Vanguard, a four-wheel electric automobile. [Club Car Inc.]

in an electric power-generating station that burns fuel and produces pollutants. In other words, the use of the electric car simply shifts the emission site from the auto exhaust to the power-generating station. Yet there are advantages to this shift. The power station can be built far from population centers so the pollution will not cause great damage to the majority of the population. Also, in the future we may hope to have power-generating centers that do not produce pollution. Besides, some of the energy sources, such as hydraulic energy, tidal energy, fission energy, and fusion energy, cannot be used directly in an automobile. These energies have to be produced and distributed from a central station. If they are to be used for the automobile at all, the electric motor is the simplest way to do it.

Recently a small company, Club Car of Augusta, Georgia, announced the first mass production of electrically powered automobiles in this country. Its Vanguard weighs 980 lb (Fig. 9.7) and is powered by six batteries and an electric motor. It has a top speed of 28 mph and a range of 53 miles before recharge of batteries. Built on a lightweight aluminum frame and fiberglas body that is impact-resistant, the Vanguard is able to make use of the battery power for an automobile because of its drastically reduced weight. Priced at $1,986, it will run 20,000 miles on one set of batteries with a total electric cost of only $108.80, about 0.5¢ per mile, which is much cheaper than gasoline-burning cars (2¢ to 7¢ per mile). Maintenance cost is extremely low. The batteries are recharged overnight by plugging into the house current. The car is not meant to replace gasoline cars but to be used as a second or third car in the family for commuting and running household errands. Because 54.1 per cent of auto trips are under 5 miles and 98 per cent are under 53 miles, the Vanguard can fulfill many auto needs. On inner-city streets high speed has no use anyway and so the 28-mph limit is no hindrance. For stop-and-start driving, such as mail delivery, ordinary cars waste gasoline when standing idle, whereas the electric car does not. As the gasoline shortage becomes more acute, the electric car will become more desirable.

How did this enterprise develop? Club Car is the manufacturer of the four-wheeled electric golf car Caroche. The very appearance of Vanguard betrays its origin in Caroche. As the pollution problem and fuel shortage become more serious, it is only natural for the golf car manufacturer to make the metamorphosis and turn itself into an automobile manufacturer.

9.4 The Turbine Car

The gas turbine is an engine in which air, as the working substance, is heated by burning fuel to produce high pressure to propel the blades of a turbine wheel. The air-to-fuel ratio is high, so combustion can be more complete than that in the auto engine, and there will be less hydrocarbon and carbon monoxide pollution. On the other hand, nitrogen oxides are produced just the same. Although the turbine has the advantage of a gasoline engine (quick start) with the additional advantage of rotary action (less vibration and less noise), there are engineering disadvantages—poor acceleration, for instance. Also, the turbine requires more costly precision tooling. Chrysler once built a turbine car, but no mass production resulted, apparently for economic reasons. Industrialist William P. Lear of Lear Jet airplane fame, who abandoned hopes for developing a steam engine car in 1968 after a crash program, announced in 1971 that he had achieved a breakthrough in developing a nearly pollution-free vapor turbine car. The advance involves the development of a new fluid (called "Learium") to replace water in the engine's vapor generator.

9.5 The Wind-Up Car

Before the battery was widely used in the toy industry, mechanical toys usually were run by winding up a spring. Why not take a cue from this old principle for the biggest toy of our civilization—the automobile? A wind-up car[3] is the newest idea brought up in the industry. It is so drastically different from other concepts that despite its many attractive features, it was not immediately accepted. Yet it remains a serious contender for the future. For our purposes, this new approach provides another excellent example for learning the workings of many basic principles and appreciating the versatility and promise of science and technology.

As mentioned many times already, any mechanized vehicle needs energy to operate. We have learned about many kinds of energy. We also have learned that although energy may be transformed from one form to another, the total amount remains always the same. Therefore we may use many forms of energy in a vehicle. The energy may be transformed in a number of ways to meet various technical conditions before finally being put to use to propel the vehicle. At present, only a few kinds of fuel are actually used in the transportation industry. This is largely determined by technical and economic limitations, but with the energy shortage and the pollution problem demanding solution, a large number of other possibilities may be opened for consideration.

In a wind-up toy, be it a model car or an animated creature, the kinetic energy that makes it run comes from the potential energy of the wound-up spring. This potential energy ultimately comes from the energy of the muscle that winds up the spring. The gear in the toy transforms the energy and puts it at the right place in the right form at the right time. In principle, we may imagine an automobile run by a giant spring wound up by giant fingers. Although such a straight analogy is fantasy, the principle involved may be carried out by making proper variations.

Even if a giant spring were technically workable, it would not be desirable, because it would lack the flexibility to be useful. When wound tight, it could not be wound any further. On the other hand, the spring symbolizes the storage of energy readily convertible to the kinetic energy of vehicle motion. When the energy is exhausted, it could be replenished by a process of "winding."

In searching for another energy source that works like a spring, one comes upon the idea of the kinetic energy of a flywheel. When a disk-shaped, flat wheel spins around an axis passing through its center, it

[3] Charles B. Camp, "A Scientist Comes Up with a 'Perfect Car'; Detroit Just Yawns," *Wall Street Journal,* March 15, 1971, p. 1.

possesses kinetic energy. The amount, according to mechanics, equals $\frac{1}{2}I\omega^2$ joules, where ω is the *angular speed* in the unit of radian/sec and I is the *moment of inertia*. The value of I equals one half of the mass of the disk in the unit of kilogram times the radius of the disk in meter squared $\frac{1}{2}mr^2$. Usually kinetic energy involves motion and thus involves a change of position; it is difficult to confine the energy bearer in a limited space. Rotational motion overcomes this limitation. Thus a flywheel spinning at high speed may be used as an energy source. Furthermore, the faster it spins, the greater the energy. Unlike a spring, it does not wind to a point beyond which it can be wound no further. The energy stored in a flywheel can be tapped for various uses. In particular, it may be used as the power source to propel a vehicle. The basic principle of a wind-up car is just that simple. Because no fuel would be burned in the vehicle, there would be no polluting emissions. The air pollution problems could thus be solved. There are two problems: how to wind up the flywheel and how to tap the energy from the flywheel to propel the vehicle.

A direct mechanical coupling of the flywheel to the four wheels of the automobile is obviously too simple-minded. Yet once the energy is there, it is merely a matter of finding a suitable transformation for putting it to use. According to the law of conservation of energy, it may be transformed without loss (except through friction and other energy drains); the total amount of energy always remains the same. Current thinking is to use a kind of magnetic coupling to change the kinetic energy of the flywheel into electrical energy to operate a dynamo. There are many detailed technical problems in this regard that have not been completely solved yet. But in principle there is no insurmountable difficulty.

In this connection we may mention that the flywheel has been used for energy storage in modern atomic research. In the Berkeley accelerator Bevatron, for example, atomic particles are accelerated in batches in 4-sec periods, during which the magnetic field of the main magnet increases to a maximum value and then drops to a low level to repeat the cycle again. During the drop the sizable amount of magnetic energy of the main magnet is not thrown away but is transferred and stored in the form of the kinetic energy of a giant flywheel, which may be reused in the next cycle of acceleration.

How do we "wind up" the flywheel? Simply use an electric motor to spin it faster and faster. One may imagine a wind-up car pulling into the garage, being plugged into the household electric outlet, and having its flywheel wound up, ready for use next morning. Here we come to an important point. The energy necessary to operate the vehicle ultimately comes, through the household electric current, from the power-generating plant. Though pollution-free, the wind-up car, like the battery-driven car, actually shifts the polluting emissions from the auto exhaust to the chimneys of the power plant. Still, we can hope that the power plant is located far from population centers or that in the future it will use nonpolluting energy sources.

Is the wind-up car practically workable? In order to operate it for a long enough time without "recharging," the flywheel must be spinning at an extremely high speed. The current estimate is that a small car carrying two or three persons could run 100 miles without stopping, at the end of which it would need 20 minutes of "rewinding." The top speed would be about 70 mph. The rim of the 222-lb flywheel would be spinning at three times the speed of sound, about 2,000 mph. This high level of speed entails many problems. The air friction loss at such a speed is considerable; to reduce loss of energy from air friction the flywheel would have to be confined in a vacuum. The appearance of vacuum technology on the scene complicates the engineering problem. For example, any mechanical coupling to tap the energy would cause a leak in the vacuum. A magnetic coupling, however, could be used to circumvent this problem.

Another problem of the high speed of the flywheel is the hazard it would pose in a crash. If the energy in the flywheel were unleashed without control in a crash, it could cause great destruction. Some kind of crash-proof system, at least to protect the flywheel, must be worked out.

The flywheel presumably would be placed flat at the

bottom of the car. It could not be placed in a vertical position, because it also works as a gyroscope and the car would not be able to make a turn. In the horizontal position the gyroscopic action would help stabilize the car against overturning in a crash.

The flywheel mechanism has been used in buses built by a Swiss manufacturer. The buses ran in Swiss and African cities in the 1950's. They took about half a minute to recharge, which was done while they were loading and unloading passengers at the bus stops, every half mile or so.

Aside from the difficulties, the wind-up car in overall design is a simple machine. It has relatively few moving parts and thus is more reliable and less liable to break down than the conventional car. The engine weight is very small, and many of the accessory systems, such as transmission, clutch, fuel, cooling, and ignition systems are unnecessary. It promises to be light, silent, fast, and cheap. It is cheap because the cost of energy provided by electric power stations is cheaper than that obtained from gasoline. The operating cost is probably in the range of 0.2¢ per mile, about 20 times lower than the gasoline cost of the present automobile. It seems to be an ideal urban car, even though it will not replace the large family automobile for making cross-continental trips. Many technical problems remain to be solved, and research projects are in progress. For example, the flywheel will have to be made of high-strength material to withstand the high centrifugal force of spinning.

At any rate, the example shows the wide-ranging possibilities open to us through modern science and technology.

9.6 Automobile, Society, and Humanity

Beyond technical improvements in the automobile, the question may still be raised in connection with energy, pollution, and congestion problems as to whether we need the automobile at all, or as many automobiles as we now have. In America the average is about one car for every three persons. In Western Europe the average is about one for every five persons. From a simplistic point of view, the energy, pollution, and congestion problems would disappear if the automobile vanished from the scene. Environmentalists have been advocating abolition of the automobile. Of course the function of the automobile would have to be taken over by substitutes. Before the appearance of the automobile, we rode bicycles and used public transportation systems. Perhaps the same may be done in the future. With the development of a modern mass transit system in urban centers—San Francisco was the first city to build such a system—commuter use of the automobile may be reduced and pollution pressures relieved. Yet the whole problem is not that simple. The large population of automobiles is a result of the automobile's interaction with the society; the present level of use reflects a social equilibrium in much the same way that the population of any living species reflects an ecological equilibrium. A drastic reduction of the automobile population would have the same disturbing effects as those that result from upsetting the balance of nature.

Consider the effect on employment. Full employment has been an undisputed social goal espoused by all political parties from the left to the right. The present structure of the American economy is such that just about one out of five workers in the national labor force is working directly or indirectly in automobile-related jobs. This conclusion is drawn from the fact that the average family spends one fifth of its budget on automobile-related expenses. Once the automobile is eliminated, 20 per cent of the population would be unemployed. Of course part of the population would be employed in substitute industries, such as public transportation, but the fraction would be small. It is of no use to say that just because man once lived without clothing, clothing can be disposed of. This cannot be done, because a new equilibrium relationship has been established in which clothing has become indispensable. A similar situation has developed with the automobile. To appreciate the significance of the change, we have to look back into the development of modern society.

Before the industrial revolution, economic problems were relatively simple. Production was largely agricultural in nature, confined mainly to food, clothing, and shelter. Each nation was nearly self-sufficient; production and consumption were almost in balance. The industrial revolution increased the rate of production of goods, such as textiles in the beginning, and thus broke the balance. Readjustments took place: the population shift from the country to cities, the importation of food from abroad, the export of finished products overseas, and so on. Problems of overproduction and unemployment arose. The classic solution in the nineteenth century was to open up overseas markets, providing a significant stimulus for the development of imperialism and colonialism and leading to a succession of wars with disastrous consequences. (Colonization began before the steam engine. One of the most important crops produced by an agricultural society is the human population; its export has been going on since the beginning of agriculture.)

However, a new solution has been successfully demonstrated by the American experience in the twentieth century. Instead of finding an overseas market to absorb the excess industrial products of the old-line commodities (such as textiles), the extra resources are applied to develop *new* products for the domestic market. The new products are generally luxury items; they are luxury because man has lived without them before. For them to be incorporated into the economic system, it is necessary that labor and material resources be apportioned to them and that extra consuming power be made available to absorb them. The increased technology has provided both. The fraction of the population living on farms necessary to produce food for the whole population has been decreasing continuously in the United States. It had dropped to about 20 per cent three decades ago and now is down to 5 per cent. Although there are city workers—such as those engaged in fertilizer and tractor production—that are indirectly engaged in food production, the total food-producing workers constitute a small fraction of the total labor force. The relieved manpower has become available for developing the new industrial products, and the profit of increased productivity has been distributed widely over the population and made available for the purchase of the new products. Psychologically the economic change has required a readjustment of the standard of value and the ethics on spending for luxuries. Happily the whole transformation has worked out well with the help of many other social factors.

The first new product that had a significant impact on the economy was the automobile. It brought forth a host of new industries and many new opportunities for employment. The increased use of automobiles led to a dramatic increase in the gross national product and a remarkable enhancement of affluence. Yet even the market for a product like the automobile may eventually become saturated because of improved production techniques. In the 1930's Americans achieved the goal of one car in every family. From there on, the demand decreased and the classic problem of overproduction and unemployment occurred again. Yet, drawing on past experience, manufacturers directed their attention to the development of further new products, and no one thought of an overseas market as a principal solution. Every new product would relieve the pressure of overproduction and unemployment and would boost the gross national product by a quantum jump. In the 1930's attention was directed to appliances, such as the refrigerator, washing machine, and so on. Their impact was not as great as that of the automobile. After World War II attention was turned to television (and later color television). Again the economic impact was limited.

Thus at the end of World War II economists from the left to the right viewed the American economy with gloom. It seemed inevitable that overproduction and unemployment, which troubled the world for a decade in the 1930's, would recur. However, to the surprise of everybody, not only did the curtain of doom never fall, but America's economy and so the world's economy climbed to unprecedented heights. The gross national product increased by leaps and bounds, tripling in 25 years. It was a phenomenon unparalleled in the past, and perhaps never again to be seen in the future. Moreover, the resources and technology available now are not essentially different from those at hand during the 1930's. Why was the whole world plagued by depression then? The political system was not the main reason. Germany and Italy were then under dictatorial rule and were in a position to do things that could not be done in a free economy.

It seems that the postwar boom was ignited by an accident; that is why it was not foreseen. Right after the war, American cities were plagued with a housing shortage. As a temporary relief, houses were built in the outlying countryside to relieve the shortage. Because it placed people far from urban centers in areas devoid of all conveniences and compelled them to spend long hours commuting, suburban living was thought to be a stopgap measure instead of a permanent solution. It was made possible by the private automobile, which provided transportation. Yet as more people moved to the suburbs, conveniences were developed to improve suburban living. Thus shopping centers were built, superhighways were constructed, and so on. While the negative side of suburban living was being improved, the positive side began to attract attention—more spacious living quarters, less noise, cleaner air, and so on. Soon the balance was tipped over to the other side, and suburban living became the desired way to live. An exodus began, and now a large fraction of the urban population has been transferred to the suburbs. In two short decades, the increased needs for housing, furniture, and appliances in the suburbs, now harboring about one third of the population, stimulated the economy greatly. Suburban living made a second car a necessity; thus the demand on the automobile increased greatly. The economic impact of a switch from the one-car-per-family economy to the two-cars-per-family economy is as great as that from a no-car economy to a one-car economy. No wonder the gross national product increased by leaps and bounds.

The development of an economy is thus not entirely determined by resources and technology. It takes a tricky, devious path of evolution to circumvent or to minimize the obstacles (the potential barriers) that prevent the development of the economy from one state of equilibrium to another. The economic operation dictated by the maximization of profit (or its equivalent) leads to a state of stable equilibrium corresponding to a potential minimum surrounded by potential barriers. Yet the potential minimum often is not the absolute minimum, and the potential barrier prevents it from developing into another lower minimum even though such a change is in accordance with the general laws of economic development. The situation is like water trapped in a mountaintop lake, unable to flow to a lake on a lower level. It takes a syphon or a tunnel to make the flow possible. In economic development we also need syphons and tunnels.

We may also use the analogy of biological evolution, of which economic evolution is actually an extension. The human body is made of the 92 chemical elements, but 92 bottles of 92 elements in the right proportion do not mix to make a man. It took 2 billion years of biological evolution, millions of enzymes to overcome the potential barriers, to transform the inorganic materials into complex living organisms. The same economic resources and technology did not mix to give rise to a two-car economy in the 1930's. It took a series of intervening steps and some good luck to bring it about in the 1960's. To disrupt the economy by any drastic change (such as a great reduction of the automobile population) is like dismembering a human body. It would have serious consequences.

In understanding the dramatic increase of the gross national product in recent years and the associated improvement in general well-being, the special role played by the automobile cannot be ignored. Once the economy had reached the stage characterized by suburban living, the automobile had become so entangled in the social, economic structure that its elimination was impossible.

The extensive use of automobiles created new problems: congestion, energy source depletion, and pollution. Preoccupied with these maladies, we find the hero of yesterday becoming the villain of today. As in biological evolution, we may expect science and technology in the future to develop remedial measures in the course of social evolution. Evolution is often a path of no return; once man has lost his claws and body hair, he is not expected to go back to the jungle again. He can only move forward by using newly developed capacities to solve new problems. The automobile is here to stay: there is no turning back. It is an unpremeditated product of an unusual historical drama. Only the naive would think man created this many automobiles merely out of greed and lust. Only the naive would think modern city problems can be solved by moralizing and by burying the automobile.

In addition to the introduction of new products, the American experience has created another method of increasing consumption and thus of coping with the problem of overproduction and unemployment: *accelerated obsolescence.* When material goods were scarce, the quality of a product was judged largely by its durability. Styling, trimming, accessories, and so on, were secondary. Once material goods became abundant, durability seemed to become a liability. If the lifetime of a product is reduced by one half, the demand for the product will be increased twofold. In order to increase consumption (and so reduce unemployment), why not cut short the useful lives of the products? Indeed, there are many ways the obsolescence of a product can be accelerated: less durable construction, more frequent styling changes, endless addition of trivial novel features, and so on. All have been tried, and all are successful in increasing consumption and boosting the gross national product. All are manifested in the automobile industry. On the other hand, accelerated obsolescence represents a waste of personal savings and a waste of natural resources. Much of the attack on the auto culture is directed toward those features related to accelerated obsolescence. In an overall assessment of the role played by the automobile in our society, the issue of accelerated obsolescence has to be carefully examined. Is it an unforgivable sin, a necessary evil, an economic expedient, or a miracle cure?

The current practice of bringing about economic progress by accelerated obsolescence is viable as long as resources are endless and environment is infinite. There is nothing intrinsically wrong with it from economic, philosophical, and moral points of view. The crux of the matter is the rapid advance in technology through which all the necessities can be produced by a small fraction of the labor force, thus threatening to render the rest of the population idle. One way to solve the problem would be just to put the idle population on the public payroll, or as an equivalent, to employ them for a token service, such as sweeping the parks, their necessities having been already produced. Yet to live off the society without meaningful work is to become a parasite, which is certainly not the goal of human aspiration. Ironically, this "socialistic" approach is being carried out piously in capitalistic nations through the welfare system; nearly 10 per cent of the population of big cities in America is now on welfare rolls.

What can a self-respecting society do in the way of finding jobs for the idle population? Whatever they can produce is by nature not urgently needed, because all necessities are already being produced. It has to be in the category of luxuries. It is human nature that once a man has a serviceable automobile, he begins to want a more attractive one—so much so that even if the added feature is trivial, he is willing to work overtime to get it. It is on this psychological basis that more demand may be generated, and this demand may be enlarged greatly by accelerated obsolescence. The extra amount of production to fulfill this extra luxury need is considered (subjectively) worthwhile and will be paid for by consumers. The workers involved in this part of the production effort are considered to be gainfully employed, not leading a parasitic existence by living off society through performance of a token service such as sweeping the parks. In reality, their functions are no different from that of sweeping the parks; their work is trivial but it results in a redistribution of national wealth. Nevertheless, the capitalistic society is able to take advantage of this frivolous human trait to make trivialities appear essential, and thus to give the workers engaged in such trivialities a sense of self-respect and purpose.

As technology advances, more and more people will be thrown into this category. It is not difficult to feed them, because all the necessities have been produced already. It is more difficult to satisfy their aspirations. Ironically, we find a miracle cure in accelerated obsolescence. As luxuries turn into necessities in the evolutional process, the last trace of parasitism will be wiped out. In this connection it may be mentioned incidentally that parasitism as an ecological phenomenon is in itself natural. The whole animal kingdom is a parasite to the vegetable kingdom. And man is the biggest parasite of all.

Although accelerated obsolescence brought about by deliberate manufacture of less than perfect items is a waste of resources, that engendered by frequent changes in style is a different story. Frequent style changes make an automobile appear older than its technical age, and thus worth less. This costs a new car owner more in depreciation. Yet there is a major difference between the two kinds of obsolescence, with significant social consequences. The loss incurred because of style changes by a new car owner when he trades in for another new car a few years later actually becomes the gain of the low-income used-car buyer, who can now afford a serviceable vehicle at low cost. In fact, it is the styling obsolence that makes the price level of used cars so low that every American can afford one regardless of his income. Although everybody complains that a haircut in America is the most costly in the world, it is not generally realized and appreciated that only in America can a car, albeit second-hand, be purchased at a cost as low as that of a suit. A car for every family is an attractive social goal. (Hitler promised a car for every German worker, and that was the origin of the Volkswagen.) From a utilitarian point of view, the new-car purchasers are actually subsidizing the used-car buyers, and the whole affair works out to be a process of equalization of wealth. Yet this is accomplished in America not by socialism or socialization, but by the "free will" (perhaps in this case snobbery) of individuals in a free society.

The desire for better cars, better housing, and better clothing may be considered selfish. Economic progress may thus be regarded as being spurred by selfish motivations. Some thinkers may consider this state of affairs deplorable. Yet in social development some superficially unselfish undertakings may turn out to be selfish collectively or even individually. Advances in medical care have always been regarded as an unblemished good thing in political opinion from the right to the left. Yet it is partly the reduction of the death rate that causes the imbalance between birth and death, resulting in population problems. Man never considers medical research undesirable, for the simple reason that the life that may be saved by the research may be his own. The idle manpower produced by the advance of technology may certainly be employed in unselfish pursuits. In biology there are unselfish instincts and motivations. In a society there are unselfish undertakings, such as the building of cathedrals, spaceships, and atom smashers. From an economic point of view these are all ways to employ idle manpower. Yet desirable undertakings are hard to come by. Even the enthusiasm over space exploration has faded quickly. The solution to the problem of employing idle manpower brought on by technological advances will still have to be found in the conventional way.

In supposedly unselfish socialist societies the construction of huge hydroelectric plants in the early years resembled more a ritual exercise than an economic enterprise. The economic return did not then justify the capital investment. Yet the projects were carried out enthusiastically. Unlike the pyramids of the Egyptians and the cathedrals of the Middle Ages, the hydroelectric plants sowed the seeds of secularization. Eventually electric power was produced in large abundance and a place to use it had to be found. What could be a better use than making more and better consumer goods? Thus the same problems that have afflicted the capitalistic countries will crop up in socialist societies, and probably the same measures will have to be taken. This transition has already taken place in the Soviet Union.

Thus accelerated obsolescence is a universal phenomenon inseparable from technical advances. It can be tolerated as long as the environment and the resources can tolerate it. Therefore considerations of environment and resources are fundamental to all future economic, political, and social problems. Technical developments to preserve the environment and to conserve resources are important not only for their immediate purposes, but also for lifting the limitations on the growth and development of a modern society.

Besides the economic and utilitarian considerations of the automobile, there is another aspect worth mentioning. There is a kind of degradation of life as technology advances. In the hunting and herding society, man roamed for hundreds of miles, sometimes from one continent to another. He was engaged in a variety of interesting activities with close contact to nature. Life, though hazardous, was never dull; it was full of adventure and novelty. Once technology brought forth the agricultural society, man became tied down to a small parcel of land, and life became the repetition of an established routine. It was far less romantic and much more boring. The contrast is sharp and obvious and is manifested in various stories, from the biblical account of Abel and Cain to the modern-day musical "Oklahoma." Still, agricultural people could enjoy a pastoral environment enlivened by the rolling hills, bubbling brooks, singing birds, and blooming flowers. The life of a modern city dweller is even more deprived. His living space at best is limited to a fraction of an acre, often a few hundred square feet of apartment space tucked in the cliffs of concrete canyons. The daily activity of an industrial worker has become extremely monotonous and dull, often involving a single act performed thousands of times a day. It is much worse than that of a farmer or an artisan in the agricultural society. Work is no longer an avenue of self-fulfillment; it has become an unbearable drudgery. The introduction of the automobile brings to the city dweller a new life. Once again he can roam over the country and plunge into the great environment. It enables him to break the monotony of daily routine and engage in refreshing activities that partly compensate for the degradation of human life by the advance of technology.

Human history has been dominated by two forms of life, the nomadic and the sedentary. Social, political, ethical, religious, and ideological developments are closely related to them. Americans have embarked upon a neo-nomadism brought about partly by the use of the automobile. This is a crucial point for understanding many of the fascinating features of modern America that do not fit into many of the old theories.

Problems

9.1 Devise a way to tap the kinetic energy of a rotating flywheel encased in a vacuum chamber. (*Hint:* Electromagnetic induction.)

9.2 Hydrogen can be used as a fuel for an internal combustion engine. Can it be used as a gasoline substitute in automobiles? Discuss all favorable and unfavorable considerations. Pay particular attention to fuel storage (in automobile) and pollution effects.

9.3 Bottled gas can also be used as an auto fuel. What is the major difference between bottled gas and hydrogen for this purpose?

9.4 As far as the engine and fuel are concerned what is the prospect for the automobile in the near future (10 years), median future (30 years), and distant future (100 years)?

9.5 A wind-up car with enough energy to run 50 miles crashes head on with a gasoline car of the same weight having enough gas to run 50 miles. During the crash the flywheel breaks and the gasoline explodes. Which car suffers more damage?

9.6 Discuss the advantages and practicality of a dual-fuel automobile. [Note: At a cost of $350, an automobile can be modified for dual operation of gasoline and natural gas; this has been done by municipal fleet operators (Sec. 17.2). A recent project aims to add a flywheel system to a gasoline car for commuting purposes, keeping the gasoline system as a backup for long trips. The flywheel system need provide energy for only about 30 miles and therefore many of the high-speed problems discussed in Sec. 9.5 may be eliminated. Perhaps the most promising dual-fuel system would be one permitting trucks to operate on both gasoline and gas—not natural gas but the old-fashioned municipal gas generated by a coal gas producer (Sec. 18.2) carried on board. During World War II many Chinese trucks were powered by a charcoal gas producer—a cylinder 5 ft high and 1 ft in diameter installed next to the driver's cab. Coal gas is not very powerful; the gasoline system is needed for acceleration and for climbing hills. But the dual-fuel system permits the more abundant and cheaper fuel coal to be used directly for automotive purposes and 2 million barrels of oil a day (12 per cent of U.S. oil consumption) could be saved.]

10 Chemical Transformations of Matter

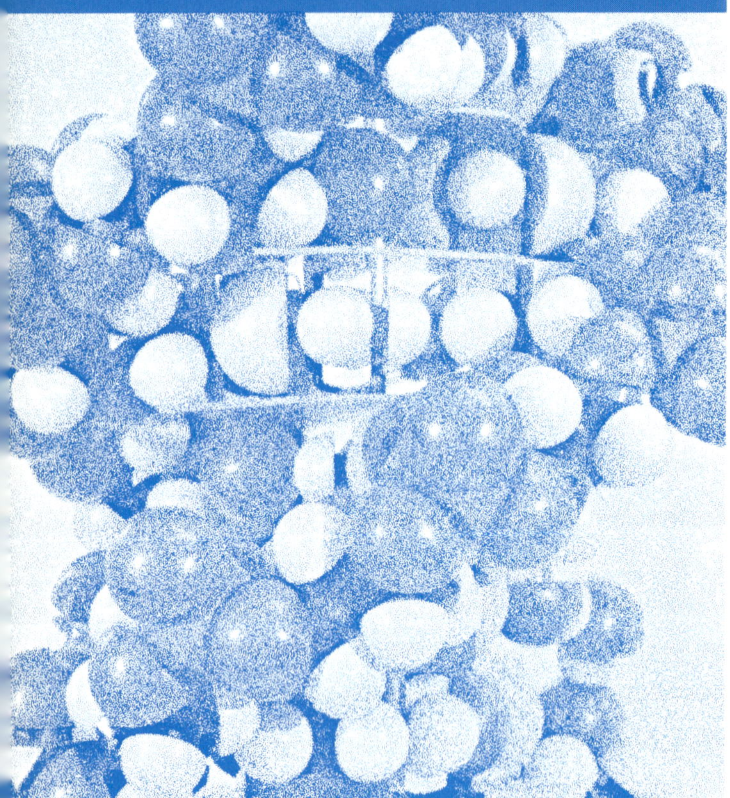

10.1 Dalton's Atomic Theory

The material substance we deal with in everyday life is usually a mixture of a number of pure substances. A mixture of several ingredients can be made in any proportion (e.g., mixing of cocktails to individual taste). It can be separated by physical means (such as fractional distillation). Moreover, each component retains its own property in the mixture.

Yet a pure substance may be broken down further into a number of simpler substances by chemical means such as burning (not by physical means). Also, simple substances may combine chemically to form *compounds*. These processes are called *decomposition* and *combination*. A compound differs from a mixture in that the proportions of ingredients in the compound are not arbitrary but definite. Moreover, each ingredient loses its own properties in the compound. Obviously combination involves a new process different from mixing.

As mentioned before, most substances can be broken down by decomposition into simpler substances. Each of the simpler substances in turn can be broken down further until we reach substances that cannot be broken down any further by chemical means; these substances we call *elements*. It was found that there are only a limited number of elements in nature. The numerous other substances in the world are made of the elements by various combinations. Because the ways of combination are endless, there is no surprise that out of a few basic elements a very large number of compounds may be formed.

After a long period of investigation, as already mentioned, it was established that all material substances naturally occurring on the earth can be reduced to 92 elements (see footnote 1 of Chapter 4, which applies throughout the book). The discovery of a new element is such an important event in the history of science that the newcomer is usually properly christened to honor a memorable person or place. By studying the spectra of light from the sun and stars, it was found that most of the elements on earth are also present in the sun and in other stars. An element first found in the sun (by its light spectrum), thus named helium, was later also found on earth. Another element expected to exist but not found on earth was found in some stars (the element can now be made artificially in nuclear physics laboratories on earth and was thus christened technetium). Thus the 92 elements are universal. The whole physical universe is made of them.

Gold is an element. It rarely forms a compound; that is why gold cannot be made of any other things. The alchemists in the Middle Ages tried to make gold from less expensive materials. Their failure suggested that gold is an element.

Out of the 92 elements many ways of combination are possible. But not all elements can combine with all others and combination is not by arbitrary amounts. The question is thus what combines with what, and by what amount. Empirical rules were established early in chemistry, from which we may learn further about the basic properties of the elements.

Two empirical rules stand out concerning the proportions of elements in compounds. First is the *law of definite proportion*. Each compound is made of elements in a definite weight ratio. Thus water is made by eight parts of oxygen and one part of hydrogen (by weight). If nine parts of oxygen were allowed to combine with one part of hydrogen, only eight parts of it will combine and there will be one part left over. This is not the case in a mixture.

The second is the *law of multiple proportions*. In special cases two elements may combine to form more than one compound with different, though definite, proportions. But in such cases the difference in proportions is usually in a simple ratio, such as 1 to 2, 2 to 3, and so on. For example, oxygen and hydrogen may combine to form another compound, called hydrogen peroxide, a common item in a medicine chest, with the weight ratio of 16 to 1. Compare hydrogen peroxide with water; the amounts of oxygen that combine with the same amount of hydrogen in the two compounds are in the simple ratio of 2 to 1 (obtained from 16 to 8).

These laws provided the basis for Dalton to expound an atomic theory in the early years of the nineteenth century (1808), which is a major step in the development of chemistry. The idea of an atom as a discrete basic unit of material substance can be traced back to the ancient Greeks. Dalton's concept of the atom is more definite and refined. The following is the essence of his atomic theory.

All substances have a granular (discontinuous) structure. The basic unit for an element is a particle called the *atom*. Thus there are 92 kinds of atoms corresponding to the 92 naturally occurring elements. Atoms in definite numbers may combine to form a composite particle called the *molecule*, which is the basic unit of the compound formed by the corresponding elements. Atoms are indivisible. Atoms of the same element are exactly the same. Atoms of different elements are different, independent, and immutable.

These assumptions are quite reasonable. On this basis the two laws of definite and multiple proportion may be explained easily. Because atoms are indivisible a molecule can contain only an integral number of a given kind of atom but not a fraction of it. Because all atoms of the same kind have the same weight, the weight proportion in a compound is determined by the numbers and the weights of the atoms and therefore is definite and not continuously variable. If two compounds are formed by the same two elements the difference can only come from the number of atoms that enter into the molecule. Because the numbers of atoms are integers the ratio can only be a simple ratio of whole numbers.

It was on this basis that modern chemistry was developed. In later years we learn from nuclear physics that these notions are not absolutely correct. All atoms of the same element are not exactly the same. (There are a few different kinds called *isotopes*.) Atoms are eventually divisible but not by chemical means and the subunits are elementary particles, which are entities completely different from the atoms and molecules. Finally, atoms of different elements are related and transmutable by *nuclear reactions*. However, as far as *chemical reactions* are concerned, Dalton's notions are still practically correct and atoms may be regarded as indivisible and immutable.

10.2 The Molecular Formula

In the atomic theory a molecule is associated with each compound, consisting of an assortment of atoms combined together. The numbers and kinds of atoms in the molecule may be put down in an expression called the *molecular formula*. For example, the water molecule is represented by H_2O, where the chemical symbols H and O stand for hydrogen and oxygen atoms and the number indicates that two hydrogen atoms combine with one oxygen atom (the number 1 is always omitted). Whenever a new compound is discovered the problem to work out first is usually the determination of the molecular formula.

If we know the relative weights of the atoms, then from the relative weight proportions of elements in a compound we can figure out the relative number of atoms of each element in the compound. Actually we do not know the relative weights of atoms beforehand. They are to be worked out together with the determination of the molecular formulas. It is like a jigsaw puzzle. Many pieces of information have to be fitted together at the same time. We start by trial and error, making use of all clues available to us.

Dalton thought that the molecular formula of water was HO and molecules of gases like hydrogen and oxygen are made of individual atoms. Then Gay-Lussac studied the volume relationship in gas reactions (Fig. 10.1) and found that the volumes of reactants and products are always in simple ratios (with pressure and temperature kept constant). As simple ratios are suggestive of ratios of atoms and molecules, this led *Avogadro* to propose the *hypothesis* that equal volumes of gases contain the same number of molecules so that the volume ratios Gay-Lussac observed are also the molecular number ratios in the reaction. For example, Gay-Lussac observed that it took two volumes of hydrogen to combine with one volume of oxygen to form two volumes of steam. This would mean, according to Avogadro, two molecules of hydrogen combine with one molecule of oxygen to form two molecules of water. According to the molecular formulas first proposed by Dalton, one would expect one volume of hydrogen to combine with one volume of oxygen to form one volume of steam, and so the Dalton molecular formulas are not right. According to Avogadro the experimental volume ratios can be explained if we assume both hydrogen and oxygen molecules are made of two atoms and the water molecule is made of two hydrogen atoms and one oxygen atom. In other words, the molecular formulas proposed are H_2, O_2, and H_2O. Then the reaction may be represented by the equation

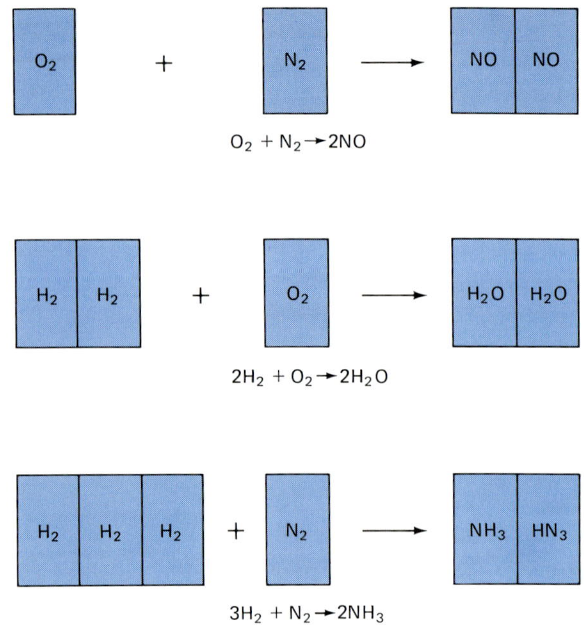

Figure 10.1 Illustration of Gay-Lussac's finding on volume relations in gas reactions.

$$2H_2 + O_2 \longrightarrow 2H_2O$$

with molecular numbers 2, 1, 2 consistent with Gay-Lussac's observation.

By a number of other gas reactions similar molecular formulas and reaction equations were established (Problems 10.1 and 10.2). Invariably the same molecular formula appears for the same compound in different reactions; this shows that the Avogadro hypothesis does not contradict itself in all tests and thus is likely to be correct. For example, the molecular formulas of oxygen and hydrogen determined from the preceding reaction are the same as those determined from the oxygen–nitrogen reaction $O_2 + N_2 \longrightarrow 2NO$ and hydrogen–nitrogen reaction $N_2 + 3H_2 \longrightarrow 2NH_3$, respectively. Moreover, the nitrogen molecular formulas determined in the last two reactions are the same. Thus the formulas are consistent among themselves. The molecular formulas we accepted now are established on this basis. On the other hand, they are not logically the only ones. If we multiply all the numbers of atoms in all molecular formulas by 2, or 3, or any integer, we would obtain another set of molecular formulas equally valid. The set we are using now is merely the simplest possible. However, many years later by other methods in molecular physics (specific heats, spectra), we can actually prove that the hydrogen, oxygen, and nitrogen molecules are made of two atoms. Thus the set of molecular formulas we are using now is actually proved to be correct.

Atomic Weight. Once we have a set of molecular formulas such as H_2O, NO, NH_3, and so on, giving the numbers of each kind of atom in the molecules, we can then use the relative weight proportions of the elements in the compounds to determine the relative weights of the atoms (Problems 10.3 and 10.4). It turns out that hydrogen is the lightest. Therefore hydrogen was chosen as the reference in the early years of chemistry. The relative weights of all other atoms referring to hydrogen as unity were called the *atomic weights*. The atomic weight of oxygen, for example, was nearly 16 and that of carbon was nearly 12. Once the atomic weights are established, the molecular formula of a new compound can be determined from the empirical weight proportions of the constituent elements of the compound (Problem 10.5). Chemistry is developed on this basis and all results obtained are consistent with one another.

Later it was found that it is more convenient to choose oxygen as the reference. The atomic weights were redefined by referring to the *element* oxygen as standard; its atomic weight is defined as exactly 16. This way the atomic weight of hydrogen turns out to be 1.008. After 1961 a new standard was used in which the *isotope* carbon-12 (not element carbon, which is 98.89 per cent carbon-12) is assigned an atomic weight of exactly 12. Atomic weights in Table 10.1 are on this standard. The value of oxygen changes from 16.0000 to 15.9994. For our purposes the differences of the three systems are very small and insignificant.

Molecular Weight. Adding up the atomic weights of all atoms in a molecule, we obtain a number defined as the *molecular weight*. Thus for water the molecular weight is $2 \times 1 + 16 = 18$. It means that the water molecule is 18 times heavier than the hydrogen atom. For carbon dioxide, CO_2, the molecular weight is $12 + 2 \times 16 = 44$. Molecular weight and atomic weight are pure numbers with no physical unit. (It is a ratio.)

The Unit Mole. The unit mole, which we have used already, is based on the molecular weight. Because the molecular weight of water is 18, we define 18 g of water as 1 mole of water. Because the molecular weight of carbon dioxide is 44, we define 44 g of carbon dioxide as 1 mole of carbon dioxide. In general 1 mole of substance is an amount in gram

Table 10.1 Atomic Weights

H	1.008	O	15.9994	Ca	40.08
He	4.003	Ne	20.183	Fe	55.847
C	12.011	Na	22.99	Au	196.967
N	14.007	Cl	35.453	U	238.03

units equal to the molecular weight of the substance. Contrary to other units, which are fixed once and for all, this unit is different in size for different material substances. Yet in certain circumstances, it is convenient to use such a unit. For example, the quantity n in the equation of state of an ideal gas is the amount of gas in the unit of mole. When n is defined this way, the other constant R in the equation becomes a universal constant. If instead n is defined as the amount of gas in gram units, the constant R is not universal but varying from one substance to another. Therefore the unit mole is preferred for n. The significance of this unit is that 1 mole of any substance contains the same number of particles (atoms or molecules), as is evident from its definition. This number is called the *Avogadro number* N_o, a basic universal constant in chemistry and physics. In atomic physics we can determine the number by several methods. The value is

$$N_o = 6.025 \times 10^{23} \text{ molecules/mole}$$

The mole is thus a unit measuring the number of particles; one mole means N_o atoms or molecules.

Valence. From the molecular formulas we notice that when hydrogen combines with other elements it always takes one or more hydrogen atoms to combine with just *one* other atom. It never occurs that more than one other atom combines with one hydrogen atom. It appears that different atoms have different capacities to combine with hydrogen. Thus we may assign an integral number, called the *valence,* to each atom equal to the number of hydrogen atoms it combines to indicate this capacity. Thus from H_2O we assign to oxygen a valence 2, meaning that each oxygen atom can combine with two hydrogen atoms. From NH_3 (ammonia) we assign to nitrogen a valence 3; from CH_4 (methane) we assign to carbon a valence 4; and so on. Moreover, this capacity represented by valence turns out to be one not only applicable to hydrogen but also applicable to other atoms. For example, from valence 2 of oxygen and 4 of carbon we surmise that carbon has a capacity twice as large as oxygen and therefore when these two combine we would expect one carbon atom to combine with two oxygen atoms. This indeed is the case and the compound is carbon dioxide, CO_2. Thus valence is not merely a property related to chemical combination with hydrogen but is a general property of an atom governing all chemical combinations. It follows that for those atoms that do not combine with hydrogen, valence may be determined through another atom of known valence with which it combines (Problems 10.11 and 10.12). Occasionally an element may have more than one valence, as exemplified by the molecules involved in the law of multiple proportion.

Chemical Bond. The same idea may be expressed by another concept, namely, the *chemical bond.* An atom of valence n is said to have n bonds, each bond representing the capacity to combine with one hydrogen atom or its equivalent. In chemical combination each bond of each atom must be associated with a bond of another atom. Graphically each bond may be represented by a short line emanating from the chemical symbol, thus —O—, —C— (with vertical bonds above and below), and so on. In a compound each bond of each atom is supposed to be connected with either a hydrogen atom or its equivalent. A compound may thus be represented symbolically by atoms connected by short lines representing bonds. The short line connecting two atoms thus represents a bonding relationship. At the same time the valence relationship is brought out by the number of bonds. Thus the molecules H_2O, CO_2, and CH_4 may be represented as follows:

```
                              H
                              |
H—O—H      O=C=O         H—C—H
                              |
                              H
```

These graphical representations are called *structural formulas.* They bring out not only the information contained in the molecular formula but also the additional information of which atom combines with which. For simple molecules such as those above, this information is inconsequential. But for complex molecules made of a large number of atoms, the same atoms may be arranged in different bonding relationships and the structural formula becomes more useful and more important.

10.3 Classification of Chemical Elements

From the molecular formulas we learn which atom can combine with which. We also learn that some atoms do not combine with some others. Among the atoms already mentioned some are very active; hydrogen combines with many kinds of atoms readily. Others are inert; gold hardly combines with any others. Some atoms have properties very similar to one another and others have very different properties. These may be taken as the basis to classify the chemical elements and to study the relations among them.

The first property distinguishing various atoms is the valence. The valence can be 1, 2, 3, 4, and so on, and atoms can be grouped accordingly. Among atoms of valence 1, many are indeed similar to one another, but not all. A close examination shows that there are two groups in them. One, consisting of the elements lithium (Li), sodium (Na), potassium (K), rubidium (Rb), and cesium (Cs), is called the *alkali metals*. They are similar to one another in the sense that what one can do the others can do also. For example, sodium (Na) can combine with chlorine (Cl) to form sodium chloride (NaCl), which is the chemical name of the table salt we ingest every day. The other alkali metal elements can also combine with chlorine to form similar compounds LiCl, KCl, RbCl, CsCl. Thus they are chemically alike. At the same time they are also physically similar. They are all metals. They are all conductors. Thus they are rightly considered as a group. They do not combine with one another. Because chlorine has valence 1 (from the compound HCl) they are all assigned valence 1 (from LiCl, NaCl, KCl, RbCl, CsCl).

The other group, consisting of fluorine (F), chlorine (Cl), bromine (Br), and iodine (I), is called the *halogens*. They combine with hydrogen in the same way; the compounds HF, HCl, HBr, HI are similar to one another. They combine with each of the alkali metal atoms and form similar compounds, e.g., NaF, NaCl, NaBr, NaI. They are all insulators. Thus they are physically and chemically similar to one another and naturally belong to one group. We may consider the valence of alkali metals to be $+1$ (the same as hydrogen) and that of halogens to be -1, to distinguish the two groups. In this sense we may say that chemical reactions take place between atoms of opposite valences. However, there are exceptions.

Next we consider atoms of valence 2. Among them is a group called the *alkaline earths,* including such elements as calcium (Ca) and magnesium (Mg). They are metals and are conductors. They can substitute for alkali metal atoms in compounds such as NaCl except that the numbers of atoms involved will be different to reflect the different valence. Thus we have such compounds as $CaCl_2$, $MgCl_2$, and so on.

In this way all elements may be classified into a number of groups, each having a set of common properties. Another group is the *noble gases,* helium (He), neon (Ne), argon (Ar), Krypton (Kr), and Xenon (Xe). They do not combine with all others[1] and with themselves. Thus they are all monatomic gases. We shall not consider the details of the other groups.

Acid-Base Reactions. The compounds formed by alkali metals and halogens are the products of a basic type of chemical reaction, the *acid-base reaction* between two groups of compounds. One group is called *acids,* exemplified by the halogen compounds of hydrogen, such as HCl. The compound HCl is a gas. When dissolved in water the solution is called hydrochloric acid. The other halogen com-

[1] Recent studies show that some of the noble gases do form compounds, albeit very unstable ones.

pounds of hydrogen, when dissolved in water, also form acids. Acids are a class of compounds formed by hydrogen with other atoms or atomic groups (called *radicals*) that will produce hydrogen ions (see next section) in solution. The most-used acids, besides hydrochloric acid HCl, are sulfuric acid H_2SO_4 and nitric acid HNO_3. The other group of compounds is called *bases,* formed by the atomic group OH with metal atoms or other atomic groups, such as NaOH, $Ca(OH)_2$, and so on. An acid and a base react readily. The H of the acid combines with the OH of the base to form a water molecule H_2O, whereas the metal of the base and the other group of the acid form a compound called a *salt.* There are many acids, many bases, and many acid-base reactions resulting in many salts. One example is

$$HCl + NaOH \longrightarrow NaCl + H_2O$$

The salt formed is the familiar table salt.

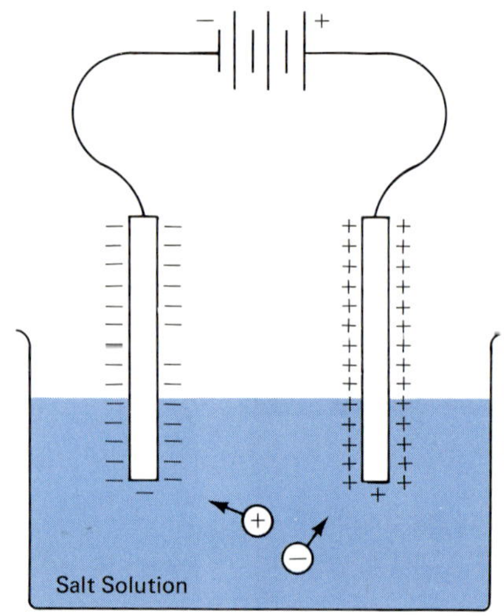

Figure 10.2 Electrolysis of a salt.

10.4 Electrolysis

Salts have a special property. When dissolved in water, they can be decomposed by passing an electric current through the solution. This process is called *electrolysis.* When two electrodes are dipped in the solution and a battery is connected to them as shown in Fig. 10.2, the salt becomes decomposed, with the metal depositing on the negative electrode. This is a way elements may be recovered from a compound. The smelting of aluminum is based on electrolysis. Another industrial application is electroplating. Silver-plated wares are made by this process, in which a thin layer of silver is deposited on the surface of a metal ware acting as an electrode in the electrolysis of a silver salt.

Its industrial significance notwithstanding, our interest in electrolysis lies in the fact that it brings out a connection between chemistry and electricity. Previously chemistry and electricity were considered two independent phenomena, not related to each other. The phenomenon of electrolysis tells us, however, that they are related and that there must be a basic connection between them. Because like charges repel and unlike charges attract each other, the very fact that the metal atoms are deposited at the negative electrode suggests that the metal atoms before being deposited at the negative electrode must in some way carry a positive electric charge, so that they are attracted to the negative electrode. On the electrode the positive charge of the atom is neutralized by the negative charge of the electrode and the atom becomes an uncharged normal atom. Likewise, the other group, before being released at the positive electrode, must carry a negative charge to be attracted and neutralized by the positive electrode. When the two elements appear separately at the two electrodes, the compound is decomposed.

How do we get the electrically charged atoms from the compound? Arrhenius in 1887 hypothesized that when a salt such as NaCl is dissolved in water, the molecule breaks up into two parts: one is a positively charged sodium atom called the sodium *ion* (Na^+) and the other is a negatively charged chlorine atom called the chloride ion (Cl^-). The process is called *dissociation*. Once this happens the two electrodes in the solution sort out the two kinds of ions and neutralize them. They are returned to the elemental form; the compound is decomposed.

It should be noted that the breaking up of the molecule into ions is not chemical decomposition, because a chloride ion is not a chlorine atom or a chlorine molecule. Because it carries a negative charge, it may have different properties. Indeed elemental chlorine is poisonous, whereas the chloride ion is not; on the contrary, it is a vital ingredient in human body fluids and serves many physiological functions.

By this theory chemical decomposition by electrolysis can be explained. Yet the question is passed on to that of the breaking up of the molecule into ions in solution. This question has to be answered based on a knowledge of the mechanism of chemical combination. In chemistry we used to speak of *chemical affinity* of atoms. (Each atom has a specific disposition to combine or not to combine with other atoms because of some kind of mysterious force.) However, based on the facts of electrolysis, we may hypothesize a mechanism of chemical combination as follows that reduces the mysterious chemical force to the well-known electric force in physics.

Because all neutral substances may be electrified, we may regard a neutral body to contain in itself positive and negative electricity in equal amounts. It is only natural to suppose that each atom, being the physical unit of matter, contains in it equal amounts of positive and negative electric charges. The chemical combination of a sodium atom and a chlorine atom may be visualized as a process in which the sodium atom transfers a part of its negative charge (actually an electron) to the chlorine atom. The sodium atom thus becomes positively charged and the chlorine atom becomes negatively charged. Because unlike charges attract each other, the two charged atoms join to form one composite particle, which becomes the NaCl molecule. Note that the molecule is again neutral electrically. In this view the force responsible for the combination of sodium and chlorine is nothing but the familiar electric force in physics and a molecule is in reality but two charged particles sticking together.

In water the electric force between two electric charges is reduced by a factor equal to the dielectric constant of water, the value of which, 81, is quite large. The electric force between the two charged particles is thus greatly reduced, so much so that they sometimes do not stick together any more. When they do separate they become two ions, Na^+ and Cl^-. The electrodes can then sort them out and the compound can be decomposed.

Thus the phenomenon of electrolysis tells us a lot about the mechanism of chemical combination. Indeed in later developments in atomic physics, a more complete theory of chemical combination is developed that contains essentially the same elements as those just described. The basic force responsible for chemical combination is thus reduced to the electric force; this is a point we have already mentioned in our discussion of the four basic forces of nature.

10.5 The Periodic Table of Elements

When Dalton introduced the atomic theory, atoms of all elements were regarded as independent and unrelated. Yet elements do fall into groups; atoms in the same group do have similar properties. Perhaps there is more to be learned from the similarities and differences among the atoms, from which we can understand better the relations among the atoms. The most significant step in this direction was taken by Mendeleev in the establishment of the *periodic table* of the elements, which brings out a remarkable "grand design" among the atoms.

Mendeleev first arranged all elements in the increasing order of the atomic weight, starting with the lightest atom, hydrogen, the number of the element in this order being called the *atomic number* Z (modified later), thus $Z = 1$ for hydrogen, $Z = 2$ for helium, $Z = 3$ for lithium, and so on. He found that as Z increases one unit at a time, the properties of the atom change drastically, corresponding to the change from one element group to another. But after a certain number of increases in Z, the properties of the atom return and repeat the same cycle of change again. In other words, the property of an atom is a periodic function of the atomic number Z. For example, lithium ($Z = 3$) is an alkali metal of valence 1; the next element, beryllium ($Z = 4$), is an alkaline earth of valence 2; the following element, boron ($Z = 5$), belongs to yet another group of valence 3; carbon ($Z = 6$) has valence 4; nitrogen ($Z = 7$) has valence 5 or 3; oxygen ($Z = 8$) has valence 2; fluorine ($Z = 9$) is a halogen having valence 1; neon ($Z = 10$) is a noble gas that does not combine with any other elements and thus has valence 0; the next element, sodium ($Z = 11$), returns to the alkali metal; the following element, magnesium ($Z = 12$), returns to the alkaline earth; aluminum ($Z = 13$), silicon ($Z = 14$), phosphorus ($Z = 15$), sulfur ($Z = 16$), chlorine ($Z = 17$), and argon ($Z = 18$) have valences 3, 4, 5/3, 2, 1, 0, in the same pattern as before (Z from 3 to 10), and the last two again are a halogen and a noble gas. From lithium ($Z = 3$) to neon ($Z = 10$) the eight elements, representing eight different element groups, are said to form a period; this one in particular is called the first *short period*. The next eight elements, from sodium ($Z = 11$) to argon ($Z = 18$), repeat the same pattern (from alkali metal to alkaline earth . . . , to halogen, and to noble gas) and thus form another period, the second short period. Because the second short period is similar to the first, its eight elements can be listed in a row below the row of the eight elements of the first period, such that elements of the same *group* fall in the same *column*. Further rows of elements from $Z = 19$ on may be added below to form a table such that elements of the same group always fall into the same column. Thus in the first column are the alkali metals, in the second are the alkaline earths. . . . , in the seventh are the halogens, and in the eighth are the noble gases. Such a table is called the periodic table of the elements (Fig. 10.3). This regularity among the properties of the atoms is truly remarkable. It appears as if all the elements are created according to a grand design laid out in the periodic table. Each element is assigned a definite place to fill.

The rule is so strict that whenever an element is missing from the list of elements, it can easily be checked out from the periodic table. Indeed, in working out the periodic table Mendeleev found it necessary to leave a few vacant spaces on the table and thus also in the atomic number sequence to represent the yet undiscovered new elements, in order to fit all elements into the above pattern. These missing elements predicted by the periodic table were actually found later and their properties are found to be actually what was expected of them from the periodic table. One such element is technetium ($Z = 43$), which we have mentioned earlier. This element does not exist on earth (all isotopes of it are radioactive). Yet it can be and has been made artificially in nuclear physics laboratories. As mentioned previously, it has been observed to exist in some stars, direct evidence that nuclear reactions take place in the stars.

Now all the available spaces on the periodic table are filled and there can be no more missing elements in between them. In other words, all elements in between them have been discovered and any new elements to be discovered must come from beyond the end of the table, i.e., beyond uranium ($Z = 92$). The search for the *transuranian elements* thus began. In recent years, indeed, more than a dozen new elements beyond uranium ($Z = 93$ to 105) have been created in nuclear physics laboratories. Because no new elements are expected below uranium, the assignment of the atomic number Z will not be changed any further. Thus Z represents a basic property. Later we will see that it is indeed related to a basic property of the atom.

The preceding discussion of the periodic table based on the sequential order of atoms according to the atomic weight needs modification. In addition to the missing elements, which Mendeleev had to skip in the table, there was the complication of three exceptions in which the order of two elements ac-

	Group I	Group II	Group III	Group IV	Group V	Group VI	Group VII	Group VIII
Period 1	1 H							2 He
Period 2	3 Li	4 Be	5 B	6 C	7 N	8 O	9 F	10 Ne
Period 3	11 Na	12 Mg	13 Al	14 Si	15 P	16 S	17 Cl	18 Ar
Period 4	19 K 29 Cu	20 Ca 30 Zn	21 Sc 31 Ga	22 Ti 32 Ge	23 V 33 As	24 Cr 34 Se	25 Mn 35 Br	26 Fe 27 Co 28 Ni 36 Kr
Period 5	37 Rb 47 Ag	38 Sr 48 Cd	39 Y 49 In	40 Zr 50 Sn	41 Nb 51 Sb	42 Mo 52 Te	43 Tc 53 I	44 Ru 45 Rh 46 Pd 54 Xe
Period 6	55 Cs 79 Au	56 Ba 80 Hg	57 La * 81 Tl	72 Hf 82 Pb	73 Ta 83 Bi	74 W 84 Po	75 Re 85 At	76 Os 77 Ir 78 Pt 86 Rn
Period 7	87 Fr	88 Ra	89 Ac **	104 Ku	105 Ha			

*Lanthanons	58 Ce	59 Pr	60 Nd	61 Pm	62 Sm	63 Eu	64 Gd	65 Tb	66 Dy	67 Ho	68 Er	69 Tm	70 Yb	71 Lu

**Actinons	90 Th	91 Pa	92 U	93 Np	94 Pu	95 Am	96 Cm	97 Bk	98 Cf	99 Es	100 Fm	101 Md	102 No	103 Lw

Figure 10.3 The periodic table of elements.

cording to increasing atomic weight has to be reversed in order to fit the elements into their rightful positions in the table. Yet the regularity shown by the table is so remarkable that the three exceptions were considered to be of secondary importance. Later in atomic physics the exceptions were explained; the atomic weight is not the fundamental quantity to correlate the properties of atoms and so exceptions may occur. It is an approximation of a more fundamental quantity, which determines the atomic number. Historically, the periodic table was first established by using the readily accessible, though less fundamental, atomic weight. (The fundamental quantity is the number of electrons in the atom, which may be established by X-ray spectroscopy; when the periodic table is arranged by the sequential order established according to this number, there are no exceptions.) Thus the definition of the atomic number Z should be modified. We may say that the atomic number is the position number of atoms in the periodic table (or the sequential number established by X-ray spectra).

There are additional complications. Although the sequence of atoms forms periods, there are minor deviations from the periods. The second short period is an exact replica of the first short period. But after the second short period there follows from $Z = 19$ to $Z = 36$ a period of 18 elements correlated to the previous periods of eight elements in a more complicated way, as shown in Fig. 10.4. This first *long*

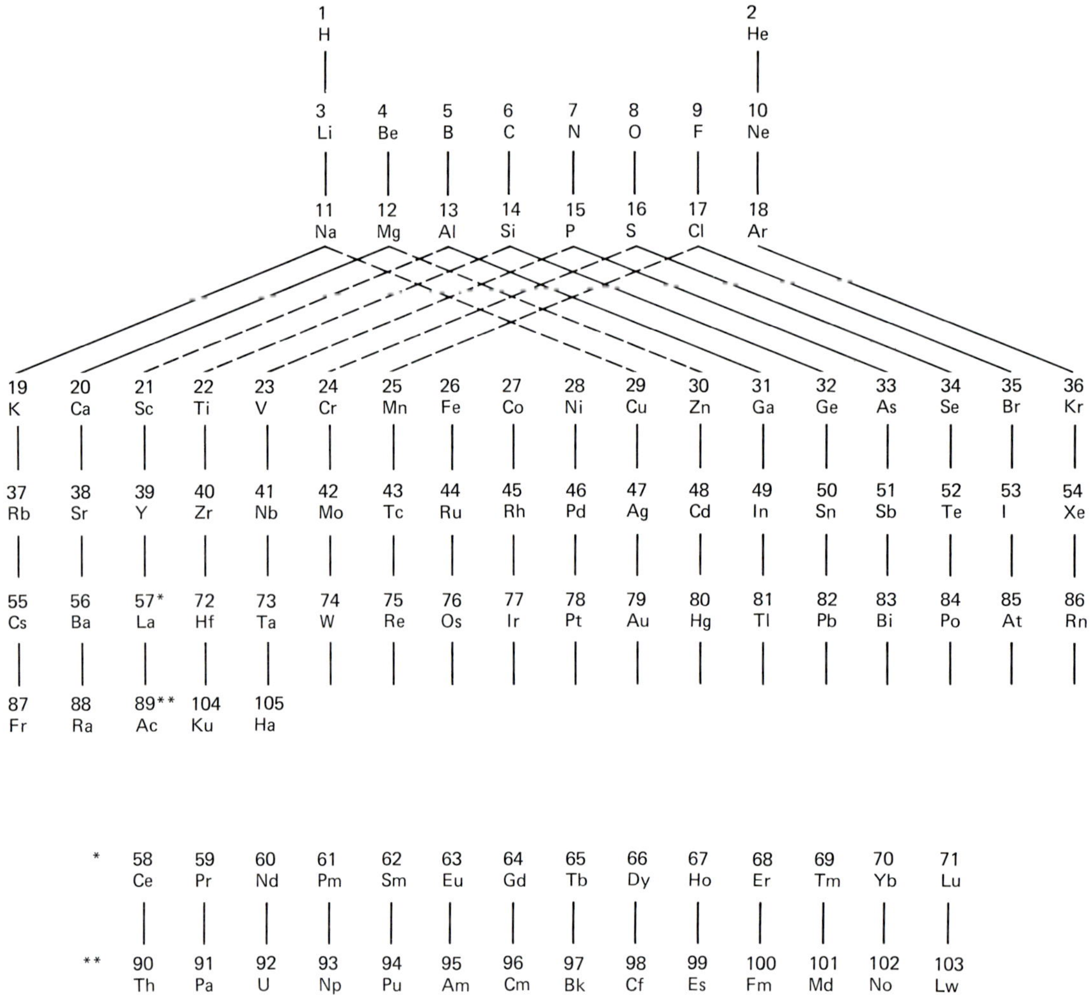

Figure 10.4 Another form of the periodic table of elements.

period is followed by the second, third, and fourth long periods of 18 elements each, except that the last long period is incomplete at the end of the table ($Z = 92$, uranium). Moreover, in the third long period it was found necessary to delete from the sequence of elements a section of 14 elements from $Z = 58$ to $Z = 71$ in order to fit the sequence into the table. These 14 elements are the *rare earth* elements with chemical properties very similar to one another. They do not follow the pattern of periodic changes and thus do not fit into the table. They are thus taken out of the sequence and out of the table and put at the bottom of the table as a separate appendage (the lanthanons). Similarly, there is a second appendage of 14 elements (the actinons) taken out of the fourth long period, which includes the new transuranian elements. Finally, before the first short period we still have two elements left, hydrogen and helium; they may be regarded as forming a supershort period in themselves, different from all the other periods, this being the very *first* period of the table. Thus even though the periodic recurrence is quite definite and striking, there are many deviations and ramifications.

Why are the elements related in such a fashion? It is obvious that atoms of elements are not independent and unrelated as Dalton thought. The close relation betweem them is likely to originate from a plan by which all atoms are built from some other, more fundamental entities. The problem of atomic structure thus presents itself. Indeed in atomic physics, discussed later (Chapter 12), we do have a theory of atomic structure by which not only the regularity of the periodic table can be explained, but also the complications and the various ramifications.

⋆10.6 Organic Compounds

Among the 92 chemical elements, carbon is very special. It forms a chemical bond with itself, and this can be repeated many times, resulting in many carbon atoms connected to form a long chain molecule. Not only is the number of compounds formed by the element increased greatly, but also the variety and complexity of the compounds are increased tremendously. In fact, the molecules that all living organisms are made of are nearly all carbon compounds; many of them are highly complex. It is not too much to say that life is based on the chemistry of carbon. Thus there is a special interest in carbon compounds, and a branch of chemistry called *organic chemistry* has been developed to deal with them. Excluding a few inorganic compounds of carbon (such as Na_2CO_3, sodium carbonate or washing soda), the organic carbon compounds occur in living materials and were once thought to occur only in living organisms. Wöhler in 1820 first converted an inorganic compound, ammonium cyanate, into the organic compound urea, and the separation of the inorganic and organic broke down. Now many organic compounds can be made in factories. Furthermore, many experiments have shown that physical processes in the primitive atmosphere of the earth (such as lightning) can produce organic compounds that are the building blocks of life and thus provide the physical basis for origination of life from inorganic material.

The carbon atom has a valence of 4. It has four chemical bonds. Each bond can be used to combine with one hydrogen atom or another atom of valence 1; two bonds may be used to combine with another atom of valence 2; and so on. These possibilities are illustrated in Fig. 10.5.

One of the simplest organic compounds is methane, CH_4. Each bond of the carbon atom is used to combine with a hydrogen atom. Each of the hydrogen atoms can be replaced by an atom of valence 1, such as chlorine, thus generating compounds such as CH_3Cl, CH_2Cl_2, $CHCl_3$, and CCl_4. These are named methyl chloride, dichloromethane, trichloromethane, and carbon tetrachloride. These names provide an idea of the naming system in organic chemistry. The system is such that from the name, one should be in a position to write down the structural formula.

Figure 10.5 Some simple organic compounds of carbon.

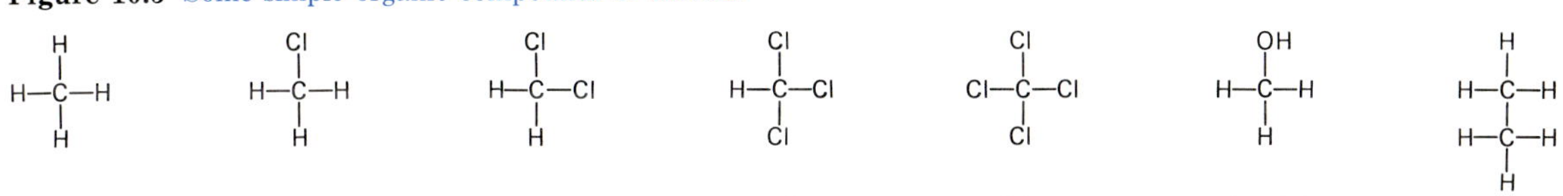

CH_4 Methane; CH_3Cl Methyl Chloride; CH_2Cl_2 Dichloromethane; $CHCl_3$ Trichloromethane (Chloroform); CCl_4 Carbon Tetrachloride; CH_3OH Methyl Alcohol; C_2H_6 Ethane

Methane is a combustible gas. Trichloromethane is used in many ways in medicine by the name *chloroform*. Previously it was used as a general anesthetic, but is now replaced by other, safer substitutes. Carbon tetrachloride is a very good solvent and therefore is used commercially for dry-cleaning purposes.

Oxygen is an atom of valence 2. It may combine with two bonds of the carbon atom. Or one bond of it may combine with one bond of carbon with all the other unused bonds free to combine with other atoms, such as hydrogen. An example is CH_3OH, methyl alcohol. Finally, in the last example in Fig. 10.5, two carbon atoms are linked by one bond and the three other bonds of each carbon atom are used to combine with hydrogen. The resulting compound is called ethane, another combustible gas.

The preceding examples of organic compounds demonstrate that a large variety of compounds may be formed of carbon and hydrogen (called *hydrocarbons*) and many other organic compounds may be looked upon as derived from hydrocarbons by substitution. Thus we may begin by studying hydrocarbons.

Saturated Hydrocarbons. Because all organic compounds are characterized by a chain of carbon atoms, hydrocarbons may be classified into two major groups: those with an open-chain structure, called aliphatic (fatty) hydrocarbons, and those with a closed-chain structure, called aromatic (fragrant) hydrocarbons.

In Fig. 10.5 we have two open-chain hydrocarbons, methane and ethane. They are actually the first two members of a series of hydrocarbons called *alkanes* (*saturated hydrocarbons*), shown in Fig. 10.6. The figure shows obviously the rule by which the compounds in the series are related. The series is generated by lengthening the open chain of carbon atoms. Each additional carbon atom brings with it two additional hydrogen atoms. Thus for a chain of n carbon atoms there are altogether $2n + 2$ hydrogen atoms. The general formula for the molecules in the series is thus C_nH_{2n+2} with $n = 1, 2, 3 \ldots$.

As n increases from 1 to 2 to 3, and so on, the compound is named, respectively, methane, ethane, propane, butane, pentane, hexane, heptane, octane, and so on ($n = 30$, triacontane; $n = 32$, dotriacontane; $n = 60$, hexacontane; and so on). The regularity of structure also reflects itself in the regularity of properties. The boiling point and melting point increase with n. Thus the earlier members are gases. Natural gas, largely methane, is a mixture of them, covering n from 1 to 4. The next dozen or so members are liquids. Gasoline is a mixture of them, covering the range of n from 6 to 12. Kerosene covers n from 12 to 18. Above octadecane ($n = 18$) the compounds become solid (e.g., wax).

These compounds, naturally present in petroleum and natural gas, are the principal ingredients of the liquid and gaseous fuels used today (gasoline, kerosene, natural gas, liquefied gas). The terms *octane* and *propane* are already familiar from these uses. These compounds figure importantly in our discussion on energy resources and environmental pollution.

Isomers. When the number of atoms in a molecule becomes large another complication appears; the same number of atoms may be arranged in different ways of combination, thus giving rise to different structural formulas for the same molecular formula. For example, in pentane, C_5H_{12} there are five carbon

Figure 10.6 The alkanes.

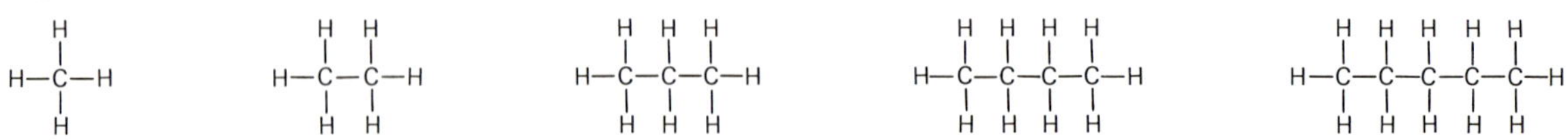

CH_4 Methane | C_2H_6 Ethane | C_3H_8 Propane | C_4H_{10} Butane | C_5H_{12} Pentane

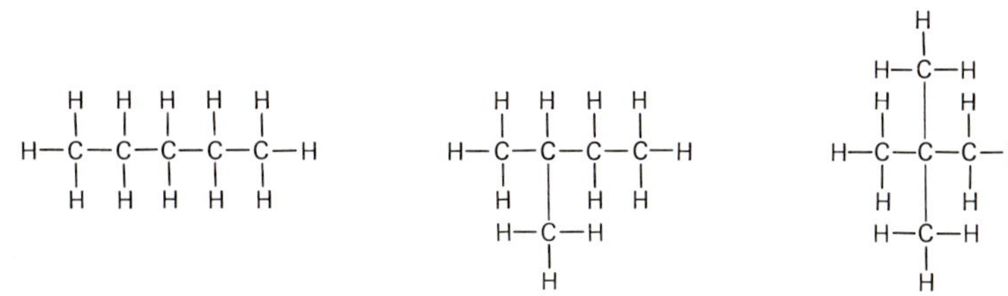

Figure 10.7 The isomers of pentane.

n-pentane — Isopentane — Neopentane

and 12 hydrogen atoms. They can be arranged in three forms according to the bonding relations between carbon and hydrogen as shown in Fig. 10.7. (These three are the only possible forms; other combinations can be shown to be in reality the same as these three. See Problem 10.14.) In spite of the fact that the molecular formulas are the same C_5H_{12}, their structures are different and as a result the physical and chemical properties (boiling point, heat of combustion, and so on) are slightly different. They are in reality different compounds; the fact that they have the same formula, C_5H_{12}, is coincidental. (This also illustrates the point that the structural formula is the more important and the molecular formula has limited significance for complex molecules.) Such compounds are called *isomers*. Thus pentane has three isomers named, respectively, n-pentane (normal), isopentane, and neopentane. For more complex molecules, more complex isomers are possible. A more systematic naming convention is introduced in which a branch chain is considered a substitute of hydrogen in a straight chain. The branch chain may be considered an alkane less a hydrogen atom and is named accordingly (see Fig. 10.8). Using the names of the branch chains a complex isomer may be named systematically as illustrated in Fig. 10.9. (The numbers refer to the position of carbon atoms in the straight chain.)

Any of the hydrogen atoms of a hydrocarbon may be substituted by a monovalent atom, such as chlorine. This way a large number of compounds may be generated from the hydrocarbons. These compounds can be named by the same system of nomenclature as that illustrated in Fig. 10.10. A familiarity with the system of nomenclature will enable one to read the drug labels more intelligently. Most drugs are organic compounds, the chemical names of which are determined by the structure according to the naming system.

Figure 10.8 The alkyl groups.

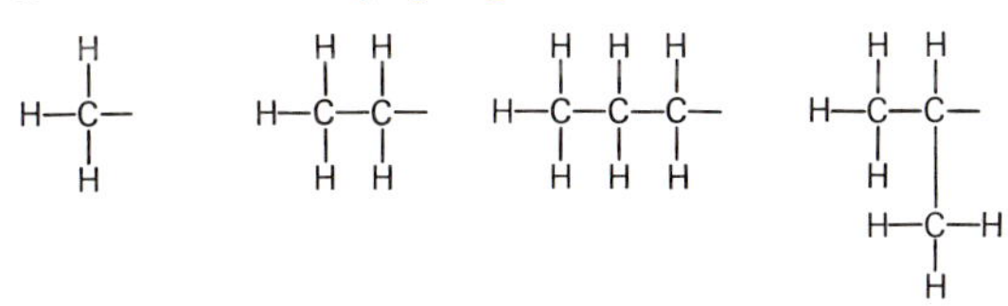

Methyl Group — Ethyl Group — Propyl Group — Isopropyl Group

Figure 10.9 Naming of the isomers.

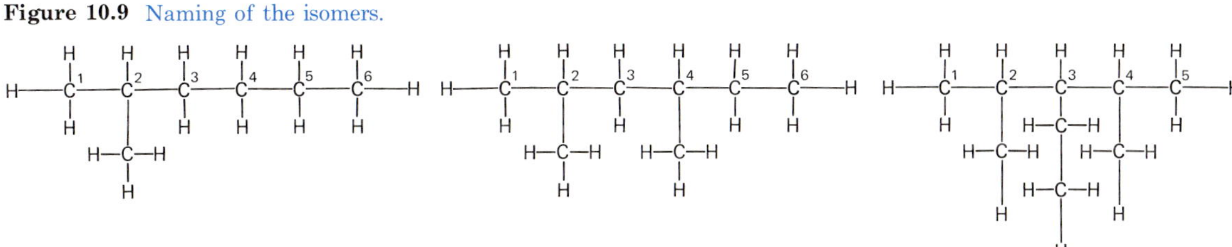

2-methylhexane — 2,4-dimethylhexane — 3-ethyl-2,4-dimethylpentane

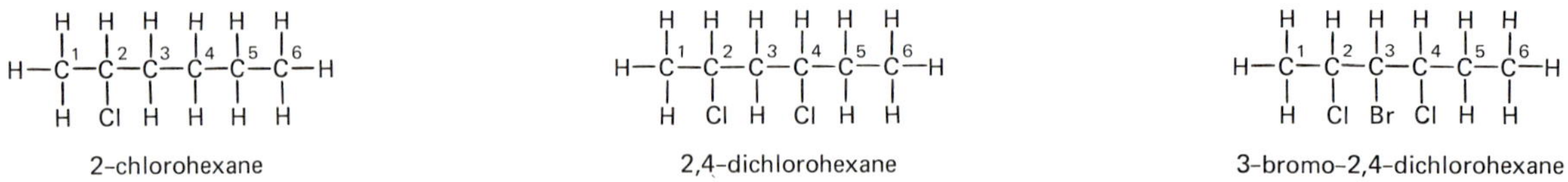

Figure 10.10 Naming of the substitution compounds.

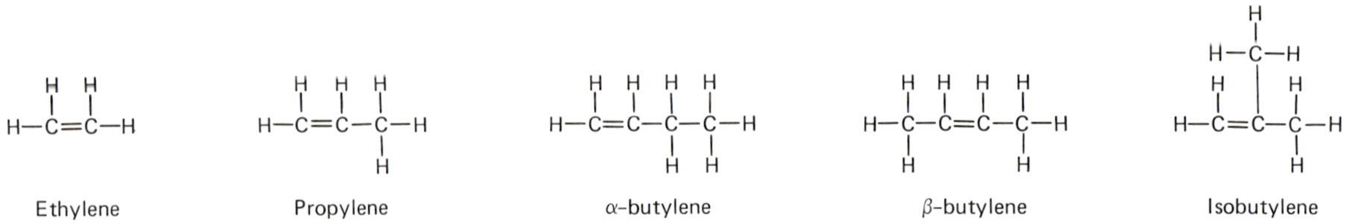

Figure 10.11 The alkenes.

Unsaturated Hydrocarbons. In all the preceding compounds carbon atoms are linked together by a single bond. It is also possible for two carbon atoms in an organic molecule to be linked by two bonds (a *double bond*). Thus another series of compounds may be generated with the general molecular formula C_nH_{2n} (two hydrogen atoms less than the corresponding alkanes). These compounds are called *alkenes* (see Fig. 10.11). Notice that the position of the double bond may be changed to give rise to isomers. These are distinguished by the designations α, β, and so on.

The properties of alkenes are generally similar to the corresponding alkanes (about the same boiling point, but lower melting point). Alkenes react with hydrogen and change into alkanes by changing the double bond into single bonds with the addition of two hydrogen atoms. Energy is released in the reaction.

In addition to the double bond there is also the possibility of a triple bond. Another series of hydrocarbons, called *alkynes,* results, having the general formula C_nH_{2n-2} (Fig. 10.12). The properties of alkynes are similar to alkanes and alkenes of the corresponding member of the same n. The boiling point is a little higher. Alkynes react with hydrogen easily to form alkenes—the triple bond is changed to a double bond by the addition of two hydrogen atoms. Because of the possibility of addition of hydrogen, alkynes and alkenes are called *unsaturated hydrocarbons.*

Figure 10.12 The alkynes.

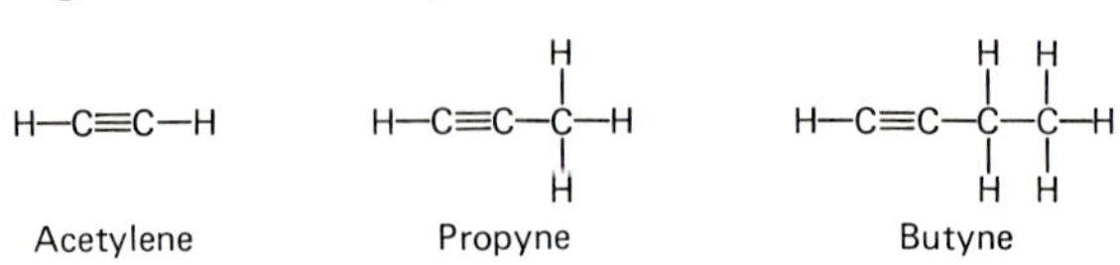

Ring Compounds. We now turn our attention to the second group, the closed-chain hydrocarbons. The simplest example is benzene, C_6H_6. For a long time it was not known how to use up the four bonds of each carbon atom in a molecule having only six hydrogen atoms with six carbon atoms. Finally the problem was solved by letting the carbon atoms form a closed chain (see Fig. 10.13). Each carbon atom forms a double bond with one neighbor and a single bond with another. The two kinds of bonds alternate with each other in the chain. This way all the bonds of carbon atoms in C_6H_6 are employed.

In spite of the existence of double bonds, the ring structure is very stable (unlike the alkenes and alkynes). This was difficult to understand but finally the problem was solved by quantum mechanics, another triumph of the theory. The benzene ring forms a unit that may participate in chemical reactions as a group, called the *phenyl group*. It may substitute a hydrogen in open-chain hydrocarbons, may combine

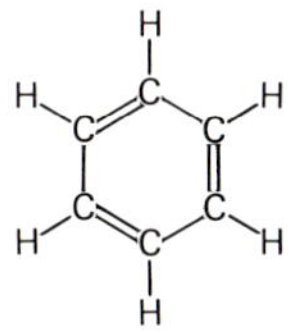

Figure 10.13 Benzene.

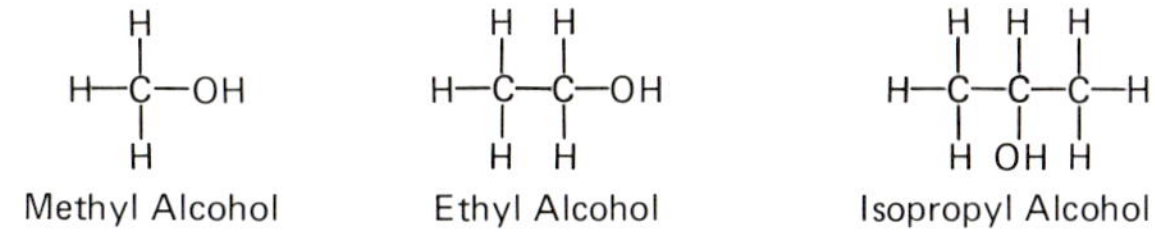

Figure 10.15 The alcohols.

with itself, or may have one of its hydrogen atoms substituted by other atoms or groups, thus forming a large variety of ring compounds. Figure 10.14 illustrates some of them. For convenience the letters *C* and *H* in the ring are omitted and the alkyl groups are written as a molecular group.

Alcohols, Aldehydes, Ketones, Carboxylic Acids. We now consider compounds formed by substitution of atoms or groups for hydrogen in hydrocarbons. The substitution by halogens, such as chlorine, has been mentioned before. Among the molecular groups that can substitute hydrogen, an important one is the hydroxyl group, OH. The compound it forms with an alkyl group is called an *alcohol* (Fig. 10.15). The alcohol in all alcoholic drinks is the second member of the series, *ethyl alcohol.* The first member is wood alcohol, which is poisonous. Isopropyl alcohol is sold in drugstores as rubbing alcohol.

The hydroxyl group can be attached to benzene by replacing a hydrogen atom and the product is called *phenol.* Phenols are similar to alcohols in some properties but different in others. Many important chemicals are derived from phenols.

Another important group is the carbonyl group, >C=O. Two classes of compounds follow: the *aldehydes,* R(H)C=O, and the *ketones,* R(R)C=O, where *R* is a radical such as the alkyl group. Some members of aldehydes are shown in Fig. 10.16. Aldehydes are readily oxidized to *carboxylic acids,* characterized by the *carboxyl group,* COOH. The first few members of carboxylic acids are shown in Fig. 10.17. Formic acid was first found in ants. Acetic acid is the main ingredient of vinegar and was produced from alcohol thousands of years before anything was known of chemistry. These acids were known earlier and formaldehyde was named after them.

Figure 10.16 The aldehydes.

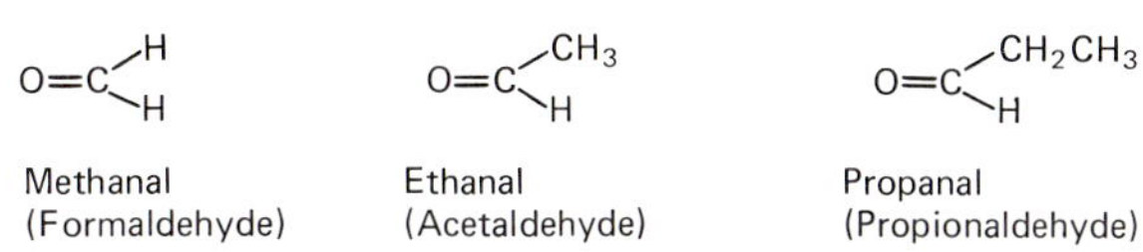

Figure 10.14 Ring compounds.

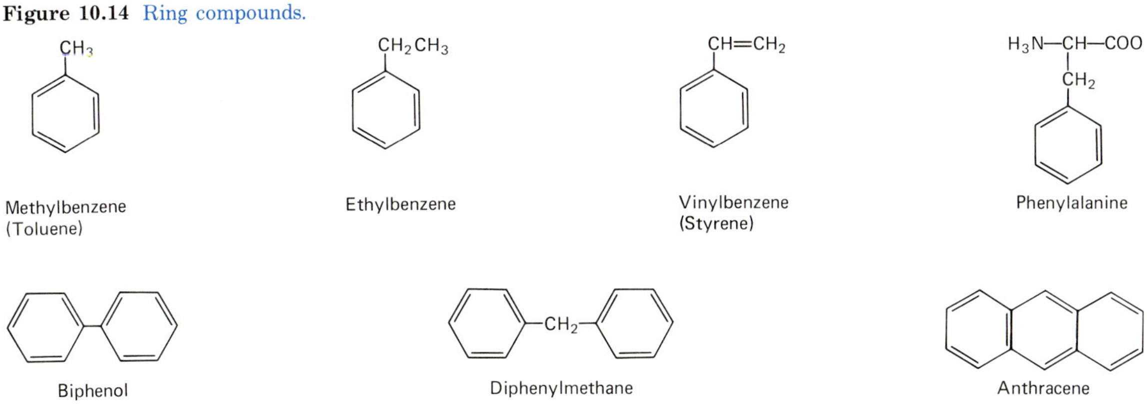

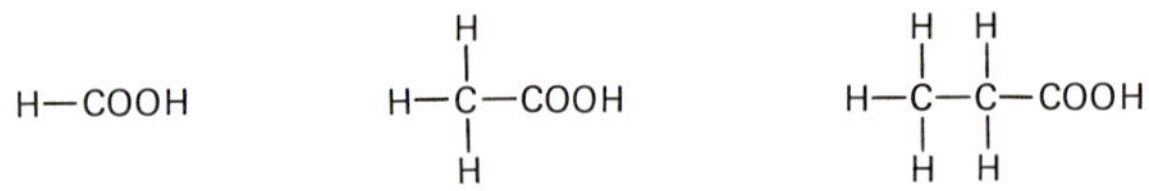

Formic Acid Acetic Acid Propionic Acid

Figure 10.17 The carboxylic acids.

The simplest ketone is CH_3COCH_3, which is commonly known as acetone, a familiar industrial chemical. The other ketones are named after the two radicals. Thus $CH_3COCH_2CH_3$ is called ethyl methyl ketone.

This survey provides a general background on organic chemistry. In the next section we will briefly describe the complex organic compounds that are the building blocks of living organisms and the active agents of life processes.

*10.7 Biochemistry

Living organisms are made of material substances—the same material substances of the physical world. The physical body of organisms can be reduced to the 92 chemical elements and their compounds. It would be interesting to have discovered some substances beyond the 92 elements in living things (or in the human body) and then to claim "this is the source of life." Yet the fact of science is that nothing of this kind has ever been discovered. This does not, though, prove that life is nothing but a machine and that humans are mere robots. Yet the material stuff in living bodies is indeed an assortment of the 92 chemical elements combined in a large variety of ways.

Major Groups of Biological Compounds. Complexity notwithstanding, the compounds living organisms are made of fall in a few general classes. The most important is the *proteins*. Proteins are the brick and mortar, nuts and bolts of the body of an organism. The proteins so used are called *structural proteins*. There are thousands of different proteins in a complex organism and they differ from one species to another. The large variety of living species is due to the variety of proteins. Yet brick and mortar alone do not make for a functioning building. A functioning building is equipped with electricity, plumbing, heating, lighting, and air-conditioning systems. In a living organism many of the physiological functions of the body are mediated by *enzymes,* which are also proteins. They do not form a part of the structure of the body. They may be called *functional proteins*.

But bricks, mortar, pipes, and wires alone are merely a heap of materials. They have to be put together according to a master plan to form a building. Likewise, there must be a master plan of life. Here we come to the second important class of compounds: *nucleic acids*. They are the physical repository of genetic information. Their molecular structure is less varied than that of proteins. In fact, as far as physical conformation is concerned, there are only a few forms. The complexity is in the permutation of the molecular units. (The large amount of information of a book is encoded in the permutation of only 26 letters.)

Indeed, the simplest organism, a virus, is made up of only one nucleic acid molecule and a few protein molecules that form a coat surrounding the nucleic acid molecule. When placed in a proper environment, given a nutrient medium, the virus particle reproduces itself into many identical virus particles. Reproduction is an outstanding feature of life and may well be *the* feature of life. The virus particle is considered a living organism, not a system of chemical compounds, because of its ability to reproduce. Yet this organism does not think, move, see, yell, eat, excrete, and so on. And it does not even get old and die. Virus particles may form a crystal that, like a piece of rock, lasts unchanged forever. As long as the blueprint is there to direct the assembly of raw material into an organized form, there is life. Not even energy is a must; energy is available in the environment. The virus is a complete parasite, depending on the environment for all necessities.

More sophisticated living organisms, even the bacteria, have a built-in power plant for their own energy needs. (From the molecular biology point of view the complexity of life is nearly complete in the level of bacteria; from bacteria to man the advance is a relatively small step.) Here enters the third class of compounds, *carbohydrates* (sugar, starch, and so on). They are simple organic compounds of carbon, hydrogen, and oxygen. Their main purpose is to provide energy for all needs of the organism, though they are also used in the structure (cellulose). The point of view we take for the physical world—considering all natural phenomena in terms of matter, energy, and their transformation—is thus also applicable to living organisms.

The fourth class of compounds is *lipids* (fats and oils). They are formed by derivatives of hydrocarbons. In addition to being used as an energy storage, they are also used in the structure—in making the membranes. The biological membranes perform many functions of life not directly controlled by genetics. The transmission of the nerve signal (electric impulses) is dependent on the permeability of the (axon) membrane to Na^+ and K^+ ions, which change in a coordinated fashion. Without the membranes life would be much duller, no matter how complex the nucleic acids.

The last class is inorganic material, mostly salts. We have already mentioned the K^+ and Na^+ ions in nerve impulse transmission. They come from the salts we ingest. Last but not least the largest fraction of the body weight is taken up by the most familiar inorganic compound—water.

Animals do not breed their own fuel molecules. (Plants do, however.) Man does not synthesize nine of the 20 amino acids necessary to make proteins. He depends on other animals and plants to supply most of the preceding compounds. Thus the preceding list also provides a diet guide. However, even with a "balanced diet" of these ingredients, the recipient still suffers from malnutrition. Something still is missing, substances that are indispensable, though only a small amount is needed. The missing ingredient is the *vitamins*. They play an important role in the control of living processes. This brings us back to the basic nature of life.

We may consider matter and energy as two basic elements of life. But we have already introduced something beyond them—the blueprint, the organization, the control systems, and so on. These may be summarized in the concept of *information,* which may be considered as the third element of life. The blueprint is encoded in nucleic acids. Like a book, it provides detailed instructions. It directs activities for the housekeeping of the organism. Although a book is composed of letters and words, these have to be organized by a system of punctuation. Two books may have different contents but the same punctuation. In fact, from English to German to French, the words are different but the punctuations are the same. Similarly, in the book of life there are punctuations performing routine controls that are the same for all living species. Some of the control mechanisms of life are mediated by a set of simple organic molecules. Unlike the enzymes, which differ from one species to another, these control molecules are the same in many species. They are the vitamins. In fact, the vitamin pills we ingest are often made of extracts from other animals and plants.

Another group of control molecules secreted internally by glands is the *hormones*. Some hormones are small protein molecules and are interchangeable among the species (like insulin).

Some animals can make some vitamins themselves. The vitamins and other compounds men cannot make themselves have to be supplied by diet. As already mentioned, man cannot make nine of the 20 amino acids. In comparison, the simple organism neurospora (red bread mold) can synthesize all amino acids as well as vitamins. (The only chemical it needs is biotin.)

Chemistry of Vitamins, Lipids, and Carbohydrates. After this general introduction we may go into the chemistry of these compounds in a little more detail. Vitamins are formed of carbon chains and rings with substitutions. For example, a simple one, vitamin B_6, has a structure shown in Fig. 10.18. The structure of the hormone insulin, being a protein, will be discussed later. The structure of the hormone adrenalin, which prepares a person to fight, is given in Fig. 10.19. The structural formulas of many vitamins, hormones, and drugs, such as penicillin, are now known, from which many can be manufactured by chemical synthesis. However, how they work and why they work the way they do are still largely unknown.

Natural fats and oils are usually formed by the combination of a carboxylic acid [for animal fats, palmitic acid $CH_3(CH_2)_{14}COOH$ or stearic acid $CH_3(CH_2)_{16}COOH$] with a special alcohol glycerol [glycerine $C_3H_5(OH)_3$; it has three OH groups substituted in the hydrocarbon propane]. Reaction of this kind is analogous to acid-base reaction in inorganic chemistry—the H of the acid COOH and the OH of the alcohol (or phenol) combine to form water. The rest of the two molecules combine to form an *ester*.

The carbohydrates ingested by humans largely take the form of starch $(C_6H_{10}O_5)_x$, where x is a large number. Starch, under the action of an enzyme in the saliva, breaks down into a sugar called *maltose:*

$$2(C_6H_{10}O_5)_x + xH_2O \longrightarrow xC_{12}H_{22}O_{11}$$

Maltose, under the action of the enzyme *maltase,* changes into two molecules of *glucose* by hydrolysis (decomposition by water):

$$C_{12}H_{22}O_{11} + H_2O \longrightarrow 2C_6H_{12}O_6$$

Glucose is a sugar naturally present in many fruits. In the human body glucose is obtained by digestion of food; it is present in the bloodstream and can be

Figure 10.18 Vitamin B_6.

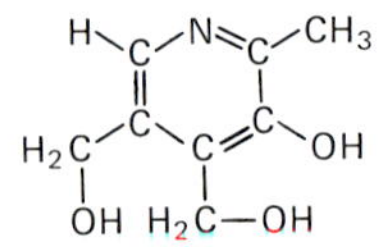

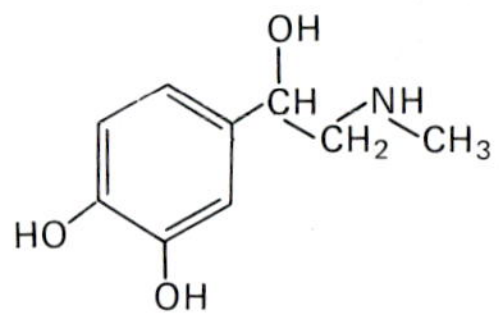

Figure 10.19 The hormone adrenalin.

transported to any part of the human body to be used as an energy source. When used as an energy source, the molecule is oxidized by combining with oxygen inhaled into the lung, carried into the bloodstream, and transported to the other parts of the body by the protein molecule called *hemoglobin* in the red blood cell. The oxidation products are carbon dioxide, CO_2, and water, H_2O; CO_2 is exhaled from the lung. In the reaction, which is like combustion, energy is released and may be used to carry out physiological functions, to perform mechanical work, or to maintain body temperature. This provides a general idea of how energy is used in living organisms.

The oxidation of sugar can be carried out by burning in the air. A cube of cane sugar, with some solid material such as a little cigarette ash rubbed on it, can be lighted by a match and burned completely into CO_2 and H_2O. The energy generated, according to the principle of conservation of energy, is the same as that obtained when oxidized in the human body. The burning in the air takes place at a high temperature. There is no such hot spot in the human body, yet the "burning" is carried out just the same. This is made possible because of the action of the enzymes—the *metabolism* of glucose consists of a set of complicated processes, taking place in a number of steps, each mediated by an enzyme, with a net result equivalent to burning in air. The structural formula of glucose is given in Fig. 10.20, which shows that the molecule has five hydroxyl groups and one aldehyde group.

Structure of Protein. Proteins are macromolecules made of smaller units called *amino acids.* Let us first consider the structure of the amino acids

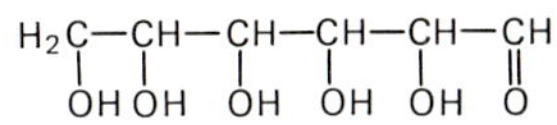

Figure 10.20 The sugar glucose.

themselves. Start with a methane molecule, CH_4, with a carbon atom (called the α carbon) in the center, surrounded by four hydrogen atoms. Keep one of the four and substitute for the other three, respectively, a carboxyl group, COOH; an *amino* group, NH_2; and another group, a radical designated by R. Then we have an amino acid, $CHR(NH_2)COOH$. In water solution the hydrogen ion H^+ of the carboxyl groups is transferred to the amino group and the structure changes to that shown in Fig. 10.21. The radical R can assume many forms; the 20 amino acids, of which proteins are made, correspond to 20 different R's. The simplest amino acid, *glycine,* is one in which R is simply a hydrogen atom. The more complex *tryptophan* has an R containing a double ring. The 20 amino acids are shown in Fig. 10.22.

The amino acids are connected together to form a protein by linking the amino group (NH_2) of one to the carboxyl group (COOH) of another. Two hydrogen atoms and one oxygen are removed to form a water molecule and the NH of one is bonded with the CO of the other. The chemical bond so formed is called a *peptide bond.* The peptide bond may be broken by hydrolysis under proper conditions and the protein is then broken up into amino acids. In the stomach, the enzyme pepsin does just that. The food protein is hydrolyzed into amino acids, which can pass through cell membranes and enter the bloodstream to be distributed to all other parts of the body and to be used as building blocks to synthesize proteins for structural (growth, repair) and functional (enzymatic activities) purposes.

Figure 10.21 Amino acids.

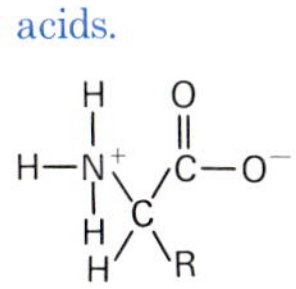

Proteins are made of 20 kinds of amino acids, as already mentioned. A protein molecule consists of a large number of amino acids connected into a long chain by peptide bonds. It is the number and sequence of the amino acids that determine the nature and property of the protein. For example, the molecule of the hormone protein insulin (the life-sustaining medicine for diabetics) is made of 777 atoms grouped into 51 amino acids connected in two chains (30 and 21 amino acids, respectively) bonded together by two *disulfide bonds* (bonds between two sulfur atoms). There is another disulfide bond within the short chain (see Fig. 10.23). The three-letter symbols in Fig. 10.23 are abbreviated names of the amino acids. It took Sanger 10 years to work out chemically the sequential structure of the molecule—the first protein structure so worked out (Nobel prize 1958). However, how the two chains coil to form a three-dimensional conformation cannot be ascertained by chemical methods. It has to be determined by a physical method, i.e., by X-ray diffraction (Hodgkin, Nobel laureate for her work on vitamin B_{12}). The function of insulin, presumably determined by the three-dimensional conformation, remains unexplained. Nevertheless, the known sequential structure makes it possible to synthesize this molecule in laboratory. The race to achieve this goal by many scientists in several nations (United States, China, Germany) is a significant event in recent scientific history (a future Nobel prize?).

There are 20,000 to 30,000 enzymes in the human body, each performing a precise physiological or biochemical function. Yet, being proteins, they are all made of the same 20 basic units. The variety of functions thus can only be related to the number and sequence of the amino acids in the protein, which can vary greatly. The number and sequence of the amino acids in the protein determine the shape, the folding pattern, or the three-dimensional conformation of the molecule, which can vary greatly. Presumably it is the shape that determines its specific function. Thus much interest is placed on determining the structure of the molecule.

The basic information in this regard is the amino

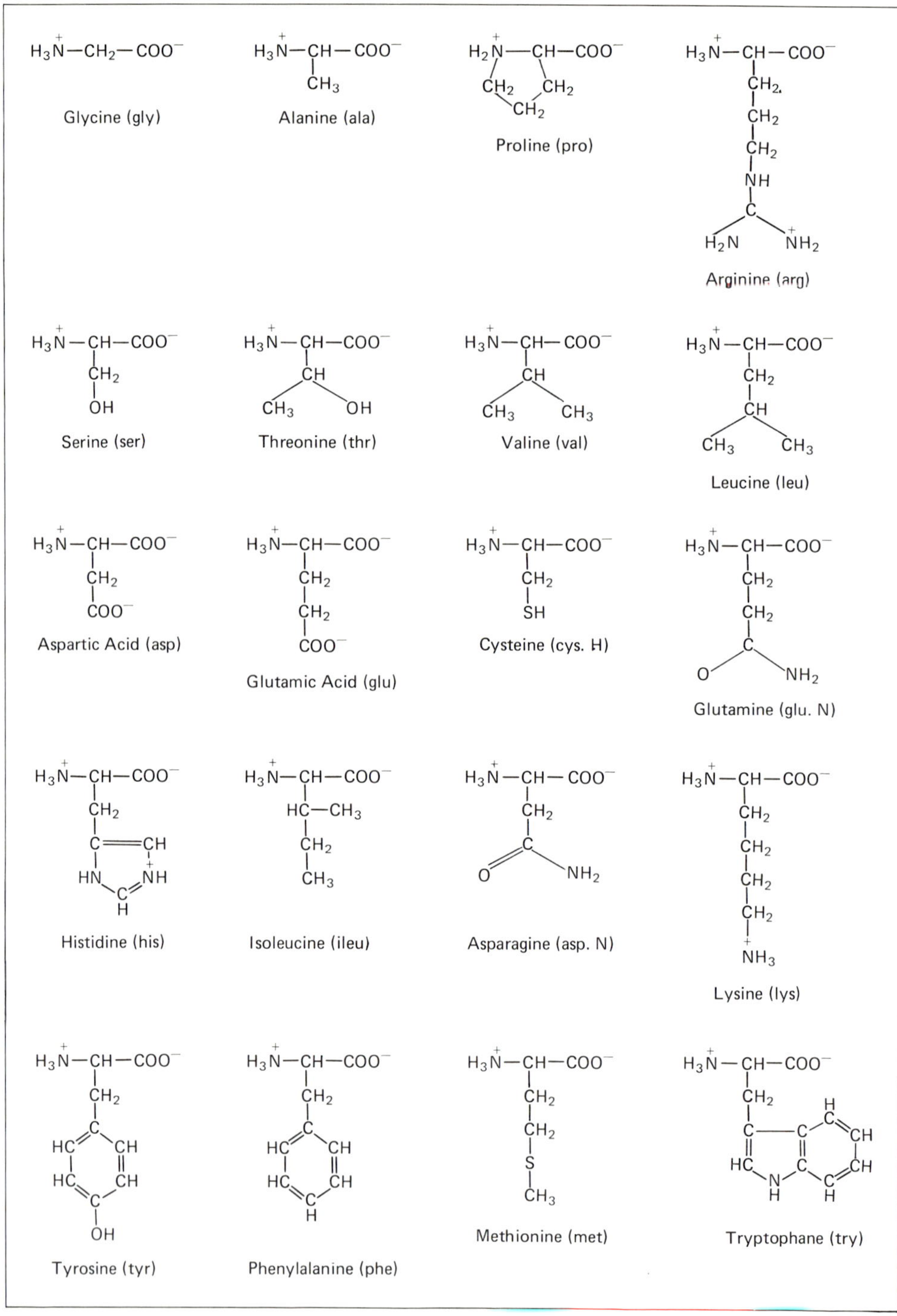

Figure 10.22
The 20 amino acids of protein.

acid sequence. This is called the *primary structure* of the protein. It can be determined, although this is difficult and time-consuming, by chemical means. After the insulin molecule, several others have been so determined. One is hemoglobin, which is found to consist of four chains: two identical α chains (141 amino acid residues[2]) and two identical β chains (146 residues). The amino acids in the chains are quite precise. A mistake of one single amino acid residue may cause malfunctioning of the molecule. This indeed is the case in the genetic disease of sickle-cell anemia, in which the genetic book of hemoglobin is misprinted for one single amino acid. Mistakes of this kind can happen by chance, resulting in defective offsprings. Sickle-cell anemia, on the other hand, has a more involved story. It originated in a part of Africa where malaria is rampant. Oddly, the person with sickle-cell anemia is immune to malaria and was picked by natural selection as the "superior" type to survive. The amino acid sequences of hemoglobin chains of some animals have been found. That of a gorilla differs from a human's by one or two residues only. That of a horse differs from ours by about 18 residues. This shows the close relation of the gorilla to man, much closer than the horse. Here we see that molecular studies (on hemoglobin structure) shed light on many aspects of biology and may be used as a tool to study pathology, evolution, and phylogeny.

The *secondary structure* of protein concerns the geometric configuration of the residues in the chain—the shape of the backbone. The first significant advance in this problem is the establishment of the *α-helix* as one basic form of the structure (Pauling et al.). In this structure (Fig. 10.24) the α carbon of the residues form a helix with 3.6 residues per turn of the helix. Each residue is linked with the residues above and below in the helix from the preceding and following turns by *hydrogen bonds*. The hydrogen bond is a weak chemical bond linking molecules together into a group but the group is not tightly bonded to be considered as a single molecule. It represents an attractive force like the intermolecular force but is specific; it acts only

[2] The amino acids in proteins, with an H_2O molecule eliminated from them, are called residues.

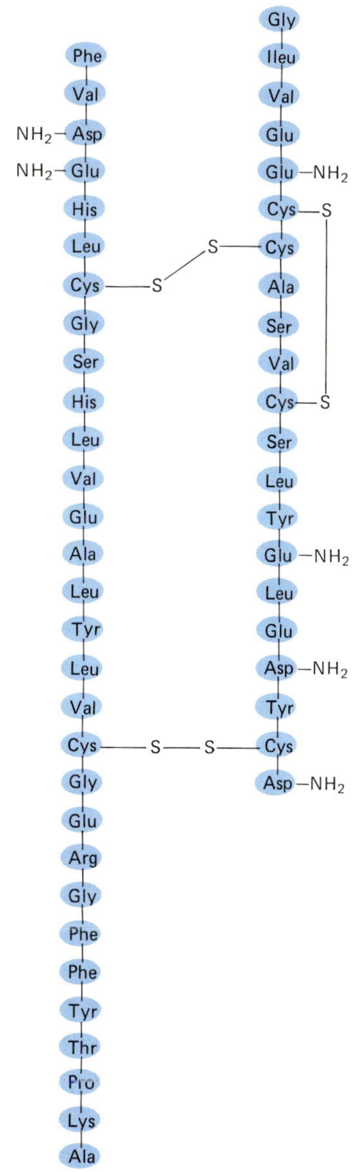

Figure 10.23 The amino acid sequence representing the primary structure of beef insulin.

between specific atoms. (Thus the hydrogen bond is in between the specific chemical bonds, which appear in the molecular formula, and the nonspecific intermolecular forces, which exist in gases, liquids, and solids but do not appear in the molecular formula. The hydrogen bond is specific but does not appear in the molecular formula.) In this case it is between the H of the NH group of one residue and the O of the CO group of another (dashlines in Fig. 10.24). These side links, though weak, help stabilize the helical structure. The radicals R of the residues project outward from the backbone. There is plenty of room for R. Changing R from one to another does not essentially affect the helical configuration. Thus many proteins can assume the α-helix structure.

Figure 10.24 Two possible forms of the α-helix, the left-handed in the left and the right-handed in the right. The right-handed helix is found in many proteins. [From *College Chemistry,* Third Edition, by Linus Pauling. W. H. Freeman and Company, Copyright © 1964.]

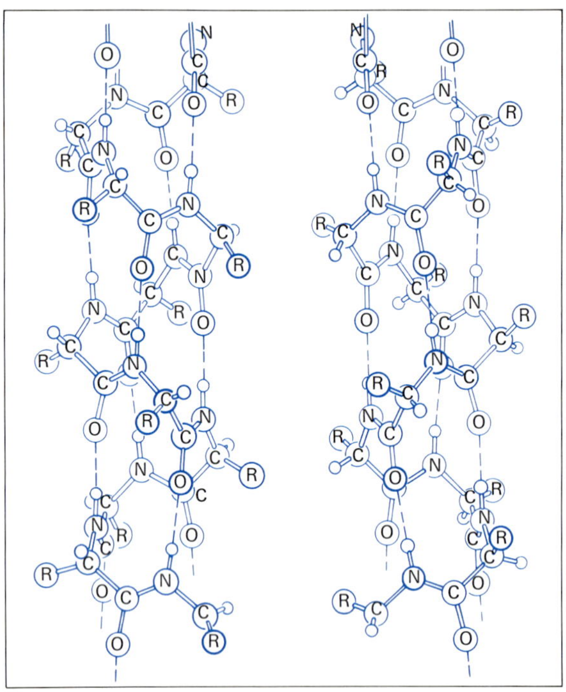

Besides the α-helix another form of secondary structure is the β-structure. The chains are fully stretched (not coiled into a helix) and hydrogen bonds are between two parallel chains. Such a structure is found in silk proteins.

The *tertiary structure* of protein concerns the folding of the backbone chain into a three-dimensional configuration. If a protein is made of one kind of amino acid only, then, all R's being the same, there is no reason to believe the molecule will assume any other structural form than a perfect α-helix. The helix will be straight, with no bending or folding. However, in most cases the R's are different from one residue to another and there are a number of ways the difference in R may change the structure slightly or abruptly. When accumulated over all the residues, the slight bendings combine to form foldings.

One such change happens when the chain passes over a *proline* residue. Proline does not have an amino group and therefore cannot form a hydrogen bond to stabilize the helix at that point. The helix may thus be bent abruptly. The protein chain may thus be bent at the proline residues into a few sections folded together, each one of which is still an α-helix. The tertiary structure of *myoglobin* worked out recently, a milestone in molecular biology, shows that the protein is globular in shape, resulting from the folding of the chain into seven short sections, each of which remains an α-helix. Another factor affecting the bending of the helix is the intermolecular forces between the R's and between R and water molecules in solution. The water-soluble R's (*hydrophilic,* i.e., "water-loving") want to make as much contact with water as possible. The water-insoluble R's (*hydrophobic,* i.e., "water-hating") tend to stick together to reduce contact with water. The helix is bent to fulfill these wishes and coils into a definite three-dimensional configuration. These configurations are often stabilized by disulfide bonds between two *cysteine* residues at two different parts of the chain, thus tying down the folding. In cases where there are several disulfide bonds, the bonds may be broken and reformed to show that the reformed bonds are between the same pairs of residues. This indicates that the folding brings the pairs to close proximity, making the formation of the bond possible. Thus the disulfide bond is not the cause of folding. It is formed after the folding and it stabilizes the folding.

Eventually we depend on the knowledge of the tertiary structure for an understanding of the function of the molecule. For example, the ability of the *antibody,* a protein in human body, to fight off infection is presumably due to its three-dimensional shape—a shape related to the shape of the *antigen* in a way like a lock to the key. The antibody thus combines with the invading antigen in lock-key fashion and thus eliminates its virulence.[3] One antibody thus works with only one antigen, just as one key works with only one lock. During the first infection the antigen is used as a mold to make antibodies. The information of making this antibody is kept in the body (in one cell in the spleen). Upon the second infection the information is called upon to make antibodies to stop the infection immediately. This is the general idea of the molecular basis of *immunology.*

The work to determine the tertiary structure (by X-ray diffraction, for example) is difficult and tedious. The work on *ribonuclease* took a decade.

Structure of Nucleic Acids. There are two kinds of nucleic acids: DNA and RNA. As far as the genetic information is concerned, DNA corresponds to the master copy of the blueprint kept in a safe (cell nucleus) and RNA corresponds to the Xerox copies of it used in the field office (cytoplasm) for day-to-day operation. We consider only the structure of DNA here.

In the early 1950's, after scientists had learned in microbiology that DNA is the genetic material, the question of the molecular structure of DNA was brought to the center of attention. Again one was guided by the thought that an understanding of the molecular structure may lead to an answer to the function of the molecule. This expectation was fulfilled in a most spectacular way. No sooner was the structure understood than a mechanism of the function was proposed. After a dozen years of intensive work, the mechanism turns out to be the very basic mechanism that is able to explain the whole complex, mysterious phenomena of genetics in terms of the physics and chemistry of the nucleic acid molecules.

The DNA molecule, like the protein, is a long chain molecule (Fig. 10.25). It is made of units called *nucleotides.* Each nucleotide is made of three parts linked together: a *sugar,* a *phosphate,* and a *nitrogenous base.* There are four kinds of nitrogenous bases in DNA corresponding to the four kinds of nucleotides: adenine (A), thymine (T), cytosine (C), and guanine (G). The sugars and phosphates of the nucleotides link alternatively by chemical bonds to form the backbone of the DNA chain, whereas the bases of the nucleotides attach on the sugars to form short side chains (Fig. 10.25). The whole structure is like a very long charm bracelet. The primary structure of DNA concerns the nucleotide sequence. The secondary structure concerns the geometrical configuration of the units forming the chain. This is the question that turns out to be related to the genetic function of DNA. The secondary structure turns out to be so rigid that there is no significant tertiary structure.

Pauling, after his success in working out the structure of the α-helix for proteins, took up the problem of the secondary structure of DNA and was guided naturally by the winning ideas of his previous accomplishment. He considered the DNA chain to be a helix. Also he thought the backbone chain to be

[3] Virulence is not due to any occult force. It is just due to a particular molecular structure—primary, secondary, and tertiary—that may cause havoc in body chemistry. Once the molecule is locked tight with the antibody, the active sites are insulated and cannot play havoc any more. Thus a mystery of life is reduced to a matter of mechanics, a typical Newtonian viewpoint. The workings of drugs and antibiotics are similar; they interrupt some vital physical or chemical processes of the bacteria, just as a nail sticking into the tire interrupts the operation of an automobile.

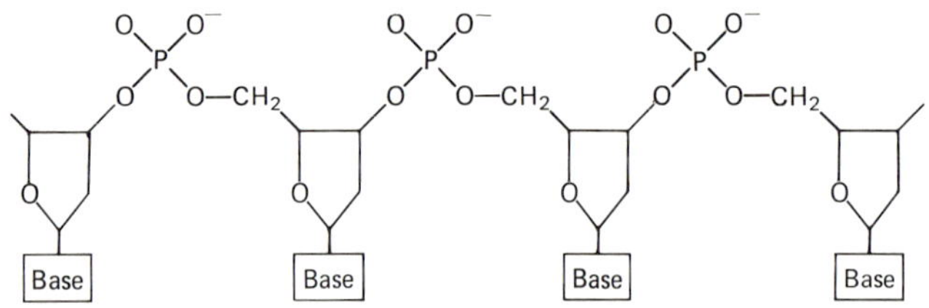

Figure 10.25 The DNA molecule.

tightly wound in the inside of the helix with the side chains (the bases) projecting outward, like the R's of the protein. That is where he erred. Otherwise he would have won another Nobel prize, for a total of three.

The correct answer was given by Watson and Crick: the molecule is indeed a helix, albeit a loose one, but the backbone is wound outside and the bases are folded inside the helix. They further introduced an extremely potent idea—the specific pairing of the bases, which turns out to be the basic mechanism of genetics. The discovery is surely one of the epoch-making events in the history of science. The affair is a great detective story, not only from a scientific point of view but also in human terms. Before we can discuss the structure, we have to explain the structure of the bases first.

The bases are pancake-shaped molecules. The structure of the four bases is shown in Fig. 10.26. They are either single-ring molecules (T, C; pirimidines) or double-ring molecules (A, G; purines). They contain quite a few nitrogen atoms and thus the name. The purine adenine and the pirimidine thymine can combine in a specific way, as shown in Fig. 10.26 by two hydrogen bonds (represented by the dotted lines). So can guanine and cytosine by three hydrogen bonds, as shown in the same figure. When so paired, the two pairs, A–T and C–G, are of about the same dimension (see the figure). The specific pairing of A with T and C with G is called *Watson-Crick pairing*.

Watson and Crick proposed that the DNA molecule is actually made of two polynucleotide chains, the base sequence in one of which can be arbitrary but that in the other is determined by the first, according to the pairing relation A–T and C–G. In other words, if the first chain has a sequence ACCCTTG, then the second chain will have a sequence TGGGAAC, so that the bases of the two chains can be paired one by one, Watson-Crick fashion by hydrogen bonding (see Fig. 10.27). Because the two pairs A–T and C–G are of the same dimension, such a paired double chain will have a constant distance of separation and thus will form a straight ladder with the base pairs acting as the rungs. Watson and Crick further assumed that this straight ladder is then twisted into a spiral staircase (explained in Sec. 12.8) with 10 rungs per turn of the spiral. The resulting *double-helix* configuration is what Watson and Crick assumed of the DNA molecule (Fig. 10.28). The model was constructed by fitting the molecular groups of DNA together into a three-dimensional configuration subjected to the

Figure 10.26 Molecular structure of the four bases A, T, C, and G and the hydrogen bonds (---) between A and T, and between C and G showing Watson-Crick base pairing.

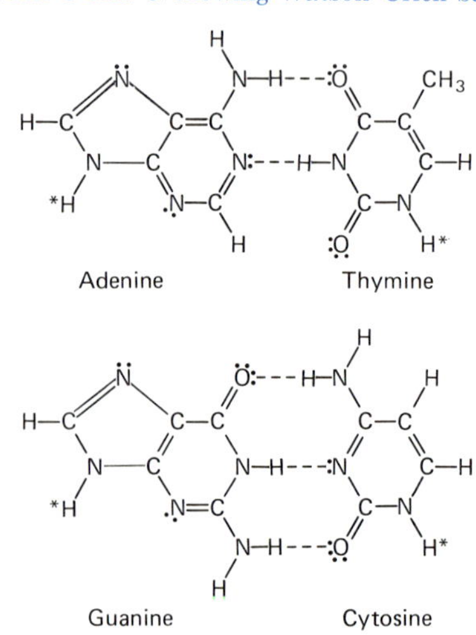

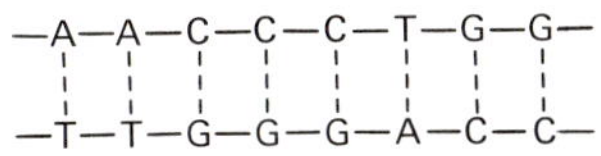

Figure 10.27 Watson-Crick pairing of the bases in DNA.

Figure 10.28 Model of the DNA molecule. Only two turns of the double helix out of the thousands are shown. This space-filling model brings out the fact that the molecule is closely packed. [Courtesy of Thomas R. Henderson]

restriction of the known bonding distances and angles of the atoms involved. There are a number of possibilities to satisfy these conditions, but they were guided to the double-helix idea by several important clues. The model was later verified by X-ray diffraction studies of the DNA molecule.

Genetic Function of DNA. At the same time they proposed the structure, they advanced a theory on how the molecule functions genetically. The sequence of the bases in one of the strains of the double helix may act as a coded message, somewhat like a telegraph message, to convey genetic information (more later). During cell division the double helix is split into two single strains. Each one acts as a template to assemble a new complementary strain; each base will attract a complementary base, A for T, T for A, C for G, G for C, and the new nucleotides will then be linked together to form a new strain (see Fig. 10.29). This way, out of one original DNA molecule, we have two new identical

Figure 10.29 Schematic representation of the replication of DNA.

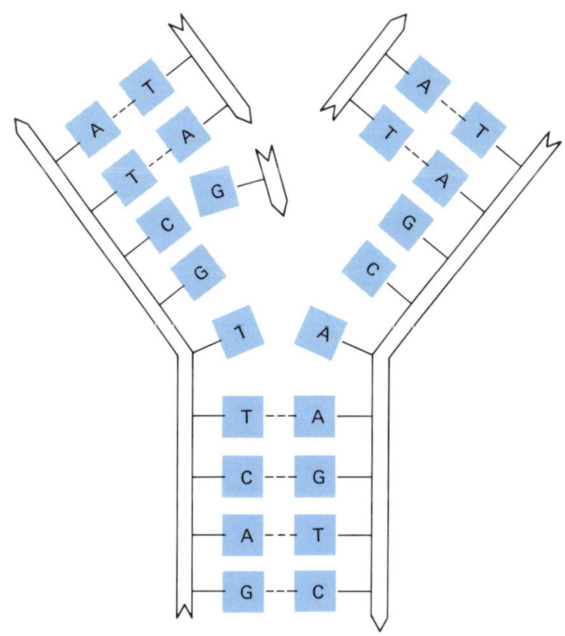

molecules. The genetic information can thus be transmitted from one generation to the next through cell division.

This replication procedure is *semiconservative*—the original DNA molecule disappears but each half of it remains intact. Theoretically replication could be totally conservative, semiconservative, or non-conservative. Experiments using radioactive tracers (Sec. 15.6) showed that replication is indeed semi-conservative, corroborating the Watson–Crick theory.

How is the genetic information expressed in detail in terms of the sequence of bases? The germ cells (sperm and ovum) are supposed to contain all the information to make all proteins, structural and functional. Each protein is specified by a sequence of 20 amino acids, like a sentence written by 20 letters. The DNA message is one written with four bases, like a telegraph message written in four symbols. For the information of the amino acid sequence of the protein to be encoded on DNA, there must be a system of translation similar to the Morse code of telegraph. It was hypothesized that there exists a *genetic code* by which three nucleotides in the DNA sequence code for one amino acid (just as three dot-and-dash symbols code for one letter). Then the base sequence of the DNA may be read as an amino acid sequence and the structural information of the protein may be encoded on the DNA. One may even consider the base sequence to act as a template for protein synthesis. Three nucleotides attract and bind a specific amino acid. The sequence of amino acids is then linked together to form the desired protein. With some modifications these speculations were verified later. (The protein synthesis template is not the DNA itself but a "Xerox copy" of it—the messenger RNA; the binding of RNA with amino acids is not direct but is mediated by another RNA—the transfer RNA.) The genetic codes of amino acids were broken in the early 1960's, another major triumph of molecular biology. Thus the nucleotide triplet GGA codes for glycine (GGU,[4] GGC, GGG also code for glycine; the code is four fold degenerate); GCA (also GCU, GCC, GCG) codes for alanine; and so on. The exciting story of RNA, protein synthesis, and genetic code, we have no time to discuss here. Altogether the intriguing problem of genetics is now nearly completely explained as molecular manifestations based on the structure of DNA and RNA together with the Watson–Crick pairing relation.

[4] U stands for "uracil." RNA's are formed of the four bases A, U, C, and G.

Problems

10.1 From the Avogadro hypothesis and the fact that one volume of nitrogen and three volumes of hydrogen combine to form two volumes of ammonia, find the simplest possible molecular formulas for these gases. Based on this information alone are other formulas possible? Give an example.

10.2 From the preceding and the additional information that one volume of nitrogen and two volumes of oxygen combine to form two volumes of nitrogen dioxide, find the simplest possible molecular formulas of oxygen and nitrogen dioxide.

10.3 From the information that 1 g of hydrogen combines with 8 g of oxygen and the reaction equation given in the text, determine the atomic weight of oxygen.

10.4 From all the preceding and the information that 7 g of nitrogen combines with 16 g of oxygen, determine the molecular weight of nitrogen.

0.5 An amount of nitric oxide is decomposed into nitrogen and oxygen in the ratio of 8 g of oxygen to 7 g of nitrogen. Find the simplest possible molecular formula of nitric oxide.

0.6 Based on the preceding molecular formula of nitric oxide, how many volumes of nitric oxide would be formed from the combination of one volume of oxygen and one volume of nitrogen?

0.7 What is the molecular weight of nitric oxide?

0.8 What is the molecular weight of nitric acid, HNO_3?

0.9 The atomic weight of carbon is 12. What is the molecular weight of alcohol, C_2H_5OH? And of butane, C_4H_{10}?

.10 How many grams is 1 mole of nitric acid, 1 mole of alcohol, and 1 mole of butane?

.11 From the compound B_2O_3 determine the valence of boron (B).

.12 From the compound $BeCl_2$ determine the valence of beryllium (Be).

.13 Which compounds can be formed from one carbon atom with various numbers of hydrogen and bromine atoms?

.14 Write down the possible structural formulas for C_5H_{12} different from those in Fig. 10.7 and show that each of them is equivalent to one of the three in Fig. 10.7.

.15 Write down all possible structural formulas of C_6H_{14}.

.16 Write down all possible structural formulas of C_5H_{10}.

.17 Write down the structural formula of the gasoline antiknock additive tetraethyl lead [$Pb(CH_2CH_3)_4$], and the additives to clean up the lead so produced, ethylene dichloride ($ClCH_2CH_2Cl$), and ethylene dibromide ($BrCH_2CH_2Br$).

0.18 Show that each amino acid can have two isomeric forms, one being the mirror image of the other (thus they are two different molecules). Are their chemical properties different? Are their physical properties different?

0.19 When a DNA molecule is analyzed for its base composition, what relations do you expect to exist among the percentages of the four kinds of bases A, T, C, and G? These relations were discovered before the Watson–Crick theory and were used by Watson and Crick to support their theory.

0.20 Do you expect these relations or other relations to exist for the base composition of RNA?

11 Chemical Energy

11.1 Oxidation–Reduction Reactions

Chemical reactions lead to the formation of new compounds and release (sometimes absorption) of energy. Both are significant from a practical point of view. Chemical reactions are involved in the production of useful materials and in the generation of energy. In this chapter we are concerned with the energy aspect of chemical reactions.

Take a typical example of chemical reaction, e.g., combustion. Heat and light are emitted, indicating the release of energy in the process. In fact, burning fuel for heat is the earliest form of energy utilization by man. Combustion was one of the intriguing problems in the history of science and in the ancient times fire was considered as one of the basic elements. It was through chemical studies, largely because of Lavoisier, that combustion was understood as a process of chemical transformation of matter in which the fuel combines with oxygen in the air. On the other hand, the heat and light emitted in the process, with all their power, are to be related to the concept of energy. Why energy is released when matter undergoes a chemical transformation is to be answered in the microscopic theory of atomic structure. In the macroscopic study here we recognize the release of energy as a fact and study its various manifestations.

Chemical reactions involving the reaction of an element or compound with oxygen are the simplest kind of *oxidation reaction*. Those involving the removal of oxygen from a compound are the simplest kind of *reduction reaction*. Besides combustion such reactions as the rusting of metal are also oxidation reactions. (Rusts are oxides of metals.) Metal ores usually are in the form of oxides of metals. The smelting process is usually a reduction reaction in which a *reducing agent* is used to remove the oxygen from the ore so as to recover the pure metal. In reduction the oxygen is combining with the reducing agent and therefore the reducing agent itself is being oxidized.

In chemistry the terms *oxidation* and *reduction* are also used in a more general sense to indicate a class of chemical reactions similar to those discussed, but without the involvement of oxygen. Similar to the burning of the metal iron (in the wire form) in (pure) oxygen, the metal sodium burns in the gas chlorine. In both cases the term *burn* is used because heat and light are emitted. The reaction is represented by the equation

$$2Na + Cl_2 \longrightarrow 2NaCl$$

We also say metallic sodium is oxidized in the reaction and chlorine is the oxidizing agent. At the same time we say chlorine is reduced in the reaction and sodium is the reducing agent. Also in electrolysis of NaCl we say that the sodium ion is reduced by the negative electrode and the chloride ion is oxidized by the positive electrode. In general, oxidation corresponds to the removal of electrons from an atom and reduction corresponds to the addition of electrons to an atom.

Even without going into atomic structure, it is easy to see why energy is involved in oxidation and reduction. When an electron is taken out of an atom, work has to be done because the electron is ordinarily kept inside the atom by an attractive electric force from the positively charged nucleus. When an electron is added to an atom, energy is released because there is a force pulling the electron into the atom. Thus during an oxidation–reduction reaction, energy is absorbed in one atom and released in the other. The net total is released in the form of heat. When the energy release is sufficiently large, light is also emitted, as in the process of combustion.

The release of energy in chemical reactions is important in two ways. From a practical point of view, it provides a source of energy that may be utilized in various ways. From a purely scientific point of view energy release or absorption determines in part in which direction a chemical reaction proceeds. We will first discuss the practical aspect and then return to the other point in later sections of the chapter.

11.2 Chemical Energy Sources

The earliest form of human use of energy is the burning of firewood. The heat energy released is used for cooking and heating. Firewood is essentially cellulose $(C_6H_{10}O_5)_x$, a carbohydrate $C_x(H_2O)_y$ from a chemical point of view. In burning, the molecule combines with oxygen in the air to form carbon dioxide and water vapor:

$$C_6H_{10}O_5 + 6O_2 \longrightarrow 6CO_2 + 5H_2O$$

The reaction products disappear in the air. (The small amount of ashes after burning is the residue of mineral contents of the firewood.)

Substances that can burn and produce heat are called *fuels*. Besides firewood a number of other fuels are in use now mostly from mineral sources. They come in solid (coal), liquid (petroleum), and gas (natural gas) forms. These are all *fossil fuels* because they had an organic origin. At the present time fossil fuel is the main source of energy.

There is another way chemical energy may be utilized; the energy released in a chemical reaction may be converted into electric energy first. Electric cells and storage batteries are devices to carry out this conversion. Let us consider one example. When zinc is placed in a solution of $CuSO_4$, a chemical reaction takes place in which zinc changes into $ZnSO_4$ and copper plates out, zinc being chemically more active than copper. In this reaction heat is evolved. For 1 mole of each reactant the heat produced is 50,130 cal:

$$Zn + CuSO_4 \longrightarrow Cu + ZnSO_4 + 50{,}130 \text{ cal}$$

This heat is not very useful for conversion to mechanical energy (see Chapter 8). On the other hand, in the *Daniell cell* this heat is wholly converted into electrical energy. The electric cell thus provides a more efficient way of using chemical energy even though not all chemical energy may be utilized this way. The Daniell cell is described in the following.

The cell has a zinc plate acting as the negative electrode and a copper plate acting as the positive electrode. The two are dipped in two solutions, $ZnSO_4$ and $CuSO_4$, respectively, the two solutions being separated by a porous partition so that ions can pass through but the two solutions will not mix immediately (Fig. 11.1). When the two plates are connected by a wire, an electric current flows from copper to zinc. At the same time zinc is dissolved into the solution, whereas copper is deposited on the positive electrode. The net chemical reaction is just the substitution reaction given earlier, in which one zinc atom replaces one copper ion in the solution. According to the principle of conservation of energy the same amount of energy, 50,130 cal/mole, should be released. But in this case this energy is not released in the form of heat but in the form of electric energy.

In electrolysis we learned that the deposition of 1 mole of substance on an electrode corresponds to the flowing through the circuit of a total amount of electric charge equal to a universal constant, called the *Faraday constant,* 96,522 coulomb multiplied by the valence of the substance (*Faraday's law of electrolysis,* Sec. 12.1). In the present case the deposition of 1 mole of copper (valence = 2) corresponds to the flowing through the circuit of an amount of electric charge equal to $2 \times 96{,}522$ coulomb. If the electromotive force of the cell in the unit of volt is designated by E, the energy evolved in the circuit will be $2 \times 96{,}522 \times E$ joules. On the other hand, the deposition of 1 mole of copper corresponds to the completion of the chemical reaction of one mole of copper with the release of 50,130 cal of energy. If this energy is converted completely into electric energy and if there is no other energy added, then the two quantities of electric and chemical energy must be equal:

$$2 \times 96{,}522 \times E \text{ joules} = 50{,}130 \times 4.18 \text{ joules}$$

We can use this equation to calculate the emf of the cell. The preceding equation gives

$$E = 1.087 \text{ v}$$

The experimental value of the voltage of the Daniell cell is 1.096 v, which is very close to the preceding prediction, justifying essentially the argument given. Still there is a small difference, of 0.009 v. This slight discrepancy has been explained by the fact that during the discharge of the cell (drawing current from the cell) the cell absorbs a small amount of heat, $H = 420$ cal, from the surroundings. Thus besides the chemical energy there is a small amount of heat energy added so that the resulting electric energy is slightly higher, represented by a slightly higher voltage, 0.009 v higher than expected. Thus conservation of energy is exactly fulfilled and convertibility of chemical energy to mechanical (electric) energy is demonstrated.

This convertibility of chemical energy to mechanical energy is not always guaranteed. A necessary condition is that the reactions involved be *reversible.* The reversible reaction is defined as one that may proceed in the reverse direction, with all physical quantities involved in the reaction remaining the same except for a possible change of sign. (Intensive quantities such as pressure and temperature do not change; extensive quantities such as heat, energy,

Figure 11.1 Daniell cell.

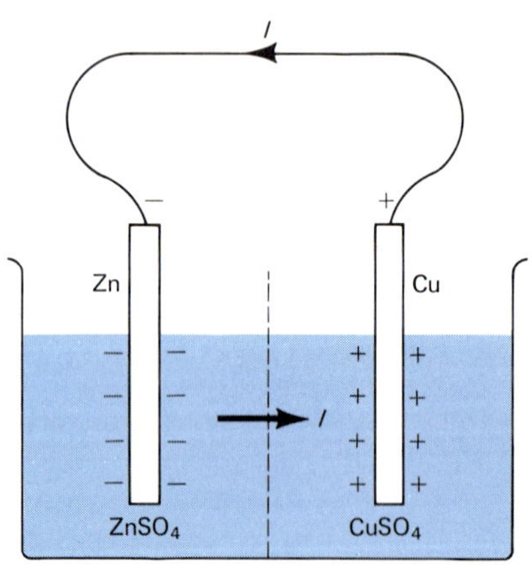

work, and charge change sign.) The reactions in the Daniell cell are reversible, the cell being thus called a *reversible cell*. In reverse operation, an external electromotive force is inserted into the circuit, driving a current in the opposite direction through the cell (Fig. 11.2). Copper is dissolved into the solution and zinc is deposited on the other electrode. The net chemical reaction is the reverse of the previous one:

$$Cu + ZnSO_4 \longrightarrow Zn + CuSO_4 - 50{,}130 \text{ cal}$$

Energy is absorbed in this reaction and the source of energy is the electrical energy of the external electromotive force. Again there is a slight discrepancy, accounted for by the heat exchange with the surroundings. This time the cell *releases* a slight amount of heat, exactly equal to H, to the surroundings. In the reverse operation, the extensive quantities 50,130 cal and 420 cal are exactly the same as before but change sign. (Absorption changes to emission.) The intensive quantities, the temperature, and emf under which the cell operates remain exactly the same (no change of sign).

In the direct reaction energy is released and may be used in the electric circuit. Once the amount of $CuSO_4$ is exhausted, the energy source is depleted and no more chemical energy is available. In this sense all chemical energy sources will be depleted when the active chemicals are used up.

Figure 11.2 Charging of a Daniell cell.

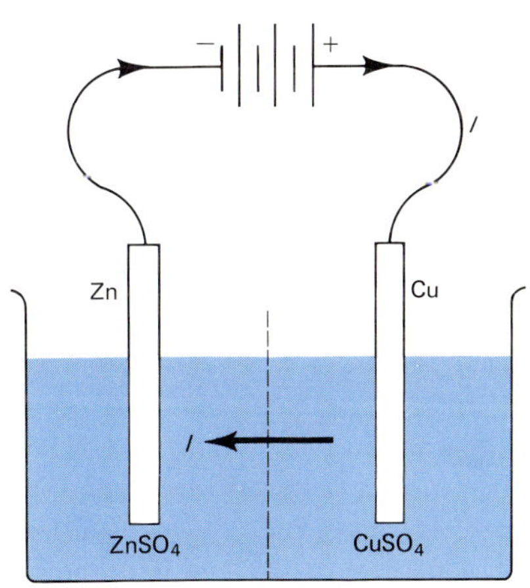

In the reverse reaction, the depleted $CuSO_4$ is replenished and may be used as an energy source again. But we do not get anything for nothing; energy has to be supplied by the external electromotive force, i.e., by the charging current. Even though we do not create any new energy by discharging and recharging repeatedly, the reversible cell does serve a useful purpose in that it can store energy from other sources—e.g., from the household current—and use it in a more convenient way, such as in a boat, where there is no socket to plug in. Used in this way, a reversible cell is not a generator or source of energy but merely a storage of energy from other sources. In a single discharge the energy produced is indeed of chemical origin but in repeated operations of charging and discharging the ultimate source of energy is the external electromotive force, such as the household current. Chemical energy is merely used as a device to store that external energy source for more convenient use. The storage battery used in an automobile does just that; the external emf charging the battery comes from the generator powered by the car engine.

When we speak of using chemical energy we are essentially using the term in this sense. Actually this applies also to some other forms of energy, such as hydraulic energy. In a watermill the mechanical energy generated is indeed derived from the gravitational potential energy of the water. But the ultimate source of this energy is solar energy, as discussed in Chapter 3. The gravity energy of water is just an intermediary for converting solar energy into a useful form. Electric power in the household circuit works in the same way; the ultimate source of energy is the heat energy generated by burning fuel in a thermal power plant or its equivalent in other plants. In speaking of energy sources, we must distinguish the immediate source and the ultimate source. In dealing with energy resources of a nation or of the earth we always deal with the ultimate source.

11.3 Energy Flow in Biosphere

Most of our present energy sources—firewood, coal, petroleum, and natural gas—are chemical in nature. Significantly, they are all related to living organisms on earth. Firewood comes from plants. The fossil fuels, though considered minerals because they issue from under the ground, are actually remains of past plants and animals. Thus our main source of energy, chemical energy, is closely related to the biological world.

In the burning of firewood, cellulose $[(C_6H_{10}O_5)_x]$ combines with oxygen in the air and forms carbon dioxide and water. In the process a large amount of heat energy is released, which can be used for cooking, heating, and generating mechanical energy. In the animal body, glucose $(C_6H_{12}O_6)$ is oxidized into carbon dioxide and water. Again energy is released for biochemical, physiological, and biomechanical uses. These substances, all carbohydrates, may be considered as chemical energy sources.

If not replenished, the carbohydrates on earth will be gradually depleted and eventually no more chemical energy will remain. They are replenished by living organisms, formed in green plants in a set of complex chemical reactions called *photosynthesis*. The net change is the conversion of carbon dioxide from the air and water in the plant into carbohydrates. The reaction is just the reverse of the preceding combustion reactions, and according to the principle of conservation of energy the reaction requires the external supply of energy. The energy necessary comes from the sunlight, thus the name *photosynthesis*. Another important element is the presence of the green stuff, *chlorophyll*. That is why it takes place only in green plants. (In this sense all animals are parasites.) How the chlorophyll and sunlight work to put CO_2 and H_2O together to form carbohydrates $[C_x(H_2O)_y]$ is a complicated chemical problem that has been studied for years. The important point that concerns us is that the energy comes from the light energy of the sun.

Thus in biological processes CO_2 and H_2O enter and leave the biological systems in cycles, just as water passes through a watermill in cycles. Solar energy is thus converted into chemical energy stored in carbohydrates and used in the biological systems.

The chemistry in living systems is largely directed to two purposes: utilization of energy and creation of body material. It concerns the building up of complex molecules (*anabolism*) and their breaking down (*catabolism*). The two together are called *metabolism,* which is a principal feature of life.

As far as plants are concerned, the raw material used in anabolism is essentially carbon dioxide and water. Both energy molecules such as glucose and structural molecules such as cellulose are organic molecules formed by reducing carbon dioxide. Both kinds of molecules can generate energy in catabolic processes.

Used as fuels, firewood gave way to coal and petroleum. Civil War locomotives used firewood, later ones used coal, and the present ones (most diesel locomotives) use oil. Yet coal is formed from vegetable remains hundreds of millions of years ago. Its energy source, like that of firewood, is still the sunlight. Oil is probably formed from animal remains of ancient times. Because the energy supply of the animal kingdom originates in the vegetable kingdom, the ultimate energy source is again sunlight.

The energy flow in the biological world works itself out in a balanced cycle in which living species can perpetuate themselves. The energy needs of the animal kingdom are supplied by the plants, the energy source of which is replenished by sunlight. However, the present industrial use of energy, which began only 200 years ago in the whole biological history, is so accelerated that it threatens to break the cycle and bring about the extinction of industrial civilization itself. Industrial society consumes fuels (coal, oil, and gas) too fast for nature to replenish them. The coal and oil deposits that take hundreds of millions of years to accumulate are being used up lavishly in a matter of several hundred years. When all the available fossil fuels are used up, which will happen in a few centuries, and if no new fuels are developed, there will be no energy to power the machines and industrial society will grind to a halt. The new fossil fuels to be formed, if any at all, will not be available until millions of years later. We cannot wait that long for our natural resources to be "recharged." Thus energy resources are a major problem for the future of mankind.

*11.4 The Le Chatelier Principle

We have seen that energy is released or absorbed in chemical reactions. We now want to show that the absorption or release of energy is not merely a passive side effect but a crucial factor that determines which way the reaction will go. From a general point of view if chemistry is reducible to mechanics, the important mechanical quantity energy should play a significant role. Similarly, other mechanical quantities, such as pressure and volume, may also affect chemical reactions. This is indeed the case. The effects of all these factors on chemical reactions may be summarized in a very useful qualitative principle formulated by Le Chatelier.

Consider a chemical system already in equilibrium. If the conditions (pressure, volume, concentration, temperature) are changed, then the equilibrium will shift in such a direction so as to reduce the changes. This is the *Le Chatelier principle,* established by empirical observation.

Let us illustrate it by the reaction of the synthesis of ammonia from nitrogen and hydrogen:

$$N_2 + 3H_2 \rightleftharpoons 2NH_3 + 22{,}000 \text{ cal}$$

It is an *exothermic* reaction—i.e., heat is evolved in the synthesis—when 1 mole of nitrogen combines with 3 moles of hydrogen to form 2 moles of ammonia, 22,000 cal of heat is evolved. The reaction is reversible; that is, ammonia may also decompose into nitrogen and hydrogen. When 1 mole of nitrogen and 3 moles of hydrogen are mixed, under proper conditions, some of them will combine to form ammonia, but not all of them. Some of the ammonia formed will again decompose into nitrogen and hydrogen. At a given temperature and pressure the two reactions in opposite directions eventually reach an equilibrium; both of them still proceed but one is compensated exactly by the other. The mixture is thus composed of three gases—N_2, H_2, and NH_3—each with a definite molar number n_1, n_2, and n_3. Such a mixture represents an equilibrium state.

When the conditions of the equilibrium state change—say, the temperature is increased or the pressure is increased—then the equilibrium concentrations represented by the molar numbers n_1, n_2, and n_3 will change in such a direction as to offset the increase of temperature or pressure. Let us first consider the effect of increasing temperature. Temperature is increased by adding heat to the system. To reduce the increase of temperature corresponds to taking away some heat. This can be accomplished by having the reaction proceed in the reverse direction—the decomposition of ammonia absorbs heat energy. According to the Le Chatelier principle, the effect of increasing temperature (or the effect of heating) is that the reaction will go backward (more decomposition of ammonia than synthesis) until a new equilibrium is reached, represented by a new set of molar numbers n_1', n_2', n_3'. Compared with the original set n_1, n_2, n_3, we find the amount of ammonia is reduced. Therefore, in the commercial process of manufacturing ammonia, temperature is kept as low as possible to increase the yield of ammonia.

Next we consider the effect of changing pressure. When pressure is increased on an equilibrium mixture, the equilibrium should shift, according to the Le Chatelier principle, in a direction that reduces the pressure. This can be accomplished by letting the reaction proceed forward—four molar volumes of reactants (one N_2 and three H_2) yield only two molar volumes of product NH_3, so that in the forward reaction, volume is reduced and therefore pressure is reduced. Indeed, under increased pressure, the yield of NH_3 is increased. In commercial production of ammonia, a very high pressure, as much as 1,000 atmospheres, is used.

Finally, we consider the effect of changing the concentrations. Suppose that in the previously described equilibrium condition we add an extra amount of N_2 to the equilibrium mixture of n_1, n_2, n_3 moles of N_2, H_2, and NH_3. What will happen? According to the Le Chatelier principle, the equilibrium will shift in such a way as to reduce the extra concentration of N_2. This can be accomplished by letting the reaction proceed forward; more N_2 will be consumed in the process. Indeed the reaction proceeds to the right.

The Le Chatelier principle is a convenient rule to figure out the outcome qualitatively. Why does nature behave in such a fashion? How much is the change quantitatively? These questions are answered in the law of mass action to be discussed in the next section. It can be shown that the law of mass action can be derived theoretically from the basic laws of thermodynamics and therefore this part of chemistry is accountable by the statistical theory of physics. The Le Chatelier principle, discovered much earlier, may now be looked upon as a convenient rule based on such a fundamental theory.

*11.5 The Law of Mass Action

Let us again take the manufacturing of ammonia as an example. The chemical reaction involved is

$$N_2 + 3H_2 \rightleftharpoons 2NH_3$$

This means that one volume of N_2 combines with three volumes of H_2 to form two volumes of NH_3. Ammonia will always be formed by this definite ratio.

Suppose that in a cylinder we mix hydrogen and nitrogen in an arbitrary ratio, say, four volumes of H_2 and three volumes of N_2. When the proper conditions of the chemical reaction are provided, we find the reactions taking place, resulting in a mixture of three compounds H_2, N_2, and NH_3 with specific proportions. Let us say, for example, only one volume of N_2 is used in making two volumes of NH_3 and there are two volumes of N_2 and one volume of H_2 left over, mixed with the two volumes of NH_3.

Suppose that into this mixture we add from outside an extra amount of H_2. The result is that an additional amount of NH_3 will be formed even though no N_2 is added. This conclusion has been arrived at by the Le Chatelier principle. The formation of the additional amount of NH_3 will deplete N_2 further; the amount of N_2 in the final mixture will be changed from two volumes to, say, one volume.

Suppose that instead of adding H_2 we now add an extra amount of N_2. The result is again that additional NH_3 will be formed even though no H_2 is added. The H_2 in the mixture will be depleted further.

Suppose that instead of adding H_2 or N_2 we now add an extra amount of NH_3 from outside. The result is that more H_2 and N_2 appear in the final mixture (again the Le Chatelier principle). This means that some NH_3 is decomposed and changed back to H_2 and N_2.

The relative proportions of the three gases at equilibrium depend on the initial proportion. Let us say that there are C_H per cent of H_2, C_N per cent of N_2, and C_A per cent of NH_3 at equilibrium. The three numbers C_H, C_N, and C_A represent the *equilibrium concentrations*. If we change the initial proportion of H_2 and N_2, the equilibrium concentrations will be changed accordingly. (Temperature and pressure are kept constant.) How they change is an important question to be investigated. The experimental study on this point leads to the following relation:

$$\frac{C_A^2}{C_H^3 C_N} = \text{constant}$$

Because these changes take place under constant temperature and pressure, the constant in the preceding equation refers to this particular temperature and pressure. At a different temperature and pressure the constant may assume a different value. In other words, the constant is a function of temperature and pressure. It is a constant with respect to the variation of the initial concentration of the mixture.

This result is a special case of a general relation in any reversible reactions. If the reaction involves compounds A, B, C, and D by the following equation:

$$aA + bB \rightleftharpoons cC + dD$$

meaning that a molecules of A combine with b molecules of B to form c molecules of C and d molecules of D, then the equilibrium concentrations C_A, C_B, C_C, and C_D are related by the following equation:

$$\frac{C_C^c C_D^d}{C_A^a C_B^b} = \text{constant}$$

the right-hand side being a constant with respect to variations of initial composition of the mixture under the same temperature and pressure. This equation represents the *law of mass action.*

The law of mass action may be understood easily from the microscopic theory point of view. In a reaction involving the combination of one molecule of A with one molecule of B, the rate of reaction depends on the chance of these two molecules coming together, i.e., the rate of collision of A and B, which is proportional to the concentration of A and the concentration of B and thus the product of the two. If the reaction involves the coming together of three molecules A, B, and C, the rate of reaction is expected to be proportional to the product of the three concentrations. If two of the molecules are the same, say, $A = B$, then two of the concentrations are the same and the rate is proportional to the square of this concentration. Thus in the reaction involving a molecules of A and b molecules of B, the rate is proportional to $C_A^a \cdot C_B^b$. The rate of the inverse reaction is likewise proportional to $C_C^c \cdot C_D^d$. At equilibrium these two rates must be equal; therefore

$$(\text{Constant})_1 C_A^a \cdot C_B^b = (\text{constant})_2 C_C^c \cdot C_D^d$$

In other words

$$\frac{C_C^c \cdot C_D^d}{C_A^a \cdot C_B^b} = \text{constant}$$

Problems

11.1 In the Clark cell the following chemical reaction is involved:

$$Zn + Hg_2SO_4 \longrightarrow ZnSO_4 + 2Hg + 81{,}130 \text{ cal}$$

If all the energy released were converted into electrical energy, what would be the emf of the cell? Actually the emf is 1.43 v. How do you explain the difference?

11.2 Hydrogen can be burned to generate heat and has been suggested to be used as a fuel. To generate hydrogen we decompose water by electrolysis. Where is the ultimate source of energy of the hydrogen fuel?

11.3 When water vapor is heated to a very high temperature what reaction will take place according to the Le Chatelier principle?

11.4 In the preceding problem if in addition we increase the pressure, in which direction will the reaction proceed?

11.5 In the preceding problem with the temperature and pressure fixed, for the same percentage change of the concentration of hydrogen and oxygen, which will lead to the generation of more water vapor?

12 Atomic Transformations of Matter

12.1 Faraday's Law of Electrolysis

The atomic theory of Dalton is based on the idea that atoms of different elements are immutable, independent, and unrelated. Yet we have already seen in the periodic table that the various atoms are related in a grand scheme. Electrolysis brings out the relation between atoms and electricity. There is one piece of information suggesting that atoms are not fundamental but have a structure. When atomic weights of the familiar elements were first determined, many of them turned out to be nearly integers. Thus the atomic weights of carbon, nitrogen, and oxygen are nearly 12, 14, and 16, respectively. This immediately suggested that perhaps all other atoms are made of one basic atom, the hydrogen atom, so that the weights are integral multiples of that of hydrogen (the Prout theory). However, later accurate measurements of atomic weights of many elements, such as chlorine, showed that they are definitely not integers. The Prout theory was abandoned. Nevertheless, after the discovery of isotopes the Prout theory saw a revival; the weight information of the atomic species (isotopes) does tell us that atoms are made of other more fundamental building blocks.

The first important advance on subatomic structure comes from the *law of electrolysis* established by Faraday, which we introduced previously. He found that in electrolysis it takes the same amount of electric charge, 96,522 coulombs, flowing through the current circuit, to liberate 1 mole of any monovalent (valence 1) substance, such as the alkali metals. Because 1 mole of any substance contains the same number of atoms (the *Avogadro number* N_0), it follows that in electrolysis it takes $96{,}522/N_0$ coulombs of charge to neutralize each monovalent ion. In other words, each monovalent ion carries the same amount of electric charge $96{,}522/N_0$ coulombs. Moreover, Faraday's law says that for bivalent (valence 2) substances, the amount of charge required is twice as much and that for trivalent (valence 3) substances, the amount of charge required is three times as much. This means that the electric charge carried by a bivalent ion is twice as large as $96{,}522/N_0$ coulombs and that of a trivalent ion is three times as large. The remarkable conclusion obtained is that there seems to exist a basic unit of electric charge $96{,}522/N_0$ coulombs and any atom, when changed into an ion, can carry only an integral multiple of it. In other words, electricity seems to have a granular structure (or "atomic" structure) the basic unit of which, $96{,}522/N_0$ coulombs, is called an *electron*. A monovalent ion acquires one electron, a bivalent ion acquires two electrons, and so on.

The facts of electrolysis tell us only the existence of a basic unit of electricity. To associate this unit with a material particle (also called *electron*), which is to become a constituent subunit of an atom, was a hypothesis. This plausible hypothesis turned out to be quite fruitful; for on this basis physicists were able to explain a phenomenon of light called the *Zeeman effect* and from this explanation were able to ascertain the charge of the electron to be negative (not positive). Finally, in 1897 this hypothetical particle was actually discovered by Thomson in an experiment concerning *cathode rays*. The particle is a material particle having a definite mass in addition to carrying a negative electric charge. It is a common constituent of all atoms and can be broken loose from the atoms. (See the next section.) Thus it was established experimentally that the atoms are not indivisible but have a structure. (At least we know that one of the constituent parts is the electron.) The problem of atomic structure was thus opened for investigation. The discovery also settled one basic question concerning the relation between electricity and the Newtonian matter. Electricity is not a separate substance beyond matter. It is a basic attribute of matter. Matter is made up of particles each one of which, besides having a mass, also has an electric charge. Thus mass and electric charge are two basic attributes of the same entity matter.

Figure 12.1 Cathode rays in a discharge tube.

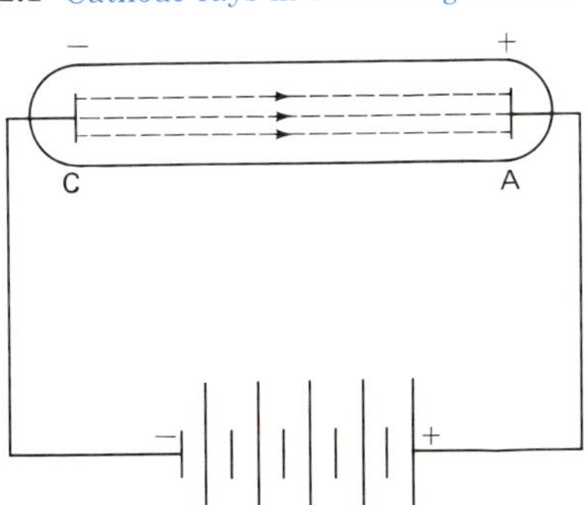

12.2 Thomson's Discovery of the Electron

Thomson's experiment establishing the existence of the electron is so fundamental that it deserves further discussion. It involves the use of a *cathode ray tube,* which is an evacuated tube with two electrodes sealed inside (Fig. 12.1). When a proper voltage is applied across the electrodes, the rarified gas inside becomes a conductor and an electric current passes through the tube. (Ordinarily gas is a good insulator and current cannot pass through it.) When this happens a greenish ray is seen to emanate from the cathode, this ray being called the *cathode ray*. It projects a sharp shadow of any obstacle placed in its way, which indicates that the ray propagates along a straight line. Yet the direction may be changed by applying an electric field or a magnetic field. This shows that the ray is not a light ray (electromagnetic waves), which is neutral, but is a stream of electrically charged particles. From the direction of deflection by an electric or magnetic field, we learned that the electric charge is negative. Moreover, by a quantitative study of the amounts of deflection produced by electric and magnetic fields, we can determine one important quantity of the cathode ray particle; that is, the ratio of the electric charge to the mass of the particle, e/m. The value of this ratio so determined is 1.76×10^8 coulombs/g. Even though we cannot determine the values of e

and m separately in the cathode ray experiments, the ratio is sufficient to enable us to establish significant conclusions.

Let us compare the value of e/m with those of the more familiar particles, i.e., the ions. Take the hydrogen ion as an example. One mole of hydrogen ions (N_0 ions) weighs 1.008 g and carries a total charge of 96,522 coulombs. Thus the ratio $(e/m)_H$ of the hydrogen ion is $96{,}522 \div 1.008 = 9.57 \times 10^4$ coulombs/g. This is smaller than that of the cathode ray particle by a factor of 1,840. Other ions, with their masses greater than that of hydrogen, have the ratio e/m even smaller. Thus judged by the value of e/m the cathode ray particle is distinctively different from the ions we are familiar with and thus represents a new particle heretofore unknown. From our previous study of electrolysis we learned that electricity has a basic unit, which is the amount carried by a hydrogen ion. If we *assume* that the same unit is carried by the cathode ray particle, then the mass of the cathode ray particle is 1,840 times smaller than that of the hydrogen ion. It is thus a very small particle compared with all atoms.

Because the cathode ray particle is produced in a cathode ray tube, it is only natural to suppose that it is split off from atoms in the cathode or atoms in the gas. When the experiment is repeated with a different cathode material or a different gas, the value e/m of the cathode ray particle determined remains always the same. This indicates that the cathode ray particle is a *common* constituent particle of all atoms.

Thus was discovered a substructure of the atom. Atoms can be broken down further, contrary to Dalton's concept of indivisibility of the atom. And we have found a common building block of all atoms—the cathode ray particle. In spite of the fact that all atoms have different properties, some of the building blocks are the same. This opens up the following questions on atomic structure. What are the other building blocks? How are the building blocks put together to form atoms? We will return to these questions later.

The cathode ray particle, being a small subunit (in terms of weight) of the atom that carries electric charge, thus meets the description of the hypothesized electron—a small material particle carrying the elementary unit of electric charge. Therefore it is identified with the electron, and Thomson was credited with the discovery of the electron. That the electronic charge is actually the smallest amount of electric charge ever observed and that any other electric charge a body can carry is an integral multiple of it is confirmed later by Millikan in the famous *oil drop experiment* (1909), which determines the electronic charge e directly, independently, and accurately: $e = 1.6021 \times 10^{-19}$ coulombs. Once the value of e is determined, the value of m can be calculated from the known ratio e/m. From the value of e and the Faraday constant 96,522 coulombs, the more accurate value of the *Avogadro number* $N_0 = 6.0248 \times 10^{23}$ is obtained. From N_0 and the atomic weights the absolute masses of all atoms are determined, e.g., the mass of the hydrogen atom is $1.00813 \div (6.0248 \times 10^{23}) = 1.6733 \times 10^{-24}$ g. This is the way the microscopic quantities are determined by a set of experiments in a very accurate way. (Half a century before, we already had a rough estimate of N_0 based on the Loschmidt method using the kinetic theory of gases.)

The use of the value of e/m to identify a particle is a technique also applied in many other occasions. In the *photoelectric effect* it was found that a beam of light may cause the ejection of negatively charged particles from a metal surface. The value of e/m of these particles was measured and was found to be the same as that of the cathode ray particle. Thus these particles were identified with the electrons and we conclude that light causes the emission of electrons from metal surfaces. In *thermionic emission* it was found that heat causes the emission of negatively charged particles from a metal surface. The value of e/m was measured and the particles were again identified with the electrons. Thus heat also causes the emission of electrons from metal surfaces.

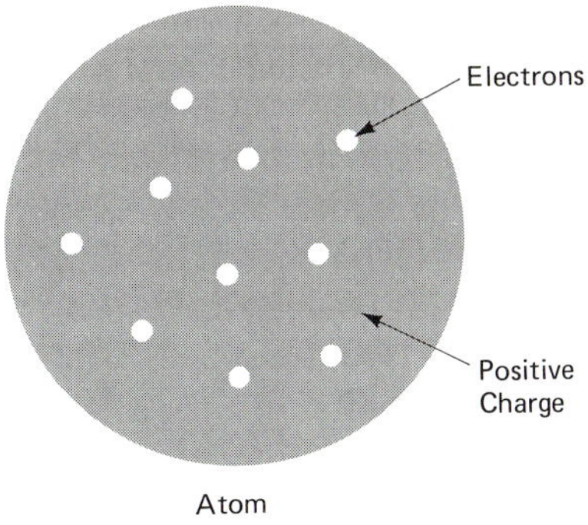

Figure 12.2 The Thomson model of atomic structure (the pumpkin model).

The Thomson Atomic Model. Up to this point all we know of the structure of atoms is that all kinds of atoms have a common constituent particle, namely, the electron, which can be split off in a cathode ray tube presumably by mechanical action, in the photoelectric effect by light, and in thermionic emission by heat. The next question that naturally arises is, "How many electrons are there in each kind of atom?" Another question is, "What is the other substance in the atoms besides the electrons?" The other substance in the atoms must carry positive electric charge so that the whole atom is electrically neutral as it should be. Because the electron weighs much less than the atom, the positively charged part must carry the main part of the weight. Many speculations followed.

Thomson first proposed a model of atomic structure that is dubbed as the *pumpkin model*. The positively charged substance is assumed to form a sphere with the small electrons embedded inside like the seeds in a pumpkin (Fig. 12.2). The electrons are assumed to be able to move about inside the atom and sometimes to move out of it. Thus as far as the electrons are concerned, the positively charged sphere must not be solid, but rather fluidlike.

This model was successful in a limited way in dealing with the problems of the atom at that time but was discarded later.

12.3 The Rutherford Atomic Model

The Thomson model held the attention of scientists for about a decade. Then in 1911, Rutherford carried out a set of experiments that revealed significant information on the structure of the atoms. The experiments involved the use of a small atomic particle called the α-particle as a probing projectile. The α-particle is the one that constitutes the radioactive ray, called the α-ray, emitted by radioactive material such as radium. It has been identified with a helium atom stripped of two electrons. When a beam of α-particles is directed on a thin foil of gold, for example, it was found that most of them go right through as if nothing is in their way, but a small fraction of them is scattered at wide angles (Fig. 12.3). The rationale of the experiment is somewhat similar to that of shooting bullets into a bale of cotton to find out if there is any weapon concealed inside. If not, the bullets will go right through. Otherwise, when the bullets hit the obstruction, they will be deflected sidewise at a wide angle. According to the Thomson model, a thin foil consisting of many

Figure 12.3 Schematic representation of Rutherford's α-particle-scattering experiment.

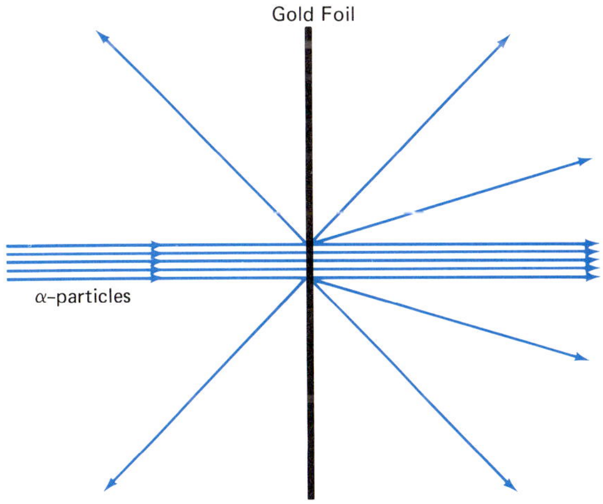

layers of atoms would be like many layers of pumpkins stacked together and the α-particle bullets would shoot through the pumpkins without wide-angle scattering. The fact is, however, that a small number of α-particles are scattered at wide angles, indicating that these α-particles have hit something hard inside. On this basis Rutherford proposed a new model of atomic structure known as the *planetary model,* which replaced the pumpkin model and is the model accepted today.

Instead of assuming the positively charged substance of the atom to be uniformly distributed over the atom, Rutherford assumed it to be concentrated in a very small region in the center of the atom, called the *nucleus*. The electrons are now left floating outside the nucleus and the situation is somewhat like that of the solar system, with a massive sun at the center surrounded by a number of smaller bodies, the planets, moving in a set of orbits (Fig. 12.4). The whole atom is thus largely empty space with a few particles moving about in it. Thus most of the α-particles can go through a thin foil as if nothing is in their way. On the other hand, once in a while an α-particle may score a direct hit on the nucleus and be scattered at a wide angle. On the basis of such a model Rutherford was able to calculate how many α-particles scattered at what angles. This prediction agreed very well with experimental results and therefore the existence of the nucleus was definitely established.

Figure 12.4 Rutherford's model of atomic structure (the planetary model).

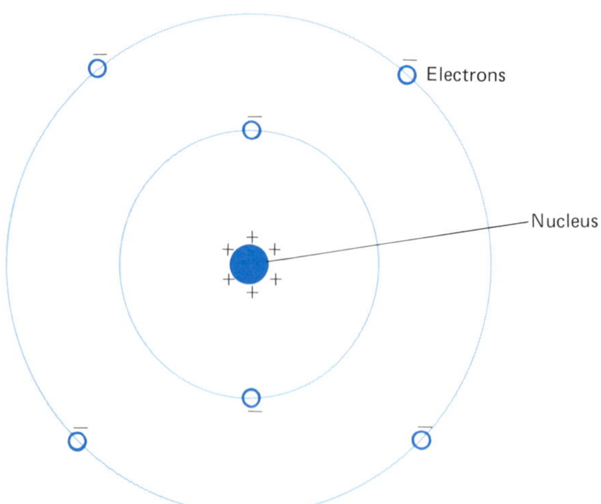

Moreover, Rutherford was able to figure out, based on the experimental information on α-particle scattering, the size and the electric charge of the nucleus and thus established some of the basic properties of the nucleus. He found that the radius of the nucleus, slightly varying from one atom to another according to atomic weight, is of the order of 10^{-12} cm, which is much smaller than that of the atom, of the order of 10^{-8} cm. The charge of the nucleus should be equal and opposite to the total charge of all the electrons in the atom and therefore it tells us the number of electrons in the atom. Rutherford's determination of nuclear charge thus also determines the number of electrons in the atom. Besides Rutherford's method there are other methods of determining the number of electrons in an atom (such as by X-ray scattering). All methods lead to the same important conclusion that *the number of electrons in an atom equals the atomic number Z* determined by the periodic table. The periodic table brings out the relations among the 92 elements. Now we have the remarkable conclusion that the characteristic difference among the 92 atoms is just a difference in the number of electrons. The simplest atom, hydrogen ($Z = 1$), has only one electron; the most complex atom, uranium ($Z = 92$), has 92 electrons. For each number of electrons Z there is a corresponding element in the periodic table with atomic number equal to Z. Because Z is an integer and all elements from $Z = 1$ to $Z = 92$ have been discovered, there can be no other elements between any of them.

The major problems of atomic structure are now the following: what is the nucleus made of and how do the electrons arrange themselves outside the nucleus in the atom? The separate studies on these two aspects led, respectively, to two branches of science: *nuclear physics* and *atomic physics*. Atomic physics in the narrow sense here used deals with only the arrangement of the electrons outside the nucleus even though, used in a broad sense, it also includes nuclear physics as the physics of the whole atom. Atomic physics will be the main concern for the remainder of this chapter and Chapter 13. Nuclear physics will be left to Chapters 14 and 15.

The differences among all atoms, according to what we have learned of the atomic structure so far, can only come from the differences among the nuclei and the differences of the number of electrons. It turns out that chemistry is essentially determined by electrons outside the nucleus and thus the differences in properties of all atoms brought out by the periodic table should be accountable by the differences in the number of electrons. How the different numbers of electrons and the different arrangements of them in atoms are related to atomic properties is the main concern in atomic physics.

12.4 The Bohr Theory of the Hydrogen Atom

The planetary model, though extremely successful in explaining the α-particle-scattering experiment, was not without serious difficulties. Unlike a planet, which can revolve around the sun indefinitely, an electron revolving around the nucleus, according to the electromagnetic field theory of Maxwell, will slow down and curve in, in much the same way as an earth satellite curves in and plunges back to earth. A circulating electric charge will radiate electromagnetic waves and thus lose energy. (This conclusion is derived from the basic laws of electromagnetic force.) Thus all electrons circulating about the nucleus would eventually curve in and plunge into the nucleus; there would be nothing left to fill up the size of the atom. Anyway an atom is supposed to be stationary (the atom of tomorrow should not be changed from the atom of today), and this model is not. The difficulty is a serious one because it involves a basic element in electromagnetism and the conflict cannot be avoided.

The problem was not completely solved until many years later when quantum mechanics was fully developed. Nevertheless, a significant step was taken in 1913 by Bohr, who proposed a theory of atomic structure, which, even though not complete and not wholly satisfactory, did bring out a number of significant features of atomic physics. Later, quantum mechanics was to develop these features and to fit them into a complete coherent theory. Because quantum mechanics is highly mathematical and difficult to present in terms of everyday language, we take the expedient of presenting atomic physics in terms of the Bohr theory, keeping in mind that what is said is more or less figurative and is justified only by a more subtle, complex treatment in quantum mechanics.

Bohr first considered the structure of the simplest atom, hydrogen ($Z = 1$), which has only one electron. He assumed first that the electron revolving about the nucleus cannot stay on any arbitrary orbit allowed by Newtonian mechanics, but can stay on only a few selected circular Newtonian orbits that satisfy a particular condition he worked out, called the *quantum condition*. This condition requires that the product of the electron mass m, the electron velocity v, and the radius of the electron orbit a (which is the angular momentum of the electron) be an integral multiple n of a universal constant called the *Planck constant*, $h = 6.625 \times 10^{-34}$ joule-sec, divided by 2π:

$$mva = n\left(\frac{h}{2\pi}\right), \qquad n = 1, 2, 3, \ldots$$

Between two selected orbits (Fig. 12.5), say those of $n = 1$ and $n = 2$, there are no other intermediate orbits the electron can occupy.

Moreover, Bohr assumed the following: contrary to the law of electromagnetism, the electron circulating on these selected orbits will not lose energy gradually by radiating electromagnetic waves but will remain on the same orbit without curve-in for a while. Then, suddenly, the electron can jump from one orbit, say, $n = 4$, to an inner orbit, say, $n = 3$,

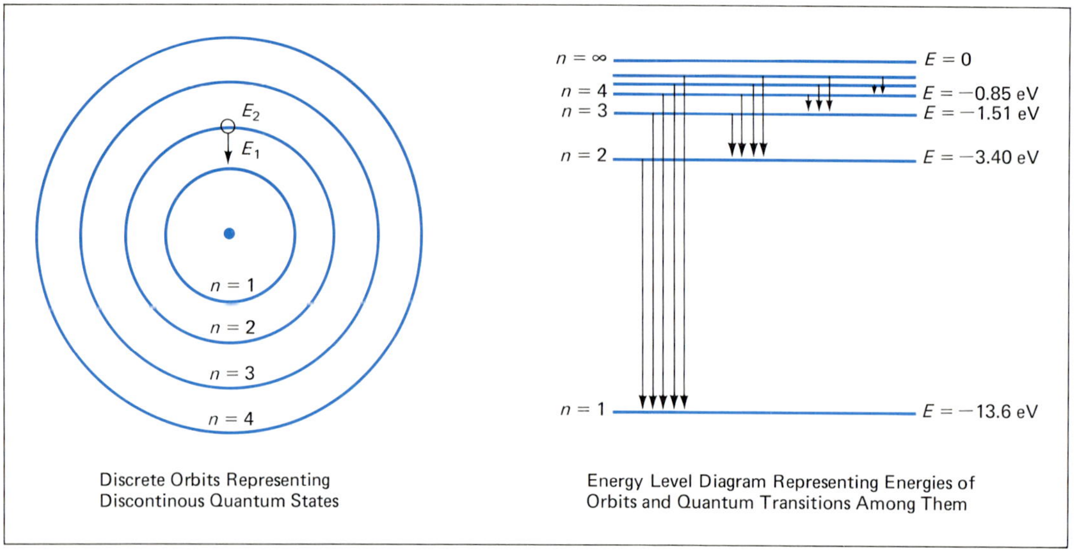

Figure 12.5 Diagrams illustrating Bohr's theory of the hydrogen atom.

a process called a *quantum jump* (a term that today has already been accepted in popular English usage), and lose a parcel of energy equal to the energy difference of the two orbits $E_2 - E_1$ all at once by emitting electromagnetic waves (light) of a particular frequency ν proportional to the total energy released $E_2 - E_1$. This second assumption is called the *frequency rule;* the proportionality constant is again the Planck constant:

$$E_2 - E_1 = h\nu$$

In other words, the continuous curve-in process of the previous theory (Fig. 12.6a) is now replaced by a series of discontinuous jumps (Fig. 12.6b). The innermost orbit ($n = 1$) has no other orbits to jump into; once the electron arrives at this orbit it will remain there indefinitely. The normal hydrogen atom is in this state (called the ground state). However, under stimulation the electron may be raised to an outer orbit (called an excited state) and light emission will follow as the atom returns to the normal state. Based on these two assumptions Bohr was able to calculate and predict the frequencies of light emitted by the hydrogen atom and this prediction agrees strikingly well with what was actually observed. Actually the two assumptions are contrived to explain the frequencies. When a theory such as this predicts several dozen numbers accurately to a minute fraction of 1 per cent, the success is so impressive that it leaves no doubt that the theory, even if not final and definitive, must contain some basic truth. The Bohr theory thus represents the first step toward solving the mystery of atomic structure.

The theory contains some elements of inconsistency and thus cannot be final. It makes use of the Newtonian mechanics for the calculation of orbits but arbitrarily limits the theory by accepting a selected group of orbits and rejecting all the others. Its assertion on light emission is contrary to the law of electromagnetism. These are the problems to be ironed out later in quantum mechanics.

One consequence of the Bohr theory that follows naturally is the prediction of the size of the hydrogen atom. The electron in the innermost orbit ($n = 1$) represents the hydrogen atom in its normal state. So the radius of the innermost orbit is a measure of the radius of the atom. The predicted value of this radius is 0.5×10^{-8} cm, which agrees well with the order of magnitude 10^{-8} cm of the atomic size.

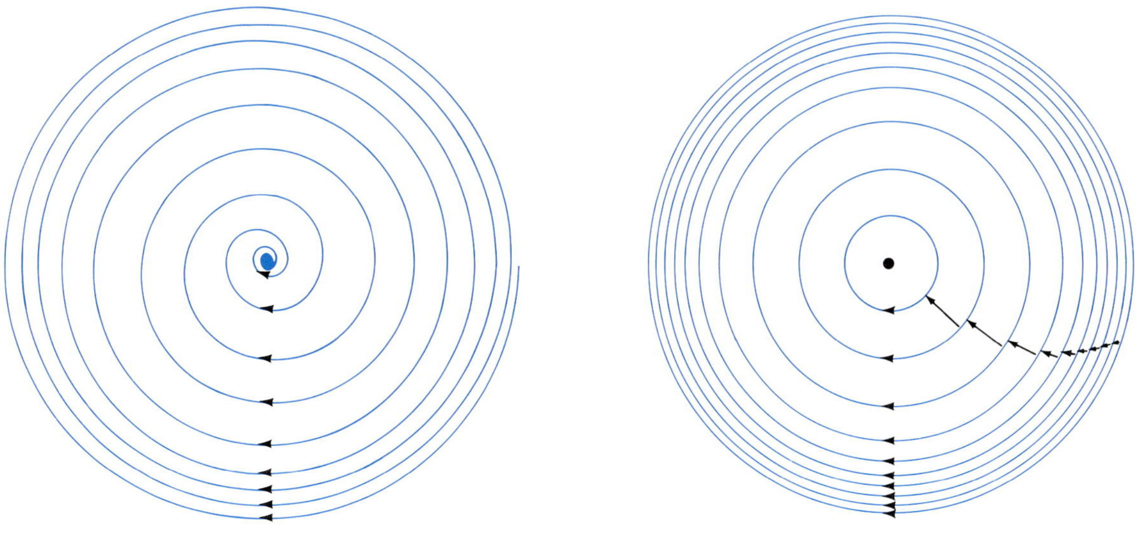

Figure 12.6 (a) In the classical theory the electron of the hydrogen atom curves in continuously. (b) In the quantum theory the electron jumps down from one orbit to another discontinuously.

12.5 The Bohr Theory of Atomic Structure

Once the theory is successful for the hydrogen atom, it is only natural to try to extend it to other atoms with more than one electron. Based on the concepts introduced in the theory of the hydrogen atom, we would expect electrons in other atoms to circulate about the nucleus on a set of selected orbits, and in the normal state all electrons would have descended onto the innermost orbit. If this were so, then, when we compare atoms from $Z = 1$ to $Z = 92$, the structure would change in a *monotonic* way and there would be no basis to explain the *periodic* changes of the properties of elements exhibited by the periodic table. Therefore a drastic revision is necessary.

Bohr a few years later worked out a theory of atomic structure based on an additional idea that each orbit can accommodate only a definite number of electrons. Each orbit is looked upon as a row of seats in a theater that can accommodate only a definite number of guests. When the quota is filled, latecomers will have to be seated in back rows one after another. Bohr's theory of atomic structure is essentially a seating plan of electrons in the allowed orbits. By filling up orbits one after another a periodicity may be brought to the fore. Indeed, the seating plan is worked out in a way aimed to explain the periods of the periodic table. Moreover, on the basis of such a seating plan and the resulting structure of the atoms, the physical and chemical properties of all atoms, including the periodic properties shown in the periodic table, can be naturally and satisfactorily explained. This seating plan, though an ad hoc assumption in the Bohr theory, comes out of the later quantum mechanics naturally and automatically without ad hoc assumptions.

The seating plan is a little complicated and some of the problems involved were not solved until the development of quantum mechanics many years later. The electrons filling up an orbit are called a shell. Thus the electrons filling up the first orbit $n = 1$ are called the K shell; the second orbit $n = 2$, L shell; the third orbit $n = 3$, M shell; and so on.

It was found necessary to assume that each shell of electrons is made up of a number of subshells. Thus the K shell ($n = 1$) has only one subshell, designated by 1s; L shell ($n = 2$) has two subshells, 2s and 2p; M shell ($n = 3$) three, 3s, 3p, and 3d; N shell ($n = 4$) four, 4s, 4p, 4d, and 4f; and so on. The letters s, p, d, f, g, h, i, . . . distinguish subshells. The subshells may be regarded as elliptical orbits of the same major axes (and thus the same energy) but different minor axes (and thus different angular momentum). The original Bohr theory of the hydrogen atom is limited to circular orbits but a generalization to elliptical orbits has been incorporated into it (Fig. 12.7). The quota of each subshell is determined by the designation s, p, d, f, g, h, i, ... Each s subshell can accommodate only 2 electrons; p subshell, 6 electrons; d subshell, 10 electrons; f subshell, 14 electrons; and so on. The subshells are to be filled in the increasing order of energy and angular momentum. In general, the seating plan is to fill the complete shells in the order of K, L, M . . . and within each complete shell the subshells are filled in the order of s, p, d, f . . . , but there are exceptions to be noted shortly.

With these quotas assumed, a grand scheme of seating the electrons is worked out as follows: the first atom hydrogen ($Z = 1$) has only one electron; this electron is seated in the 1s orbit. The electronic structure of this atom is designated by 1s. The second atom helium ($Z = 2$) has two electrons; both are seated in the 1s orbit and this completely fills the 1s subshell, which is also the K complete shell. The electronic structure of this atom is designated by $1s^2$. The filling of the K shell corresponds to the completion of the first, supershort period of the first two elements in the periodic table.

The third atom, lithium ($Z = 3$), has three electrons; two of them are seated in the 1s subshell and the third is relegated to the first available back row, the 2s orbit. The electronic structure of this atom is designated by $1s^22s$. The fourth atom, beryllium ($Z = 4$), has four electrons to fill both the 1s and 2s subshells. The electronic configuration is designated by $1s^22s^2$. The fifth atom, boron ($Z = 5$), has five electrons. With four filling the 1s and 2s subshells, the fifth is relegated to the next available orbit 2p. The electronic configuration is designated by $1s^22s^22p$. Because the 2p orbit can accommodate 6 electrons, the filling of the 2p subshell corresponds to elements from $Z = 5$ to $Z = 10$. We note now that the filling of the L shell ($n = 2$), which consists of two subshells 2s and 2p, corresponds to the completing of the first short period of 8 elements from $Z = 3$ to $Z = 10$ in the periodic table.

Figure 12.7 Extension of the Bohr theory to include elliptical orbits.

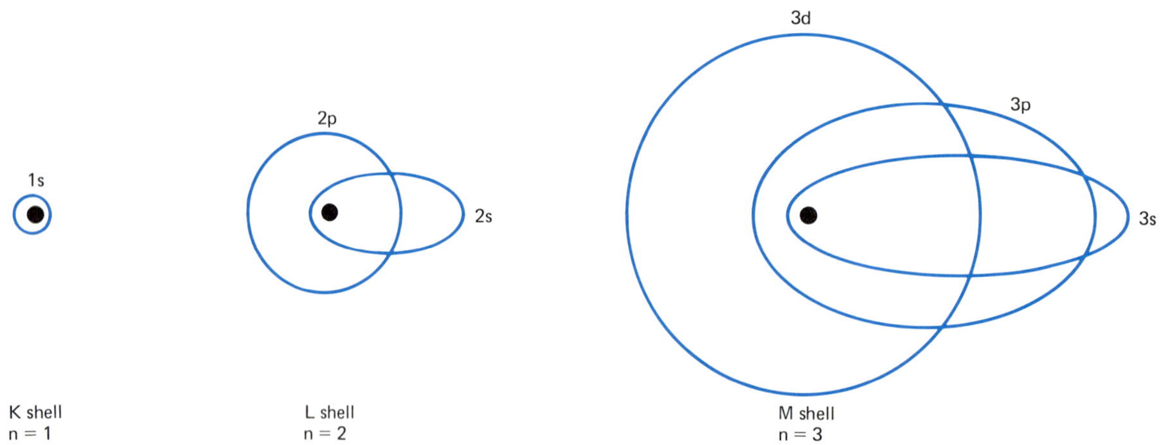

Subshells	1s	2s, 2p	3s, 3p	4s, 3d, 4p	5s, 4d, 5p	6s, 4f, 5d, 6p	7s, 5f, 6d
Periods	I	II	III	IV	V	VI	VII

The next 8 elements from $Z = 11$ to $Z = 18$ will have their extra electrons over and above those filling up the K and L shells seated in 3s and 3p subshells in a way similar to the filling of 2s and 2p subshells. These eight elements from sodium to argon thus form the second short period similar to the eight elements forming the first short period.

At this moment one might suppose that the next available orbit is 3d. Here a reverse of order takes place because of energy considerations; the 4s subshell (2 electrons), having lower energy, is being filled up first before 3d (10 electrons), which is then followed by the 4p subshell (6 electrons). A total of 18 elements is involved in filling the $4s^2 3d^{10} 4p^6$ vacancies, with electronic configurations somewhat related to the 8 elements in the short period (s and p subshells) but not exactly the same (because of the d subshell). They thus form the first long period of 18 elements with a rather complicated correlation with the short period as shown in the periodic table (Fig. 10.4).

The next 18 elements from $Z = 37$ to $Z = 54$ fill up subshells 5s, 4d, 5p in order. This part is similar to the filling of 4s, 3d, 4p subshells in the first long period and thus we have the second long period of 18 elements correlated closely with the 18 of the first long period. The subshell 4f is skipped because its energy is higher than 5s and 5p.

We would have a third long period of 18 elements corresponding to the filling of 6s, 5d, 6p subshells beginning $Z = 55$ exactly correlated with the previous long periods, but this time we find the energy of the 4f subshell falling in between 6s and 5d and therefore the subshell 4f (14 electrons) is to be filled between 6s and 5d. Thus we have a "superlong" period of $18 + 14 = 32$ elements. In the periodic table, the 14 elements corresponding to the filling of the 4f subshell are identified with the rare earth elements (lanthanons), which have very similar properties among themselves and thus do not fit in any periodic scheme, as mentioned previously. This group of atoms is usually taken out of the periodic table and put in the bottom as a separate appendage. With this excision the superlong period of 32 elements becomes a regular long period of 18 elements and can fit into the periodic table neatly as the third long period except that there is a gap of 14 of the atomic numbers following $Z = 57$.

After this period the next atomic number is 87, which is the beginning of the last and unfinished period of the table. After the filling of the 7s subshell (2 electrons), we have a similar situation in the filling of the 5f subshell (14 electrons) before 6d, resulting in the second appendage of 14 elements (actinons), which includes not only the heaviest natural element, uranium, but also most of the man-made transuranian elements up to $Z = 103$. These elements are also put in the bottom of the periodic table separately. The last known man-made element is hahnium, $Z = 105$, in the incomplete 6d subshell. The order of filling the subshells is summarized at the top of the page.

12.6 Explanation of the Properties of the Elements

The seating plan is not merely a game of numbers to fit the periodic table. The periodic table is based on the chemical and physical properties of the elements. The seating plan would not mean much unless it provides a basis to explain the periodic physical and chemical properties of the elements. Indeed the main task of atomic physics is to explain the physical and chemical properties of the atoms in terms of the arrangement of the electrons in the atoms. Because the periodic filling of shells corresponds to the periods in the table, the periodic properties can be explained as follows.

Let us consider a sodium atom and a chlorine atom. Their respective electronic configurations are $1s^22s^22p^63s$ for Na and $1s^22s^22p^63s^23p^5$ for Cl (Fig. 12.8). The Na atom has one extra electron outside a complete shell (L shell) and the Cl atom is one electron short to complete a shell (M shell). Because the seating order of subshells is the increasing order of energy, the atoms with a completed shell are particularly stable, because the next electron will be seated in the next shell of higher energy. Therefore those with an extra electron beyond a complete shell, such as Na, are expected to have the last electron loosely bound to the atom and easily detachable from it. Those with one electron short of a complete shell, such as Cl, are expected to have a strong tendency to pick up an additional electron to complete the shell. Thus when a Na atom and a Cl atom are brought together, it is natural for the Na to give up its outermost electron and for the Cl atom to pick it up. After that, the Na atom, losing an electron and becoming positively charged, changes into a positive ion, Na^+, and the Cl atom, picking up an electron and becoming negatively charged, changes into a negative ion, Cl^-. Positive and negative charges attracting each other, the Na^+ ion and the Cl^- ion attract together by electric force to form a composite system. This explains the formation of the NaCl molecule from the two atoms.

The idea of charge transfer has been mentioned previously in the dissociation theory in connection with electrolysis, but now it is worked out in a more complete theory on the electronic structure of the atoms. The phenomenon of electrolysis can thus be explained in terms of the Bohr theory of atomic structure. When NaCl is dissolved in water, the large dielectric constant of water reduces the electric attraction between Na^+ and Cl^- and the two ions separate from each other (dissociation). When electrodes are inserted, they can be sorted out and neutralized. The compound is thus decomposed by electrolysis.

In this way the chemical combination of two atoms leading to the formation of a molecule is explained and, as mentioned previously, the chemical

Figure 12.8 Chemical combination of a sodium atom and a chlorine atom by transferring one electron from the sodium to the chlorine. The two ions thus combine to form the sodium chloride molecule.

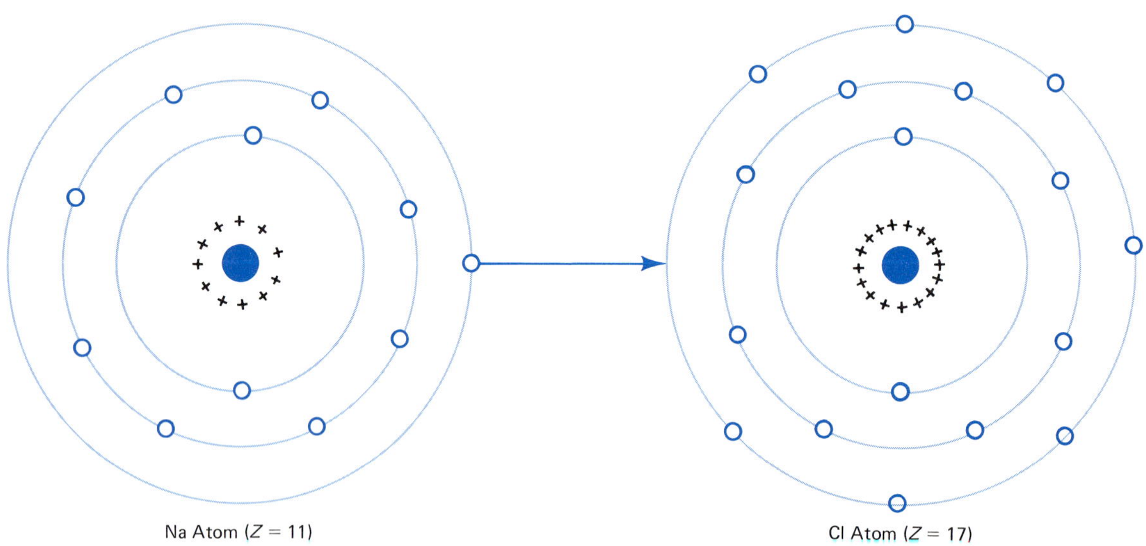

affinity that used to be associated with a mysterious force of the atom is now explained on the basis of the well-known electric force. Moreover, the property of valence can now be explained naturally. When hydrogen combines with chlorine, it has only one electron to lose and chlorine can pick up only one electron. The transfer of the electron gives rise to one H^+ ion and one Cl^- ion and the resulting compound is one made of one H atom and one Cl atom. Thus the fact that H and Cl have valence 1 is related to the fact that one electron is involved in the transfer. The formation of NaCl with one Na to one Cl is also determined by the fact that Na has one loose electron to donate. On the other hand, when chlorine is to combine with magnesium, Mg ($Z = 12$, $1s^2 2s^2 2p^6 3s^2$), which has two electrons beyond a complete shell, we have a slightly different situation. The Mg atom is disposed to lose two electrons and become a doubly charged positive ion Mg^{++}. The two electrons are picked up by two Cl atoms to form two Cl^- ions. The Mg^{++} ion will attract two Cl^- ions and thus form a neutral molecule $MgCl_2$. Once neutral, no other ions can be added to it and therefore the molecules formed by Mg^{++} and Cl^- ions are always $MgCl_2$. Phenomenologically we say that Mg has a valence of 2 so it combines with two chlorine atoms of valence 1. Now we see that this is due to the fact that Mg has two electrons outside the complete shell. Thus for such atoms valence is just the number of electrons outside the closed shell. (This corresponds to atoms of positive valence.) By a similar argument, for those atoms a few electrons short of a closed shell, valence is just the number of electrons short of the closed shell (corresponding to negative valence).

Thus we have explained not only the valence but also the two kinds of valence ($^+$ and $^-$) corresponding to excess or deficiency of electrons over a complete shell. Only atoms of opposite valences can combine in this mechanism of chemical combination, one donating and the other accepting electrons. Li and Na have valence 1 but they do not combine because both are electron donors, and no chemical bond can be formed by this mechanism. This explains why alkali metals combine with halogens, one atom to one atom, but alkali metals do not combine among themselves.

We are now in a position to explain the grouping of elements according to their chemical and physical properties. Each group of elements, such as the alkali metals, actually consists of those atoms having the same electronic configuration in the outermost shell, e.g., having the same number of extra electrons outside a complete shell. It follows that they have the same valence and similar chemical properties in that they can form similar compounds (e.g., NaCl, KCl, LiCl). Elements in the same group have similar physical properties, e.g., all alkali metals are conductors. The easily detachable electrons become *free electrons* in a solid and can easily move about to transport electricity and heat. The solid thus becomes a good electric and heat conductor. On the other hand, halogens and other elements in the right-hand side of the periodic table are short of electrons for a complete shell. They do not have easily detachable electrons and so they are electric and heat insulators. Thus practically all chemical properties and many physical properties of the atoms are determined by the electronic configuration of the *outermost* shell. The electrons in the inner shells are inactive and do not correspond to any observed properties (exception: X rays). They are buried deeply inside and are not "seen" in ordinary physical and chemical processes.

The group of alkali metals consists of atoms with one electron in the outermost shell. Thus Li has one electron beyond the completed K shell. Na has one beyond the completed L shell, K has one beyond the M shell, Rb has one beyond the N shell, and Cs has one beyond the O shell. The group of halogens consists of atoms with one electron short of a complete shell. Thus F, Cl, Br, I correspond to one electron short of L, M, N, O shells. (Here the M, N, O shells are not complete in the sense that the d and f subshells are to be completed later. They are "complete" in the sense used here that a new shell is to be started after them.)

The element between a halogen and the next alkali

metal is one that has the outermost electron shell completely filled. When the shell is complete there is no tendency to donate or accept electrons and therefore there is no tendency to enter into chemical combination with other elements. The valence, according to the number of electrons, is 0, which is a proper representation of the fact that it combines with no other elements at all. Indeed, the elements between the halogens and alkali metals do form a group, the noble gases, characterized by their chemical inertness. They not only do not combine with other elements chemically to form compounds, but also do not aggregate with themselves physically to form a solid or a liquid. They are thus all gases (He, Ne, Ar, Kr, Xe, and Rn). They have electronic configurations corresponding to the completion of the K, L, M, N, O, P shells.

The atoms following the alkali metals have two electrons in the outermost shell and form the group alkaline earths having similar properties among themselves. They include Be, Mg, Ca, Sr, Ba. They all have valence 2 and are all metals. They can be substituted for one another chemically.

As the atomic number Z increases after a complete shell, we first have an atom in Group I of the periodic table (Fig. 10.3), the alkali metals, which have valence 1; then the next in Group II, the alkaline earths, which have valence 2; then the next in Group III, the boron or aluminum group, which has valence 3; then the next in Group IV, the carbon and silicon group, which has valence 4; then the next in Group V, the nitrogen or phosphorus group, which has valence 5 (positive) or 3 (negative); then the next in Group VI, the oxygen group, which has valence 2; then the next in Group VII, the halogens, which have valence 1; and finally the next in Group VIII, the noble gases, which have valence 0. This completes a period after which another period follows. Thus we have explained the periodic changes of properties of elements and the arrangement of the elements in the periodic table according to rows and columns representing periods and groups, respectively.

The exceptions to the periodic changes of properties are represented by the two appendages at the bottom of the periodic table, corresponding to the rare earth elements, the lanthanons ($Z = 58$ to $Z = 71$), consisting of the 14 elements filling the 4f subshell, and the actinons ($Z = 90$ to $Z = 103$), consisting of the 14 elements filling the 5f subshell. Why do the periodic changes of properties of elements, as the atomic number Z increases, suddenly come to a stop when Z runs through these two gaps with elements in each gap very similar to one another? These exceptions can also be explained based on the electronic structure. Because the seating in the 4f and 5f subshells are each superseded by an outer shell, these subshells are deeply buried inside and the change of number of electrons in these subshells as they are being filled does not change the electronic configuration of the outermost shell and thus does not essentially change the chemical properties of the elements. Thus both the regularities and the exceptions of the periodic table can be explained on the same basis of the electronic structure.

*12.7 Chemical Bonds

The preceding theory of atomic structure explains chemical combination of elements, such as alkali metal and halogen, leading to a compound that can be decomposed by electrolysis. Yet there are compounds that cannot be decomposed by electrolysis. Sugar is an example. Among them are all the compounds made of two identical atoms, such as H_2, O_2, Cl_2, and so on. The two atoms being identical, there is no reason why one should become a positive ion and the other negative. It thus appears that there are two kinds of compounds formed by two different mechanisms associated with two different kinds of chemical bond. The mechanism of electron transfer leads to the formation of *ionic compounds,* based on the *ionic bond* (or *polar bond*). There must exist another mechanism not based on electron transfer yet having the same effect of combining two atoms but this time to form a nonelectrolytic compound. The corresponding chemical bond is called the *covalent bond* (or *homopolar bond*). The resulting compound is called a *covalent compound.* Our knowledge of electricity and electronic structure of atoms makes the ionic bond easily understandable but does not shed any light on the covalent bond. For a molecule made of two identical atoms, the two must be on the same footing and so the molecule cannot be polarized. With both atoms neutral, how can they stick together? This problem was not solved until the full development of quantum mechanics. It was then found, in studying the motion of electrons between two atoms, that sometimes two electrons, one from each atom, move together in such a way that the two atoms become bound together. This kind of motion of a pair of electrons explains the covalent bond. This is a *quantum effect,* meaning that it comes only with the quantum theory but not with the Newtonian theory. It comes out of quantum mechanics naturally and automatically and can be predicted on a quantitative basis even though it cannot be described verbally and pictorially. The explanation of the covalent bond is thus a great triumph of the quantum theory; it tied up one of the last loose ends between physics and chemistry. It demonstrated once again that the phenomena of chemistry, like many other natural phenomena, are ultimately reducible to the problem of motion (of the electrons) and thus are reducible to mechanics. Moreover, the force involved in the covalent bond, in the final analysis, is again the familiar electric force (manifested in a quantum effect). Thus all the "forces" associated with chemical affinity of atoms are eventually reduced to the familiar force of electricity as we have stated earlier. In the following we shall see this statement also holds for the other interactions, i.e., hydrogen bond and intermolecular force. Thus chemistry is based on one single force—the electric force.

The Hydrogen Bond. Chemical compounds are formed either by an ionic bond or by a covalent bond, these being the two basic chemical bonds that appear in the molecular formula. Two atoms combined by such bonds have a definite distance between them. Three atoms combined by covalent bonds have a definite angle between the two bonds. (Bond direction is the direction connecting the two atoms.) For example, the two OH bonds of the water molecule H—O—H do *not* lie along a straight line but make an angle of 104.5°. The molecules formed thus have not only a definite chemical composition but also a definite geometrical configuration. However, besides these bonds there is another, much weaker, chemical bond representing another kind of molecular interaction. It is the *hydrogen bond* mentioned in the Watson-Crick theory. For example, water molecules in liquid water are not single molecules H_2O. A number of H_2O molecules are connected by linking the H of one to the O of another, thus forming a cluster (Fig. 12.9). This link, much weaker than the covalent bond within the H_2O molecule (weaker by a factor of 10), is due to the hydrogen bond. Because it is weak, the links are not permanent. The H_2O molecules associate and dissociate constantly. The cluster does not have a definite composition (the number of H_2O molecules increases and decreases in time), nor a definite shape. The cluster is thus not considered as a single large molecule and hydrogen bond links do not appear in the molecular formulas. Nevertheless, it represents a form of molecular interaction that is important in a number of ways.

The origin of the hydrogen bond may be regarded as the residue of the primary chemical bonds. Thus the hydrogen bond may be considered as a secondary bond. In the molecule H_2O the electron of the hydrogen atom is shared with the oxygen atom and thus the hydrogen becomes slightly positive and the

Figure 12.9 Water molecules forming a cluster by hydrogen bonds.

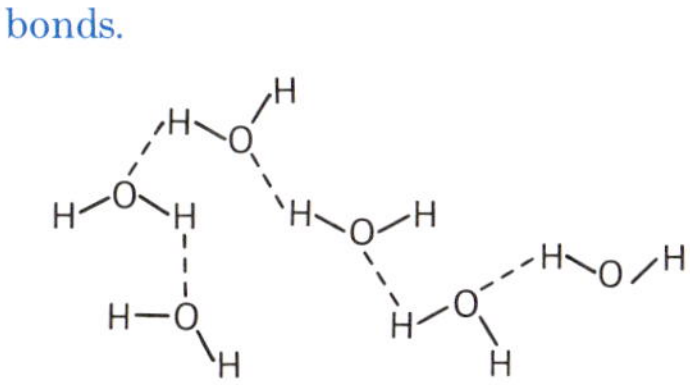

oxygen slightly negative; the water molecule thus is electrically polarized and appears as a *dipole* (two equal and opposite charges q separated by a distance d); it has a permanent *electric dipole moment* (the product qd). The hydrogen thus may be attracted to the oxygen of a neighboring H_2O molecule, forming a hydrogen bond. Thus the force of the hydrogen bond is again the electric force.

If there were no hydrogen bond, life would not be possible. The primary bonds are too strong; once a compound is formed it does not easily break up. Everything would be deadly set. Life means activities; there must be a way association and dissociation can be made to proceed more easily so that a lot of changes can occur. The hydrogen bond fulfills this requirement. Moreover, the hydrogen bond is specific; it does not bind blindly with anything coming its way. The specificity is exactly what is required in the Watson–Crick pairing; C pairs only with G and A only with T. This specificity makes possible the storage of a large amount of biological information in DNA. The weakness of the bond makes the use of this information (in replication, transcription, and translation) possible on a physical base. Thus the specificity and the weakness are the two features of the hydrogen bond that make life possible.

*12.8 Intermolecular Force

Besides the hydrogen bond, on an even weaker level, there is a third kind of molecular interaction—a very weak and nonspecific attractive interaction between all molecules. It represents a blind, indiscriminate attraction force among all molecules. The origin of this force is again the familiar electric force. The nature of this interaction is basically not different from the primary and secondary bonds. However, because it is very weak and nonspecific, it does not (and cannot) enter into the molecular formula and it is not even considered as a chemical bond. It is simply called the *intermolecular force*.

The quantum theory of the force that is responsible for the chemical combination of atoms also provides the explanation of the intermolecular forces between the molecules. From the study of macroscopic properties of matter we already know the existence of the intermolecular force. In solids such a force must exist among molecules of the same kind so that the molecules stick together to form a solid. In liquids the intermolecular force manifests itself in such phenomena as *adhesion* and *surface tension*. In gases, where the molecules are far apart, the existence of the intermolecular force is ascertained from the van der Waals equation. Without knowing the structure of the atoms and molecules, the intermolecular forces would appear just as mysterious and as unrelated to the other forces as the forces of chemical bonds. Once we know the structure of atoms and molecules in terms of a definite arrangement of the electrons around the nuclei, then the intermolecular force must have its origin rooted in the forces among these particles and in the arrangement of these particles. In other words, intermolecular force should be explainable in terms of the atomic and molecular structure.

For one thing the positive charge of the nucleus and the negative charge of the electrons in the atoms and molecules do not exactly cancel their effects. They would if they occupied exactly the same position but electrons are *outside* the nucleus at some distances. Therefore even though the major parts of the electric forces from the nuclei and from the electrons of a molecule cancel each other out, there may still be a small amount of residual force left, called the *dipole force,* that can act on other electric charges. Thus a molecule having a dipole moment (such as water) may exert a dipole force that, being residual, is very weak. When the dipole force is acting on the electrons and the nucleus of another molecule, the positive charge of the nucleus and the negative charge of the electrons experience opposite but nearly equal forces. The two forces nearly balance out, but not exactly. Thus there is a net residual force left that becomes the intermolecular force between the two molecules. Therefore the ultimate origin of this intermolecular force is again the familiar electric force.

The intermolecular force just discussed is one existing between two molecules, each having a permanent electric dipole moment. It is called *dipole-dipole interaction.* There are two other kinds of intermolecular force. One exists between a molecule with a permanent dipole moment and another without. The first molecule exerts a dipole force to polarize the second molecule, producing an *induced-dipole moment.* Then an interaction takes place between the two dipoles. This kind of intermolecular force is called *dipole–induced-dipole interaction.* This force as well as the previous one is understandable from the Newtonian standpoint.

The third kind is one existing between two molecules both without a permanent dipole moment. There is no dipole–dipole interaction. Because no one exerts a dipole force to polarize the other, there is no induced dipole either and thus no dipole–induced-dipole interaction. Thus from the Newtonian point of view they should not attract each other, yet they do. This is another puzzle, like the covalent bond, that is eventually solved by quantum mechanics, again naturally and automatically without adding any new assumption. Roughly speaking, the origin of this interaction may be visualized as follows. One molecule for no reason suddenly becomes slightly polarized by itself. It thus generates a dipole field to polarize the other. The other thus generates a dipole field to polarize the first further. This action goes back and forth many times until both assume induced dipoles. Then a force appears between these dipoles. The first step is crucial. It is like borrowing money to get the business started. Later, not only can the loan be repaid, but also an ongoing business may be established. Such "borrowing" is not possible in the Newtonian theory but is permitted in the quantum theory (because of the uncertainty principle). Thus this *induced-dipole–induced-dipole interaction* can be explained in the quantum theory. (Another piece of the jigsaw puzzle falls into place!) The origin of the force is again the familiar electric force. In general, between any two molecules the intermolecular force is a combination of the three kinds of interactions.

The last interaction is also referred to as the *London force.* It has a peculiar property that may have a significant bearing in biology. Two identical molecules under the action of the London force will attract in such a way that they will overlap as much as possible. Two spade-shaped molecules will thus stack one upon the other, tip to tip and stem to stem. This follows the quantum mechanical calculations under the same dynamical laws and force laws without any new assumption. In biology, identical macromolecules do line up or stack up in structures (tissues) and in functional processes (replication, transcription, and translation). If there is no supernatural influence, what makes the molecules line up or stack up before they can be processed in biochemical processes? Perhaps it is the London force.

The chemistry of life is peculiar in that it needs both stability and changeability. Also both variety and regularity. (The role of the hydrogen bond in providing changeability has already been discussed.) It needs variety to accommodate the large volume of genetic information. Yet it needs regularity to process the molecules in biological functions the simplest way possible. The two requirements are contradictory but they are both fulfilled in a way that is most ingenious indeed. Take the DNA molecule as an example. The variety of genetic information is reflected in the variety of the base sequence of the chain. Ordinarily a chain of such a complexity (like a protein) would mean a very complex molecule, presumably with a very complex structure, like the Gordian knot, which would be very difficult to process (e.g., to unravel in replication). Even a tangled garden hose is enough of a headache for an intelligent human being. A tangled nucleotide chain, a million times more complex than a tangled garden hose, would require an omnipotent intelligent being to handle. Yet DNA carries out its function efficiently, even in such lowly creatures as a bacterium or a virus. (Should God waste His creative energy in tending the daily chores of multiplying bacteria and viruses?) We can now understand this by recognizing that in spite of the variety of the base sequence, the structure of DNA is very simple and very regular; in fact, it is merely a double helix. This is made possible first by adding a complementary chain so that the two chains form a ladder of uniform rungs. Then because of the intermolecular force between the base pairs—largely the London force, which tends to stack the base pairs as close as possible—the base pairs are drawn together to stack closely one upon another. As a result the straight ladder is twisted into a spiral staircase. A double helix is a simple and regular geometric structure and can be processed by robots (replication enzyme, transcription enzyme). Thus in the DNA molecule both variety and regularity are present. The physical basis for this unusual feat is the hydrogen bond and the London force.

Whereas chemistry takes up the lion's share of molecular interaction (ionic and covalent bonds), biology deals with the residue of it (hydrogen bond and London force). Yet it is in the seemingly trivial residue that the most interesting thing—life—happens. And all this is reducible to one single force—the electric force.

Problems

12.1 In the electrolysis of a copper salt how many coulombs of electric charge are needed to liberate 1 g of copper? The atomic weight of copper is 63.54 and the valence is 2.

12.2 Write down the electronic configuration of the structure of fluorine, $F(Z = 9)$. What are the properties that can be predicted from this structure?

12.3 Write down the electronic configuration of the structure of calcium, $Ca(Z = 20)$. What are the properties that can be predicted from this structure?

12.4 Explain the chemical combination of calcium and fluorine.

13 Electromagnetic Energy

13.1 Electromagnetic Waves

Up to this point all considerations of energy—kinetic, gravitational, thermal, and electric—have been related to material systems. This is natural from the Newtonian point of view, in which matter is the central concept and energy is a manifestation of matter. However, as science advanced, we came to know a new entity, the electromagnetic wave, and found it to possess energy as well. An electromagnetic wave is not a material system; it is a state of space. Empty space has energy! Indeed, the theory for it—the field theory—represents a revolution against the classic Newtonian ideas. Now the field theory is generally accepted and we send energy (radio signals) routinely to spacecraft through empty space without a material medium.

From the practical point of view, electromagnetic waves have entered modern life in more than one way—radio, television, radar, X rays, and so on. Some of our environmental problems are related to them.

Let us introduce the concept of the electromagnetic wave, beginning with a discussion of waves. We are already familiar with sound waves. The simplest kind of wave is the water wave. When a stone is dropped into a pond, it produces a wave, a series of ripples emanating from where the stone is dropped outward in all directions. The ripples propagate outward with a definite velocity; this is referred to as the *wave velocity*, v. The ripples consist of a series of concentric circles, some representing the peaks of the wave, some the troughs (Fig. 13.1). The distance between two neighboring peaks or troughs is called the *wavelength*, λ. If we fix our attention on one particular point, we see the water level changing up and down and we may speak of a *wave frequency*, ν, representing the number of vibrations per second. The wave propagates in such a way that when one cycle of vibration is completed, the *wave front* advances one wavelength. In one second the front ad-

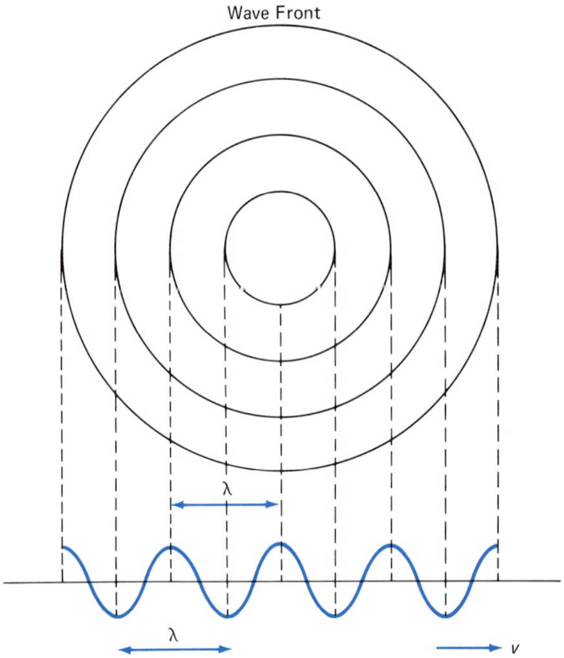

Figure 13.1 Water wave.

vances a distance of $\nu\lambda$, which by definition is the wave velocity. Thus

$$v = \nu\lambda$$

The idea of the wave first introduced in such examples may be extended. Whenever a physical quantity changes periodically in space and time, we called it a *wave*. In a water wave, the physical quantity that changes periodically is the height of the water level, which moves up and down in the course of time for a fixed point in space, and undulates in space for a fixed time. When some other quantity changes in a like manner, another wave has manifested itself. Thus the air wave is a periodic change in space and time of the air density. This air wave is the physical basis of sound, and we can speak of sound velocity, wavelength, and frequency. Sound velocity under normal conditions is 1,130 ft/sec. The frequency of the middle C note is 254 cycles/sec. A higher pitch by one octave corresponds to the doubling of the frequency.

Not only the water level and the density of air, but also many other physical quantities may change periodically, giving rise to other kinds of waves. When the physical quantities that change periodically are the electric field and magnetic field (the two usually accompany each other because of electromagnetic induction), the wave is called an *electromagnetic wave*.

The production of an electromagnetic wave may be visualized as follows: Suppose that an electric charge Q is vibrating up and down, say, along a metal wire (Fig. 13.2). At a distant point P in space the Coulomb electric field produced by Q, along the line joining P and Q, also changes direction periodically. According to electromagnetic induction, a periodic change of electric field induces a periodic change of magnetic field, which in turn also induces periodic changes of electric field. Thus the fields produced by a vibrating electric charge are more complex than that of a static charge. They include not merely a Coulomb field, $E = Q/r^2$, but an electromagnetic wave, a periodically changing electric field and an accompanying magnetic field that changes in a similar fashion.

Although electricity and magnetism have been studied separately from ancient times, the electromagnetic wave (Fig. 13.3) is a rather recent discovery. Its existence was first predicted by the electromagnetic field theory of Maxwell in the latter part of the nineteenth century. The theory also predicts the velocity of electromagnetic waves to be 3×10^8 m/sec. Not long afterward, Hertz discovered electromagnetic waves experimentally and verified many theoretical predictions. Soon after that

Figure 13.2 Generation of an electromagnetic wave by an oscillating charge Q.

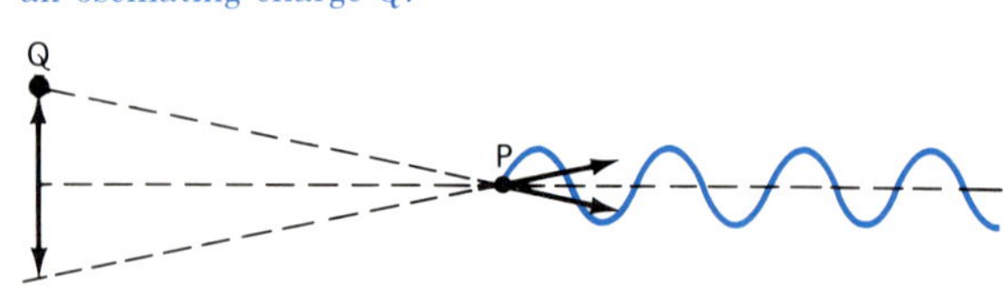

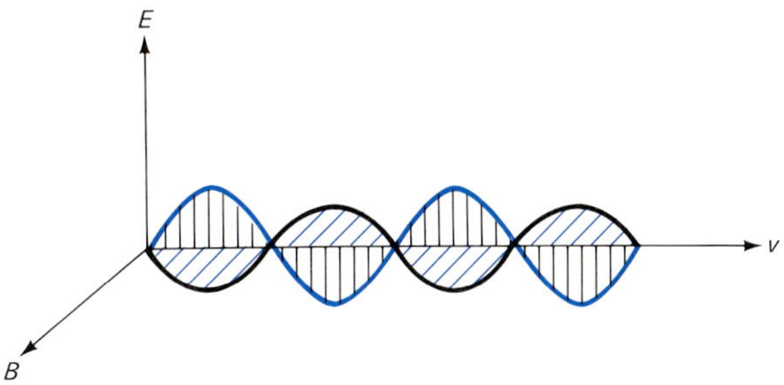

Figure 13.3 The electric and magnetic fields of an electromagnetic wave related to the direction of propagation v.

the discovery was put to practical use in radio communication. In fact, the antenna of a broadcasting station is essentially a metal wire in which electric charges can vibrate, as shown in Fig. 13.2, to generate electromagnetic waves.

One significant development of the electromagnetic theory is the identification of light waves with electromagnetic waves. Light has been studied for a long time. Early in the nineteenth century it was established by experiments on *interference, diffraction,* and *polarization* of light that light is a wave. But what kind of wave? What is the physical quantity that changes periodically in the light wave? This question remained unanswered for a long time. The discovery of electromagnetic waves provided a possible answer, which was supported by the fact that the velocity of light c, which has been known for some time, is exactly the same as the velocity of electromagnetic waves:

$$c = 3 \times 10^8 \text{ m/sec}$$

After much investigation, it is now firmly established that light is an electromagnetic wave.

Although all electromagnetic waves have the same velocity c, their wavelengths vary and indeed change over a wide range. The frequencies, equal to c/λ, also vary over a wide range. A mixture of electromagnetic waves of various wavelengths may be separated into a *spectrum,* e.g., the white sunlight may be separated into a spectrum of colors from red to violet. The study of the spectrum is called *spectroscopy.* One technique in optical spectroscopy is to use a prism to separate the light waves. The principle and the procedure may be extended to other electromagnetic waves.

The other kinds of electromagnetic waves besides light enter our modern life one way or another and thus are practically important. The following is a systematic description of them throughout the electromagnetic spectrum beginning with the long-wavelength, or low-frequency, end.

The household electric current is an alternating current of 60 cycles/sec. When this current flows in a wire loop it means that the electric charges rush back and forth in the wire loop 60 cycles each second. In other words, the electric charges are vibrating with a frequency of 60 cycles/sec. Consequently, an electromagnetic wave of 60 cycles/sec (corresponding to a wavelength of $3 \times 10^8 \div 60 = 5 \times 10^6$ m) will be generated. It is a fact of nature, which is also predicted by the electromagnetic field theory, that the power of electromagnetic wave emission increases rapidly with frequency, other factors remaining the same. At 60 cycles/sec, which is an extremely low frequency for electromagnetic waves, the power is so weak that this wave emission can generally be ignored. On the other hand, the electric disturbances resulting from the wave emission generated by the operation of electric appliances (shavers, can openers) are usually strong enough to interfere with the reception of radio and television programs.

The radio waves emitted by standard AM broadcasting stations are in the wavelength range of several hundred meters. The AM radio receiver covers a frequency range that can accommodate a dozen stations of different wavelengths. These wavelengths are still long in electromagnetic wave standard. *Long-wave* radio signals attenuate quickly over a distance. Standard broadcast usually covers just a metropolitan area. For transcontinental radio communication shorter wavelengths of the order of meters are used and a different radio (*short-wave* radio) is required.

During World War II even shorter waves were used

in the development of *radar*. The wavelengths of radar are in the range of centimeters and the waves are called *microwaves*. Besides their use in radar, microwaves are used in quick-heating microwave ovens. In such applications the practical advantage may be offset by environmental hazards. The oven may leak harmful doses of radiation. This is the subject of radiation pollution already discussed in Sec. 2.6.

Proceeding to even shorter wavelengths down the spectrum, we come to *infrared* waves, *light* waves, and *ultraviolet* waves, in that order. These are usually associated with the phenomena of light and were studied in *optics* before the waves were used in communication. Light waves are of course the most familiar. Their wavelengths are in the range of 10^{-5} cm. The length scale becomes so small that we need a new unit and the angstrom, equal to 10^{-10} m, is introduced, abbreviated A. The wavelength range of the visible light, from red to violet, is 7,000A to 4,000A. Beyond this range the eye cannot see.

Light waves not only have a visual effect, but also give rise to a heat effect as well as to a chemical effect, such as affecting a photographic plate. The light waves from red to violet in the visible spectrum can be tested for these effects. It was found that the heating effect extends beyond the red light in the spectrum, indicating that there exists some "light" with a wavelength longer than red light, invisible to the eye but detectable by heat-sensing devices. This "light" is *infrared*. In a similar manner, the chemical effect on photographic plate extends beyond violet, indicating the existence of a shorter-wavelength, invisible "light," i.e., *ultraviolet*. The waves are all present in a light source and are basically alike, but our eyes are not good enough to see them all and thus only the middle portion of this spectrum was first recognized as light. There are indications that some animals, such as rats, can see infrared rays and thus can see things in the dark.

Proceeding to waves of even shorter wavelengths, we come to the *X rays,* which are made familiar to us by their medical applications. The wavelengths are in the range of angstroms. The X ray was discovered in 1895. Its nature was a mystery for some time (hence its name), but its penetration power and its use for examination of the human body attracted attention. It was later identified as a very short-wavelength electromagnetic wave.

The shortest wavelengths belong to the *γ-rays* first discovered in radioactivity. They appear frequently in nuclear physics. Their wavelengths are in the range of a small fraction of an angstrom, down to 10^{-3}A, or 10^{-13} m. Ultraviolet rays, X rays, and γ-rays, in addition to the physical and chemical effects, also show significant biological effects and figure importantly in the problem of radiation pollution. The various waves are summarized in Fig. 13.4.

*13.2 Thermal Radiation

We now consider a point related to heat, namely, thermal radiation. Earlier we have discussed that a hot object may lose heat even though it is surrounded by space and that the rate of energy outflow is proportional to the fourth power of its absolute temperature. How is heat energy transferred through a vacuum? This question can now be answered. It is through the emission of electromagnetic waves, which carry away energy. For an ordinary hot object, such as a cup of hot coffee, the waves emitted that carry away heat energy, and thus cause the coffee

Figure 13.4 Electromagnetic wave spectrum.

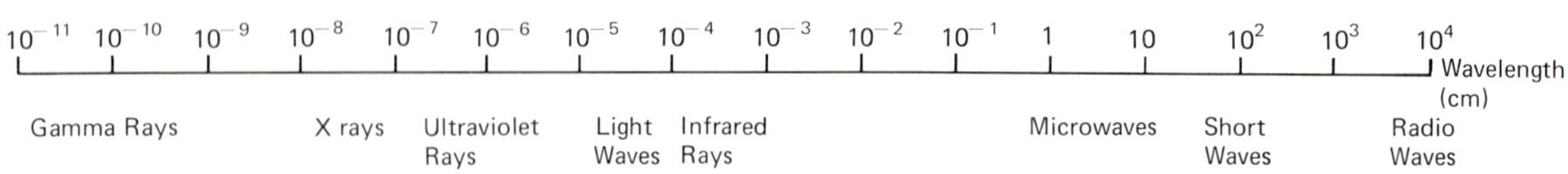

to cool, are in the infrared range and are invisible. We may also call them heat waves. The emission of heat waves from a hot body is different from the emission of waves from atoms, molecules, and nuclei to be discussed in Sec. 13.4. It is due to the collective action of all atoms (or molecules) the hot body is made of. Thus the emission spectrum is not discrete but continuous. Moreover, the spectrum is not specific to the emitting substance. It is essentially determined by the temperature of the emitting body, thus the term *thermal radiation*. To study this phenomenon we concentrate on the *blackbody*, the body of maximum emissive power.[1] The radiation it emits is called *blackbody* radiation.

The spectrum of blackbody radiation is given in Fig. 13.5, which shows the intensity of the wave radiated as a function of the wavelength λ. It depends only on the temperature T (independent of chemical constitution). As the temperature increases gradually, we find the maximum of the distribution curve λ_m shifts to lower wavelengths. The shift is such that the product $T\lambda_m$ remains a constant (Wien's law). This means that, at a higher temperature, the majority of the radiations emitted have shorter wavelengths or higher frequencies. As the temperature of a hot body is increased, the majority of the heat waves emitted shifts from the infrared to visible light. Thus a piece of iron heated to about 500°C (red hot) begins to glow with the emission of red light. As the temperature increases to 1,000°C (white hot) the maximum wavelength λ_m shifts toward higher frequencies and the light emitted becomes white. At even higher temperatures the color of light emitted becomes bluish. The change of color thus is closely related to temperature, a fact known to the artisans of ceramics for a long time. As the temperature increases further, the peak centered at λ_m moves out of the visible region into the ultraviolet and the light emitted by such an extra-hot object is invisible again! This does not happen in everyday experience because everything, including the furnace, evaporates when temperatures reach higher than 3,000°C. Thus no fire can be built hotter than this limit. On the other hand, in cosmic events (and in atomic explosions) this may well happen.

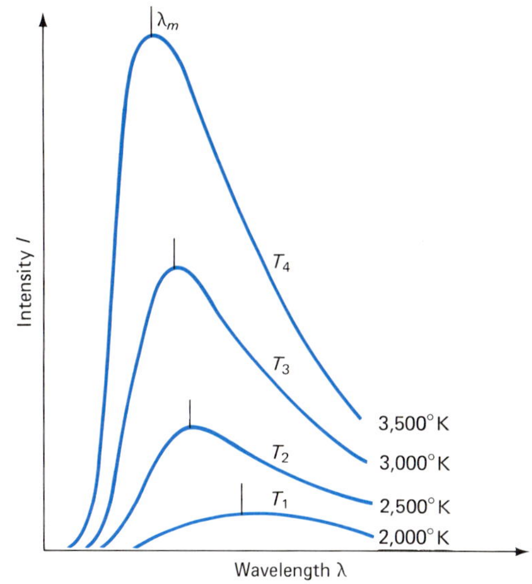

Figure 13.5 Spectral distribution of the blackbody radiation for a number of temperatures.

At a high temperature not only the wavelength of maximum intensity λ_m shifts to a low value, but also the total energy, the sum over the distribution curve shown in Fig. 13.5, increases rapidly. In fact, it increases to the fourth power of the absolute temperature. This is the *Stefan–Boltzmann law* of thermal radiation we mentioned in heat transfer (Sec. 7.1). The spectral distribution of Fig. 13.5 merely shows the breakdown of the total energy according to wavelengths.

The theory of blackbody radiation is the starting point of the development of the quantum theory (next section), and thus is a topic of considerable theoretical and historical interest. It is also involved in the study of the beginning of our universe (the primeval fireball), to which we will return in Chapter 21.

[1] It also has maximum absorptive power, thus black.

*13.3 The Quantum and the Photon

The spectral distribution curve shown in Fig. 13.5 was known at the end of the nineteenth century. The problem that naturally arose was to explain this curve in terms of the existing laws of physics. The problem involves mechanics, electromagnetism, and heat. At the end of the nineteenth century the laws of mechanics, electromagnetism, and heat in macroscopic physics had indeed been developed to a state of completion. (Microscopic physics of atomic structure had not started.) These laws have been applied to study the problem of spectral distribution and they led to a definite conclusion (the Rayleigh–Jeans formula). Unfortunately the result did not agree with the experimental facts. Thus there arose a crisis in theoretical physics.

Physics before this point is called classical physics. It deals with macroscopic phenomena and is nearly complete and consistent. Many physicists at the time thought that all basic discoveries in physics had been made and that future progress would be limited to refinements of instrumentation. [This optimism, as physicists learned later, proved to be quite premature. The discovery of the electron (1897), together with the discovery of X rays (1895) and radioactivity (1896), opened up a new, unexplored field of microscopic physics.] Yet classical theory predicted wrongly the law of blackbody radiation. This indicated that classical theory might need revision. Indeed, this is the starting point of the quantum theory, and later we are to learn that quantum theory is the basic theory for microscopic physics.

The impasse was overcome and the quantum theory was initiated by Planck (1901). Although the details of Planck's theory are rather involved, the basic idea can be stated simply. He assumed that a particle oscillating with a frequency ν (such as a pendulum or a vibrating atom) cannot take on any arbitrary amount of energy but can assume only a selected set of discrete values given by the formula

$$E = nh\nu, \qquad n = 1, 2, 3, \ldots$$

where n is an integer and h is a universal constant, now called the Planck constant, which we have already used in the Bohr theory. The value of h is very small, given by

$$h = 6.625 \times 10^{-34} \text{ joule-sec}$$

In other words, if the energy of the particle is to increase (by receiving external work done on it) or decrease (by losing energy one way or another such as through friction), it cannot change continuously but can change only by discontinuous jumps. Each jump involves a parcel of energy $h\nu$ (change of n at least one unit). This unit of energy is called a *quantum* of energy. Based on this esoteric assumption, together with the other laws of classical physics, Planck was able to derive theoretically a spectral distribution formula in good agreement with the experimental curve.

The assumption is esoteric because it is alien to the Newtonian theory, which is based on the concept that all physical quantities are continuous. The argument between continuity and discontinuity has been going on among philosophers for a long time without conclusion. Whether nature is actually continuous or discontinuous is a matter of fact, not a matter of logic. Actually matter is discontinuous with the atom as the unit. Electricity is discontinuous with the electron as the unit. There is no reason why energy cannot be discontinuous with the quantum $h\nu$ as the unit. On the other hand, the concept of continuity that was fostered by the Newtonian theory is so deeply rooted that the Planck assumption appears quite unusual. Actually, the modern concept of continuity is a mathematical concept created for mathematical analysis, a purely mental construct that has not been proved to represent physical reality.

Indeed, Planck himself spent the rest of his life trying to reconcile his esoteric assumption with the existing theory. With the benefit of hindsight, we now know that this approach is a blind alley. Nature is actually discontinuous, energy not excepted. Newtonian theory is in reality an approximate theory; it smoothes over the discontinuities. What has to be done is to reintroduce the discontinuity and other features related to it that were whitewashed away in the Newtonian theory. Such a theory, to be developed for the discontinuous physical systems, is the *quantum theory*. Its mechanics is quantum mechanics.

The next important step in the development of the quantum theory was the *photon theory of light* developed by Einstein (1905). He had the audacity to recognize that the quantum $h\nu$ is not to be explained away but to be affirmed and developed further to explain other phenomena. In the photon theory he assumed that light is discontinuous and each unit, the light corpuscle called the *photon,* has a parcel of energy

$$E = h\nu$$

where ν is the frequency of light and h is the Planck constant. This is the same as the quantum of energy $h\nu$ in the Planck theory dealing with a material particle but is now applied to a completely different entity, the electromagnetic wave. In a way, Einstein revived the ancient corpuscular theory of light but with a modern slant—the light particle is related to the wave frequency and thus cannot be dissociated from the wave theory.

The photon theory was introduced to explain a phenomenon the classical theory cannot explain, the photoelectric effect—the ejection of electrons from a metal surface by light (Fig. 13.6). It was found that there is a *threshold of frequency* below which light cannot induce photoelectric effect. According to the classical theory the ejection of photoelectrons should be determined by the energy of the light wave, not by the frequency. The photon theory is introduced just to explain the frequency dependence. If the metal surface is to absorb energy from light, it must absorb either a whole photon or none at all, not a fraction of it. The parcel of energy $h\nu$, if greater than the minimum energy ϕ necessary to release an electron, will give rise to photoelectric effect. The energy left over, $h\nu - \phi$, will become the kinetic energy of the electron $\frac{1}{2}mv^2$. Thus we have the Einstein equation

$$h\nu = \tfrac{1}{2}mv^2 + \phi$$

for the photoelectric effect. This equation has been verified by experiment. The conflict between the corpuscular theory and wave theory of light was eventually resolved in the *quantum theory of radiation* (circa 1930). For our purpose the significance of the photon theory is the confirmation of the existence of the energy quantum $h\nu$ in another, independent phenomenon.

The quantum idea was further developed in the Bohr theory. As already stated, the theory is based on two assumptions: the quantum condition and the frequency rule. The quantum condition is a generalization of Planck's idea. Planck assumed that the energy of an oscillating particle is discontinuous. Bohr extended the discontinuity of energy to electrons in atomic orbits. Only a selected set of orbits is possible, each one corresponding to a discrete energy. Thus only a selected set of energy values is possible. On the other hand, in this case the selected energy values are not spaced at equal intervals, but are related in a more complex way. The other assumption, the frequency rule, is more or less an extension of Einstein's idea of the photon. When the electron jumps from one orbit to another with the release of an amount of energy $E_1 - E_2$, the frequency rule says that the energy is converted into a photon $h\nu$ of frequency ν,

$$h\nu = E_1 - E_2$$

Figure 13.6 Photoelectric effect.

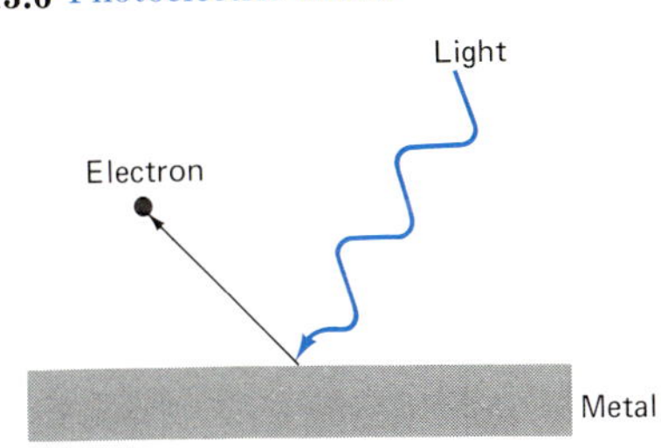

The Bohr theory and its extension (the *old quantum theory*) were able to give a reasonable account of many atomic phenomena, indicating that the two assumptions do represent some basic truth in atomic physics.

The final completion of the theory is left to *quantum mechanics* (1925). For our purpose we will satisfy ourselves by relating the atomic phenomena of interest to the two basic assumptions of Bohr. The atomic structure we discussed in the last chapter is largely based on the first assumption. The emission of electromagnetic radiation we are going to discuss in the rest of this chapter is largely based on the second assumption. One deals with the material system; the other deals with the wave field. The two together provide a reasonably good account of the quantum nature of the atomic world.

*13.4 Emission of Electromagnetic Waves by Atoms

In Sec. 13.1 we surveyed the spectrum of electromagnetic waves. In Sec. 13.2 we discussed the radiation of electromagnetic waves by a material body of high temperature. Now we consider the dynamical mechanism of the emission of electromagnetic waves by atoms. As already mentioned, we will do this largely by proceeding on Bohr's second assumption. (Actually this mechanism is also needed in the blackbody radiation at high temperature but the temperature effect overshadows the mechanism.)

As mentioned earlier, electromagnetic waves may be produced by an oscillating electric charge. Indeed, this is the way the communication waves (radio, radar, TV) are produced. The classical theory of electromagnetic waves gives a good account of this process. It is only natural to try to think of light waves being produced in a similar fashion. After the discovery of the electron in 1897, it was supposed that light waves are produced by electrons oscillating inside atoms. However, this approach was only moderately successful, and soon it was clear that a major overhaul of the whole radiation theory was required. The Bohr theory was the first step.

In the Bohr theory the frequency rule asserts that an electron circulating in one orbit of energy E_1 may "jump" into another orbit of a lower energy E_2 and that the energy so released, $E_1 - E_2$, will be converted into the energy of an electromagnetic wave now emitted by the atom, the frequency of which is given by $(E_1 - E_2)/h$, where h is the Planck constant. Based on this assumption as well as the quantum condition, Bohr was able to make a prediction of the frequencies of light emitted by the hydrogen atom.

The experimentally observed light spectrum of the hydrogen atom as seen through a prism spectroscope is not a continuous band of rainbow colors but consists of four discrete lines colored orange, blue, lavender, and violet, each with a definite wavelength. They are called H_α, H_β, H_γ, and H_δ lines and are shown in Fig. 13.7. Beyond the H_δ line in the ultraviolet region there are additional lines that, together with the four, form a *series* with a regular pattern. This series is called the *Balmer series*. In the far ultraviolet there is another series, called the *Lyman series*. In the infrared there is a *Paschen series*. The wavelengths of all the lines in all series have been measured accurately. The prediction of the Bohr theory agrees with them remarkably well. All quantum jumps to the innermost orbit ($n = 1$) result in the Lyman series; to the next orbit ($n = 2$), the Balmer series; to the third orbit ($n = 3$), the Paschen series. Thus a breakthrough was made in the understanding of the mechanism of emission of light (including ultraviolet and infrared) by atoms.

Because light emission is discontinuous, an intense light source is one with a large number of quantum jumps taking place at the same time. On the other hand, each of the quantum jumps is independent of the others and the wave phases of all the quanta are unrelated (each quantum is a *wave* of energy $h\nu$). Such a beam of light is considered *incoherent*. Later

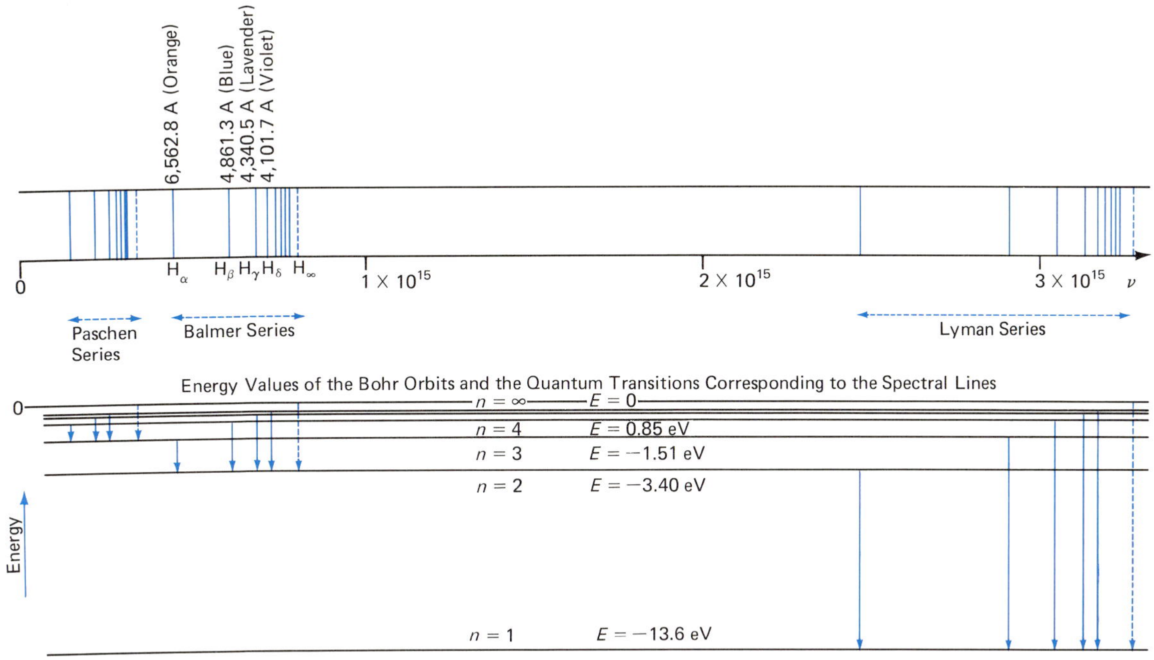

Figure 13.7 Spectrum of the hydrogen atom.

when we discuss the *laser* we will introduce *coherent* light in which the phases of all the quanta are the same. We may use an analogy with a set of identical swings hanging on a bar: if the swings start swinging at staggered times, they swing incoherently; if they swing in unison, they do so coherently.

If the emission of electromagnetic waves by atoms is discontinuous, what about the emission of electromagnetic waves by a radio antenna, which seems to be continuous and is well accounted for by the classical electromagnetic theory, which is continuous? Once we have established the quantum theory, we consider all emission processes to be discontinuous, including the emission by a radio antenna. Only the photons are so numerous that the discontinuity is smoothed over and it appears continuous. The quantum theory in this case reduces to the classical theory as a special case. Thus we have one consistent theory explaining all emission processes.

We now consider the emission of X rays and we limit ourselves to a particular kind, the *characteristic X rays,* which have a discrete spectrum characteristic of each element (analogous to atomic spectra). X rays are produced in a cathode ray tube when the electrons are stopped by the anode (Fig. 13.8). A part of the X rays emitted form a discrete spectrum characteristic of the anode material, indicating that they are emitted from the atoms of the anode (Fig. 13.9). The characteristic X-ray spectrum is rather simple (compared with atomic spectra,

Figure 13.8 Generation of X rays in a cathode ray tube.

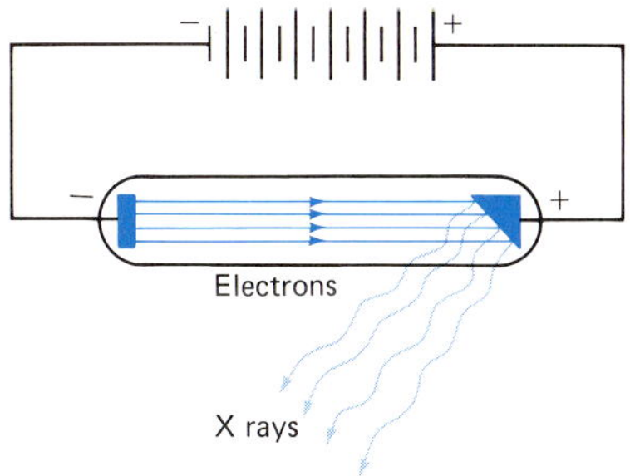

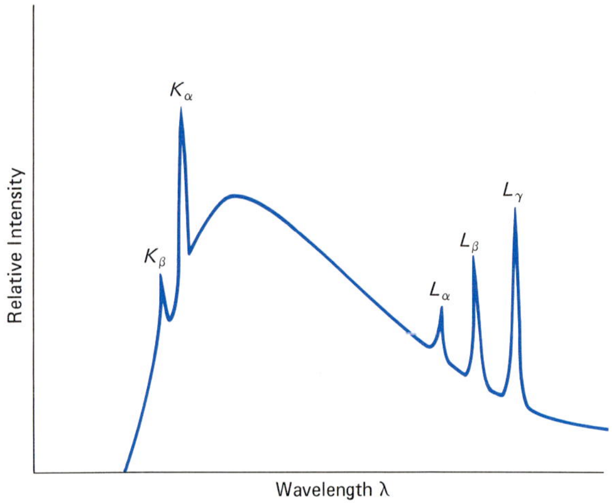

Figure 13.9 The spectrum of X rays from a particular anode shown here is made of two parts: a line spectrum superimposed on a continuous spectrum. The line spectrum is composed of the characteristic X rays: lines K_α and K_β forming the K series and lines L_α, L_β, and L_γ forming the L series. The continuous spectrum is independent of the anode material but is determined by the voltage across the cathode ray tube.

which are complex). Moreover, it is similar from one atom to another except for a change of scale of the wavelength (Fig. 13.10). This shift is so precise that the spectra of elements can be arranged in a very regular pattern in the increasing order of frequency (Fig. 13.10). This ordering of elements turns out to be nearly the same as that by atomic weight. Moreover, as far as the periodic table is concerned, the three exceptions of wrong ordering according to atomic weight become right according to the X rays. Furthermore, this shift of the spectrum is so precise that the sequence of a particular line, say, the first line of the K series K_α, appears like the rungs of a ladder; any missing rungs can be spotted right away. In fact, the missing elements in the periodic table correspond to the missing rungs in the sequence. Thus the sequential ordering of elements by their characteristic X rays is much more meaningful and precise than that by atomic weight, and this ordering may be used to define the atomic number Z and to construct the periodic table. This way we are freed of Mendeleev's difficulties.

The mechanism of X-ray emission is explained as follows. The cathode ray particle (electron) knocks out an electron in an *inner* shell of an atom in the anode, thus creating a vacancy (Fig. 13.11). Electrons in outer orbits may jump in to fill the vacancy. The electromagnetic wave emitted according to Bohr's second assumption manifests itself as the X ray. The frequency so calculated agrees with the experimental

Figure 13.10 Comparison of characteristic X ray spectra of elements in the order of atomic number.

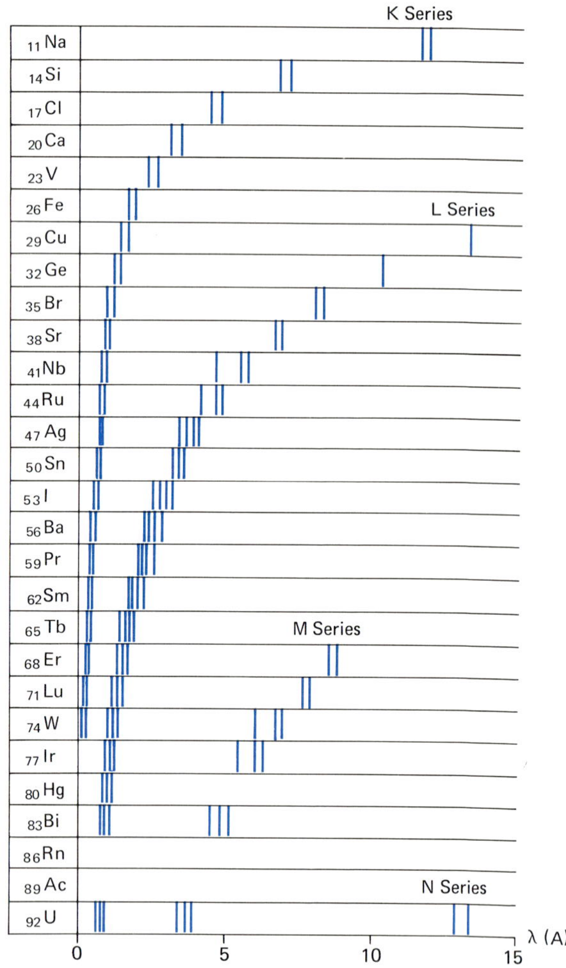

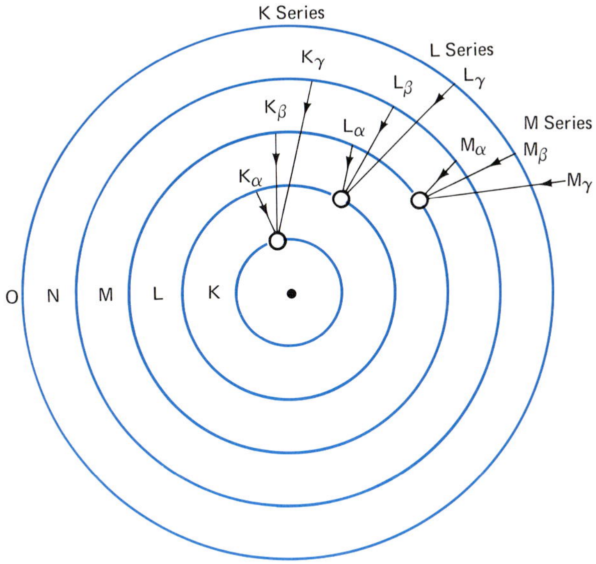

Figure 13.11 Explanation of the emission of characteristic X rays.

result. The X rays originating from a vacancy in the K shell form the K series; from the L shell, the L series; and so on.

When atoms combine to form compounds, the spectra of light emitted by the molecules are completely different from those of the constituent atoms. However, the characteristic X rays of the molecules are simply those of the constituent atoms put together. These are so because chemical combination involves the change of electrons in the outermost orbit and we associate light emission with electrons in the outermost orbit and X-ray emission with electrons in the inner orbits which are not disturbed in chemical combination. In this view the X-ray spectra provide information on electrons in the inner shells. On the other hand, the information on electrons in the outer shell is constructed based on chemical and optical (light) properties. The two together provide a more complete picture of the structure of the atom. In fact, the terms *K, L, M . . . shells* of electrons were derived from the terms *K, L, M, . . . series* of the characteristic X rays. The mechanism also explains why the X-ray spectrum is simple (because only a few higher orbits are occupied by electrons) and varies with the atomic number Z monotonically instead of periodically as in the optical spectra (because all atoms have the same *inner* electronic configuration).

The emission of γ-rays is similar to that of light except that the particle that does the jumping in orbit is not an electron but a proton and the stage on which the emission takes place is not the atom but the atomic nucleus. It is a nuclear process and thus the γ-ray is related to radioactivity and nuclear reactions. The energy involved in nuclear processes is generally a million times higher than that in atomic processes and therefore, according to Bohr's second assumption, the γ-rays' wavelengths are generally a million times shorter than those of light waves, hence in the 10^{-3}A range.

Thus the emission of light (infrared, visible, ultraviolet) and hard radiation (X rays, γ-rays) are all related to the change of orbits, or more appropriately, the transition of quantum states of the atom or the nucleus. The electromagnetic waves emitted are waves of definite frequencies discrete from one another. The spectrum is thus a *discrete,* or a *line, spectrum* (Color Plate), consisting of a number of bright lines instead of a continuous band of colors (as the spectrum of white light resolved by a prism). Each atom, as well as each nucleus, has its own characteristic spectrum that is so specific that it can be used for chemical identification. Trace chemical impurities that can hardly be detected by conventional chemical analysis can be detected by observing their characteristic lines in the light spectrum or X-ray spectrum. Chemical analysis by X rays is only a decade old; the constituents of a gold ring can be ascertained in a matter of a few seconds.

Besides the atomic and nuclear spectra there is also a molecular spectrum originating from the transition of molecular quantum states. All spectra, molecular, atomic, and nuclear, are related to the structure of their emitters and thus provide a means to study the structure of molecules, atoms, and nuclei. For example, the molecular spectrum is related to the rotation of the molecule as a whole and the vibration between atoms in the molecule. The study of the molecular spectrum thus can give us important information concerning the distance between the constituent atoms in a molecule.

The process of light emission may be reversed. In the *absorption* process, a light quantum comes in and is absorbed by an atomic or molecular or nuclear system together with a transition of quantum states, which is the reverse of the emission process. This puts the system in a higher energy state, which has a tendency to return to the lower energy states by light emission. Thus emission usually follows absorption.

Just as the light emitted is characteristic of an atom, the light absorbed is also of specific frequencies. In fact, the frequencies of the absorption lines are the same as the corresponding emission lines because they are both determined by the same equation based on Bohr's second assumption:

$$\nu = \frac{E_1 - E_2}{h}$$

Thus all atomic, molecular, and nuclear systems absorb selectively. For example, the molecule ozone, O_3, absorbs ultraviolet rays strongly. The ozone in the upper atmosphere thus acts as a blanket shielding the earth from the lethal ultraviolet rays from the sun. This is an important point relevant to the environmental problems of the supersonic transport airplane and the spray-can propellent, which we have discussed earlier.

Figure 13.12 Laser light. [RCA Laboratories]

*13.5 Lasers

The last topic to be discussed in connection with electromagnetic waves is the *laser* (Fig. 13.12), one of the significant inventions in recent years. It began with a similar device involving microwaves, called a *maser*. When extended to light waves, the name *laser* is given. Now there are speculations that it may even be extended to X rays. Theoretically the same principle may be applied across the spectrum of electromagnetic waves, though practical limitations and difficulties often determine engineering feasibility.

The principle on which all these devices are based is a phenomenon called *stimulated emission*. An atom or molecule in an *excited state* E_1 (high energy orbit) may return to a lower energy state E_2, usually the lowest energy, the normal state of the atom, i.e., the *ground state*, by emission of a photon of frequency $\nu = (E_1 - E_2)/h$. This process takes place of the atom's own accord and is called *spontaneous emission*. In addition, if the excited atom of energy E_1 is exposed to a beam of light of the same frequency as the emitted photon ν, then the rate of radiation, or the rate of transition from the state E_1 to state E_2, will be greatly increased. The additional radiation mechanism induced by the radiation already present is called stimulated emission, first discovered theoretically by Einstein.

It happens that some atomic (and molecular) systems have a few unusual excited states that have a very low rate of spontaneous emission. An atom excited to such a state may remain there for a moment, sometimes up to a second or so, which is very, very long compared with other usual excited states (10^{-8} sec). These states are called *metastable states.* Such an atomic system may have a sizable fraction of atoms excited to a metastable state by shining light on it to induce absorption. This process is called *optical pumping.* Once this is done, a single photon coming into the system from outside may induce stimulated emission of many photons of the same frequency. Thus the device accomplishes the action of *amplification* (similar to a triode and an n–p–n junction transistor), which is very useful in many practical applications. The word *maser* is an acronym of the expression *microwave amplification by stimulated emission radiation.* The word *laser* originated by substituting *l* (*light*) for *m* (*microwave*).

An important feature of the laser light is that it is *coherent.* In spontaneous emission all atoms do their own emitting independent of others and the waves emitted from the many atoms are unrelated in phase (See Fig. 13.13a). The light so generated is incoherent. On the other hand, in stimulated emission all photons emitted from the many atoms have the same phase (see Fig. 13.13b) and the light so generated is coherent. A beam of coherent light is exactly parallel and thus can propagate to a long distance without much weakening in intensity. The laser light can also be made very intense. These properties are the basis of many useful applications of the laser developed in recent years.

Laser light has been sent to the moon and reflected back with sufficient intensity to be detected. From the time the light takes to travel to the moon and back the distance from earth to the moon can be determined. This is the most accurate way to measure a large distance and can be used to improve the surveying of the earth's surface.

The intense, concentrated light of a laser beam delivers a large amount of energy to a small spot and can be made to perform a number of useful functions, from surgical operation on the eye to experimental production of nuclear fusion reactions (Sec. 15.4). Now lasers are being considered for such

Figure 13.13 (a) Incoherent light. (b) Coherent light.

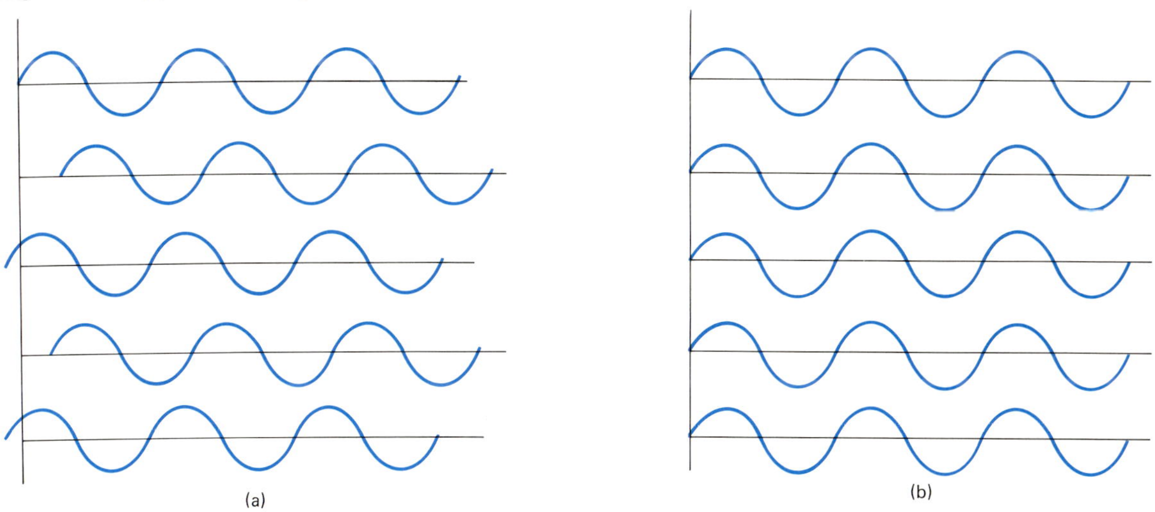

mundane functions as cutting paper, meat, and cloth.

A beam of coherent light, like a radio-wave beam, may be used (by amplitude modulation or frequency modulation) to transmit wave signals for communication purposes. Because the frequency of light is much higher than that of the radio wave, the number of channels (telephone calls, TV programs) that can be carried is much greater. It may revolutionize the communication industry. But the technical problems of this application have not all been worked out yet.

An intense coherent light beam! A novelty with many unusual features, the laser may become a general tool that may be used to perform many services. Like the stick of a caveman, the wrench of a handyman, the computer of a spaceman, it may well be the most versatile tool for future man.

Problems

13.1 Find the frequency of the green light of wavelength 5,000A. Find the wavelength of the radio wave of 900 kilocycles/sec.

13.2 In an atomic explosion the initial temperature is of the order of 10^8°K. If the explosion takes place in a vacuum it is dark in the beginning and is also silent. Explain why. Nevertheless it is extremely hot, so hot that in a vacuum (above the atmosphere) it can burn up objects miles away. Explain why. This is the principle of the antiballistic missile. (The explosion in a vacuum becomes visible later as it cools. On the other hand, an explosion in air ionizes the surrounding air to about 150 yd in 1 sec and this produces a fireball of that size, which is visible. There is also a loud noise.)

13.3 The energy values of the Bohr orbits of the hydrogen atom is given by the formula $-13.6/n^2$ eV, where n is an integer 1, 2, 3, ... Calculate the wavelengths of the hydrogen lines H_α, H_β, H_γ, and H_δ.

13.4 In the calcium atom one of the electrons in the K shell is knocked out by a photon, say. Draw a diagram and show which X rays will be emitted.

13.5 Once the K series X rays appear, the L, M, . . . series appear simultaneously. However, if only the L series is made to appear, then only the M, N, . . . series will appear simultaneously, but not the K series. Why?

13.6 Not all emission lines are present in the absorption spectrum. Why? Which of the lines are present?

13.7 For a clock pendulum oscillating with a frequency of 1 cycle/sec, how large is the discontinuous unit of energy? Is this discontinuity large enough to be detectable?

13.8 What is the energy of the photon of the green light and of the radio wave in Problem 13.1? Is the discontinuity of the radio wave large enough to be detectable? And the green light?

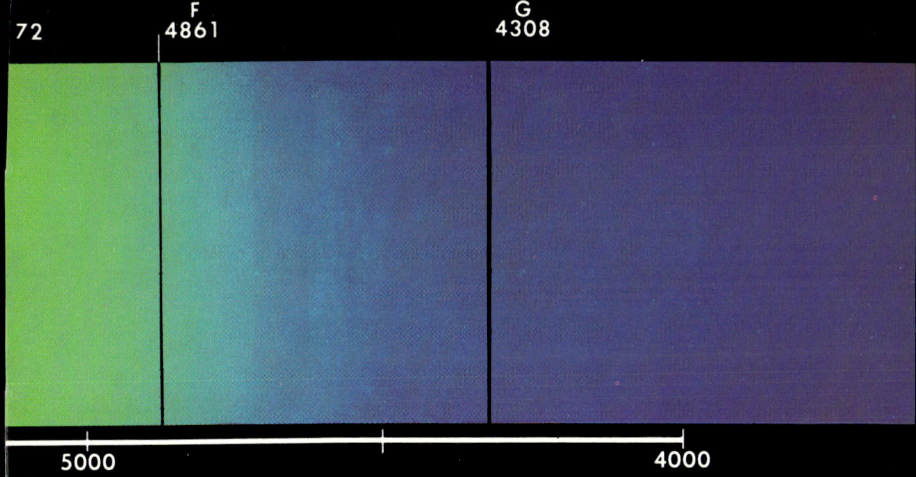

Such diverse and fundamental information on the nature of matter as the composition of distant stars and the structure of atoms and molecules has been obtained by analysis of the light emitted from any incandescent substance.

In the SPECTROSCOPE, such light, passed through a slit and a prism, is broken up into its component wavelengths, which are observed as colored lines, or light of different energies, characteristic of the differences between the various electron energy levels of the atoms. This EMISSION SPECTRUM is CONTINUOUS when the images of the wavelengths are uninterruptedly overlapping; it is a LINE spectrum when only certain specific wavelengths are emitted, as shown here for the elements hydrogen, helium, mercury, and uranium.

On the solar spectrum across the top of this plate appears a series of dark lines — FRAUNHOFER LINES — forming an ABSORPTION SPECTRUM: Some of the light from the intensely hot interior of the sun is absorbed by the cooler gases of its outer layers as the light energies raise the atoms in these layers to higher energy states; bright lines are not, therefore, seen for these changes.

The spectra are calibrated in angstroms ($1\ A = 10^{-10}$ m); the letters are arbitrary designations introduced by Fraunhofer for lines important in spectroscopy.

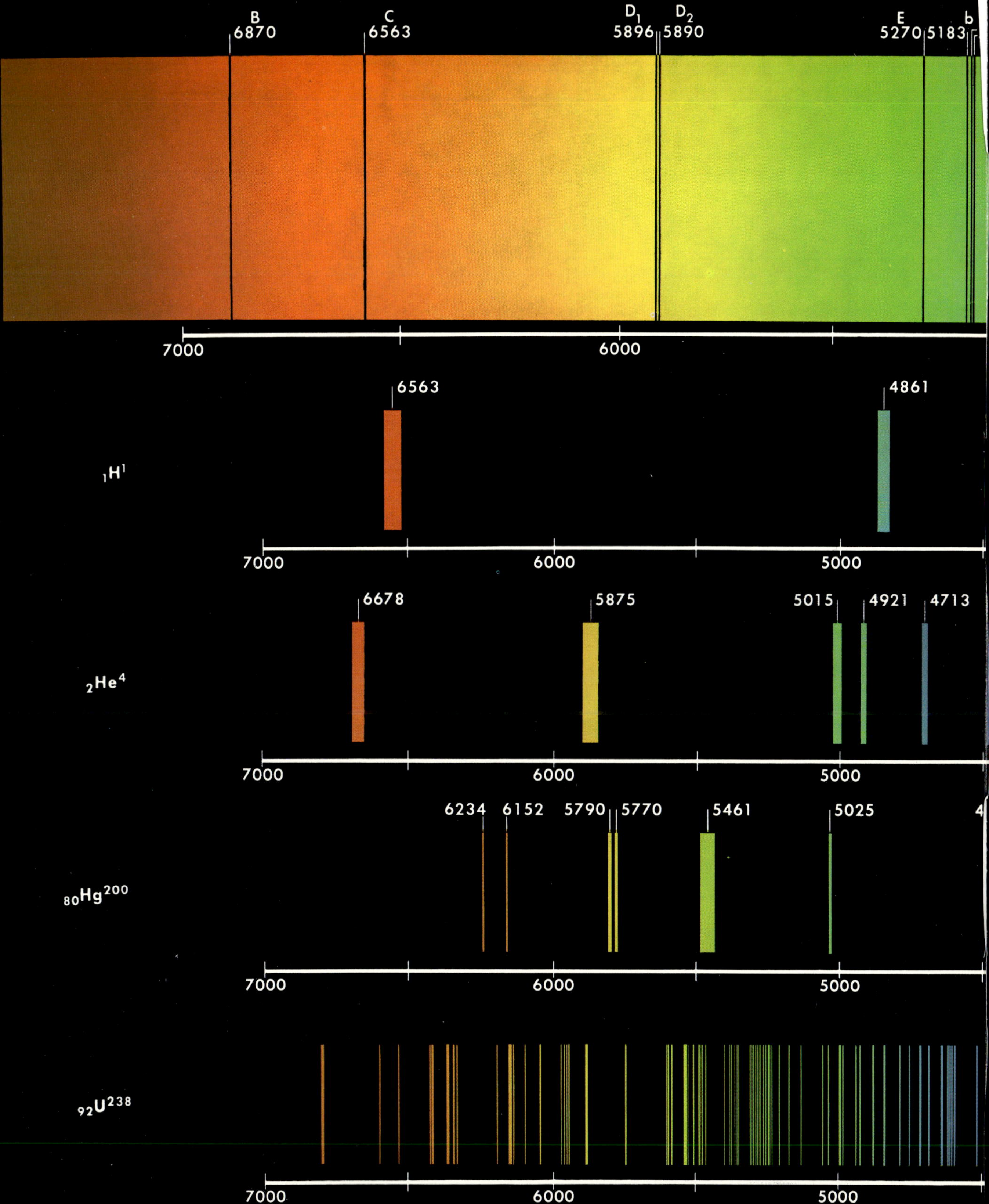
B
6870
C
6563
D1
5896
D2
5890
E
5270
5183
b
7000
6000
6563
4861
1H1
7000
6000
5000
6678
5875
5015
4921
4713
2He4
7000
6000
5000
6234
6152
5790
5770
5461
5025
80Hg200
7000
6000
5000
92U238
7000
6000
5000

14 Nuclear Transformations of Matter

14.1 Nuclear Constitution

We now turn our attention to the structure of the atomic nucleus. As mentioned before, we learn from Rutherford's experiments the sizes and the charges of various nuclei. The mass of the nucleus is practically the same as that of the atom, which we already know. The next question is what the nucleus is made of.

Isotopes. As early as 1910 Soddy, based on the study of radioactivity, suggested that the atoms of an element may assume several forms, differing in weight but otherwise identical chemically. This was a drastic departure from Dalton's concept that all atoms of the same element are exactly the same. The atomic weights used in chemistry, in this view, are actually average values over the several naturally occurring atomic species. Soddy called them *isotopes*. In 1913 Thomson found, by determining the value of e/m of the neon ion, that neon ions indeed consisted of two groups with different values of e/m, corresponding to atomic weights of 20 and 22, respectively. The chemical atomic weight of 20.2 is actually the average of a mixture of the two atomic species in the ratio of 10 to 1. Thus the existence of isotopes was confirmed. After that all elements were investigated for isotopic composition and it was found that most elements are not pure but a mixture of a few species. More important, the atomic weight of each isotope invariably turns out to be very nearly an integer.

This was the situation that led to the Prout theory, which was abandoned because some atomic weights are definitely not close to integers. Now we know that, after all, the nonintegral nature is due to averaging and that the real atomic weights of the isotopes are nearly integers. Naturally the Prout theory was revived. Yet atoms cannot be made of hydrogen atoms as Prout first thought. The hydrogen atom has a structure of itself—a nucleus surrounded by an electron with the weight concentrated in the nucleus. The fact that weights of all other atoms are nearly an integral multiple of that of hydrogen thus would suggest that all other nuclei are made of the nucleus of the hydrogen atom, which we call the *proton*.

This assumption accounts for the weights of the nuclei, but not for the charges. Each proton carries one electronic unit of positive charge. If the nucleus were made of protons exclusively, then the atomic number (which equals the number of electronic units of the nucleus charge) would be the same as the atomic weight (which, in this case, equals the number of protons). Actually, the atomic number is about one half of the atomic weight.

This problem may be overcome superficially, as was first proposed, by assuming that the nucleus is made not of protons exclusively but of protons and electrons. The addition of electrons, which are light in weight, does not upset the weight relation already established, but has the desirable effect of reducing the nuclear charge. For example, the neon isotope of atomic weight 20 is assumed to have a nucleus made of 20 protons and 10 electrons, thus giving rise to an atomic weight of 20 and an atomic number $20 - 10 = 10$, as desired. The other isotope of atomic weight 22 is assumed to have a nucleus made of 22 protons and 12 electrons. This way, from the atomic weight and the atomic number, we can assign the numbers of protons and electrons to account for the mass and charge of any nucleus. This theory, however, had to be abandoned later.

Neutron as a Nuclear Constituent. The assumption that electrons existed in the nucleus was helped by the observation in radioactivity that occasionally a radioactive atom emits an electron, which can only come from the nucleus (β-decay). The view that the nucleus is made of protons and electrons was accepted for two decades, but many difficulties with this theory emerged in the later years, largely from quantum mechanical considerations; the electron is simply an unlikely resident of the nucleus. In 1932 a new elementary particle called the *neutron* was discovered; it has nearly the same mass as a proton but it carries no electric charge. Considering it as a nuclear constituent, a new theory was worked

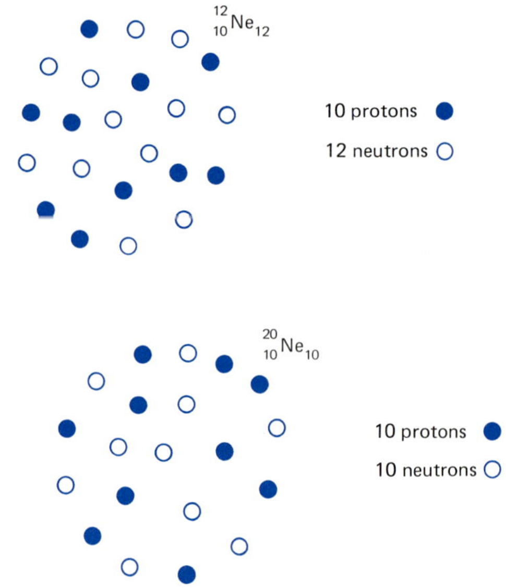

Figure 14.1 Composition of the nucleus.

out assuming that the nucleus is made of protons and neutrons. (The electron is evicted.) In this view the neon isotope of atomic weight 22 is made of 10 protons and 12 neutrons with a total weight nearly 22 times that of proton and a total charge of 10 electronic units (Fig. 14.1). Based on this theory all the quantum-mechanical difficulties of the electron in the nucleus encountered in the previous theory disappear. This is the theory accepted currently. Thus in the present view all atoms, and so all material substances, are made of three kinds of elementary particles—electrons, protons, and neutrons—the latter two constituting the nucleus of the atom.

14.2 Elementary Particles

It would be nice if the list of elementary particles were limited to the three already mentioned. Everything would be made of them. A very simple order of the physical universe would be established. Yet nature is usually more subtle than we suppose. Other elementary particles were discovered later, some desirable and others unwanted. A host of problems associated with them arose that remain incompletely understood at the present. The frontier of science, instead of vanishing, simply recedes further back. As far as elementary particles are concerned (which are the foundations of the physical world), we are still far from having a complete understanding. On the other hand, the elementary particle problems can usually be separated from the other problems in physical science, including all problems of practical significance that occur every day in the physical world. As far as the properties of the nucleus, atom, molecule, and bulk material are concerned, they can be dealt with adequately by our knowledge of the three basic particles and the basic laws of nature discussed previously. The unknowns in elementary particles in a sense can be quarantined. From a practical point of view we do have a good grasp of the principles in the physical world. We will give a brief account of the elementary particles here to provide a glimpse of the last frontier of physical science.

Let us begin with the nucleus. Once the electron is evicted, how do we account for the emission of an electron from the nucleus in radioactivity (β-decay)? We can only say that the electron is created inside the nucleus shortly before its emission. Here we come across the idea of *creation* of matter, which was alien to Newtonian physics. Yet by 1932 quantum mechanics had developed to such an extent that the idea of creation of matter was no longer strange. Besides, from quantum mechanics it can be predicted that another particle, called the *positron,* should exist that is the same as the electron except that the electric charge is positive (a positive electron). Moreover, it can be predicted that an electron–positron pair can be created out of no matter but at the expense of energy (Fig. 14.2) and also can annihilate each other, with matter disappearing and changing back to energy. The conversion between mass and energy is governed by Einstein's equation $E = mc^2$, which was firmly established by that time. The positron was actually discovered in 1932 and the processes of *pair production* and *pair annihilation* were all confirmed afterward. Thus we have one more elementary particle, the positron.

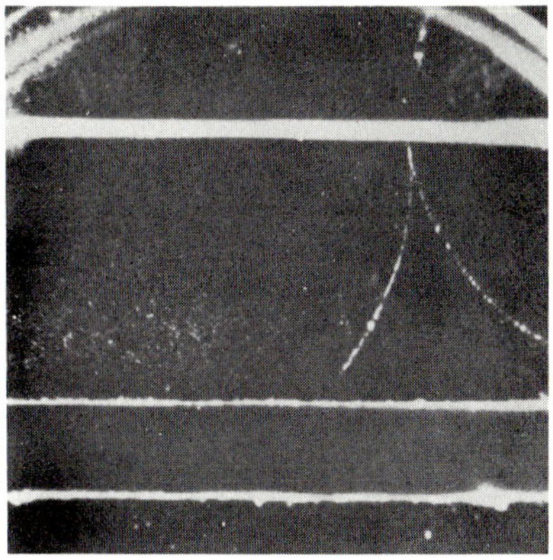

Figure 14.2 Pair creation recorded by a cloud chamber. A γ-ray photon, entering from above, hit the lead plate and converted itself into an electron and a positron whose paths curved in opposite directions because of a magnetic field applied. [From Holton-Roller, *Foundations of Modern Physical Science,* 1958, Addison-Wesley, Reading, Mass.]

In the late 1930's a new particle, called the *meson,* was discovered with a mass heavier than the electron but less than the proton. In the 1940's two kinds of meson were distinguished; they are called *π-mesons,* or *pions,* and *μ-mesons,* or *muons.* In the 1950's another group, called *K-mesons* (or kaons), was added to the list. Then another group of particles heavier than the proton, called Λ, Σ, Ξ particles, was discovered. This group is called the *hyperons.* These so-called elementary particles can be electrically positive, negative, and neutral, the charge being always one electronic unit. Usually they exist in pairs (two particles identical except for their opposite charges), like the positron and electron pair, which are called particle and *antiparticle* pairs. In the 1950's the *antiproton* and *antineutron* were discovered and the pairs were completed. Also in the 1950's the existence of a hypothetical particle, the *neutrino,* introduced in the 1930's to explain the β-decay phenomena (Sec. 14.5), was established by experiment. Altogether the number of elementary particles had grown to around 30 in the 1950's. When the number becomes that large, one begins to wonder whether they are really "elementary." Furthermore, they are related by radioactive decay, e.g., a pion decays into a muon and a muon decays into an electron (Fig. 14.3). If they are related, then they are not quite elementary.

The discovery of the positron fulfills a theoretical prediction; this is excellent. The discovery of the meson (later identified with the pion) fits into the theory of nuclear force (the third fundamental force) by Yukawa; this is fine. The confirmation of the neutrino completes our theory on the fourth basic force (and saves the principles of conservation of energy, momentum, and angular momentum); this is marvelous. However, the other elementary particles, the muons, the kaons, and the hyperons, are unwanted. It would be better if they did not exist at all, for there are no niches for them to fit in the theory. They are extra pieces of the jigsaw puzzle of elementary particle theory that do not fit the pattern just pieced together; this could only mean that the pattern just worked out is not complete. In the development of quantum mechanics in the 1930's, every piece falls into its rightful place—the covalent bond, the resonance energy, the London force, and so on. The theory is complete. On the other hand, the elementary particle theory is not. Here is the true frontier of physical science. Fundamentally new discoveries still could be made.

In the 1960's there was a population explosion of the elementary particles and the number increased to hundreds. In fact, they can be made at will by increasing the energy of the particle accelerators. Moreover, they are closely related in a complicated scheme. Obviously they are not "elementary"; in fact, we have a new "chemistry" of the newly discovered particles, a set of "reactions" representing the transformations among them. The term *elementary particle physics* is not meaningful anymore and is now changed to *high energy physics*. Presumably in the future we shall go through a history of development parallel to what happened to chemistry, which includes the discovery of the periodic table, the formulation of the Bohr theory, and the establishment of quantum mechanics by which all chemistry was eventually explained. Right now we seem to be just at the beginning. We are in the process of putting together a "periodic table" and not even a "Bohr theory" is yet in sight. How to reduce the great complexity of the "elementary" particles to more fundamental entities and more basic laws of nature is the main problem of pure science in the future.

Figure 14.3 Photographic emulsion tracks showing a π^+ meson changing into a μ^+ meson, which in turn changes into a positron e^+. [From J. Orear, *Fundamental Physics* (New York: Wiley, 1961)]

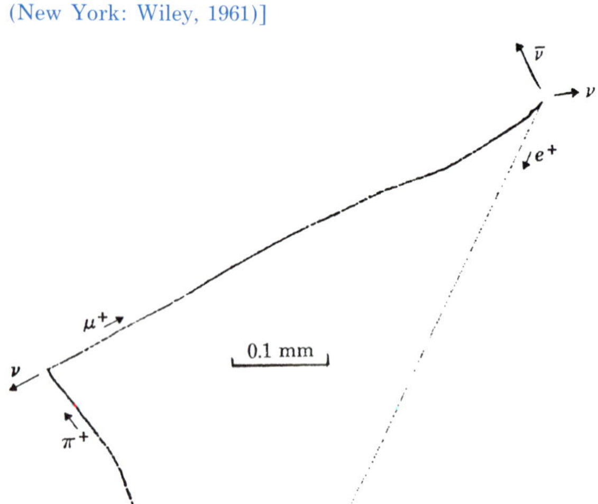

*14.3 Nuclear Structure

Once we know the nucleus is made of a collection of particles, a problem similar to that of atomic structure arises. How do the particles arrange themselves in the nucleus? We may hope that a knowledge of the structure will lead to an understanding of many properties of the nucleus.

Two differences stand out between nuclear and atomic structure. There are two kinds of particles (proton, neutron) in the nucleus instead of one (electron) in the atom. The atom has a center (the nucleus) that provides the attractive force by which the electrons are kept in the atom, whereas the nucleus has none. Nevertheless, the protons and neutrons exert attractive nuclear forces (strong interaction) among themselves. Adding the forces together, a particle experiences a net force in a way not much different from that of an electron in an atom. Thus the same situation as that found in the atom may occur in the nucleus. Inasmuch as there are two kinds of particles, it may happen to protons and neutrons separately.

In the atom we learned the existence of a shell structure of electrons. Therefore one may suspect a shell structure of protons and another one of neutrons in the nucleus. The closed-shell nuclei, like the noble gas atoms, are expected to be especially stable. A close examination of the systematics of nuclear properties shows that indeed there are a few nuclei with special numbers of protons or neutrons that are especially stable corresponding to closed shells. These numbers, called *magic numbers,* are 2, 8, 20, 50, 82, 126. Unfortunately they do not correspond to the closed-shell numbers in the structure of the atom. Where do the magic numbers come from?

The problem was solved by Mayer and Jensen. From quantum mechanics one finds that the nuclear subshells are essentially the same as the atomic subshells except that they are ordered differently and grouped into complete shells in a different way, i.e., (1s), (1p), (1d, 2s), (1f, 2p), (1g, 2d, 3s), (1h, 2f, 3p), (1i, 2g, 3d, 4s) . . . . The total number of particles corresponding to the closed complete shells is thus 2, 8, 20, 40, 70, 112, 168. These are not yet the magic numbers observed and the differences are explained by an effect called *spin-orbit coupling* by which the subshells 1g, 1h, 1i split in such a way that 10 particles of the 1g subshell go over to the previous shell, making the previous closed-shell number equal to $40 + 10 = 50$. Also 12 particles of the 1h subshell go over to the previous shell, making the closed-shell number equal to $70 + 12 = 82$, and 14 particles of the 1i subshell go over to the previous shell, making the closed-shell number equal to $112 + 14 = 126$. Thus the magic numbers 50, 82, 126 are explained. This scheme not only explains the magic numbers, but also correlates practically every property of the nuclei. A large body of nuclear information, otherwise inscrutable, now becomes organized and understandable. The nuclear shell theory is thus the cornerstone of nuclear structure physics.

Consider one example. Once the neutrons complete a closed shell the nucleus becomes more stable and thus is less disposed to absorb an additional neutron. If all nuclei in the universe were created from small ones by adding neutrons to it, the nuclei with a magic number of neutrons, not disposed to neutron addition, will survive with larger abundances. Indeed, they occur in nature more abundantly than their neighbors. In the theory of the origin of the universe, nuclear physics is applied profusely and the magic numbers play an important part (see Chapter 22).

14.4 Nuclear Reactions

Once we know the nucleus to be a collection of protons and neutrons, the possibility arises that we may change the nuclear constitution by adding or removing protons or neutrons to it. When the *neutron number* N is changed with the *proton number* Z fixed, the nucleus changes from one isotope to another. When the proton number Z is changed, the nucleus changes from one element to another. Dalton's assertion that atoms are immutable may now be violated. Indeed, in nuclear physics laboratories transmutation of elements is carried out this way every day. The dream of making gold from other substances is finally realized except that the cost of making gold this way is a thousand times more than that of buying it. Thus no economical gain is to be realized by the new discoveries.

Given N and Z of a nucleus we find the total $A = N + Z$, representing the total number of particles in the nucleus. This integral number is called the *mass number,* which equals (very closely) the atomic weight. The nucleus of a chemical element X may thus be denoted by ${}^{A}_{Z}X_{N}$, or simply ${}^{A}_{Z}X$ to show its nuclear constitution; for example, $^{22}_{10}Ne_{12}$ represents the neon isotope that has 10 protons and 12 neutrons, and an atomic weight very close to 22.

The first report of transmutation of elements was big news, overthrowing the long-established concept of Dalton. It was made by Rutherford in 1919 to explain a phenomenon in which an α-particle (emitted from a radioactive atom), moving through the air, gives rise to a proton (Fig. 14.4). The explanation is that the α-particle, which is actually the helium atom stripped of two electrons and thus is the helium nucleus, when colliding with a nitrogen nucleus in the air, enters the latter to form a *compound nucleus,* which then breaks up into a proton and another nucleus. The nitrogen nucleus, with atomic number equal to 7 and atomic weight equal to 14, is denoted by $^{14}_{7}N$. The helium nucleus is denoted by $^{4}_{2}He$. The compound nucleus is the amalgamation of the two, thus having a total of $7 + 2 = 9$ protons and $14 + 4 = 18$ particles of both kinds. Because $Z = 9$ refers to fluorine, the compound nucleus is thus an isotope of fluorine. Its atomic weight is equal to 18. It is thus denoted by $^{18}_{9}F$. Natural fluorine is 100 per cent $^{19}_{9}F$. The compound nucleus thus formed is one not naturally existent in nature. It is not stable, breaking up by emitting a proton. After that, both numbers A and Z are reduced by one and the remaining nucleus becomes an isotope of oxygen $^{17}_{8}O$. The nuclear processes may be represented by the equation

$$^{14}_{7}N + ^{4}_{2}He \longrightarrow ^{18}_{9}F^{*} \longrightarrow ^{17}_{8}O + ^{1}_{1}H$$

Such processes are called *nuclear reactions,* in which one element changes into another. They are not to be confused with chemical reactions, in which one compound changes into another.

Figure 14.4 Cloud chamber tracks showing, among the long α-particle tracks from ^{212}Po and the short α-particle tracks from ^{212}Bi in the same source, the track of one α-particle hitting a nitrogen nucleus in the air and converting the two into an oxygen nucleus and a proton, shown by the forked track (courtesy of W. Gentner, from *An Atlas of Typical Cloud Chamber Photographs,* 1945, Pergamon Press).

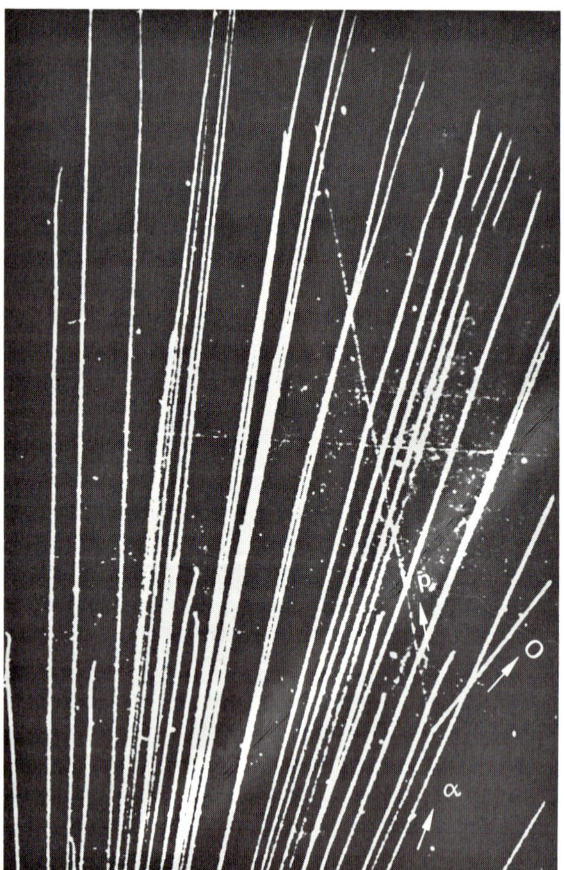

Since 1919, hundreds of nuclear reactions have been observed. Many isotopes not naturally existent in nature (such as technetium) have been created by nuclear reactions. In particular, over a dozen transuranian elements from neptunium ($Z = 93$) to hahnium ($Z = 105$) have been artificially created. Their chemical properties are exactly what is predicted from the periodic table based on the theory of atomic structure. Because all mundane isotopes that can be produced have been produced, scientists are currently looking for the esoteric, hypothetical *superheavy* elements, with Z around 114, N around 184, and A close to 300. They are predicted to exist theoretically based on the nuclear shell structure (magic numbers 114 for the proton and 184 for the neutron). If actually found, the discovery would be a great triumph for theoretical physics.

Most nuclear reactions are initiated by the bombardment of a target nucleus with a small particle, such as a proton or an α-particle. The reaction products are usually a large residual nucleus and a small particle, such as a proton as in the preceding reaction of Rutherford. The preceding reaction can be written simply as ${}^{14}_{7}N(\alpha,p){}^{17}_{8}O$. Other reactions involving small particles can be written in a similar fashion.

Nuclear Accelerator. The bombarding particle Rutherford used is the α-particle from natural radioactivity. To investigate nuclear reactions we need a large variety of bombarding particles and a large range of energy of the particles. For this purpose a number of *particle accelerators* (atom smashers) have been built, ranging in size from small "table models" to giant mile-long monstrosities. The earliest type was the one-stage linear accelerator (*voltage doubler, van de Graaf machine*). An important advance was the ingenious invention of a multistage, circular accelerator called the *cyclotron*. Multistage

Figure 14.5 Aerial view of National Accelerator Laboratory in Batavia, Illinois. Protons are accelerated to an energy up to 400 BeV in a ring 4 miles in circumference shown in the photograph. [National Accelerator Laboratory]

linear accelerators also were devised. The largest one currently in operation, at Stanford University, is 2 miles long. More sophisticated circular accelerators were constructed (*synchrocyclotron, synchrotron, proton synchroton*). The largest one in existence is at the National Accelerator Laboratory in Batavia, Illinois. [See Fig. 14.5 for aerial view and Fig. 14.6 for inside view (main ring tunnel).] It represents such a large investment that it can be done only by a national effort. The particle energy produced (400 BeV in 1973) is 100,000 times larger than that of the α-particles from natural radioactivity first used by Rutherford. One recent experimental result is shown in Fig. 14.7.

Classification of Nuclear Reactions. Nuclear reactions may be initiated by charged particles, such as protons and α-particles; by neutral particles, such as the neutron; or by light. These reactions have different characteristics. For example, charged particles are more effective at high energies but neutrons are more effective at low energies.

As the bombarding energy increases, the reaction products may involve more than one small particle. Usually as energy increases, the number of small particles emitted also increases. The process may be compared with evaporation. The compound nucleus is "hot" and evaporates several small particles.

Once the energy of the bombarding particle becomes sufficiently high, the nuclear reaction switches to a different mechanism. Instead of forming a compound nucleus first and evaporating particles later, the bombarding particle knocks out one or two protons or neutrons by direct collision. The mechanism is thus called *direct reaction,* to distinguish it from *compound nucleus formation.*

In both cases the reaction products are a number of small particles plus a large residual nucleus. The question thus arises whether a nucleus can split into a few comparable pieces. It was found later (1939) that such a reaction actually can take place, but only for the very heavy nuclei such as uranium. This reaction, called *fission,* is the source of the so-called atomic energy, an important energy source of the future, and so will be discussed more in detail in the next chapter. Also to be discussed is another nuclear reaction, *fusion,* which is another important future source of energy.

As the energy of the bombarding particle becomes even higher, the more important events that happen is the production of new elementary particles, such as the mesons, the hyperons, and the antiparticles. Nuclear reactions involving the rearrangement of protons and neutrons becomes secondary and the investigation passes from nuclear physics to high energy physics.

Figure 14.6 Shown in the photograph inside the ring of the 400 BeV accelerator are bending magnets which make the protons move in a circular path 4 miles in circumference. [National Accelerator Laboratory]

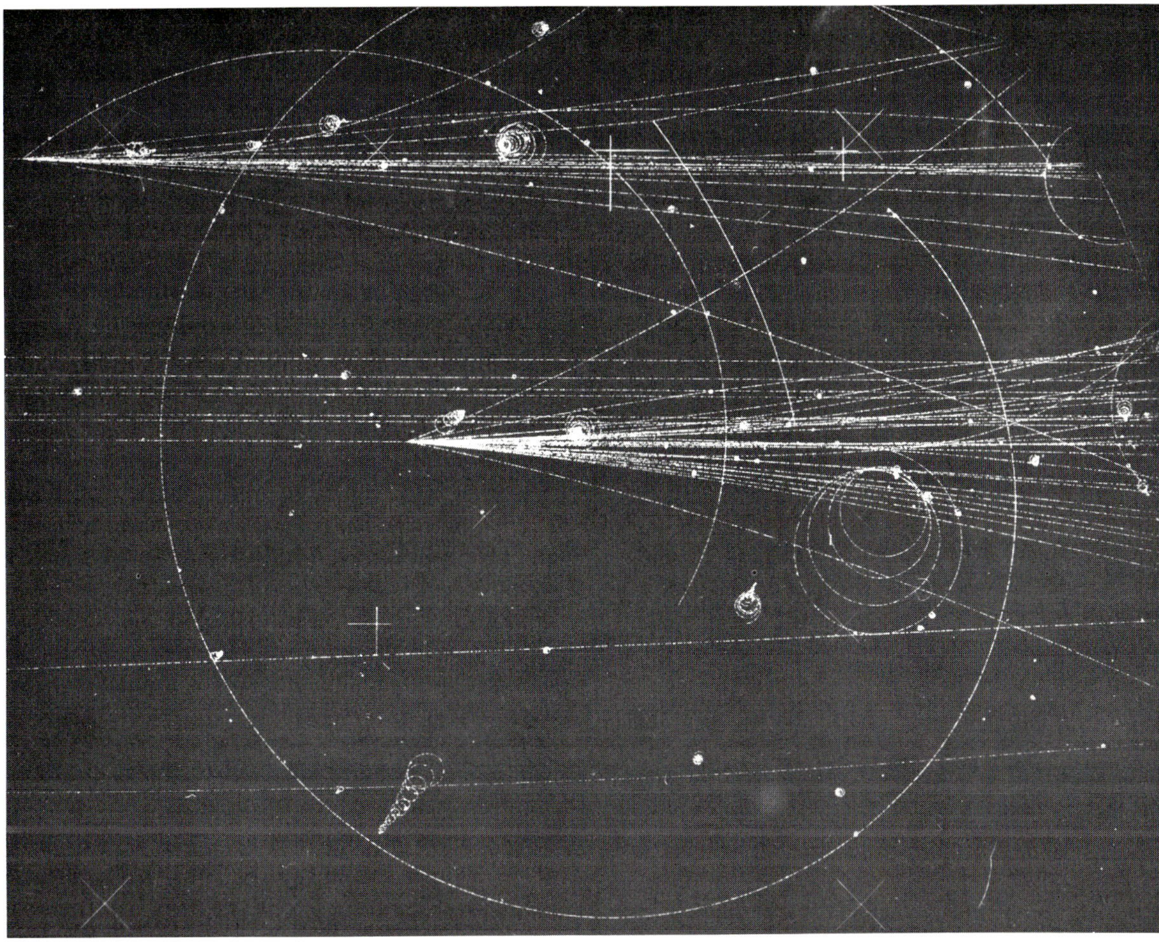

Figure 14.7 Experimental results from National Accelerator Laboratory. Two high-multiplicity events produced at 300 BeV energy recorded by bubble chamber in August, 1972. [National Accelerator Laboratory]

14.5 Radioactivity

Radioactivity is a nuclear phenomenon but its study began 15 years before the discovery of the nucleus and so for a long time it was a great mystery. It was discovered accidentally by Becquerel (1896), who found that a securely wrapped photographic plate was exposed by a piece of uranium compound placed on it. The compound must have emitted a ray, now called the *radioactive ray,* which penetrated the plate holder and affected the photographic plate in a way similar to the action of the X ray. The discovery of radioactivity in 1896 together with the discovery of the X ray the year before and that of the electron the year after opened up a new era of modern physics and launched investigations culminating in the establishment of atomic and nuclear physics.

The phenomenon of radioactivity is drastically

different from any physical and chemical transformations known before. Radioactive substances, such as radium, constantly emit radioactive rays and seemed to provide an interminable source of energy. In other transformations energy emission is never spontaneous or endless. Moreover, the radioactivity of radium, say, remains the same, whether in element form or in chemical compounds. Thus radioactivity has nothing to do with chemistry (outer electron shell). Finally, in the process of radioactive emission, new chemical elements appear and the original radioactive substance diminishes. In other words, a *radioactive transformation* takes place in which one element changes into another. The concept of Dalton that atoms are immutable was thus directly refuted by fact. The process thus must affect the nucleus of the atom.

These are strange properties, contrary to previously established rules. In the early years efforts were concentrated on learning the facts. A celebrated event was the discovery of the new element radium by Madam Curie. This substance provides an intense source of radioactivity so that investigations on radioactivity can be carried out expeditiously. The explanation of the facts has to await the development of nuclear physics.

The first question is, "What are radioactive rays?" It was found that there are three kinds, called the α-ray (alpha ray), β-ray (beta ray), and γ-ray (gamma ray). They can be separated by applying a magnetic field (Fig. 14.8). The α-ray is a stream of positively charged particles identified with the helium nuclei as already mentioned. The β-ray is a stream of negatively charged particles identified with the electrons (by the value of e/m). The γ-ray is neutral; as discussed before it is a kind of electromagnetic wave like the X ray, with wavelengths even shorter than X rays. Each radioactive atom emits one or two of the rays. Because a sample of radioactive material is usually a mixture of many radioactive atomic species, all three are present generally.

What happens to the atoms that emit the radioactive rays? It was found that after the emission of the α-ray or the β-ray the radioactive atoms change

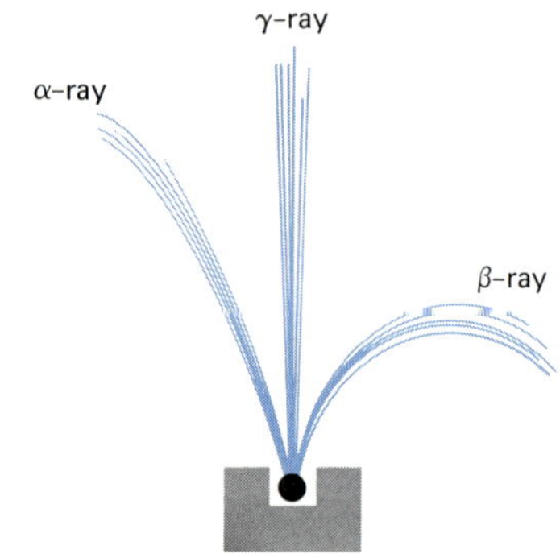

Figure 14.8 The three components of the radioactive rays separated by a magnetic field.

into different atoms, this being the radioactive transformation or *decay*. The decay product, the *daughter,* of α-decay occupies a position in the periodic table two places below that of the *parent* (Z is reduced by 2). The daughter of β-decay occupies a position one place ahead of the parent (Z is increased by 1). These are the *displacement laws* of radioactive decay. In γ-decay the atom does not change its species.

The daughter of a radioactive atom is usually also a radioactive atom itself and will emit its rays and transform itself into another atom. A succession of radioactive decays forms a *radioactive series,* starting from a grandparent, changing through over a dozen descendants one after another, and ending with a nonradioactive (stable) final product. All naturally occurring radioactivities can thus be organized into three series: the *uranium series* (starting from ${}^{238}_{92}U$, ending with ${}^{206}_{82}Pb$), the *actinum series* (starting from ${}^{235}_{92}U$, ending with ${}^{207}_{82}Pb$), and the *thorium series* (starting from ${}^{232}_{90}Th$, ending with ${}^{208}_{82}Pb$).

Artificial Radioactivity. As just mentioned, natural radioactivity stops at lead, ${}_{82}Pb$. It might be suspected that all the strange behavior associated with radioactivity might be just a freak of nature peculiar to the heavy elements with Z greater than 82. Later, however, in nuclear reactions new isotopes are formed that are radioactive and they can be formed for any atomic number. Thus radioactivity is not an oddity of the heavy elements, but a general property of all elements.

In fact, for any element Z, isotopes may be formed with a variety of neutron numbers. Except for a few stable (meaning nonradioactive) ones, all others are radioactive. Compared with the stable ones, those with an excessive number of neutrons are β-active, emitting electrons. Those with a deficiency of neutrons are radioactive, but the particles they emit are positrons. This represents a new kind of radioactive decay, which is called *positive β-decay* to distinguish it from *negative β-decay*. In a general sense the term *β-decay* includes both positive and negative processes.

Thus we may imagine a variety of nuclei formed by a variety of combinations of N and Z but most of them are unstable (radioactive). A nuclear species specified by a given N and Z is called a *nuclide*. The nuclides, each specified by two numbers N and Z, may be represented by points on a graph, called the chart of nuclides, with N and Z as the x and y coordinates. The stable nuclei generally fall on or near a curve called the *β-stability line* (Fig. 14.9). For light nuclei the stable ones are generally those with $N = Z$ and the β-stability line in this region is nearly a straight line inclined at 45° with the N-axis. On the other hand, for heavy nuclides, the stable ones usually have more neutrons than protons ($N > Z$) and the β-stability line levels off as shown in Fig. 14.9. Nuclei below the β-stability line are negative β-active, whereas nuclei above the line are positive β-active.

Figure 14.9 The β-stability line.

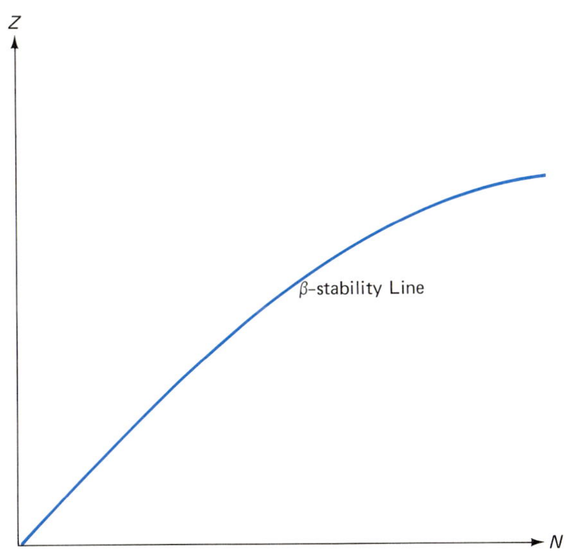

Law of Radioactive Decay. How fast does a radioactive substance emit the radioactive ray? It was found that in all kinds of radioactivity, α, β (positive and negative), γ, natural or artificial, the process of radioactive decay follows a definite law—the *law of radioactive decay*. The law specifies that a sample of radioactive substance with N atoms to begin with will always lose a definite fraction λ of it per unit time by radioactive decay. The constant λ is called the *radioactive constant,* which is a material constant varying from one nuclide to another. Notice that the constant quantity is the *fraction*. The actual number is not constant. In fact, as N decreases in time the fraction of atoms λN also decreases in number. Therefore N does not decline by constant rate and the curve representing the decline of N in time is not a straight line. For Radium $^{226}_{82}Ra$ the value of λ is 4.3 per cent per century; this means 4.3 per cent of a sample is changed into radon, Rn, in one century by emitting α-particles. Starting with a large number N, subtracting 4.3 per cent after a century, and then subtracting 4.3 per cent of what remains (95.7 per cent of N) for the next century, and so on, we construct a curve $N(t)$ representing the decline of N in time, as shown in Fig. 14.10. As already mentioned, the curve is not a straight line; it is an *exponential curve*. By the calculus method we can prove that the curve is represented by the equation

$$N(t) = Ne^{-\lambda t}$$

which is the quantitative representation of the law of radioactive decay.

As radioactive decay proceeds, sooner or later the

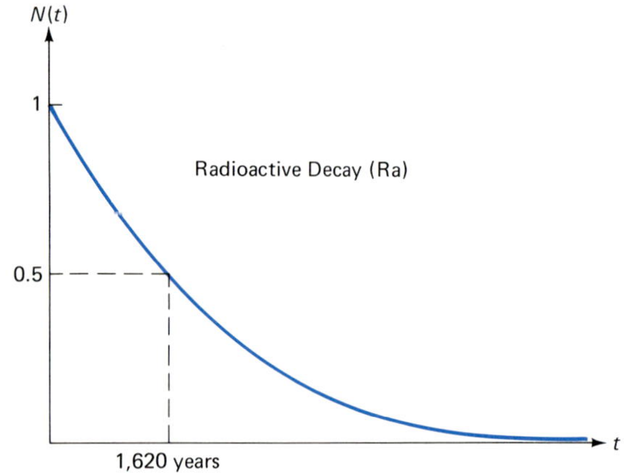

Figure 14.10 Exponential curve representing the law of radioactive decay.

time will come when the original amount is reduced by one half. This time is called the *half-life, T,* of the radioactive nuclide. Mathematically we can prove that T and λ are related by the equation

$$T = \frac{\log_e 2}{\lambda} = \frac{0.693}{\lambda}$$

Thus T is another way of representing the rate of radioactive decay and is usually the one determined experimentally. The half-life of radium is 1,620 years. This means that a given sample of radium will be reduced by one half after 1,620 years, the other half being converted to radon by emitting α-particles. Each radioactive nuclide has a definite half-life. The values of half-lives of radioactive atoms range from billions of years to a minute fraction of a second. Uranium has a half-life of 4.5 billion years. It decays so slowly that the uranium on earth inherited from the "day of creation" has not completely disappeared yet. Otherwise uranium would have disappeared from earth long ago. This also indicates that the geological history must be finite and that there was indeed a "day of creation" (at least for the uranium). Thus the phenomenon of radioactivity gives us important information on the origin of the universe and actually we can calculate the age of earth (or the age of uranium-bearing rocks) by the law of radioactive decay based on the ratio of uranium and its decay product, lead, in uranium ores.

The law of radioactive decay is interesting in another respect. Why do some radium atoms decay in a few years and some not decay after thousands of years, inasmuch as all radium atoms are supposed to be exactly identical? When a sample of radium has decayed so much that there is only one atom left, what will happen to this atom after 1,620 years? From the law of radioactive decay we can only say that after 1,620 years this radium atom has a 50 per cent probability of having decayed and a 50 per cent probability of not having decayed. What actually happens to this radium atom in the course of time cannot be predicted. If all atoms were predictable in the same way, we would have a completely different law of radioactive decay. Thus an element of probabilistic behavior is introduced into physical laws, which is contrary to the deterministic viewpoint of the Newtonian laws. After the discovery of radioactivity, scientists realized that *the law of causality had broken down.* We have already mentioned that the quantum theory takes a probabilistic point of view and here we see a concrete example for the necessity of such a viewpoint. It is to be noted that the probabilistic view is necessary because of the discontinuous (atomic) structure of matter; there exists a basic unit atom which must act as a whole. Quantum theory, dealing with discontinuous systems, thus is destined to be probabilistic.

Explanation of Radioactivity. Up to this point we describe radioactivity as a natural phenomenon. On the other hand, once we know the structure of the atom, we would expect an explanation of radioactivity in terms of the structure and the basic laws of motion and of forces. Because radioactive decay involves the change of the element and an element is determined by the charge of the nucleus, radioactivity must involve a change of the nucleus. The displacement law relating the parent and the daughter in the periodic table tells us how the charge of the nucleus is changed: in α-decay charge Z decreases by two units and in (negative) β-decay charge Z increases by one unit. These changes equal the charges carried away by the α- and β-particle and therefore the total charge is conserved. Once the nucleus is known to be made of protons and neutrons we have to conclude that the proton number decreases by two and increases by one in α- and β-decay, respectively.

Because the α-particle is identified with the helium nucleus ^4_2He, which is made of two protons and two neutrons, the emission of an α-particle is looked upon as the extraction of two neutrons and two protons from a radioactive nucleus, with the four extracted particles that come out together forming an α-particle. Or we may think of the four particles forming an α-particle inside the nucleus, which moves under the influence of the nuclear forces of the other protons and neutrons. Thus we have a problem of the motion of a particle under the influence of the force of strong interaction. This is the kind of problem that can be dealt with by the laws of mechanics. In this case it turns out that Newtonian laws are not adequate and quantum mechanics has to be applied. In fact, by applying quantum mechanics, we can predict that after the α-particle has moved inside for some time, there is a definite probability of escaping the nucleus. On this basis we can calculate the radioactive constant or half-life of various α-emitters. The calculations agree well with experimental results and this successful prediction was considered a major triumph of quantum mechanics in its early years. After an α-particle is emitted the numbers N and Z both are reduced by two; the nucleus thus changes into that of another element displaced two places backward in the periodic table. The mass number of the nuclide is reduced by four. The change of atomic weight is also checked by experiment.

In β-decay an electron is emitted and the nucleus changes into that of another element displaced one unit forward in the periodic table. Therefore the electron cannot come from those outside the nucleus and must come from inside the nucleus. The earlier theory that the nucleus is made of protons and electrons suits this situation well. One of the electrons inside the nucleus is emitted as the β-particle and the nuclear charge is increased by one unit as a result, agreeing with the displacement law. Once this theory is replaced by the proton–neutron theory, we have to assume the β-particle to be newly created inside the nucleus. Because the proton number is increased by one, we have to assume that one neutron changes into a proton with the emission of an electron. The total charge as well as the mass is conserved in this process. The "force" responsible for this change is the weak nuclear force introduced specifically for the purpose of explaining the β-decay. Moreover, the same weak interaction later also explains other kinds of radioactive decay of other elementary particles. Altogether the strong and the weak nuclear forces explain most of the nuclear processes.

A free neutron outside the nucleus is unstable and is radioactive. It emits an electron and changes into a proton with a half-life of 12 minutes. This is just what is assumed to happen inside the nucleus in β-decay. In the case of positive β-decay, a proton is changed into a neutron and a positron.

One complication appears here. If the neutron changes into two particles, the proton and the electron, then all energy, momentum, and angular momentum will not be conserved, as experiments show. To save the three conservation principles one has to assume the emission of a third elementary particle, called the *neutrino,* in β-decay, which picks up the missing energy, momentum, and angular momentum. This particle is neutral in charge and practically massless. (A "material particle" with no mass!) Many years later the existence of the free neutrino was demonstrated by experiment. The neutrino also has its antiparticle and the one emitted in negative β-decay is the antineutrino $\bar{\nu}$. Thus the complete basic equations of β-decay are

$$\mathrm{n} \longrightarrow \mathrm{p} + \mathrm{e}^- + \bar{\nu} \qquad \text{for } \beta^-\text{-decay}$$

$$\mathrm{p} \longrightarrow \mathrm{n} + \mathrm{e}^+ + \nu \qquad \text{for } \beta^+\text{-decay}$$

In γ-decay the nucleus does not change species. The emission of the γ-ray is a process very similar

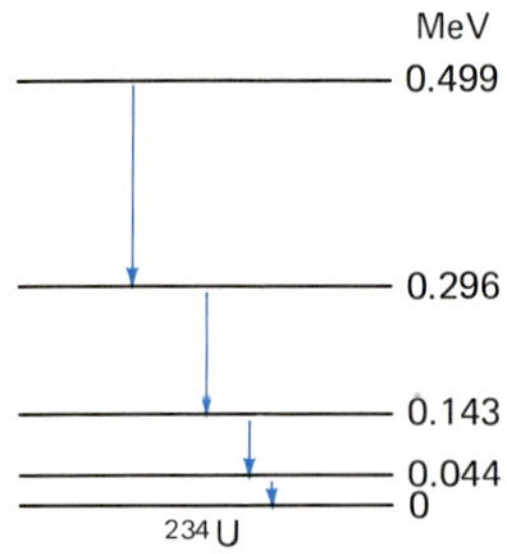

Figure 14.11 Quantum transitions of nuclear energy states and γ-ray emission of ^{234}U.

to the emission of light by the quantum jump of an electron in an atom. In Sec. 13.4 we have already discussed the mechanism of γ-ray emission. The γ-rays emitted by ^{234}U are shown by quantum jumps in Fig. 14.11. The force responsible for the process is the electromagnetic interaction. Thus based on the basic laws of motion and three of the four basic forces, all radioactive processes as well as other nuclear processes can be explained.

Uses of Radioactivity. Radioactive rays can kill living cells; from very early times radium has been used for cancer therapy to kill malignant cells. Yet these rays also kill healthy cells. When radioactive atoms escape into the environment inadvertently, they become a health hazard. Radiation pollution is a point we have already discussed in Sec. 2.6.

Another use of radioactivity in connection with the environmental problem is the use of the *neutron activation method* for monitoring air pollutants. Very small amounts of hazardous air pollutants, such as arsenic, cannot be measured by chemical analysis but can be measured by the neutron activation method. Upon irradiation by neutrons, the trace pollutant elements undergo nuclear reactions and change into new isotopes that are radioactive. The radioactive rays can be easily measured by nuclear physics methods. Because each radioactive isotope has a definite half-life, the measurement of the half-life of the radioactivity can be used to identify the nuclear species and thus the pollutants can be identified. The quantitative measurement of the radioactivity provides the basis for monitoring the amounts of the pollutants.

Other uses of radioactivity will be discussed in the next chapter (Sec. 15.6). The discussions here illustrate the two aspects of the impact of any scientific discovery, constructive and destructive. The discovery itself is neutral; whether it becomes destructive or constructive depends on how we make use of it.

Problems

14.1 How many protons and neutrons are there in ^{12}C and ^{14}C? The atomic number Z for carbon is 6. How many protons and neutrons are there in ^{238}U and ^{235}U? The Z value for U is 92.

14.2 Natural chlorine consists of two isotopes: 75 per cent of ^{35}Cl and 25 per cent of ^{37}Cl. What is the average atomic weight? This is the atomic weight determined and used in chemistry.

14.3 Tin ($_{50}Sn$) has an unusually large number of stable isotopes. Why?

14.4 Radioactive decay of naturally occurring heavy radioactive elements terminates at lead ($_{82}Pb$). Why not earlier or later in the sequence?

14.5 Complete the following nuclear reactions:

$$^{7}_{3}\text{Li} + \text{p} \longrightarrow \alpha + ?$$

$$^{27}_{13}\text{Al} + \text{n} \longrightarrow \alpha + ?$$

$$^{3}_{1}\text{H} + ? \longrightarrow {}^{4}_{2}\text{He} + \text{n}$$

14.6 In a uranium mine ^{206}Pb is found to exist, in the ratio of 0.093 g of lead to 1 g of uranium, as the radioactive decay product of ^{238}U. Find the age of the uranium-bearing rock. The half-life of uranium is 4.5×10^9 years.

15

Nuclear Energy

15.1 Nuclear Binding Energy

Just as energy is released in chemical reactions when atoms associate or dissociate, energy is also released in nuclear reactions when protons and neutrons associate or dissociate. In fact, the amount of energy released per atom in nuclear reactions is generally a million times that released in chemical reactions. The energy stored in the atomic nucleus was recognized when nuclear studies began. At first it was thought to have no practical significance because nuclear reactions did not take place on a large scale. In 1939, when the possibility of large-scale practical use of this energy was first realized, a large number of scientific and technological problems still remained to be solved. The wartime need for atomic weapons elicited a crash program. By effort and by "luck" the difficulties were overcome. Not only was the bomb developed, but a new energy source was made available, which, in the long run, is vastly more significant than the atomic weapon. Numerous beneficial "fallouts" of the program in biology, medicine, agriculture, and engineering make the program one of the most significant scientific events in history.

The atomic nucleus is made of two kinds of particles, Z protons and N neutrons. The atomic weight of the proton is 1.007596 and that of the neutron is 1.008986. The nucleus of the helium atom, which is made of two protons and two neutrons, ought to have a total atomic weight of $2 \times 1.007596 + 2 \times 1.008986 = 4.033164$. Yet the actual atomic weight of the helium nucleus is slightly smaller; it is 4.00279. Earlier when the experimental values were not accurate enough, any slight difference, if it existed at all, was ignored and it was hoped that the discrepancy would go away as improvements were made. However, more accurate experimental values, such as those listed here, show that the difference is real and will not disappear in the future. For the first time the law of conservation of mass seems violated. This would be disastrous, but the theory of relativity tells us mass can be changed to energy and so is not necessarily conserved by itself.

This difference in mass between the nucleus and the constituent particles is called the mass defect. It is always a shortage when the particles combine to form a nucleus. The theory of relativity claims that m kg of mass are equivalent to mc^2 joules of energy, where c is the speed of light 3×10^8 m/sec:

$$E = mc^2$$

Thus there is the possibility that part of the mass may be changed to energy and be radiated away when particles combine to form a nucleus. This is advanced to explain the mass defect. This explanation can be tested quantitatively as follows. Most nuclides (N,Z) can absorb a neutron and convert themselves to another nuclide $(N+1, Z)$. The mass of $(N+1, Z)$ does not equal the mass of (N,Z) plus the mass of a neutron and the amount of mass lost can be calculated. At the same time, after the absorption of the neutron, the nucleus does radiate energy in the form of γ-rays. The energy of the γ-rays emitted can be measured and its amount can be compared with the mass loss multiplied by c^2. This comparison can be carried out for many nuclear reactions of this kind and many similar reactions. In all cases the equality of the two is verified. Thus the loss of mass in nuclear reactions is indeed converted into the energy radiated.

A nucleus (N,Z) may be regarded as being built up from a proton by adding neutrons and protons one at a time until the numbers N,Z are reached. If in each step the mass lost equals the energy radiated in the process, then the total mass defect of the nucleus (N,Z) equals the total energy radiated when neutrons and protons combine to form a nucleus. As hundreds of nuclear reactions are involved in the verification of the equation $E = mc^2$, mass defect evidence is the most strong and convincing proof of the equivalence of mass and energy.

The equivalence of mass and energy combines the two previously independent principles of conservation of mass and conservation of energy into one single principle—conservation of the total of mass and energy. Each of them may increase or decrease but must be compensated by an equivalent change of the other. In nuclear physics this conversion happens frequently. Now if mass is converted into energy and lost when particles combine to form a nucleus, then in the inverse process, in the breaking up of a nucleus into the constituent particles, an equal amount of energy has to be added into the nucleus to replenish the lost mass of the particles. In the absence of this amount of energy the particles stick tightly together and therefore the energy is called *binding energy*. The larger the binding energy, the more difficult to break up the nucleus and the more stable the nucleus. Thus the binding energy, or equivalently, the mass defect, is a measure of the stability of the nucleus.

It is a basic fact in nuclear physics that the binding energy of a nucleus is nearly proportional to the number A of particles in the nucleus (sum of N and Z). In other words, the binding energy per particle is nearly a constant, which is about 8 MeV. It requires the addition of about 8 MeV energy to separate a particle, proton or neutron, from the nucleus.

The binding energy per particle is not exactly, but only approximately, a constant. There is an individual variation and also a systematic change. The binding energy per particle is largest for medium-weight nuclides; the maximum is at mass number $A = 56$ (iron). As the mass number increases to the end of the periodic table or decreases to the beginning, the binding energy per particle decreases slightly and gradually (Fig. 15.1). Thus median-weight nuclides are most stable. The heaviest and the lightest are least stable. If the heavy nuclei are broken up into median-sized nuclei, or light nuclei are combined to form median-sized nuclei, the mass defect of the system becomes larger and the additional mass lost turns into energy that will be released in these processes.

These processes are actually found in nuclear physics. The process of breaking up a heavy nucleus into two median-sized nuclei is called *fission*. The process of combining light nuclei into heavier ones

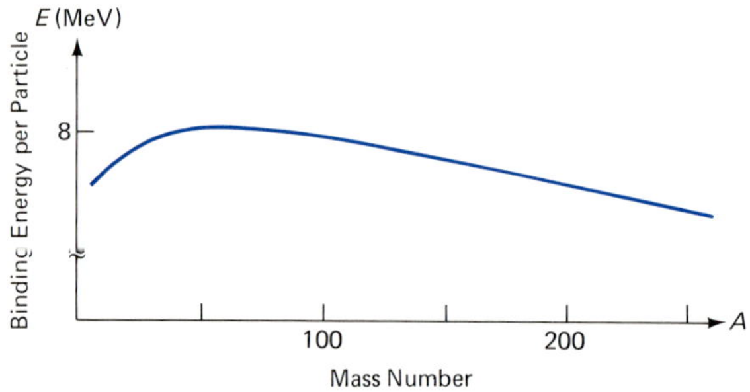

Figure 15.1 Binding energy per particle as a function of the mass number A.

is called *fusion*. In both cases energy is released. This is the physical basis of *nuclear energy*. Thus nuclear energy may be obtained in two ways—through nuclear fission and nuclear fusion.

15.2 Nuclear Fission

We first consider fission energy. Nuclear physics started around the turn of the century. Many nuclear reactions have been observed in the first four decades but all of them are concerned with the addition or deletion of a few particles (protons or neutrons) to a nucleus. Inasmuch as a nucleus is made of a large number of particles of two kinds, the possibility of its splitting into two comparable fragments cannot be ruled out. Yet such a fission process has not been observed or even imagined before 1939. And it was through a rather devious path and after considerable scientific efforts that the process was finally discovered in 1939. It was big news, a kind of nuclear reaction never known before, and the process had many unusual features that attracted attention. It was first observed in the heavy element uranium when bombarded by neutrons. What happened was that the uranium isotope ${}^{235}_{92}U_{143}$, after absorbing a neutron, became another uranium isotope, ${}^{236}_{92}U_{144}$, which then split into two fragments, X and Y, which may be designated by the respective proton and neutron numbers ${}^{A_1}_{Z_1}X_{N_1}$, ${}^{A_2}_{Z_2}Y_{N_2}$ (Fig. 15.2).

$$ {}^{235}_{92}U_{143} + n \longrightarrow {}^{236}_{92}U_{144} \longrightarrow {}^{A_1}_{Z_1}X_{N_1} + {}^{A_2}_{Z_2}Y_{N_2} $$

The splitting of the numbers is such that

$$ \begin{aligned} A_1 + A_2 &= 236 \\ Z_1 + Z_2 &= 92 \\ N_1 + N_2 &= 144 \end{aligned} $$

One of the unusual features is that there is a large variety of X's and Y's. (The fission products are not unique, but show a wide distribution.) Altogether there are some 200 isotopes that can be produced as fission products; some of them do not exist in nature and practically all are radioactive. Thus fission is the source of many new artificially created nuclides. The two fragments are generally unequally divided; this is called asymmetric fission. Why this is so is one of the outstanding problems in nuclear physics.

Figure 15.2 The fission process.

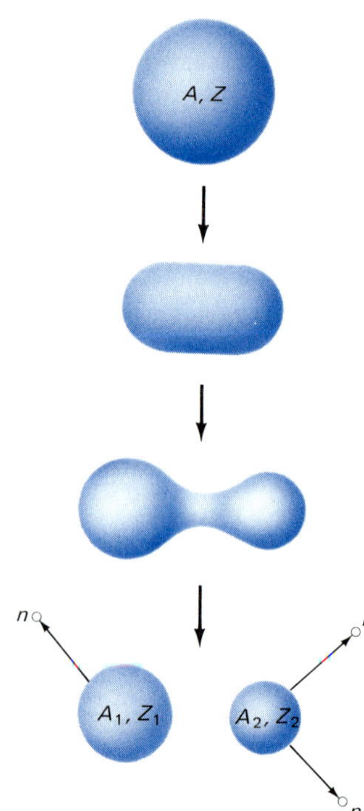

The second unusual feature is that fission is accompanied by the release of a large amount of energy. In most nuclear reactions known previously, the energy released, if any, is of the order of 5 to 10 MeV, comparable to the binding energy of one or two particles. In fission, the energy released is of the order of 200 MeV, which is much greater.

The most significant feature is that the fission products emit neutrons spontaneously. Remember fission is initiated by a neutron. As the fission products emit neutrons, these neutrons may be used to start fission of other uranium nuclei. The process may be repeated one stage after another and a "chain reaction" may result (Fig. 15.3). It only takes one neutron to start a chain reaction in which millions or billions of nuclei split and a tremendously large amount of energy is released. For the first time we see the possibility that nuclear energy in large quantity may be obtained in an economically feasible way.

The existence of nuclear energy was known for a long time, but no one realized its practical significance. The pioneer nuclear physicist Lord Rutherford considered nuclear studies to be a pure intellectual endeavor having no practical or economical applications at all. Although the energy of nuclear reactions (a few MeV) is large, about a million times larger than that involved in a chemical reaction (a few eV) atom for atom, the release of the energy is achieved by using some very expensive particles—neutrons and high-energy protons—as bombarding particles in the reaction. It costs one such particle for one nucleus to react with the release of a few MeV energy. Economically it never pays off. The chain reaction changes all this. It costs only one neutron, no matter how expensive, to start the chain reaction of millions, billions of nuclei with the release of a vast amount of energy. (In fact, the first neutron does not cost us anything at all.) For the same number of atoms, the fission energy is 10 million times larger than chemical energy. Thus 1 lb of nuclear fuel is a million times more powerful than 1 lb of chemical fuel (coal, gasoline, and so on).

Figure 15.3 Chain reaction.

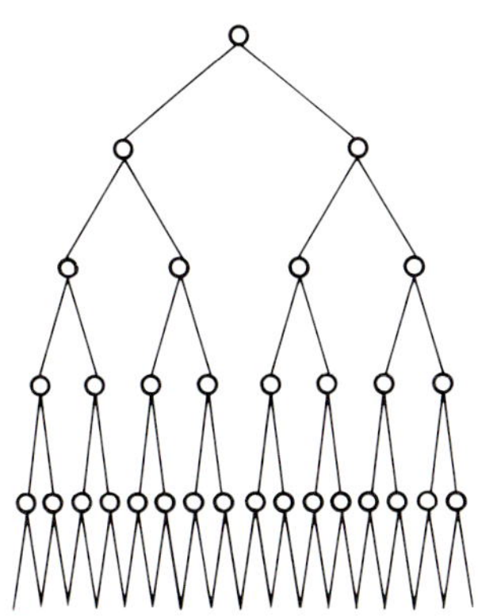

Once the emission of neutrons by fisson products was discovered, the possibility of practical use of nuclear energy suddenly dawned. This was all the more awesome because of its high potency. The year 1939 was a time of impending war. The military implication of this highly concentrated energy source was recognized by both warring parties. The military use of nuclear energy was the stimulus to initiate large-scale research and development programs. To the public, the energy is known as atomic energy and the bomb so developed is the atomic bomb. They should be referred to as nuclear energy and the nuclear bomb. The fact is that the term *atom* is familiar to the public but *nucleus* is not.

Fission Chain Reaction. Let us return to the fission chain reaction. If out of 100 uranium atoms that undergo fission, the 200 fission products together should emit enough neutrons to produce fission of 50 uranium atoms in the second generation, then we would expect 25 in the third generation, 12.5 in the fourth, and so on. The chain would proceed with diminished strength and all generations added together would yield a total of 200 uranium atoms that have fissioned, doubling the original 100. The amount of energy released would be increased by a factor of 2, which is not very significant. This kind of chain reaction is said to be *convergent.* It has little practical significance.

On the other hand, if the original 100 fissioning

uranium atoms should yield 200 in the second generation, then there would be 400 in the third, 800 in the fourth, and so on. The chain would grow in strength and soon all available fissionable material would be used up completely and all the energy would be released nearly at once. This would result in an explosion. For military uses of making a bomb, this kind of chain reaction is exactly what is desired. This kind is said to be *divergent.*

A divergent chain reaction has no use in tapping nuclear energy for peaceful purposes such as generating electric power. For such a purpose we want the chain reaction to proceed at a constant rate, not convergent, not divergent, but exactly in between. In other words, we want 100 original fissioning atoms to produce exactly 100 in the second generation, 100 in the third, 100 in the fourth, and so on. This way energy will be released at a constant rate, the "fire" will not die out or build up to an explosion. Such a chain reaction is said to be *stationary.*

Thus the first question that has to be answered is, "Can a fission chain reaction of uranium be made self-sustained (divergent or stationary)?" A necessary condition is that the pair of fission products emit, on the average, at least one neutron, at least to replenish the one neutron consumed in the reaction. The number of neutrons emitted by a pair of fission products, designated by ν, was a tightly guarded atomic secret. Now it is well known to scientists; for ^{235}U the value of ν is 2.5, which is indeed greater than 1, and by a large margin. Otherwise, no nuclear energy would be possible. This is one fortunate situation that helped the program.

But not all 2.5 neutrons can be used for second-generation fission. Neutrons may be absorbed, and thus wasted, by other material in the device. A large amount of research and development was carried out to insure that more than 1 out of the 2.5 neutrons is used in second-generation fission. Thus both stationary and divergent chain reactions are made possible.

The device in which a stationary fission chain reaction is made to occur is called a *nuclear reactor.* The first reactor was constructed at the University of Chicago campus during World War II. The first self-sustained (stationary) chain reaction was achieved on December 2, 1942, a landmark date in human history. By this demonstration the industrial use of nuclear energy became a reality, even though the original purpose of this first reactor was to demonstrate the feasibility of nuclear chain reaction for making atomic bombs.

15.3 Nuclear Fission Reactors

Because the reactor is the basic device to tap nuclear fission energy, a more detailed description of its structure and operation is in order here. It must be first stated that natural uranium consists of two isotopes, 99.3 per cent of ^{238}U and 0.7 per cent of ^{235}U. It is the rare isotope ^{235}U that is the active ingredient in the fission reaction. The majority of natural uranium ^{238}U is inert in the chain reaction; it fissions very inefficiently and moreover the neutrons emitted by fission products do not have sufficient energy to cause second-generation fission of ^{238}U (thus no contribution to the chain reaction). If natural uranium should consist of ^{238}U exclusively, there would be no atomic energy and no atomic bomb. But, happily, there does exist a trace of ^{235}U, 0.7 per cent of the natural element. This changes the situation completely. This is another fortunate situation that makes nuclear energy possible. More will be said on this point later.

The Use of a Moderator. The major isotope ^{238}U not only does not contribute to the chain reaction, but impedes it by absorbing, and thus wasting, the vital neutrons. However, an ingenious arrangement was contrived, which is the basic principle of the reactor, to circumvent this difficulty. It is based on two scientific facts: that the fissionability of ^{235}U increases when the neutron energy decreases and that the absorption of neutron by ^{238}U is greatest for neutron energy in the neighborhood of 25 eV. Neutrons emitted by fission products have an energy on the order of 1 MeV. The central idea of the reactor is to reduce the energy of the neutron from 1 MeV down, passing and avoiding the "death trap" of ^{238}U at 25 eV, all the way to the end, the thermal motion energy of 0.04 eV. Then the neutrons are used to produce second-generation fission. (The first neutron initiating the chain comes from the natural environment—from the very rare spontaneous fission of ^{238}U—and need not be artificially provided.) To achieve this end, natural uranium is made into thin, long rods (Fig. 15.4), so thin that most neutrons emitted by fission products in the rod will escape the rod and thus stay away from the death trap of ^{238}U. The rods are arranged in parallel and the rather large space between the rods is filled with a material acting as a *moderator* to reduce the energy of the neutron without absorbing them. There are only a few substances that qualify as a moderator, e.g., heavy water and graphite. When the neutrons enter the moderator they collide with moderator atoms and thus lose energy gradually. The lighter the moderator atoms, the faster is the rate of reduction of energy, but only a few light substances have the desired property of not absorbing neutrons to qualify as a moderator. The neutrons stay in the moderator while the energy is being reduced, past the vulnerable energy value 25 eV, all the way down to 0.04 eV, which is the average heat energy per particle in the moderator and thus cannot be reduced any further. Without being absorbed, the neutrons at this terminal energy wander around the moderator "indefinitely" and thus eventually wander back into the uranium rod. Now these so-called *thermal neutrons* (low-energy neutrons) in the rod will not be absorbed by ^{238}U but will react with ^{235}U to produce second-generation fission. Everything considered, after such a complicated sequence of events with various possibilities of losing neutrons, there is still more than one neutron out of the original 2.5 left that is now available to produce second-generation fission of ^{235}U (and this is an engineering feat). Thus a self-sustained chain reaction may be generated. The rods and the moderator enclosed in thick walls form a *reactor* (Fig. 15.5).

Figure 15.4 Uranium rods and moderator in a reactor.

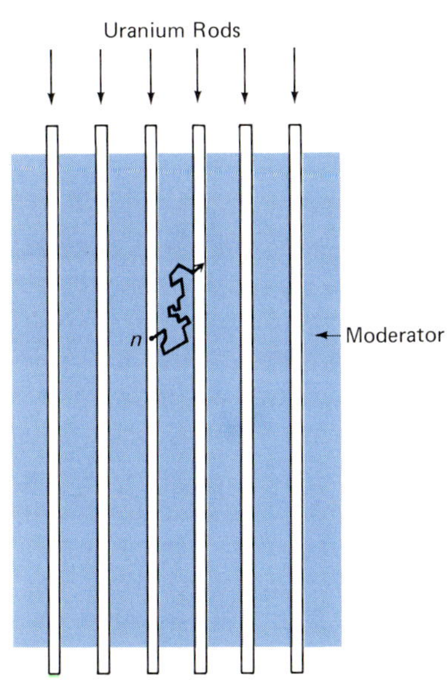

Reactor Control. If at this stage more than one neutron out of the original 2.5 should produce second-generation fission, the chain reaction will accelerate to an explosion. For industrial power production we want exactly one neutron out of the 2.5 to be used in second-generation fission. This requires careful and delicate control of the reactor to bring the number to 1 precisely. This control is accomplished by using a cadmium rod, a material that has a high power of absorbing neutrons. When the rod is inserted all the way into the reactor, it absorbs most neutrons in the moderator and the nuclear "fire" is "extinguished." By pulling the rod out gradually the number of neutrons in the moderator increases and eventually it will reach a point when the number of second-generation fissions exactly equals that of the first. Then a stationary chain reaction is achieved. If there is any sign the reaction becomes divergent, the rod can be pushed in to reduce the level to normal and to avert an explosion.

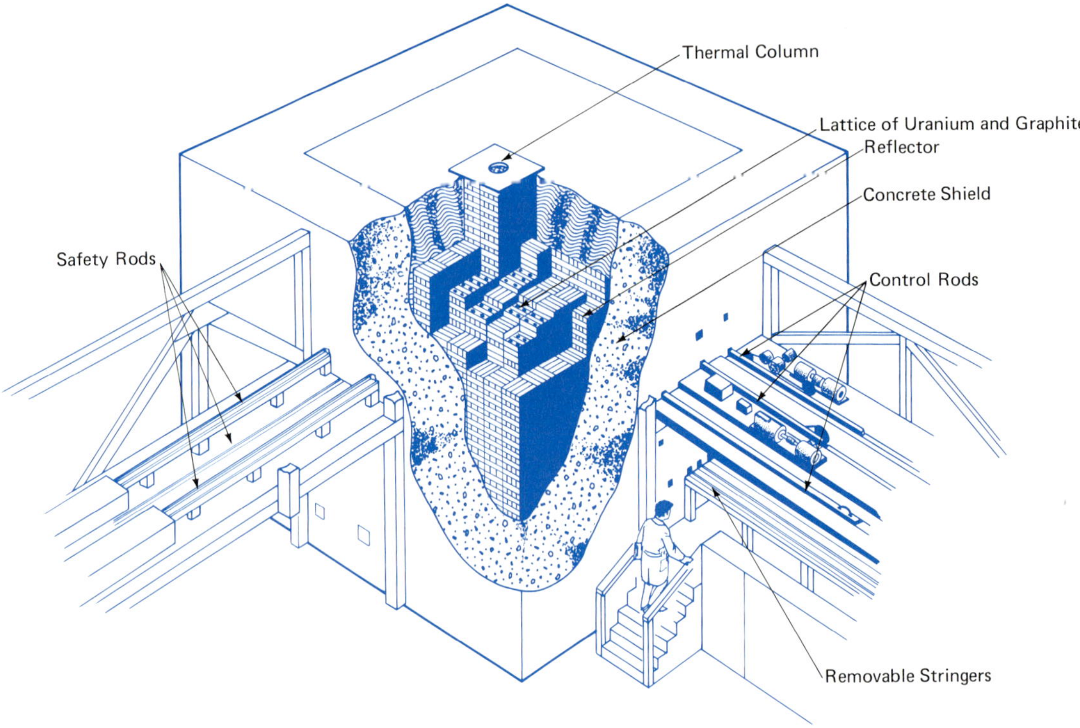

Figure 15.5 Cutaway view of the first nuclear reactor showing lattice of uranium and graphite enclosed by concrete shield. [University of Chicago]

To shut down the reactor, push the rod all the way in. To start the reactor, pull it out. A stray neutron from spontaneous fission of ^{238}U or other sources is always present to start a chain reaction. As the chain reaction intensifies, temperature of the reactor increases. When the reactor reaches a desirable level of operation, push the rod in until a constant level of operation is established. This is in principle the way a reactor is controlled. Indeed, the first reactor was controlled this way, by pushing a cadmium rod in and out manually.

On December 2, 1942, when the first reactor became operative, the rod was in the hands of a scientist operating according to instructions from Professor Enrico Fermi, the chief architect of the reactor (Fig. 15.6). On the first trial, like landing on the moon, there were all sorts of uncertainties and danger. (And the danger was quite real. A divergent reaction out of control could mean an explosion.) Besides the cadmium control rod, there were two additional control systems for emergency. Other scientists held a bucket of "water" (neutron-absorbing cadmium salt) standing on top of the reactor ready to pour it into the reactor to "quench" the nuclear "fire" should anything go wrong. Nothing went wrong that day; a historical feat was accomplished.

It would be a hair-raising experience to try to control a nuclear chain reaction [the A-bomb reaches the exploding stage in a microsecond (10^{-6} sec)] by manual control of a rod or a bucket, which would require at least a second to effect. It would be worse than letting a sleepy old man steer a supersonic jet plane. Yet the feat was accomplished without mishap. It was *not* a miracle. It was *not* by sheer luck.

It was due to another happy accident of nature—the presence of the *delayed neutrons*. About 1 per cent of the neutrons from fission products are emitted with a time delay of a few seconds. By itself it is a minor, inconsequential scientific fact. It turned out to be of crucial importance in the reactor control. Because this minute fraction is delayed several seconds, the explosion, which would require all the neutrons present at the same time, is delayed by several seconds. This delay provides enough leeway so that the reactor can be controlled by any engineering device at all, even by hand manipulation. Thanks to the delayed neutrons, the architects of the reactor were able to accomplish the seemingly impossible.

Thus the successful exploitation of fission energy depends on many factors; some are happy accidents. Without the trace isotope ^{235}U no fission chain reaction would be possible. If the number of neutrons per fission ν were not sufficiently large, chain reactions would not be possible. Without the delayed neutrons the control of a chain reaction would not be possible. We can only say that nature has been very generous to us. On the other hand, given these favorable conditions, there are still many serious technical problems that have to be solved by hard work before the chain reaction can proceed. Much skill and ingenuity were enlisted and the atomic energy program turned out to be one of the most spectacular scientific achievements of all time.

Plutonium. Up to this point the chain reaction involves only ^{235}U. The nuclear energy derived is at the expense of destroying ^{235}U. Therefore ^{235}U may be regarded as a fuel, a nuclear fuel. Yet ^{235}U accounts for only 0.7 per cent of all natural uranium. Thus only a small fraction of uranium is useful fuel.

As the nuclear research program unfolded, it turned out that the inert ^{238}U could be converted to useful nuclear fuel. This is significant because the

Figure 15.6 The historic moment on December 2, 1942—the first nuclear fire is ignited. [University of Chicago]

total nuclear fuel reserve is thus expanded a hundredfold. We have mentioned that ^{238}U may absorb a neutron but did not mention what happens afterward. When ^{238}U absorbs a neutron, it changes into another isotope of uranium, ^{239}U:

$$^{238}U + n \longrightarrow {}^{239}U$$

This isotope is unstable; it undergoes two successive stages of β-decay, changing to a stable isotope of a new element called plutonium.

$$^{239}_{92}U \xrightarrow{\beta^-} {}^{239}_{93}Np \xrightarrow{\beta^-} {}^{239}_{94}Pu$$

Previously the periodic table of elements terminated at uranium. All possible elements below uranium having been discovered, any discovery of new elements must come from beyond uranium. For many years the search for *transuranian elements* had been a major scientific effort. In fact, the discovery of fission was an unexpected by-product of such a search that was not successful. By 1940 and 1941 the preceding reactions were discovered by experiments in the laboratory and two new transuranian elements (Np and Pu) were created. It would have been great scientific news, but because of the war, the discovery was kept secret and the world learned of it only by the explosion of the atomic bomb over Nagasaki in 1945.

These reactions do take place in a nuclear reactor. The inert ^{238}U, after all, does transform itself to ^{239}Pu. It so happens that ^{239}Pu is similar to ^{235}U and can be used as a ^{235}U substitute either as nuclear fuel in a reactor or as a nuclear explosive in an atomic bomb. If all ^{238}U could be converted into ^{239}Pu by the reactor, then the inert ^{238}U would be converted into useful material.

The Atomic Bomb. During the war attention was concentrated on making the atomic bomb. A

Figure 15.7 Oak Ridge Gaseous Diffusion Plant. Originally producing ^{235}U for military use, the plant now produces enriched uranium for nuclear reactors. [U.S. Atomic Energy Commission]

Figure 15.8 Site of Electromagnetic Separation Plant originally built to produce ^{235}U for military use but shut down in 1947. The Oak Ridge Y-12 Plant now houses other units including the Biology Division of Oak Ridge National Laboratory. [U.S. Atomic Energy Commission]

sizable chunk of ^{235}U by itself will explode automatically. A stray neutron from a natural source will cause fission of one ^{235}U nucleus and the neutrons emitted by the fission products will be absorbed by other ^{235}U nuclei inside the chunk, if it is sufficiently large (over a *critical mass*) to trap more than one of the 2.5 neutrons inside the chunk, to produce a divergent chain reaction. (This time neutrons are not slowed down first.) All that has to be done to make a bomb is to keep a mass of ^{235}U over the critical mass in small separate pieces each under the critical mass. When the pieces are shot together to form a mass over the critical mass, the explosion takes place immediately. It is now common knowledge that this shooting together, or *implosion,* is done by a charge of conventional explosive TNT comparable to the amount used in a conventional bomb. The major task of making atomic bombs is to make pure ^{235}U.

Pure ^{235}U has to be separated from natural uranium. Natural uranium is an inexpensive mineral product but the separation process is costly. Unlike the separation of two chemicals, which can be done by making use of their different chemical properties, the separation of two isotopes can be done only by making use of the very slight difference of their masses, only about 1 per cent. The chemical properties of the two isotopes are exactly the same, so they respond to any chemical treatment in the same manner and thus cannot be separated from each other. Two physical methods were used to separate the isotopes: *gaseous diffusion* (Fig. 15.7) and *electromagnetic separation* (Fig. 15.8 and 15.9). Both are tedious and costly, even though the principle involved is simple—particles of different masses move differently. Because the mass difference is small, the diffusion process has to be repeated many, many times to achieve an appreciable result. Both methods were successfully developed at Oak Ridge at great cost.

The use of ^{239}Pu to substitute ^{235}U provides an alternative. Plutonium is a different element from uranium with different chemical properties. Therefore it can be separated from uranium, from which it is created in the reactor, by chemical methods, which are more efficient than gas diffusion and electromagnetic separation. Because the isotope separation programs were not guaranteed to succeed, it was decided to pursue both the isotope separation and the plutonium production programs at the same time in the hope that one of them would eventually succeed. A number of reactors were built in the state of Washington (the Hanford Works, Figs. 15.10 and

Figure 15.9 Electromagnetic Isotope Separators (Calutrons). [Oak Ridge National Laboratory, operated by Union Carbide Nuclear Division for the U.S. Atomic Energy Commission]

Figure 15.10 Operating crew readies the rear face of one of the huge Hanford plutonium production reactors for discharge of irradiated fuel elements. [U.S. Atomic Energy Commission]

15.11) for the exclusive purpose of producing plutonium. (The energy generated was thrown away, with the effect of heating the Columbia River, which was used as coolant.) The plutonium project was as successful as the Oak Ridge projects in producing nuclear explosives (Fig. 15.12). Because ^{239}Pu can substitute ^{235}U in fission chain reactions, the process of ^{239}Pu production may also be used to generate nuclear fuel for peace-time use of nuclear energy. Reactor technology was thus developed and after the war was directed to build reactors for energy production (Figs. 15.13 to 15.16).

The Breeder Reactor. The conversion of inert ^{238}U to fissionable ^{239}Pu opens the possibility that the nuclear fuel reserve may be expanded a hundredfold. To achieve complete conversion is not a simple task. This indeed is a major objective of the current nuclear energy development program. We know that each fission of ^{235}U produces 2.5 neutrons. One of them is needed to keep the chain reaction going. To produce one ^{239}Pu atom requires at least one neutron. If all other neutron losses in the reactor can be reduced to below 0.5 per fission, then there is the possibility that out of the 2.5 neutrons from the fission of one ^{235}U nucleus one neutron may be used to keep the chain reaction going and another may be used to produce a ^{239}Pu to replace the used ^{235}U. When all ^{235}U is used up, an equal number of ^{239}Pu has been produced to replace them and the process may now be repeated over and over again until the whole of ^{238}U is converted into ^{239}Pu and burned up in the reactor. Such a reactor that regenerates the fuel as it burns is called a *breeder reactor.* Right now a *demonstration fast breeder* reactor is under construction at Oak Ridge. The technology has passed the experimental stage but has not entered the commercial stage. (The demonstration is guaranteed to work but not guaranteed to make money.) The fast breeder makes use of fast neutrons, not thermal neutrons. The design is thus completely different.

If the number of neutrons per fission ν should be less than 2, then a 100 per cent breeder reactor would not be possible in principle. The fact that the number is 2.5 makes the breeder reactor in principle possible. But the margin is only 0.5 neutrons and it would take great ingenuity to make the breeder work. In fact, the breeder is designed to breed a little more fuel than is consumed, so that the margin is even smaller. Such a breeder will become a fuel supplier to conventional reactors. Breeder reactor technology in the United States is lagging behind; Britain, France, and Russia all have breeders operating at the present time.

Assume the future use of the breeder reactor. All natural uranium can then be used for energy production. (It still requires the minute fraction of ^{235}U to get things started and we are still indebted to the existence of the small amount of ^{235}U in nature.) In addition, there is another source of nuclear fission fuel, thorium. Natural thorium has only one isotope $^{232}_{92}Th$. It is inert, just as ^{238}U is. On the other hand, by absorbing a neutron in a reactor, it may be converted into fissionable ^{233}U, another active nuclear fuel. Again a kind of breeder reactor is required. Besides uranium and thorium, there is no other fissile material, directly or indirectly. Thus the total amount of fission fuel on earth is limited to the mineral resources of uranium and thorium.

The high-grade uranium and thorium ores on earth are rather limited (Chapter 3). The total energy available from burning all of them is not tremendously large. It could keep industrial society going for perhaps a few thousand years. On the other hand, low-grade ores may become economical through the breeder reactor and its energy reserve may last for 100,000 years.

We now turn to the other source of nuclear energy, fusion.

Figure 15.11 The "N" reactor on the Columbia River at Hanford, Washington, the nation's first dual-purpose reactor, producing plutonium as well as energy. The steam produced is sold to an 800-megawatt power-generating plant to generate electric energy. [U.S. Atomic Energy Commission]

Figure 15.12 "Baker Day" at Bikini, July 25, 1946. Since then Bikini has become the byword for anything daring, spectacular, and shocking, which, judged by this photograph, is fully justified. [U.S. Atomic Energy Commission]

15.4 Nuclear Fusion

Nuclear fusion provides the second source of nuclear energy. Light nuclei, when combined to form heavier nuclei, release energy. For example, when two protons and two neutrons combine to form a helium nucleus, an amount of energy of 28 MeV is released, based on the mass defect of the helium nucleus. In the 1930's it was learned that stars, including our sun, derive their energy from fusion of hydrogen into helium. Because all energy sources (except nuclear) on the earth are traceable ultimately to the sun, we have been using fusion energy all along, albeit indirectly.

Fusion takes place when two nuclei collide with high speed. High-speed particles can be produced in *nuclear accelerators* but this is too costly for fusion energy production purposes. Instead we rely on a high temperature. When fusion fuels, essentially ^{2}H, ^{3}H, and perhaps ^{6}Li, are heated to a high temperature, the average kinetic energy of a particle ($\frac{3}{2}kT$) in the gas becomes very high. The particles collide with high speed and fusion may result. Such reactions are called *thermonuclear reactions*. The fusing of light nuclei, such as that taking place in the sun,

GRAPHITE REACTOR
LOADING FACE

Figure 15.13 Oak Ridge National Laboratory's graphite reactor. The world's oldest reactor has been operating for 20 years. Formerly a top secret, it is now a National Historic Landmark visited by thousands of persons yearly. [Oak Ridge National Laboratory, operated by Union Carbide Nuclear Division for the U.S. Atomic Energy Commission]

requires a very high temperature, of many tens of millions of degrees centigrade. Because most material substances evaporate and disappear at a temperature of 3,000°C, a fusion furnace maintaining a temperature of millions of degrees is out of the question.

In the past there was no way of raising the temperature to millions of degrees. Thus the fusion process as a source of energy on earth was not seriously considered.

The Hydrogen Bomb. With the advent of the atomic bomb the fusion process began to be considered as an explosive energy source. In the explosion of an atomic bomb the temperature goes to 100 million degrees. Thus the possibility is opened of using an atomic bomb as a match to ignite the fusion fuels (^{2}H, ^{3}H). This is exactly the principle of the hydrogen bomb, which includes a fission bomb as a starter. The advantage of the hydrogen bomb is that it can be made as large as one wishes. The atomic bomb, though powerful, cannot be made as large as one wishes. The size of an atomic bomb is limited by the critical mass of fissionable material (about 10 kg). The bomb cannot be made larger or smaller. There is no such limit on a hydrogen bomb. On the other hand, one of the hydrogen bomb ingredients, tritium ^{3}H, is made in reactors. To make one ^{3}H would cost at

Figure 15.14 Lifting of the reactor head for refueling at the Connecticut Yankee Nuclear Power Plant. The 575-megawatt plant, once the largest commercial nuclear power plant in the nation, began operation in 1967. [U.S. Environmental Protection Agency]

Figure 15.15 Virginia Electric and Power Company's Surry Power Station at Surry, Virginia. Each of the two nuclear-generating units is designed to produce 788 electric megawatts (started up in 1972 and 1973). Pressurized water reactors are used. [U.S. Atomic Energy Commission]

least one neutron and thus cost about one ^{235}U atom, which can be used in the atomic bomb directly. Thus the hydrogen bomb is uneconomical and vastly wasteful as far as the use of nuclear explosives is concerned. This was one practical reason against the development of the hydrogen bomb. The only advantage of the hydrogen bomb is the concentration of a large amount of energy in one single shot, which is desired for intercontinental balistic missiles, which do not have pinpoint accuracy. However, the situation was changed completely when a cheap way of making the bomb was introduced. In the hydrogen bomb explosion the following reaction takes place among the hydrogen fusion fuels ^{2}H and ^{3}H:

$$^{2}_{1}H + ^{3}_{1}H \longrightarrow ^{4}_{2}He + n$$

The neutrons so released have an energy of 17.6 MeV. This energy is so high that it can split the inert ^{238}U and cause the release of a large amount of *fission* energy. Thus a casing made of natural uranium, which is cheap, is used to surround a hydrogen bomb. The cheap inert material ^{238}U can now be used as a nuclear explosive and the explosion power yield is raised tenfold. The so-called hydrogen bomb is in reality like a three-layer cake, a fission–fusion–fission bomb.

One environmental consideration of the atomic weapon may be mentioned here. The fission products are radioactive and therefore the fission bomb causes dangerous radioactive fallout. It causes more casualty in a wider area than the blast and leaves long-lasting deleterious genetic effects. The fusion explosion does not have radioactive fallout and was considered "clean." However, the three-stage fission–fusion–fission bomb (the original hydrogen bomb) produces the largest amount of radioactive fallout and thus is the "dirtiest" of all. Later the so-called clean hydrogen bomb (without the third fission stage) was also developed [presumably used in antiballistic missiles (Problem 13.2) which are to be exploded over our own air space hundreds of miles outside the atmosphere].

Fusion Energy. Although fusion energy has been used for military purposes, with the necessary high temperature provided by an atom bomb, use for industrial purposes has not been achieved. Researchers in many countries, using a variety of approaches, have been working on the problem for many years. Impressive progress has been made, but the final goal is still far from being reached. However, with steady progress there is hope that the difficulties may eventually be overcome, even if it takes many decades (current estimate is 1995).

The reason to tackle this difficult technological problem is that the fusion fuel reserve is quite abundant and eventually when all other energy sources on earth become depleted, there is no other recourse but to turn to fusion energy. No matter how difficult, we are forced to face the problem.

Fusion reactions involve largely the heavy isotope of hydrogen, deuterium ^2H. Thus industrial fusion fuel is essentially ^2H. Fusion fuel material must be readily available, abundant, and cheap. Of course it should fuse quickly. Although most light-isotope fusion reactions yield energy, hydrogen is the easiest to fuse. (Because the nuclear charge is the smallest, so that the Coulomb repelling force between two nuclei against fusion is the smallest.) Among the hydrogen isotopes ^1H is the most abundant but ^1H fusion reactions, such as those taking place in stars, proceed too slowly and so cannot be used in fusion reactors. The ^2H fusion reactions take place more

Figure 15.16 From atom to electricity. Heat energy generated from nuclear fission in the reactor (left) is extracted by liquid metal, which goes through a boiler (bottom) to generate steam. The steam goes through the turbine (middle right) to generate mechanical work, which operates the generator (middle left) to produce electric power. Through the transformer and transmission lines the electric power is fed to homes and factories. [U.S. Atomic Energy Commission]

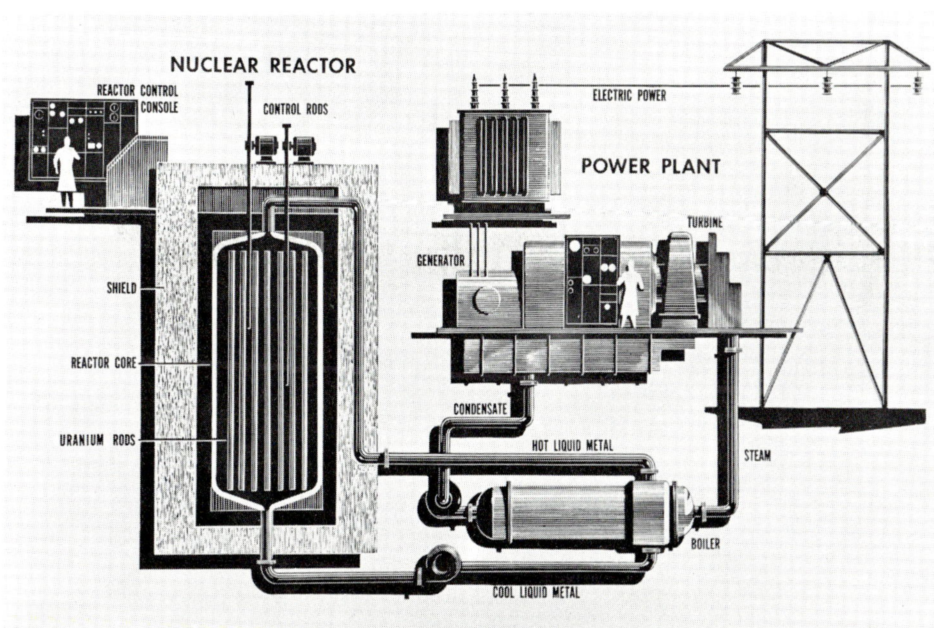

rapidly. Even though ^{2}H accounts for only 0.015 per cent of natural hydrogen, the total amount of hydrogen on earth, present in water in oceans, is sufficiently large that the energy in ^{2}H of ocean water is sufficient to sustain our energy needs much longer than any fuels we know. To get an idea of the amount of energy available from the deuterium in ocean water, it may be remarked that the fusion energy stored in 1 gal of water is equivalent to 250 gal of gasoline. The total fusion energy in oceans is enough to supply our energy needs at the present rate for 10^{10} years (to the end of the world!). The cost of separating ^{2}H from ^{1}H (in the form of heavy water) is not high at all. Thus attention on fusion energy is largely focused on ^{2}H.

Fusion Reactions. Deuterium ^{2}H may react with another deuterium at a high temperature as follows:

$$^2_1\text{H} + ^2_1\text{H} \longrightarrow ^3_2\text{He} + ^1_0\text{n} + 3.2\ \text{MeV}$$

$$^2_1\text{H} + ^2_1\text{H} \longrightarrow ^3_1\text{H} + ^1_1\text{H} + 4.0\ \text{MeV}$$

^{3}He does not react further, but ^{3}H (tritium) reacts with deuterium as follows (the same reaction takes place in the hydrogen bomb):

$$^3_1\text{H} + ^2_1\text{H} \longrightarrow ^4_2\text{He} + ^1_0\text{n} + 17.6\ \text{MeV}$$

Thus the fusion products are two stable isotopes of helium and neutrons. The energy released is on the average 5.0 MeV per deuterium atom.

Besides the *deuterium–deuterium reaction* another light-element fusion reaction has been proposed as a mechanism to tap the fusion energy. The *deuterium–lithium reaction* works with tritium and neutron as intermediaries, which are regenerated in the fusion reactions as follows:

$$^6_3\text{Li} + ^1_0\text{n} \longrightarrow ^3_1\text{H} + ^4_2\text{He} + 4.8\ \text{MeV}$$

$$^2_1\text{H} + ^3_1\text{H} \longrightarrow ^1_0\text{n} + ^4_2\text{He} + 17.6\ \text{MeV}$$

The fusion products resulting from the consumption of deuterium and lithium is the stable isotope helium, ^{4}He. The energy produced is 22.4 MeV per deuterium atom. In this case another element, lithium, is used in conjunction with deuterium as fusion fuel. However, the abundance of lithium is only 10^{-8} times that of deuterium, and so the lithium energy reserve is much smaller than that of deuterium. In fact, it is comparable to the fossil fuel energy reserve. It is worth noticing that in both cases no radioactive atoms appear as final products. However, the neutron may generate radioactivity and the intermediary ^{3}H may leak and cause radiation pollution.

15.5 Nuclear Fusion Reactors

To obtain controlled release of fusion energy, first we have the problem of heating the fusion fuel to the extremely high temperature of tens of millions of degrees. Then there is the next problem, that of confining the high-temperature material in a container for a sufficiently long time. Any solid material evaporates at a temperature of 3,000°C but now the desired temperature is higher by a factor of 10,000. (In a bomb the container problem does not exist.) These problems are tackled traditionally in the following way: For heating we may use a *plasma*—an ionized gas through which a large electric current passes, developing a large amount of energy to heat up the plasma. For confinement we may use a *magnetic bottle*—a magnetic field that constrains and confines the plasma in a limited region in space for a short time, thus working like a bottle without a material wall (Fig. 15.17). A physical bottle confines its contents by a material boundary, but the magnetic bottle confines by a force field instead of by a material substance. It is hoped that once the fusion fuel is heated to a high temperature and sustained in a confined volume for some time, fusion will take place.

These mechanisms are further explained as follows. When an electric current passes through a gas discharge tube, such as a neon sign tube or a fluores-

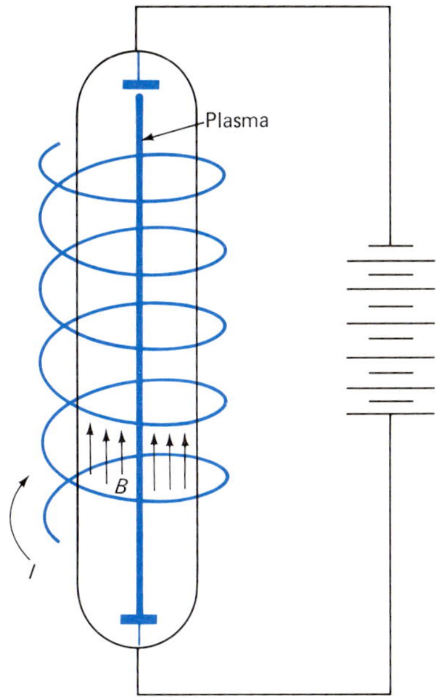

Figure 15.17 Magnetic bottle.

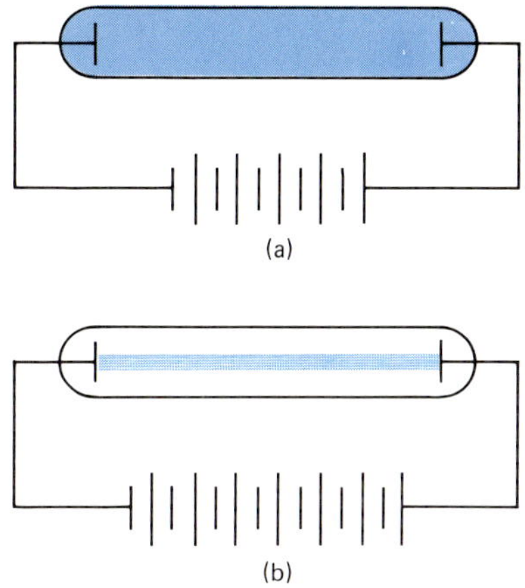

Figure 15.18 (a) Glow discharge. (b) Arc discharge.

cent light tube, the gas inside becomes ionized. (Electrons are knocked out of atoms of the gas.) The gas now consists of negatively charged ions (electrons) and positively charged ions (ionized atoms); such a state of matter is called a plasma. (It is said to be the fourth state of matter after solid, liquid, and gas.) When the current is increased to a large extent, the electric discharge of the tube changes from *glow discharge* (such as that in the neon sign tube or the fluorescent light tube; the tube glows uniformly, with light filling the entire inside of the tube as shown in Fig. 15.18a) to *arc discharge* (such as an electric arc or the welding arc; the glowing area inside the tube becomes confined to a narrow, thin filament, called the arc, spanning from one electrode to another, as shown in Fig. 15.18b). The two kinds of discharge are also familiar in lightning; the light in one is a generally uniform flash and in the other is a sharp, bright arc accompanied by loud thunder.

The formation of the arc is due to a process called the *pinch effect,* which can be easily demonstrated. Two parallel electric currents flowing in the same direction attract each other, whereas two in the opposite direction repel (Fig. 15.19). This is due to the third property of the electromagnetic force—a current produces a magnetic field, which exerts a force on the second current. The electric current flowing through the space inside the discharge tube may be looked upon as a bundle of thin current elements all flowing in the same direction. Therefore they attract among themselves, pinching themselves into a narrow filament—an arc (Fig. 15.20). The pinching heats up the plasma tremendously because the same amount of heat released is now confined in a much smaller volume (the arc instead of the tube). The pinching acts as a compression process, which heats up the contents. This way a high temperature can be generated.

Figure 15.19 Electric currents in the same direction attract each other; currents in opposite directions repel each other.

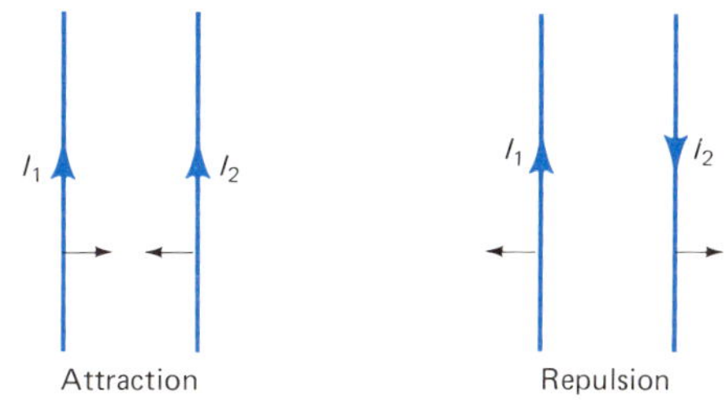

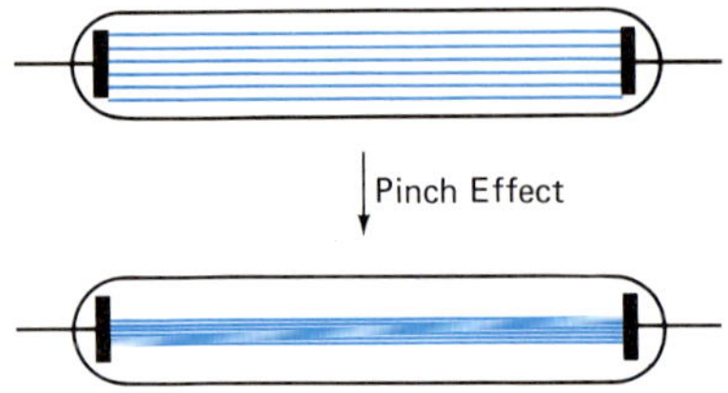

Figure 15.20 The pinch effect.

However, the high-temperature plasma is unstable. That is where the magnetic bottle comes in. A solenoid carrying a current placed over the discharge tube produces a magnetic field B parallel to the tube (Fig. 15.21). The ions in the plasma, moving with a high velocity and exposed to this field B, will execute a helical motion along the direction of B. This follows from Newton's second law, with the force on the ions given by that exerted by the magnetic field. Thus the ions are constrained to move along the direction of the field, which is the lengthwise direction of the tube (disregarding the tiny circular motion in the helix, which is of little consequence) and the plasma is thus confined (instead of breaking up by sidewise movement). At very high temperature the magnetic bottle is not leakproof and the plasma will still break up. What is needed is to confine the plasma sufficiently long and keep it sufficiently hot so that nuclear fusion will take place before the plasma breaks up. After breaking up, the process may be repeated; it is hoped that in each cycle a certain amount of fusion fuel will fuse and nuclear energy will be released. The repeated operations thus provide a controlled release of fusion energy.

Figure 15.21 Confinement of a plasma by a magnetic field.

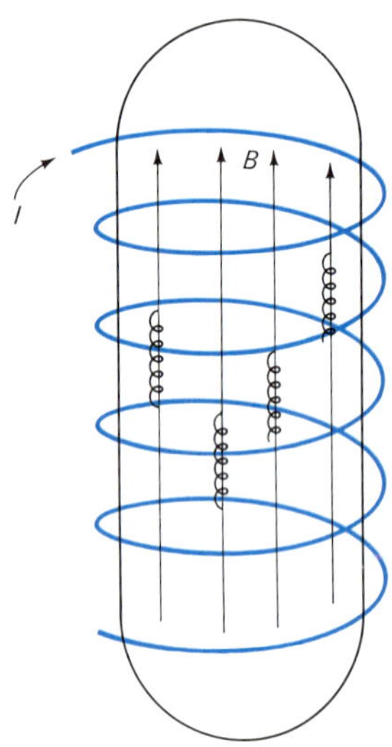

Thus the anticipated fusion reaction works in pulses. In each pulse the temperature reaches a peak T°C and lasts for a time of t sec. T must be sufficiently high (10^8°C) and t sufficiently long (10^{-1} sec) for nuclear fusion to take place. Fusion research in recent years is marked by development of devices that will produce a hot plasma with increasingly larger values of T and t. The temperature attained inches toward the million-degree mark. The confinement time has increased from microsecond to millisecond. Up to the present, with impressive progress made possible by the contributions of many people from many countries, we are still short of the final goal of achieving controlled release of fusion energy. Because all devices work on pulses, in spite of the high temperature attained, the tube itself is still cool.

The Tokamak Machine. The preceding discussion based on a straight discharge tube is convenient for illustrating the principles. In practice, other, more advantageous geometrical configurations are used. The most interesting development in recent years is the "tokamak" concept advanced by the Russians in 1969. It involves a toroidal-shaped plasma. The compression of the plasma is carried out by shrinking the major radius of the toroid (Fig. 15.22). The concept has been further developed by scientists in many nations (Fig. 15.23 and 15.24). In May, 1972, the ATC machine (Adiabatic Toroidal Compressor, Fig. 15.25) of Princeton University began operation. It uses a deuterium–tritium fusion fuel in a toroidal magnetic bottle. The heating cycle

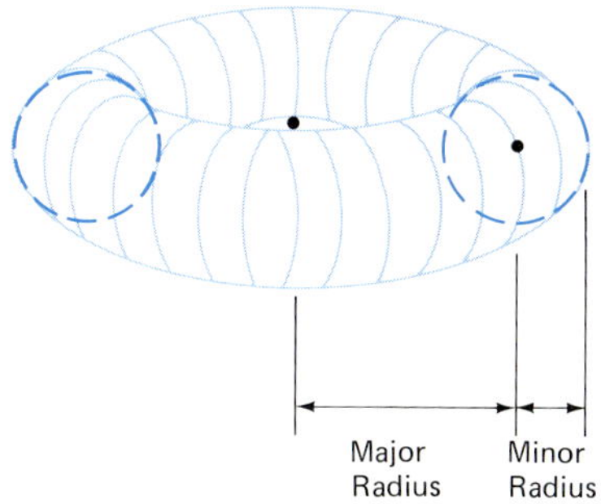

Figure 15.22 Major radius of a toroid.

starts by establishing a large current in a toroidal plasma. The ATC electron temperature rises to about 10 million °C. Then the plasma is compressed; the doughnut shrinks. The major radius is reduced by a factor of more than 2. The temperature increases threefold and density increases fivefold. The electron temperature rises to 25 million °C and the electron density is compressed to 10^{14}/cc. For the first time the density barrier for fusion is broken. However, the positive ion temperature of ATC (deuteron temperature) is still only about 7 million °C. The tokamak method is not likely to achieve the ignition temperature of 100 million °C but the success of ATC points out the way for further development.

Laser Heating and Implosion Methods. Plasma heating has been the main strategy in fusion research. Recently alternatives have been explored making use of the laser. Fusion fuels (deuterium-deuterium or deuterium-tritium) are made into small pellets. High-power laser beams are fired into the pellet. The laser light heats up the pellet so fast that the temperature might reach the 100 million °C region for fusion ignition before the heat dissipates by conduction and radiation. Because it involves a solid pellet rather than a gaseous plasma, the problem of confinement is eliminated; there is no need of a magnetic bottle. On the other hand, the laser energy necessary to ignite fusion is much too high.

Besides the laser-heating approach another alternative is the laser-implosion approach pursued by many laboratories all over the world. A number of laser beams, properly oriented and timed, are fired onto a fusion fuel pellet from all directions at the same time to compress the pellet rapidly into a very high density state to ignite fusion (Fig. 15.26). A beam of light exerts a pressure when it falls upon a surface. This pressure is so small in ordinary light that it is often unnoticed. A laser produces a beam of intense light and thus exerts a high pressure. The pressures from all directions produce an implosion (inward explosion). The necessary laser energy required in this method is much smaller than that in

Figure 15.23 Oak Ridge National Laboratory's ORMAK, a toroidal fusion research device inspired by the Russian Tokamak. The composite photo shows both the interior and the exterior with the toroidal core gleaming in the center. [U.S. Atomic Energy Commission]

Figure 15.24 A 5-meter arc segment of the eventual 15-meter toroidal fusion research device Scyllac at Los Alamos Scientific Laboratory, New Mexico. Initial experiments on the first arc shown here indicated an ion temperature of close to 10 million °C for a confinement time of 10 microseconds. [U.S. Atomic Energy Commission]

the laser-heating method and thus the difficulty is reduced. Russians in the Lebedev Institute in Moscow use nine laser beams on a spherical fusion fuel pellet in a vacuum chamber. They have observed some neutrons from the resulting fusion reaction. A new 27-beam machine is under construction.

*15.6 Other Applications of Nuclear Energy

Before we leave the topic of nuclear energy we want to discuss another important aspect of the nuclear energy program, that is, the production of radioactive isotopes by reactors. This is another major use of the reactor and its significance is by no means less than those of military applications and electric power generation. In fact, the use of isotopes has revolutionized many aspects of modern technology. It alone is sufficient to make atomic energy the most far-reaching technological advance in human history.

The reactor produces a large flux of neutrons. Any isotope placed inside a reactor will capture a neutron and convert itself into another isotope. Thus a large variety of new isotopes of practically all elements may be produced. Many of them are radioactive and, because of it, have wide applications in science, industry, agriculture, and medicine.

For scientific research, principally in the biological sciences, radioactive isotopes may be used as *tracers* to trace atomic and molecular processes that cannot be unraveled otherwise. As far as chemical properties and reactions are concerned, two isotopes of the same element behave exactly the same. Thus in all chemical and biological processes ^1H and ^3H are used indiscriminately as if they are the same thing. However, ^3H is radioactive and its radioactivity can be

Figure 15.25 The ATC machine (Adiabatic Toroidal Compressor) at Princeton University. The $1.3 million project currently in operation is the forerunner of a new-generation "Large Torus" experimental device now under construction at a total cost of $13 million and scheduled to be operational in 1975. The Princeton program is one of four supported by the AEC with a total annual budget of $40 million in fiscal 1973. [U.S. Atomic Energy Commission]

detected by such devices as photographic film and the Geiger counter. Therefore it can be distinguished from ^{1}H. Thus radioactive isotopes are often called tagged atoms, distinguishable from the stable ones. This possibility of tagging is the basis of a wide range of applications.

Biologists have been using the tagging technique for a long time. For example, they catch migratory animals, such as salmon, at one source and put a tag, usually a metal mark, on a large sample of them before releasing them. A few months later these salmon are caught over a wide-ranging area; they are counted and their distribution studied. This way their movement can be traced. The radioactive atom acts as a tagged atom, and problems similar to those encountered with migratory animals can be worked out in the same way. For example, once we have developed a new drug, we like to know how fast it moves to the desired place of the human body to act after it is taken orally. Tracing the movement of the drug molecule in the body is like tracing the movement of the salmon in the ocean, and the tagging technique may be used. We substitute one kind

Figure 15.26 Laser implosion method of producing nuclear fusion.

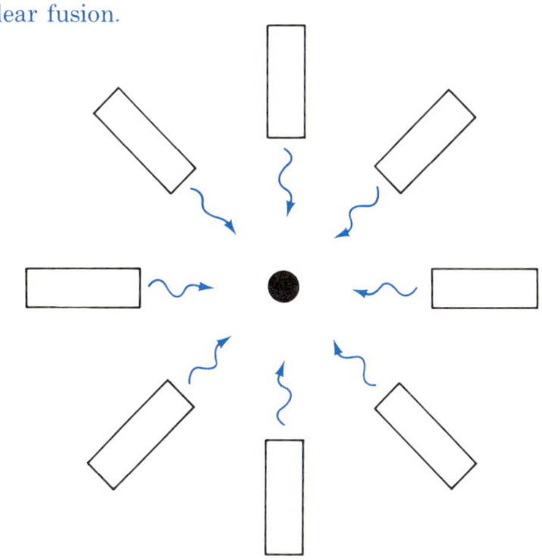

of atom in the drug with its radioactive isotope, say, substituting ^3H for ^1H. The drug works just the same but the molecule is now tagged, like the tagged salmon, and can be detected. By moving a Geiger counter over various parts of the patient's body after the drug is taken orally, we can detect the whereabouts of the drug molecules at various times and thus trace their movement in the human body (Fig. 15.27).

More important is the use of isotopes as tagged atoms to trace the detailed processes of biochemical reactions. In a natural process, many complex reactions take place at the same time. All chemistry can observe is the gross total; it cannot separate the individual reactions. The tagged atoms make it possible to unravel some of the complexities and much new knowledge can be discovered. In fact, much of the recent progress in molecular biology is based on the use of isotopes. The spectacular advances in modern biology are thus a fallout of the atomic energy program.

To give an example let us consider the experiment of Meselson and Stahl (1958) that established the semiconservative replication of DNA and thus lent strong support to the Watson–Crick theory (1953). Ordinary DNA contains ordinary nitrogen atoms ^{14}N. There is another isotope of nitrogen, ^{15}N, that is slightly heavier (15:14). We can make DNA that is composed of this heavy nitrogen ^{15}N exclusively. The way to do it is to let bacteria (E-coli) grow for many (fourteen) generations (7 hours) in a nutrient medium whose nitrogen (from NH_4Cl) is ^{15}N exclusively (96.5 isotopic purity). The DNA extracted from such a bacteria colony, called fully labeled DNA, is slightly heavier than the DNA of wild (meaning natural) E-coli, called unlabeled DNA. These two kinds, being different in mass, can be separated by the technique of *centrifugation*. If there is a DNA molecule of which half of the nitrogen is ^{14}N and the other half is ^{15}N, called half-labeled DNA, its weight should be exactly halfway between the previous two kinds, and in the centrifuge separation it should appear halfway in the mass spectrum between the fully labeled and the unlabeled DNA.

Meselson and Stahl first prepared an E-coli colony with fully labeled DNA in a medium containing

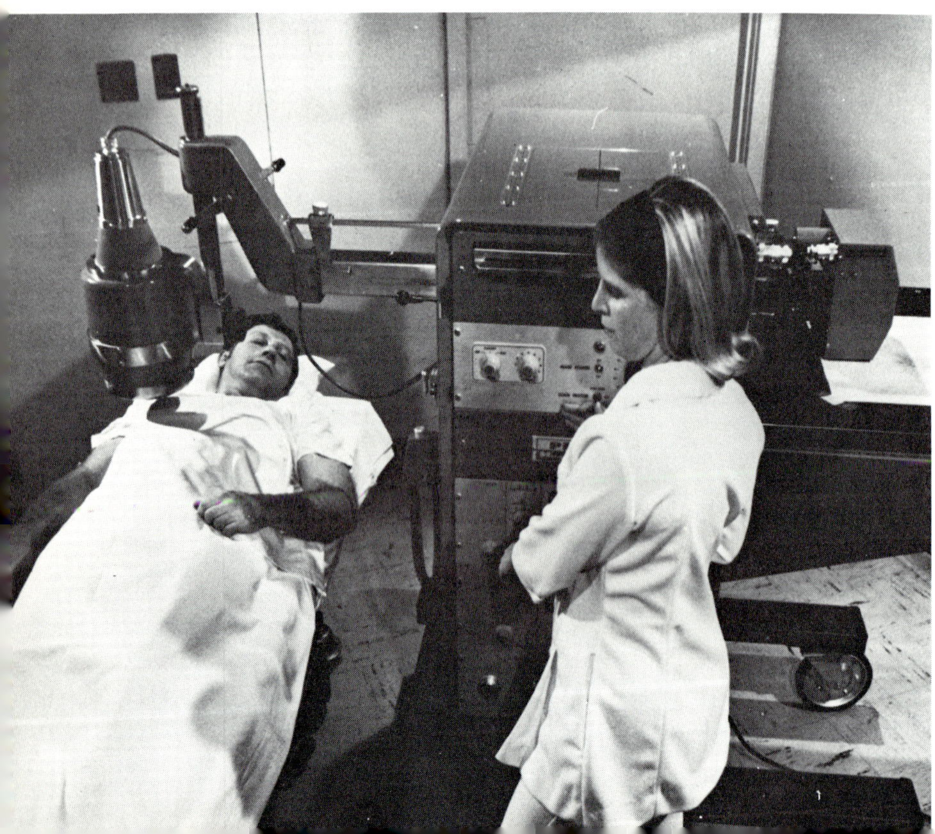

Figure 15.27 Scanner. This radiation detection scanning instrument is used to check the uptake of a radioactive substance in an organ of the patient as a means of diagnosing disorders. [U.S. Atomic Energy Commission]

$^{15}NH_4Cl$ exclusively in the preceding manner. Then they suddenly changed the nutrient medium to one of normal nitrogen, ^{14}N, exclusively (added 10 times concentrate $^{14}NH_4Cl$). From this point on newly synthesized DNA will be made of ^{14}N exclusively, whereas old DNA is still made of ^{15}N. At a time of one generation later (one-half hour) each parent cell divides into two daughters, each presumably contains some old and some new DNA. How the old and the new are distributed in the daughters can be learned on the basis of their different weights. Meselson and Stahl took a sample of bacteria after a time of one generation and analyzed its DNA by centrifuge. They found all DNA are of one kind, the half-labeled (half old and half new). Then they repeated the procedure after a time of two generations from the start and found this time the DNA are of two kinds, either half-labeled or unlabeled. They repeated the procedure some 10 times up to five generations (including fractional generation times). After the first generation the fully labeled DNA disappeared and never appeared again. On the other hand, the half-labeled appeared at the one generation time and remained for later times even though its concentration diminished gradually. This behavior is exactly what would be expected of the Watson–Crick model of replication. (Verify this yourself!) Conversely, by a judicious application of logical arguments, the experimental results obtained are sufficient to establish that DNA replication is semiconservative (not totally conservative and not nonconservative).

Where is the ^{15}N from? In other isotope experiments other isotopes are used. Where are they from? They are created in nuclear reactors. Without the reactors none of these experiments would be possible. There would hardly be any molecular biology. It is in these fundamental processes that we hope to explicate the mysteries of life, including the solution of such problems as cancer. Such efforts would not be possible without the development of atomic energy.

In the preceding experiment the isotope is distinguished because it has a different weight. In most other applications it is distinguished because of its radioactivity. One way to solve a problem by using radioactivity is to make an *autoradiograph.*

The DNA molecule contains hydrogen ^1H. Let it be replaced by the radioactive isotope ^3H. The bacteria DNA from E-coli is a threadlike molecule about 1,000 μ long (0.1 cm). Let it stretch out and put a photographic film over it. The radiation from ^3H will cause a blackened spot on the film right above the ^3H atom. All the ^3H atoms along the length of the DNA will do the same and thus will produce a dotted curve on the film, which is an image of the DNA molecule. Thus a picture of the DNA molecule can be made. A length of 1,000 μ is easily visible under the microscope. The technique is similar to those of making a track of a nuclear particle by the photographic emulsion and the Wilson cloud chamber in nuclear physics. By making use of the pictures of DNA, molecular biology problems may be solved.

Cairns has produced autoradiographs of DNA of E-coli growing in a "hot" medium (^3H-thymidine) for one hour (two generations). The DNA pictures show visibly the Y configuration of a replicating DNA, which is expected from the Watson–Crick mechanism (one parent DNA molecule joined by two equal-length daughter molecules). This is the most vivid proof of the Watson–Crick replication mechanism. After the wild bacteria complete one replication in a hot medium the two daughters are expected to have "hot–cold" hybrid DNA. In the second replication in the hot medium, which is recorded by Cairns' autoradiographs, these daughters are expected to produce granddaughters of two kinds, one with "hot–hot" hybrid DNA and the other with "hot–cold" hybrid DNA. Indeed, in Cairns' pictures one daughter track of the Y is twice darker than the other daughter and than their parent, indicating that it has twice as much radioactivity as the others (and thus is the "hot–hot" hybrid). The semiconservative nature of replication is again demonstrated.

The advance of biology depends on instrumentation of observation. Aristotle used the naked eye

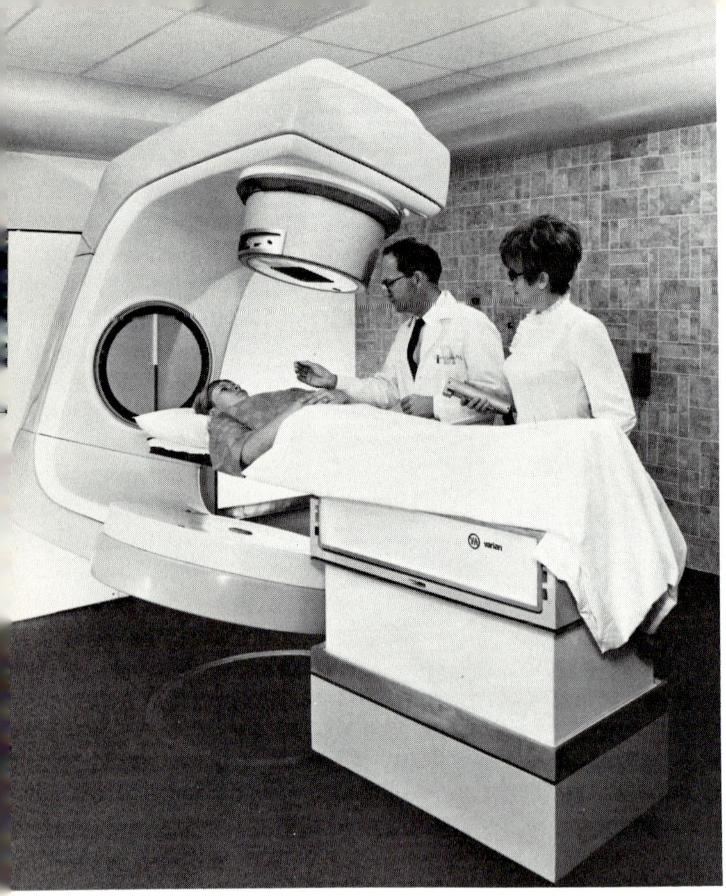

Figure 15.28 Tumor treatment by nuclear radiation. This machine, built by Varian Associates, is called Clinac 4. [U.S. Atomic Energy Commission]

for observation and his biology stopped at *taxonomy*. By the dissection technique biology advanced to *anatomy*. The introduction of the microscope brought biology to the cellular level and opened up a new field *cytology*. Recent development of the *electron microscope* made it possible to study even smaller units—the *organelles* of the cell. But a molecule can never be seen by any microscope! Nevertheless the molecular processes can be studied by the isotope technique, including the tracer and the autoradiograph. Thus biology advanced to the molecular level and we have *molecular biology*.

The isotope technique can be made very sophisticated. In the *pulse-labeling* technique the radioactive atoms are administered in a short pulse at a particular time. This way reactions taking place at this particular time can be separated from other reactions and a mixture of complex reactions can be separated into individual reactions with their time sequence worked out.

Turning from pure scientific research to practical applications we find again numerous uses of the radioactive isotopes in a variety of fields covering medicine, agriculture, and industry. One of the isotopes produced in reactors is ^{60}Co. It emits a radiation strong enough that it can be used as a substitute for radium for the treatment of cancer (Fig. 15.28). In one hospital in Chicago alone the cobalt radiation is stronger than all the radium of the world combined. Thus radiation therapy can be offered more widely. Moreover, the cost is reduced a thousandfold compared with radium. Already the number of lives saved by atomic energy through cobalt therapy has

Figure 15.29 Nuclear-powered heart pacemaker. [U.S. Atomic Energy Commission]

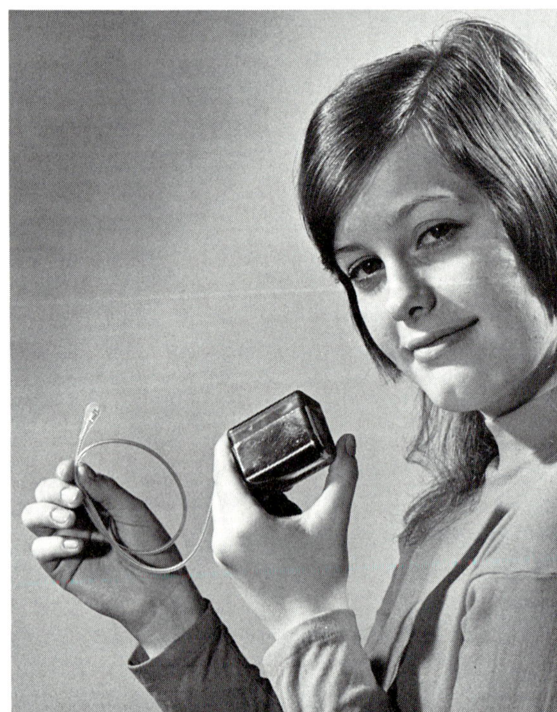

Figure 15.30 The Project Rulison 40-kiloton nuclear device is lowered into its 8,442-ft-deep emplacement hole on August 14, 1969. The device is 15 ft long, is 9 in. in diameter, and weighs about 1,250 lb. [U.S. Atomic Energy Commission]

far exceeded that killed by atomic energy through the two bombs.

A novelty in medical application is the nuclear-powered pacemaker for heart patients. Because nuclear energy is a powerful energy source, the problem of changing the battery of the pacemaker by repeated surgery can be eliminated (Fig. 15.29).

In agriculture, radiation from radioactive isotopes can be used for the preservation of food. Radiation-induced mutation helps produce new plant species and thus improve the quality and quantity of the crops.

In industry the tracer technique can also be used to solve industrial problems. Radioactive isotopes are used in leak detectors. The machine tools industry is able to use radioactivity to determine the wear and tear of machine tools and thus can save billions of dollars yearly.

Even the destructive power of the atomic bomb and hydrogen bomb can be put to constructive use. The explosive power may be used to dig a canal or excavate a harbor. Edward Teller, the "father of the hydrogen bomb," has visited Thailand to explore the possibility of using nuclear explosives to dig a canal through the Isthmus of Kra, connecting the Gulf of Siam to the Bay of Bengal. In Project Rulison (Figs. 15.30 and 15.31), carried out in Colorado, hydrogen bomb explosions were set off underground to break

Figure 15.31 The Project Rulison 40-kiloton nuclear device was detonated on September 10, 1969. Gas-bearing rocks were fractured and natural gas was released. At the end of 1970 the quality of the gas, shown burning in the photograph, is improving and its radioactivity decreasing as fresh gas flowed into the well from the surrounding fractured rock. No radioactivity above background level is found in streams and drinking water. The low tritium radiation level in the air assures safety in later production operations. [U.S. Atomic Energy Commission]

rock layers loose, in the hope of freeing trapped natural gas. There are environmental hazards, but in a time of energy crisis such undertakings are by no means trivial.

Thus out of an originally military project a vast program has been developed that encompasses many aspects of modern life. A new energy industry has grown out of it on which much of man's future energy needs depend. The radioactive isotopes produced by reactors provide a powerful tool contributing not only to pure scientific research but also to technological advancements. Yet the danger of hydrogen bomb warfare still hangs over the world like the sword of Damocles. By utilizing nuclear energy one may produce abundant energy and useful material, uplift mankind from poverty and misery, and provide affluence and leisure, or one may destroy the whole of mankind, ruin civilization, and pollute the earth with deadly radiation, making it nothing but a breeding ground for retarded and deformed monsters. Choose!

Problems

15.1 Calculate the fusion energy content of 1 g of water. How much would it be worth if sold at the rate of 3¢/kwh?

15.2 In a fission process a ^{235}U nucleus absorbs a neutron and splits into two nuclei of ^{100}Zr and ^{136}Te. The atomic weights of the four particles are, respectively, 235.11794, 1.00898, 99.94962, and 135.96089. Calculate the energy released in the process. If sold at the rate of 3¢/kwh, how much is 1 g of ^{235}U's energy worth?

15.3 Calculate the binding energy of ^{2}H and ^{3}He in MeV and compare the binding energy per particle with that of ^{4}He. Atomic weights of ^{2}H and ^{3}He are 2.01474 and 3.01711, respectively.

15.4 A person of 75 kg and an antiperson (made of antimatter) of the same mass meet and annihilate themselves. How much energy is produced? How much is the energy worth at the rate of 3¢/kwh?

15.5 A sewer line develops a leak under the basement concrete floor. Using the tracer method find a way to determine the point of leak for repair to avoid digging up the whole concrete floor.

15.6 Estimate how many collisions it takes for the moderator (assumed to be ^{1}H, which is used in light-water power reactors prevalent in U.S.) to slow a neutron of 1 MeV down to thermal energy.

15.7 If by accident the chain reaction in a nuclear reactor becomes divergent, the reactor will "explode" with a force perhaps comparable to that of a 150-mph car crash but will never explode like an atomic bomb. Explain why. The major safety consideration of a reactor is the possibility of accidental melt-down, never explosion.

16

General Laws of Transformation and Their Environmental Implications

In the preceding chapters we have discussed various forms of matter and energy, and also their transformations. The transformations are governed by three general principles, namely, the principles of conservation of matter, conservation of energy, and increase of entropy. We discuss these principles here, and in the following chapters will apply them first to problems of energy and the environment and then to problems of the origin and evolution of the universe.

16.1 The Principle of Conservation of Matter

The first general principle has been touched upon in several earlier chapters; it will be presented more systematically here. This section is partly a review.

We have mentioned the principle of conservation of mass in Chapter 4, which tells us that matter cannot be created or destroyed, even though it may change from one form to another. Matter may change in its shape, volume, color, smell, and so on, and thus may occasionally seem to disappear. To be able to say that matter cannot be destroyed or created, there must be some criterion to establish the existence of matter. Newton introduced the concept of mass as a property of matter that never changes and thus may be used as a measurement of the quantity of matter. From this point of view conservation of matter is implied by the principle of conservation of mass, which states that the total mass of a system undergoing a physical or chemical change is always the same before and after the change.

From the modern science standpoint, this principle is embodied implicitly in the view that the material world is composed of a few kinds of elementary particles; each is permanent and unchanging, and whatever changes take place in nature are merely changes of association, dissociation, and rearrangement of elementary particles. Because the total numbers of elementary particles are the same, the total mass is always the same before and after any changes. At the end of the nineteenth century, the elementary particles were thought to be the 92 atoms of the chemical elements. Even though these atoms are now known to be divisible, they act as basic units in chemistry, and thus mass is conserved in chemical changes. Later the elementary particles were thought to be the three particles the atoms are made of, that is, the proton, neutron, and electron. Even though these particles may be changed in high-energy physics, they are the basic units in atomic transformations and thus mass is conserved in atomic physics. The conservation of mass in many aspects of science is thus expressed implicitly in the microscopic structure of matter.

This view was not obvious, however, in the early days of science and even in everyday experience. A candle disappears after burning. Apparently matter can be destroyed by combustion. In all other burning processes material seems to disappear at least partly. In fact, burning is commonly considered as a way of destroying material substances, as in burning the trash. Even though Newton had recognized mass as a basic and constant property of matter, what happened during burning was not known. Combustion, a chemical phenomenon, was first studied independently of the Newton theory.

The study of combustion is a crucial step in the development of chemistry, for it concerns the basic question of what matter is. Only after the combustion process was understood was it possible to formulate the modern science of chemistry based on the Dalton atomic theory. It was nearly a century after Newton that this was done, largely by Lavoisier, as mentioned in Sec. 4.1. Before him thought on combustion was dominated by the phlogiston theory. Fuels were assumed to contain phlogiston. Upon burning, the phlogiston is released, forming the fire. The ashes are left. Many metals can be burned; the ashes are called calx. Again it was assumed that metals extracted of phlogiston become calx. Calx, which includes the metal ores, can be changed back to metal, as in the smelting processes. In such a process heating is applied. It was thought that phlogiston of the heat source enters the calx and converts it back to metal.

As early as the seventeenth century, however, it was observed that when tin is burned to form calx, its weight increases, despite the extraction of phlogiston. This contradiction was not considered seriously at the time. It was Lavoisier who, late in the eighteenth century, by a series of experiments in which weights (including those of gases produced) were measured accurately, established that calx weighs more than the metal by an amount exactly equal to that lost by the air (oxygen of the air) during the combustion process. One of the experiments involved the burning of an iron wire in oxygen-enriched air. By accurate measurement he determined that the weight of calx formed is 35 per cent greater than the iron, whereas at the same time the weight of air (oxygen) is reduced by exactly the same amount. Many other similar experiments confirmed the same relation. For example, Lavoisier decomposed the calx of mercury into mercury and oxygen (further demonstrating the role of oxygen in calx) and again the weight of the calx is exactly the same as those of the mercury and the oxygen combined.

Thus it was established that what is involved in combustion is a material substance, oxygen in the air, not a new substance, phlogiston. The "fire" is a manifestation of the chemical combination of the fuel material with oxygen, not itself a substance. The participation of oxygen is established by the exact amount of weight (mass) involved. Weight relationship thus becomes a crucial point in studying chemical changes, and in all changes the total weight of participants remains exactly the same before and after the change. Thus the principle of conservation of mass was established.

The significance of mass in mechanics had already been established by Newton much earlier. The principle just established extends the concept and its significance to chemistry as well. Even though in chemistry, material substance seems to disappear by burning, the mass is not destroyed, but is changed into another form, an invisible gas, with the total mass unaltered.

In spite of this scientific truth, the common-sense view of combustion and matter still prevails in our everyday life. In the environmental problem of solid waste disposal, we still do not think the production of waste to be a serious problem because we always think waste material can be destroyed by burning. Yet burning only converts the waste material to another material of the same weight, plus that of the oxygen consumed. The result is the heavy pollution of the air. This problem will be dealt with in more detail in Chapter 20.

Although the principle of conservation of matter was held as a basic tenet in science for more than a century, recent theoretical and experimental work have definitely established that under certain unusual circumstances matter and thus mass can be destroyed or created, even though not into or from nothing, but into or from an equivalent amount of energy. It started with the prediction of the relativity theory that m kg of mass is equivalent to mc^2 joules of energy. This relation has now been amply verified by innumerable experiments in nuclear physics. Thus in principle there is only one conservation law of mass and energy together. The two previous conservation laws of mass and energy are two special cases of this general law. On the other hand, for most of the natural processes in our daily life, the conversion of mass to energy and vice versa rarely takes place in a significant way and for practical purposes we may still speak of conservation of mass and conservation of energy separately.

16.2 The Principle of Conservation of Energy

In Chapter 5 we introduced the principle of conservation of mechanical energy. We introduced kinetic energy and potential energy and showed that in mechanical processes involving conservative forces, such as in the swing of a pendulum, the total amount of these two kinds of energy remains always the same. The principle of conservation of mechanical energy, for one thing, may be used as a convenient rule to help solve mechanical problems.

However, the principle is more important than a mere problem-solving technique. It is also related to our economic activities and thus to our material well-being. After the recognition of the existence of other forms of energy, such as heat energy, chemical energy, and so on, we established a general principle of conservation of energy involving all kinds of energy in all physical and chemical processes. Because most natural phenomena and human material activities may be analyzed into physical and chemical processes, the two overriding principles, conservation of matter and energy, governing the transformation of matter and energy in all physical and chemical changes, assume paramount importance.

Conservation of mechanical energy is a consequence of the Newton dynamical laws, which have already been established. The first step in the generalization is the inclusion of heat energy. Macroscopically, heat was shown to be equivalent to energy according to a definite conversion ratio by Joule. Empirical evidence from the study of heat (later in this section) shows that in all physical and chemical processes involving mechanical and heat energies, the total of these energies cannot be increased or decreased. Thus the conservation principle, including heat energy, is established empirically. This generalization makes it possible to include the nonconservative forces in the conservation principle—the energy lost because of the dissipative forces changes into heat energy and the grand total, including heat, is still conserved.

Microscopically, heat energy is regarded as the mechanical energy of the molecules in a material body. From this point of view the generalized principle of conservation of energy does not involve any generalization in the principle; it merely applies the same principle on a different level, that is, on the microscopic level of molecular motion.

Electric energy is the energy associated with electromagnetic force, which conserves energy. The conservation principle thus also includes electric energy.

In chemical reactions, energy, which may be heat, mechanical, or electric energy, is absorbed or released. These energies are known to be conserved and therefore chemical energy is included in the conservation principle. We have made use of this principle to solve such problems as the determination of the electromotive force of a reversible cell.

The next generalization includes the radiation energy of electromagnetic waves. This involves a generalization in the principle. Electromagnetic waves are not a material system. They are in a different world from the Newtonian system and are governed by a separate set of dynamical laws given by the electromagnetic field theory. Yet in spite of the change to the field theory the conservation principle remains intact. This is because of the symmetry principle, as we have mentioned previously.

The last generalization is to include mass as a form of energy as manifested in nuclear and "elementary particle" physics. Empirically it is a fact that the mass of nuclei may increase or decrease with a corresponding change of energy, with the conversion ratio given by $E = mc^2$. Moreover, elementary particles may be created (or annihilated) at the expense of energy, according to the same conversion ratio. Theoretically, the equivalence of mass and energy is demanded by the theory of relativity. The conservation principle again survives the revolution because of the symmetry principle.

The First Law of Thermodynamics. Let us go back to history and consider energy from a human need standpoint. The material to be presented below is related to the historical development of the principle of conservation of energy from a macroscopic point of view.

That energy is involved in economic activities is well known, and was known even in early times. The watermill is the machine that taps hydraulic energy for economic purposes. Without referring to energy, we see that a machine like a watermill produces motion and the motion performs an operation and accrues earnings. A machine that would yield endless motion would give us great profit. Thus arose the idea and began the pursuit of the so-called *perpetual motion machine*.

The pursuit of the perpetual motion machine is an activity very much like that of making gold out of base metals, an activity based on understandable human desires. Its course of development, also like that of alchemy, is that of a pseudoscience pursued by "quacks"; this is also understandable historically. Like alchemy, the pseudoscience paved the way for the development of a modern science. Alchemy led to chemistry, and the pursuit of the perpetual motion machine led to the modern science of thermodynamics.

A variety of schemes for making a perpetual motion machine have been advanced, but none have worked. Just as the failure of making gold out of base metals brought out the principle of immutability of elements, the failure of the perpetual motion machine brought out two basic principles, i.e., the *first and the second laws of thermodynamics*. The first is in reality a generalized form of the principle of conservation of energy, including heat as a form of energy, which will be discussed here. The second will be discussed in the next section.

One scheme of the perpetual motion machine, even though trivial, may be mentioned here to illustrate the basic idea. We know that heavy objects sliding down a slope gain kinetic energy, which can be utilized to do work. Yet after they have finished the run, the energy is exhausted. To make the machine run continuously and indefinitely, it is necessary to bring the heavy objects back to a high level again. Thus the following scheme was conceived: A number of heavy balls are connected into a chain and fitted over a triangle, as shown in Fig. 16.1. As the balls on the inclined side slide down the slope, the chain pulls the other balls up to start the slide again. It was hoped that the machine would run perpetually.

From what we learned in mechanics based on Newton's laws we can prove immediately that this

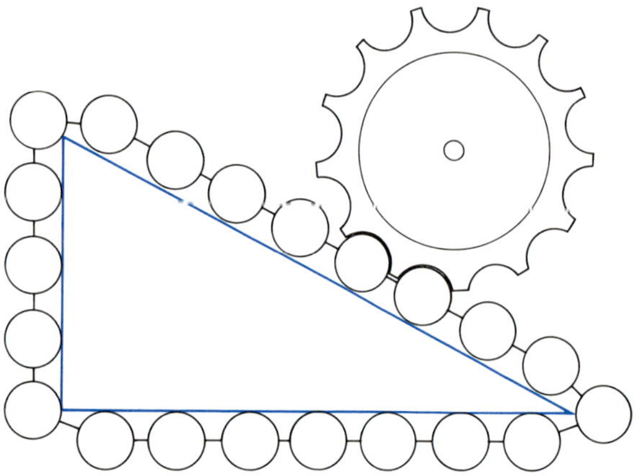

Figure 16.1 A perpetual motion machine that does not work.

machine will not work, simply by the principle of conservation of mechanical energy. The kinetic energy gained in sliding down the slope equals the potential energy lost. To put a ball back to the starting point requires putting back the lost potential energy and thus demands all the kinetic energy it produced in sliding down the slope. Thus all the energy generated in sliding down the plane will be used up exactly and completely in hauling the balls up and therefore nothing is left to be utilized to do work. Under the best condition (without friction loss of energy), the chain of balls may run around indefinitely with a constant velocity, seemingly achieving perpetual motion. But no energy can be taken out to be used to make a profit. As soon as the kinetic energy is tapped, say, by coupling it to a gear to turn a clockwork or a grindstone, the motion slows down and soon comes to a stop.

The failure of the simple mechanical device did not prevent people from pursuing other alternatives. For example, we know that electricity can be used to generate mechanical energy, using a motor. Also mechanical energy may be used to generate electricity, using a generator. Suppose we use a generator and a motor together. Let the motor operate the generator to produce electricity and let the electricity so generated drive the motor (Fig. 16.2). Some of the energy coming out of the motor will have to be used to operate the generator. But after this use if there should be a small fraction of energy left, no matter how small, it could be used to do other work and make a little profit. By repeating the operations millions of times, an endless amount of energy could be generated and the little profit would multiply into a huge fortune. The dream of the perpetual motion machine would be realized.

Such a scheme, if actually tried out, would also not work. No one has made himself a millionaire this way. And there are many other similar schemes. That none of them would work indicates the existence of a general principle. Now we can understand the situation in terms of the principle of conservation of energy. If energy is conserved, there can be no scheme that will generate new energy out of nothing. The energy needed to operate the generator is at least as much as that produced by the generator. When a generator and a motor are connected together, the energy generated and consumed at best compensate each other and there is none left over to make a profit.

Once heat was known to be able to generate energy, such as by a heat engine, other possibilities of the perpetual motion machine were opened up. For example, when steam condenses into water, a certain amount of heat is released and can be used to do work. To repeat the feat again, water must be changed back to steam but in this process a certain amount of heat has to be added to the water. If the heat to be added should be less than that released in the condensation process, then there would be a net gain of heat, no matter how small, in one cycle of operation, and by repeating the cycle many many times, endless heat might be generated. There are many similar schemes, yet no one has ever made any

Figure 16.2 Another perpetual motion machine that does not work.

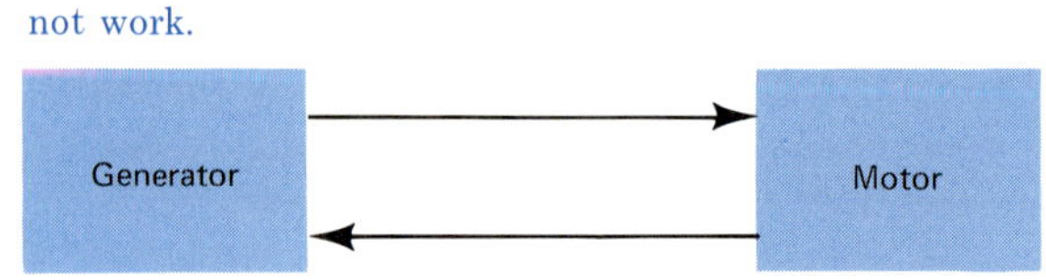

profit this way. This is true even when the cycle involves a change of the pressure and the latent heat.

One can think of many examples in endless variations. If only in one single case there is a tiny, tiny difference between the output and input to make a net gain, then by repeating the cycle many times, one can make a fortune. By trying out the endless variations, there might be a special case in which a small difference might show up. One is tempted to think this way, just as one is tempted to gamble in the hope that by trying many times one might hit the jackpot. Yet the fact is that no matter how hard one tries, the "jackpot" never shows up.

The jackpot in gambling will show up after a very large number of trials, if it is just a matter of chance. If never, then it is not a matter of chance and there must be a *law of nature* forbidding it to happen. Such a law forbidding the generation of energy out of nothing is formalized as the *first law of thermodynamics,* an empirical law established by facts. The fact that none of the schemes would work makes us realize that energy cannot be generated out of nothing and must be a conserved quantity. Therefore the law may be identified with the principle of conservation of energy.

This is the macroscopic approach to the general principle of conservation of energy. This is the way the science of thermodynamics was established; it is a macroscopic theory detached from any microscopic considerations. The microscopic approach to the principle has already been discussed previously: All energies are mechanical energies of the atoms or molecules and the general principle is merely the principle of conservation of mechanical energy applied to the atomic and molecular level.

Perpetual motion machines of this kind, intending to generate energy out of nothing, are called *perpetual motion machines of the first kind* to distinguish them from a second kind to be discussed in the next section. The first law of thermodynamics thus may be stated by saying that a perpetual motion machine of the first kind is impossible.

When speaking of conservation of energy we must include in the system all parties involved in the exchange of energy. If we consider only one body, its energy may increase or decrease because of the exchange of energy with others. When all such interacting parties are considered together, they form an *isolated system* (all interactions confined among the parts of the system exclusively). For such an isolated system the total energy is a constant.

The whole universe, by definition, is an isolated system. Thus the total energy of the whole universe is a constant, remaining always the same. This is another way of stating the first law of thermodynamics.

The implication of the principle on the energy problem is obvious. Any energy we consume must come from some energy source. The total of all energy sources is finite and thus is exhaustible.

16.3 The Principle of Increase of Entropy

The Second Law of Thermodynamics. The failure of the perpetual motion machine of the first kind did not stop the pursuit of the machine because there were other alternatives. Granted that energy is conserved and cannot be created out of nothing; there is still a large amount of it around. The amount of heat energy in the ocean is tremendous. If only we could lower the ocean temperature by a few degrees and convert the heat energy so extracted into mechanical energy, we would reap a fortune (Problem 16.1). Heat energy can indeed be converted into mechanical energy, as is done in a heat engine such as the steam engine, but the heat must come from a high-temperature reservoir, e.g., the hot boiler. Our dream would be fulfilled if only we could transport heat from the ocean to the boiler without any cost, then the engine would do work for us without cost. Devices so contrived are called *perpetual motion machines of the second kind.* They do not aim at creation of energy out of nothing; they merely attempt to transport existing energy (in the ocean, say) to a high-temperature heat reservoir. If only one of them worked, its inventor would become rich instantly.

There is indeed a device that can transport heat from a low-temperature reservoir to a high-temperature reservoir—the refrigerator, which is used in kitchen refrigerators, freezers, and air-conditioning units. But there is one hitch; it costs electric energy to operate the machine. On the other hand, one may try to use a refrigerator in combination with an engine (Fig. 16.3). The heat transported to the boiler can be used in the engine to do work. Use part of the work to pay back the work consumed in operating the refrigerator; in other words, let the engine, through an electric generator, operate the refrigerator. If only there were a small amount of work left over after paying back the advance, then by repeated operation of the two in combination, the work left over, no matter how small, could be multiplied to a huge sum. There are many kinds of refrigerators and many kinds of heat engines. In all possible combinations, perhaps by chance, we might hit one that indeed would work in this way; then we would have hit the jackpot. It does not seem terribly difficult. After all, we need only a very small amount of work left over. Out of all possible combinations the law of chance might operate in our favor.

Figure 16.3 Yet another perpetual motion machine that does not work.

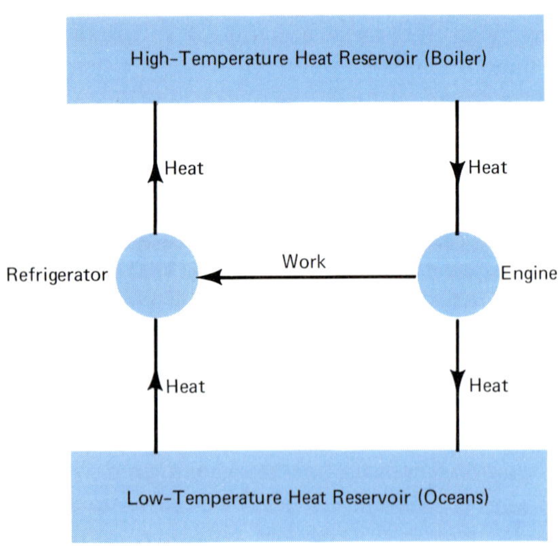

Yet nobody has made a fortune this way. No matter how you try, it will not work. It is not a matter of chance. Actually there is a law of nature forbidding it to happen. This law is formulated and called the *second law of thermodynamics.* Simply stated, the law says that perpetual motion machines of the second kind are impossible. Another way is to say that heat cannot be made to flow to a higher-temperature reservoir without cost (in work).

The Carnot Engine. What does the second law of thermodynamics mean besides telling us that there is no instant rich? This is the subject for the rest of the chapter. For one thing we can derive a theorem originated by Carnot concerning the efficiency of heat engines. It is a basic theorem and very important in engineering. It lies at the foundation of a thorough understanding of such environmental problems as thermal pollution by the energy industry and air pollution by the internal combustion engine (Sec. 2.4, Chapters 8, 9, 17).

We introduce a hypothetical heat engine called the *Carnot engine,* which converts heat from a boiler into mechanical work. It (Fig. 16.4) is composed of (1) the hardware consisting of a cylinder filled with, say, an ideal gas (used as the *working substance*) and fitted with a piston connected to a flywheel by a rod; (2) a *boiler* (a *high-temperature heat reservoir*) where fuel is burned to generate heat at a high temperature; (3) a *condenser* (a *low-temperature heat reservoir*) where a coolant, such as river water, is used to cool (take heat away) the working substance. In all heat engines these components also are present with minor variations. In all heat engines the input heat is generated in the boiler and the work output is derived from the flywheel, which may be connected by belt or by gear to operate other machines and do work. The ratio of work output to heat input is called the *efficiency* of the engine. Another aspect often overlooked is the output of waste heat removed by the coolant and eventually dumped in the envi-

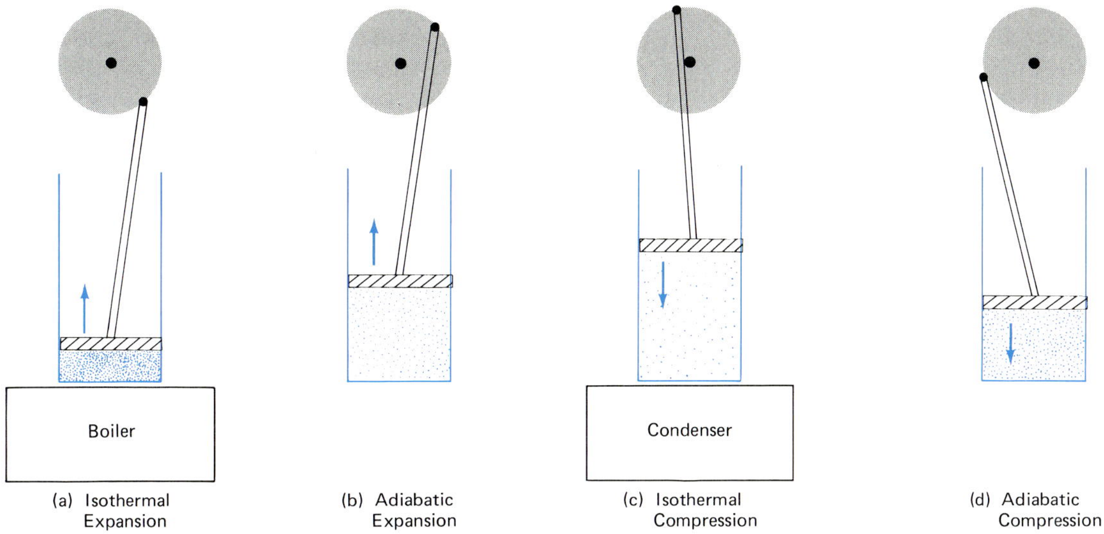

Figure 16.4 The three components of a Carnot engine, namely, the hardware, the boiler, and the condenser, and the four "strokes" of its operation cycle, (a), (b), (c), and (d).

ronment (the river, for example). This, as mentioned before, is the origin of thermal pollution.

The operation of the Carnot engine consists of four "strokes" (Fig. 16.4): (a) an isothermal expansion of the ideal gas when it is in contact with the boiler at a high temperature of T_1; (b) an adiabatic expansion of the gas when the heat contact with the boiler is disconnected; (c) an isothermal compression when the gas is placed in contact with the condenser at a low temperature of T_2; (d) an adiabatic compression of the gas when the heat contact with the condenser is disconnected. At the end of (d) the gas is brought back to the initial state and the cycle may be repeated over and over again. The four processes are required to be reversible; they can run backward with all the intensive quantities remaining the same and all the extensive quantities remaining the same in magnitude but with a change of the sign. The Carnot engine is thus called a *reversible engine*.

In expansions (a) and (b) the gas does work; in compressions (c) and (d) the gas borrows work (Fig. 16.5). The work done in (a) and (b), after paying back the borrowed work in (c) and (d), is the net work output W of one cycle of operation. At the same time, heat Q_1 is absorbed by the gas from the boiler in (a) and heat Q_2 is extracted from the gas by the coolant in (c), there being no heat exchange in (b) and (d). The net heat absorbed is $Q_1 - Q_2$. According to the first law of thermodynamics, this must equal

Figure 16.5 Change of state of the ideal gas as the working substance of the Carnot engine in one cycle of operation and the work output W of the engine.

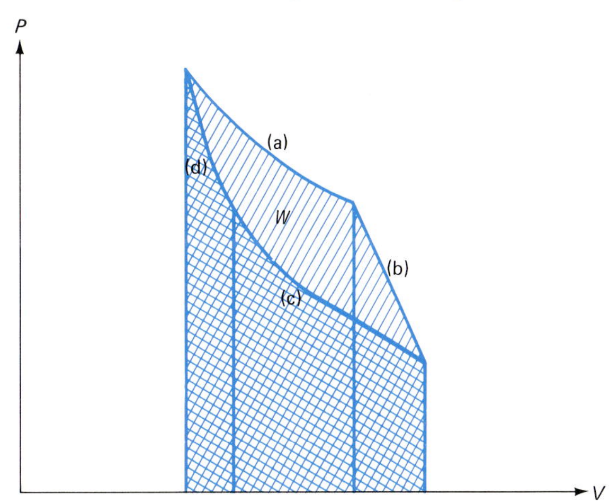

the net work output:

$$W = Q_1 - Q_2$$

The efficiency e_c of a Carnot engine is thus

$$e_c = \frac{W}{Q_1} = \frac{Q_1 - Q_2}{Q_1}$$

The total work done by the ideal gas in (a), (b), (c), and (d) (positive in a and b, negative in c and d) can be calculated by using the equation of state of the ideal gas. Thus W can be obtained (Fig. 16.5). The heat Q_1 for the ideal gas equals the work done in (a) and thus can also be calculated. From W and Q_1 we can calculate the efficiency. Because the calculation has to be done by the calculus method we cannot go through the details here. We will simply state the final result of the calculation: the efficiency is given by

$$e_c = 1 - \frac{T_2}{T_1}$$

The temperatures T_1 and T_2 are in the absolute scale. The reason we use an ideal gas is that the equation of state is exactly known and so the efficiency can be calculated exactly.

The Carnot Theorem. We now come to the Carnot theorem. It states that all heat engines operating between a boiler of temperature T_1 and a condenser of temperature T_2 (coolant temperature) have an efficiency no better than that of the Carnot engine, i.e.,

$$e \leqq 1 - \frac{T_2}{T_1}$$

This is very important in engineering because it tells us that the efficiency of any engine, such as the steam engine, internal combustion engine, and so on, has a limit $1 - T_2/T_1$, which can never be surpassed. If the actual efficiency is close to this limit, we know it cannot be improved any further and do not have to spend further effort to improve it. For a steam engine operating with a boiler temperature of 180°C (under high pressure, the boiling temperature of water goes above 100°C) and cooled by river water at 20°C the maximum efficiency is

$$e_c = 1 - \frac{20 + 273}{180 + 273} = 35 \text{ per cent}$$

It means that at best only 35 per cent of the heat energy generated by the fuel is converted into useful work. The other 65 per cent is flushed away into the river by the coolant. Not only is it wasted, but it causes thermal pollution!

We know that work can be converted into heat 100 per cent but the reverse is not true. However, it is not generally realized that the percentage is that small. Why is nature so mean to us? We will return to this point later. In the meantime let us first finish the study of the Carnot theorem.

Before anything else we recognize that the Carnot engine, being reversible, can be run backward—the four strokes are now (d), (c), (b), and (a), and compression is changed to expansion and vice versa. The gas now *absorbs* an amount of heat Q_2 from the condenser, *rejects* an amount of heat Q_1 to the boiler, and consumes an amount of work W supplied by an outside source, usually a motor. Such an operation is exactly what a refrigerator does. The kitchen refrigerator absorbs heat from the inside of the refrigerator (now acting as the low-temperature reservoir) and thus keeps it cool. It consumes work from the house current. We pay for the energy consumed in running the appliance. Finally, there is the undesirable side effect—the machine dumps heat outside of it, that is, into the kitchen (now acting as the high-temperature reservoir). (Refrigerators and air-conditioning units do not destroy heat; they simply remove heat and dump it at the neighbor's doorsteps, and with an extra amount, the work consumed, added.) All refrigerators are actually heat engines working in reverse. Thus heat can be forced to flow from a low to a high temperature at the expense of consuming an amount of work supplied from outside. It cannot so flow without cost.

By a rather involved argument deriving from the second law of thermodynamics, which we omit, it can be proved that all reversible engines working between the same temperatures T_1 and T_2 have the same efficiency. (The argument goes like this: If this were not true it would be possible to contrive a viable perpetual motion machine by using two reversible engines, one acting as a motor and the other as a refrigerator.) Because the Carnot engine is a reversible one, all other reversible engines will have the same efficiency, $1 - T_2/T_1$. By a similar argument we can prove that the efficiency of an irreversible engine is less than that of a reversible engine. Therefore no engine can have an efficiency better than $1 - T_2/T_1$, which is what the Carnot theorem states.

Interpretation. We base our proof on the basic assumption that a perpetual motion machine of the second kind is impossible. Why should a basic law of nature be formulated in human terms, particularly with a slant of economic gains? If humans were not in existence, would this law based on human terms disappear? The last vestige of a pseudoscience still remains in a scientific theory! Bridgman[1] criticized the traditional formulation of thermodynamics from this point of view, which is quite justified in view of the many superficial expositions of the subject in textbooks. On the other hand, we should not underestimate the intelligence of the founding fathers of thermodynamics. From their standpoint the second law of thermodynamics may be regarded as originating in the basic nature of heat transfer, i.e., when two bodies are allowed to interact thermally, such as by conduction, or radiation, heat *always* flows from the hot to the cold. The second law of thermodynamics is thus a statement on the nature of heat transport. We will not elaborate on this point further here.

[1] P. W. Bridgman, *The Nature of Thermodynamics* (Cambridge, Mass.: Harvard University Press, 1941).

Entropy. Heat can be transported from high temperature to low temperature spontaneously. But to do the reverse, work must be done. Clausius introduced a new concept, *entropy* (from the Greek word for *transport*), which we will not try to define, and found that the unsymmetric behavior of heat transport is equivalent to the statement that in an isolated system the total entropy always increases, at best remaining the same. This is the *principle of the increase of entropy,* which is actually equivalent to the second law of thermodynamics. This principle, derived from the Carnot theorem, is a little abstruse and we shall not go into it in detail. Nevertheless, entropy is such a seminal concept that it even appears in the title of a recent novel. A casual acquaintance with it is well within the scope of this book.

Cosmic Consequences. The universe as a whole is an isolated system and therefore the second law may be stated by saying that the entropy of the whole universe is increasing. As already stated, the first law may be stated by saying that the energy of the whole universe is a constant.

The second law implies that heat at a high temperature is useful; it can be, at least partly, converted into work. The higher the temperature, the better. When temperature is reduced to that of the environment (used as the condenser) the efficiency reduces to zero and the heat is completely useless, unable to be converted into work by any percentage. From the economic point of view, energy resources thus mean heat energy at a high temperature. Fuels can generate heat at a high temperature and thus are energy resources.

As heat in a high-temperature region always spontaneously moves into a low-temperature region (by conduction, radiation, and so on) the amount of useful energy of the universe thus keeps on decreasing. This is an important consequence of the second law of thermodynamics.

Because the total energy of the universe is finite, the total useful energy is finite. Because it keeps on decreasing, eventually it will decrease to zero. This would happen when all the fuels would have been burned and all bodies in the universe would have reached the same temperature. Then, in spite of the fact that we still have the same amount of energy, none of it would be useful. All of it would be heat energy at the same temperature and none could be converted into work. Not only would industrial civilization grind to a halt, but all lives would cease to exist. (Life is sustained by useful energy.) The whole universe would be dead. This is the state of the so-called *heat death,* which would be the ultimate fate of our universe. This state is, of course, a long way off. The sun is still shining and we can get a large amount of useful energy from it to sustain life and industry. The sun has 7 billion more years to go.

What Is Entropy? Kelvin and Clausius formulated the two laws of thermodynamics in 1850 but it took Clausius 15 years to introduce entropy. Since then attempts have been made to explain entropy. Because the Newton laws were the basic laws of science at the time, it is only natural to try to reduce the entropy principle into the Newton laws. This attempt, however, never succeeded.

Boltzmann, one of the most ingenious theoretical physicists, introduced an alternative. He interpreted the entropy principle as a probability principle on the basis of the molecular theory. It is thus a principle outside the Newtonian system and hence a new independent basic principle in science. This interpretation is accepted today and verified without any doubt. With the benefit of hindsight, it is now easy to see the point. As the entropy principle is rooted in the nature of heat transport, let us consider heat transport from a molecular point of view. The fact that heat flows from the hot to the cold is due to the equalization of the (average) molecular energies of the two bodies resulting from molecular collisions. That unequal things become equalized by random interaction is due to the law of probability. A bottle of beans, black in the bottom half and red in the top half, after being shaken randomly for a while, becomes mixed; nonuniform density distribution changes into uniform. This is due to the law of probability. Molecules from the hot body and those from the cold body interact and equalize their average energy also by the law of probability.

Nevertheless, this view was in the minority in the nineteenth century. In spite of chemistry based on the atomic theory, in spite of the kinetic theory of gases, in spite of Boltzmann's impressive theoretical work, the statistical interpretation of entropy was not accepted; even the existence of the molecule was not considered proved. Boltzmann was driven to commit suicide. A major adversary was Mach, who is remembered by posterity through the Mach number of supersonic flow. It was only after 1905, when Einstein's theory of Brownian movement demonstrated "molecular motion" to the eyes of everybody, that the existence of molecules was accepted without further argument. So was the statistical interpretation of entropy. We may now consider entropy as a quantitative expression of *randomness.* The entropy principle simply says that randomness tends to increase spontaneously.

Being a statistical principle, the second law of thermodynamics or the entropy principle covers not only natural processes involving the equalization of molecular energy, but also those involving the equalization of molecular density. Nonuniform materials tend to become uniformly mixed. The amount of iron on earth, when concentrated in ores, is not uniformly distributed. But after human utilization, the amount becomes more uniformly distributed over the earth by the law of probability. This may also be regarded as a manifestation of the entropy principle. The iron in the ore, like heat in a high-temperature reservoir, is useful, but the iron in trash dumps all over the world, like heat energy in the ocean, is useless. The entropy principle again tells us the trend from the useful to the useless.

Entropy and Life. In an isolated system the molecules will reach equilibrium and all relations of order and organization will become randomized. On the other hand, the evolution of life is to proceed from the simple to the complex, from the random and unordered to the highly organized and ordered. It seems to go against the trend of the second law of thermodynamics. The paradox originates from a misunderstanding and can be cleared up easily. The biosphere of the earth is *not* an *isolated* system. It is *not* in an *equilibrium* state. Thus there is no contradiction with the second law of thermodynamics. Rather it is in a *steady* state of energy flow; energy flows into the biosphere from the sun and exits to the outer space surrounding the earth. There is a constant *flow through,* but the configuration does not change from one time to the next (just as in the metal rod used in the experiment of measuring heat conductivity, there is a constant flow through of heat but the temperature distribution of the rod does not change in time). In a steady state more interesting things can happen. If in a system many chemical reactions take place, one leading into another, we would expect the system to change from time to time. For the system to remain steady, no change from time to time, the reactions must organize themselves into cycles. Indeed, the cycles (replication, metabolism, birth–death cycle, and so on) are the very characteristics of life. Thus energy flow in the biosphere is instrumental in the very origin of life and fundamental for an understanding of the nature of life.

Conclusion and the Environmental Problem. We have established three basic laws governing the transformations of the two basic entities in the physical world, i.e., matter and energy. Each can be transformed in a variety of ways and the transformations manifest themselves as the various kinds of natural processes. Thus matter can be transformed at the nuclear, atomic, molecular, and bulk level (Chapters 14, 12, 10). Energy can be transformed among kinetic, potential, heat, electric, chemical, electromagnetic, and nuclear energies (Chapters 5, 7, 11, 13 and 15). Indeed, we have surveyed the whole of physical science based on these transformations. But we have two conservation principles that say that the total of each (ultimately the total of both) is always the same throughout these transformations. This provides an important guideline in looking into the transformations. Still, these two principles do not tell in which direction transformations (many of them reversible) will proceed. This point can be resolved by the third principle, that of increase of entropy. Generally, transformations take place from the less probable to the more probable, from the ordered to the random. The entropy principle in the mathematical form makes it possible to figure out the direction when offhand we have no way of knowing which is more probable. Thus the three principles give us very useful guidelines and many problems can be solved by them without further information. In Chapters 21 through 25 we consider the origin and evolution of the universe. The problem will be looked upon as transformations of matter and energy. These three principles will play an important role.

If this is the way science helps us understand natural phenomena, this is also the way science will help us in dealing with human-created problems. The three principles will be the basic principles of energy and environmental discussions in the next four chapters.

The environmental problem concerns the utilization of matter and energy. The two conservation principles tell us that the amount of matter and energy is always the same even after utilization. But the third entropy principle tells us that the amount of useful material and useful energy will keep on decreasing. The depletion of material resources is due not to the destruction of the atoms but to the redistribution of the atoms from a concentrated state, such as in metal ores, which is useful, to a dispersed state, such as rusted metal cans in trash dumps, which is not useful. This redistribution is dictated by the second law of thermodynamics. This point will be elaborated further in Chapters 19 and 20. The decrease of useful material and energy casts a dark shadow on our future and is the crux of the resource and environmental problems. Notice that this follows the basic laws of physical science, disregarding industrialism, capitalism, expansionism, and environmental neglect, which only accelerate the process.

On the other hand, we know that the trend of the

entropy principle can be forced to reverse by supplying energy. Wasted material can be recovered and recycled by supplying energy. Therefore the *only* path of salvation of our future is to find a new abundant energy source. There is no other alternative. No amount of environmental ethic, no amount of nature loving, no amount of self-denial and restraint can solve the problem completely. They can only postpone the problem. It is as clear as the impossibility of the perpetual motion machine.

The only new source of energy that can fulfill this role is ultimately the fusion energy. It lasts (5×10^{10} years) longer than the sun (7×10^{9} years) and thus longer than all lives on earth. The controlled release of fusion energy has not been accomplished as yet. To harness fusion energy is a task that should be made the one paramount scientific objective having priority over all other pursuits. Even without fusion energy we can survive on fission energy for at least 100,000 years.

And this is the way science will serve humanity.

Problems

16.1 If the ocean temperature were reduced by 1°C and the heat energy so extracted were converted entirely into electric energy and sold at the rate of 3¢/kwh, how much money would be made? Consider that 70 per cent of the earth surface is covered by ocean with an average depth of 12,000 ft.

16.2 Can you cool the kitchen by opening the door of the refrigerator?

16.3 An amount of coal is used to generate electricity, which is then used for cooking. Compared with using the same amount of coal in a stove for cooking, does the energy value of the coal increase, decrease, or remain the same? Why?

16.4 A car of 5,000 lb is driven from Atlanta (elevation 1,000 ft) to Savannah (elevation 10 ft), covering 265 miles, and back to Atlanta on an alternate route with the same mileage in the same time under the same weather conditions; the wind velocity is zero. Which way consumes more gasoline? Assuming an overall efficiency of 20 per cent of converting the heat energy of gasoline to mechanical energy of the car, what is the difference of the gasoline mileages (in the unit of miles/gal) of the two trips? Heat of combustion of gasoline is 3.0×10^{7} cal/gal.

16.5 Assuming the use of air conditioning in a house and in a car, which of the following two alternatives is more energy-saving: Cooking in the kitchen or in the backyard? Parking a car in sun or in the shade? Serving hot tea or iced tea on the dinner table (remember even iced tea must first be brewed with hot water and do not forget where the ice comes from)? Ordering hot tea or iced tea at a drive-in restaurant and having it in the car?

16.6 Chickens fight for dominance, rush for food, and pursue a variety of other activities, including sex. The fighting, rushing, and so on, are undesirable from the point of view of modern chicken farming and are eliminated to achieve maximum efficiency—to produce a bird of about 4 lb in merely eight weeks with only 8.1 lb of feed. Explain why.

16.7 Chicken feathers are not edible and thus represent a waste of feed. If it were possible, would it be desirable to breed a new species of featherless chicken?

17 Environmental Effects of Energy Utilization

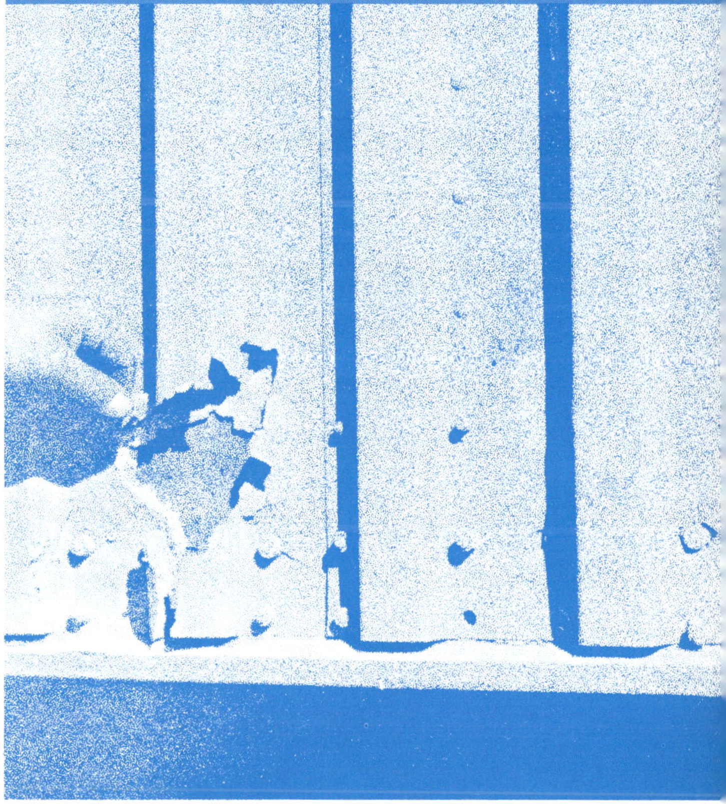

17.1 Depletion of Useful Energy and Increase of Useless Energy

Energy utilization is governed from the physical point of view by two basic principles, the first and the second laws of thermodynamics, which may be phrased thus: the total amount of energy in the universe is a constant, but the amount of useful energy keeps on decreasing. The term *useful energy* means mechanical energy or any energy that can be converted into mechanical energy. It is mechanical energy that is useful in the industrial sense. Heat obtained from burning fuels can be converted into mechanical energy only partly by a heat engine. Only part of it is useful energy. Heat in ocean water or in the earth's crust is not useful, because it cannot be converted into mechanical energy, even though it may be used sometimes in other economically profitable ways.

In the utilization of energy no amount of energy is destroyed; the total amount of energy involved always remains the same. The utilization involves a transformation of energy in which useful energy, such as electrical energy, is consumed and in its stead an equal amount of "useless" energy, usually in the form of heat energy given to the earth environment (air, water, land), is generated (in engine exhaust, in heat production by friction of mechanical parts, and so on).

The utilization of energy does have the consequence of depleting natural resources irreversibly. This point is of little significance as far as solar energy is concerned. Solar energy will not be depleted until billions of years from now, but this eventuality is irrelevant because it is certain that other natural catastrophes will occur long before this ultimate doom. On the other hand, the use of other energy resources—i.e., fossil fuels, fission fuels, and fusion fuels—is a process involving the irreversible depletion of natural resources. Our future optimism about energy resources is based on the large supply

of nuclear fusion fuels, a supply that is practically inexhaustible from a human point of view even though the technology for its use has not yet been fully developed. In the meantime we can count on the fossil and fission fuels for our consumption in the near future before the necessary technology of fusion is perfected.

Another consequence of energy utilization is the addition of useless heat energy to the environment (thermal pollution). In fact, all the fuels and power consumed have turned up as heat in the environment. This may one day become a major problem for mankind, as we have mentioned earlier.

17.2 Material Pollution by Fossil Fuels

There are side effects of energy utilization, that is, environmental pollution, which are our concern here. The energy in any kind of fuel is associated with a material substance (the fuel itself). During the process of consumption (burning), this material substance is transformed. Yet according to the law of conservation of matter, the total amount of matter remains exactly the same after transformation. The fuel does not disappear after burning, as is usually thought. It is transformed into other substances—i.e., ashes, smoke, gaseous products—which are dumped into the environment as waste products. In fact, fuel consumption accounts for a major part of environmental pollution (Fig. 17.1). The amount of food grain consumed in the world averages about $\frac{1}{3}$ ton per person per year; yet the coal consumed averages about 1 ton per person per year. After burning, this amount of coal changes into more than 1 ton of carbon dioxide, ashes, and a few other substances such as sulfur dioxide. Thus a large amount of waste products is released into the environment.

Figure 17.1 Burning of fuel in the launch of Minuteman from silo complex. [U.S. Environmental Protection Agency]

Solid Fossil Fuel. Let us consider the pollution produced by various kinds of fuels. The fossil fuels derive their energy from the carbon atom, which changes into carbon dioxide (CO_2) after combustion. Carbon dioxide is not harmful and is necessary for the maintenance of plant life. It generally is not considered a pollutant. On the other hand, most fuels contain other natural or man-made impurities that give rise to undesirable combustion products such as sulfur dioxide (SO_2), thus contributing to air pollution. Moreover, an incomplete combustion process produces additional pollutants, such as carbon monoxide (CO).

Among the pollutants produced by coal burning, sulfur dioxide and fly ash are the principal ones. As already mentioned, sulfur dioxide causes throat irritation and respiratory disorders. Moreover, after being oxidized to sulfur trioxide SO_3 it combines with water to yield sulfuric acid (H_2SO_4), which is highly corrosive and attacks many things exposed to it. Thus in a humid atmosphere it contributes to the rusting of metals (Fig. 17.2), deterioration of fabric, and corrosion of stone structures and sculptures. The famous London smog that has existed for several centuries is a combination of smoke and fog. Its unpleasant effect is due largely to sulfur dioxide produced by burning coal.

Large amounts of sulfur dioxide may be removed, for example, by wet scrubbers, such as a sodium carbonate solution. Sometimes pollution control yields sulfuric acid as a by-product (Fig. 17.3). Ironically, in a copper mine in Tennessee the pollution control by-product has turned out to be more profitable than the copper smelted there. Fly ash, consisting of small particles, may be attracted and thus removed from smokestacks by electrically charged devices (Fig. 17.4). Cleaning up the black part of the smoke is relatively easy, but the colorless part is still hazardous. The collected fly ash may be made into bricks and other concrete products, thus paying for part of the pollution control expenses. Another remedy is to limit the sulfur content of the fuel. Now even cities in the traditional coal-producing Appalachian region are using low-sulfur coals transported from 1,000 miles away in the west.

Figure 17.2 Corrosion of metal by air pollution. [U.S. Environmental Protection Agency]

Figure 17.3 Sulfur recovery plant, Union Oil. [U.S. Environmental Protection Agency]

Liquid Fossil Fuel: Primary Pollutants. The case of liquid fossil fuel is quite different. Largely used in automobiles, gasoline as a fuel produces a new class of pollutants and a new kind of smog first seen in Los Angeles. At first the smog was thought to be due to emissions from incinerators and oil refineries. But control of these sources did not reduce the smog. Finally it was found that the smog is the product of photochemical reactions of pollutants from automobile exhaust. The *photochemical smog* causes eye irritation and blight of vegetation. The once sunny and clear sky of southern California is now filled by an unpleasant, ugly, brownish air mass.

Gasoline is composed largely of hydrocarbons. When burned completely, it yields carbon dioxide and water vapor; both are harmless. Yet the internal combustion engine used in automobiles works in such a way that it burns only momentarily but the burning is repeated in quick succession. The combustion is likely to be incomplete (Sec. 8.3), and the combustion products released into the air from the exhaust include carbon monoxide and unburned hydrocarbons. The latter also enter the air through evaporation of the fuel in storage and in transit. Besides the main ingredient, hydrocarbons, gasoline often contains a number of additives to perform a variety of auxiliary functions. Most important is the additive tetraethyl lead, $Pb(CH_2CH_3)_4$, which reduces the knocking tendency of the engine and thus increases its performance. This compound is highly toxic. The combustion of this compound yields lead oxide, which is harmful to the engine. Further additives, ethylene dichloride $ClCH_2CH_2Cl$ and dibromide $BrCH_2CH_2Br$, are used to convert it to the more volatile $PbBr_2$ and $PbCl_2$, which pass out with the exhaust gas. In addition, a host of compounds are added to act as antioxidants, antirust and anti-icing agents, detergents, and lubricants. All additives contribute to the pollution. The list of *primary pollutants* from automobile engines is completed with the addition of nitrogen oxides, mainly NO_2 and NO, which are formed during the combustion process from the nitrogen of the air taken into the cylinder together with the oxygen. The quantitative breakdown of the automobile primary pollutants is as follows: carbon monoxide (77 per cent), sulfur dioxide (1 per cent), particulates (1 per cent), hydrocarbons (14 per cent), and oxides of nitrogen (7 per cent).

Secondary Pollutants: Photochemical Smog. Under the action of sunlight, hydrocarbons and oxides of nitrogen interact and yield a variety of *secondary pollutants* that give rise to the photochemical smog. The hydrocarbons do not absorb sunlight and undergo chemical changes themselves. But they react with NO_2 and its photochemical products, and are oxidized to aldehydes 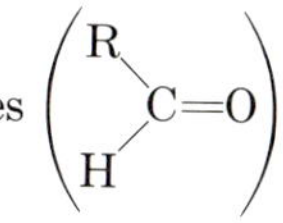 and ketones 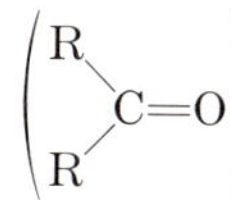 as well as converted to a variety of other compounds. Nitrogen dioxide (NO_2), the primary pollutant, absorbs ultraviolet light from the sun and decomposes to form nitric oxide NO and atomic oxygen O. The atomic oxygen reacts with molecular oxygen (O_2) to form ozone (O_3). Furthermore, molec-

Figure 17.4 The same smokestack with the pollution control device, an electrostatic precipitator for removing particulate pollutants, turned on (left) and off (right). [U.S. Environmental Protection Agency]

ular oxygen may absorb sunlight to raise itself to an excited state, O_2^*.

$$NO_2 \xrightarrow{\text{light}} NO + O\cdot$$

$$O\cdot + O_2 \longrightarrow O_3$$

$$O_2 \xrightarrow{\text{light}} O_2^*$$

All O, O_3, and O_2^* are oxidizing agents and are ready to oxidize the hydrocarbons. The hydrocarbons are thus changed to free radicals directly and indirectly;

$$\text{Hydrocarbons R} + (O\cdot \text{ or } O_3 \text{ or } O_2^*) \longrightarrow \text{free radicals } RO\cdot \text{ or } RO_2^{\cdot} \text{ or } RO_3^{\cdot}.$$

The free radicals are highly reactive and yield other compounds:

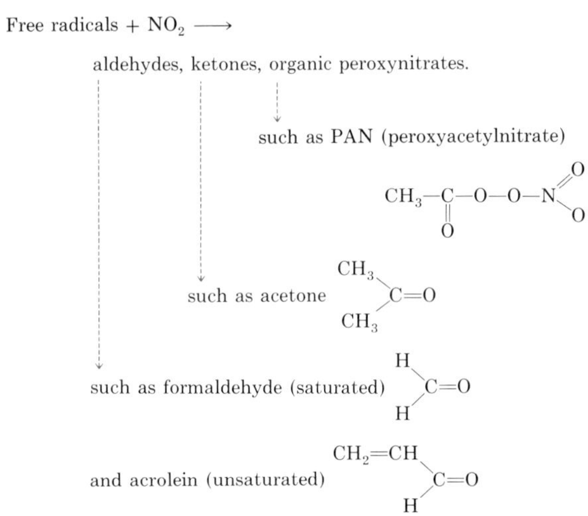

Formaldehyde, acrolein, and PAN irritate the eyes and nose. PAN destroys vegetable leaves and orange trees. Ozone irritates people (it has a peculiar odor), attacks plant leaves, and causes rubber to become brittle. Many of the products are liquids and form tiny droplets, producing an aerosol smog. Nitrogen oxides, sulfur dioxide, and carbon monoxide are either toxic or irritating or both. Thus the major features of the smog are accounted for by photochemical reactions. Incidentally, the following sequence of reactions may occur:

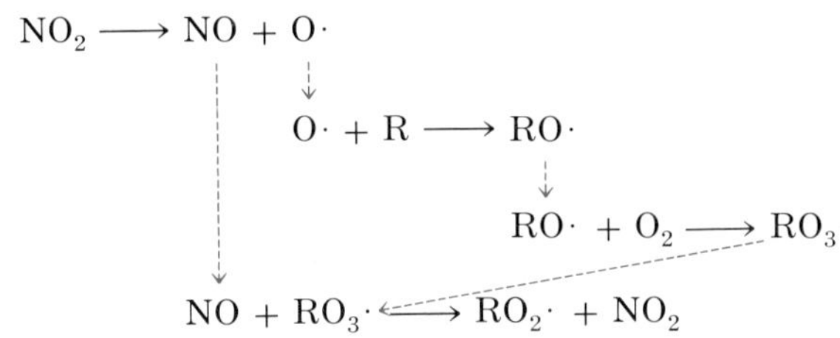

in which a hydrocarbon R changes into a free radical RO_2 with the help of NO_2, which is regenerated and thus acts as a catalyst that can be used over and over again. The catalytic cycle may start from NO, which is also a primary pollutant. Ozone also appears in cycles; it disappears when reacting with NO to produce NO_2 and O_2, and it appears when atomic oxygen reacts with O_2. Ozone may oxidize sulfur dioxide into sulfur trioxide, which yields sulfuric acid when combined with water.

The photochemical smog is a consequence of large-scale use of automobiles and is destined to persist and spread as the use of automobiles increases. It has already struck major cities all over the world because of the increase in the auto population during recent years. It first occurred in Los Angeles (Fig. 17.5) because of a peculiar geographic and weather condition. Los Angeles is a coastal city backed by soaring mountain ranges on three sides, hindering air circulation. A low-altitude ocean breeze is the only circulation of any significance. Ordinarily temperature decreases as altitude increases. The ocean breeze cools the air near the ground (1,000 to 2,000 ft), but poor circulation does not allow the quick mixing of the air, and the next few thousand feet of air may remain hot, hotter than in the lower altitude, a phenomenon called *temperature inversion*. The hot-air layer represents a high-pressure region that produces a downward draft. Air pollutants rising from the surface are thus prevented by the high-pressure region from rising any further to disperse, so they react among themselves with the help of sunlight coming through the clear, dry air

Figure 17.5 Smog formation in Los Angeles. [U.S. Environmental Protection Agency]

above to produce photochemical smog. Because the smog cannot rise and disperse vertically, it forms a blanket spreading out laterally to cover the whole basin persistently.

Smog control is a pressing environmental problem today. Because smog is due largely to incomplete combustion of auto fuel, an obvious remedy is to burn away the residual hydrocarbons. The hydrocarbon pollutants may be reduced by raising the flame temperature of the engine. Although it takes both unburned hydrocarbons and nitrogen oxides to produce smog, the former are easier to control. At the same time, health considerations demand the control of carbon monoxide, which is not related to smog. Thus stricter limitations were set by authorities on hydrocarbon and carbon monoxide emissions. To meet these demands, the auto industry, to maintain the high performance of the engine, raised the flame temperature in the hope of burning away more of these combustible products. But this led to the production of more nitrogen oxides, thus largely defeating the original purpose of reducing the smog. Right now the approach is to burn away the combustible pollutants in the catalytic converter. But again a new problem arises: the converter increases sulfuric acid pollution.

Other Pollutants. Besides smog and carbon monoxide, other pollutants from automobiles, such as lead, have detrimental effects that cannot be overlooked. Lead poisoning, such as that caused when children swallow paint chips containing lead, is well known. When it enters the human body, airborne lead may increase the lead level in the blood (tests on this point are now in progress) and in body tissue. Although the effect of airborne lead has not been ascertained completely, lead should be avoided by those persons whose bodies have already accumulated it. The source of airborne lead is usually automobile exhaust; studies have shown that from 1962 to 1969, the air lead level increased by amounts ranging from 33 per cent to 64 per cent at eight sampling sites in the Los Angeles area. Because snow is a more efficient air cleanser than rain, snow is found to contain more lead than rain water, sometimes exceeding the safety limit set by the U.S. Public Health Service tenfold. Studies also show higher lead concentration in snow near main highways, confirming the source to be automobile exhaust. Because birds in winter depend on snow for their water supply, some dead birds are found to have high lead concentration in their bones, on the order of 50 per cent of the lethal dosage.

Chronic Effects. It must be kept in mind when considering the health hazard of air pollution that besides the *acute effects* already mentioned, such as irritation, respiratory disease, and even death, there is a *chronic effect,* less obvious but not less important. The situation is somewhat similar to radiation hazard; besides the short-term *somatic effect* associated with high doses of radiation, there is also long-term *genetic effect* associated with low doses of radiation. The current radiation safety standard is set to insure safety against the low-dose, long-term effect. Yet the current safety standard for air pollution is set merely as a safeguard against the acute effects. (It is as if we were buying $1 million earthquake insurance but not buying any auto insurance at all.) The acute effects of air pollution are easily noticeable. During the London smog spell of 1952, in five days there were 4,000 more deaths than expected according to the average mortality rate. It is not difficult to set safety standards to protect against the acute effects. In 1969 California set the following limits: 0.1 ppm of oxidant (including ozone), 20 ppm of carbon monoxide, 0.1 to 0.5 ppm of sulfur dioxide, 0.25 ppm of nitrogen dioxide, and 100 $\mu g/m^3$ of particulates. Yet the long-term, low-level effects of pollution in the air are more difficult to gauge, and much additional research is needed.

It is suspected that SO_2 and NO_2 are mutagenic and carcinogenic. A *mutagen* causes changes in the DNA molecule and thus garbles genetic information. When this happens in the genetic cells, it usually results in a deleterious mutation, giving rise to defective offspring. When it happens in somatic cells, it may accelerate aging and shorten the lifespan. A *carcinogen* induces cancer. There are no obvious statistics like the 4,000 deaths in London that link air pollution to lifespan and cancer, but a detailed statistical analysis indicates some correlation between pollutants and accelerated aging.[1] Thus the low-level, long-term effect is a problem that deserves serious attention.

Gaseous Fossil Fuel. We now turn our attention to the environmental effect of burning natural gas. As a heating material it burns clean, yielding carbon dioxide and water vapor with no sulfur dioxide pollution. It is thus a more desirable fuel than coal. The replacement of coal by gas as a household heating fuel has greatly contributed to the reduction in sulfur dioxide pollution of the air. The air of many small cities is actually cleaner now than decades ago because of the substitution of gas for coal. This is an indication that technological progress is not necessarily detrimental to the environment.

To reduce air pollution some gas companies have adopted natural gas as an automobile fuel (Fig. 17.6). When natural gas is used in internal combustion engines, nitrogen oxides are expected to appear in the same way as in a gasoline engine. Yet the hydrocarbons are of a different kind (lower molecular weight), and the pollution effects may be quite different. At any rate before we have a network of natural gas stations on the highways, the use of gas as a vehicular fuel will be limited to fleet operators who have their own fueling stations. Recently some municipalities have equipped their vehicle fleets for dual operation with natural gas and gasoline ($350 per car) to meet the energy crisis and to reduce fuel cost (7¢ for the equivalent of 1 gal of gasoline). A full

[1] Richard J. Hickey, in *Environment,* William W. Murdoch, ed. (Stamford, Conn.: Sinauer Associates, Inc., 1971), p. 189.

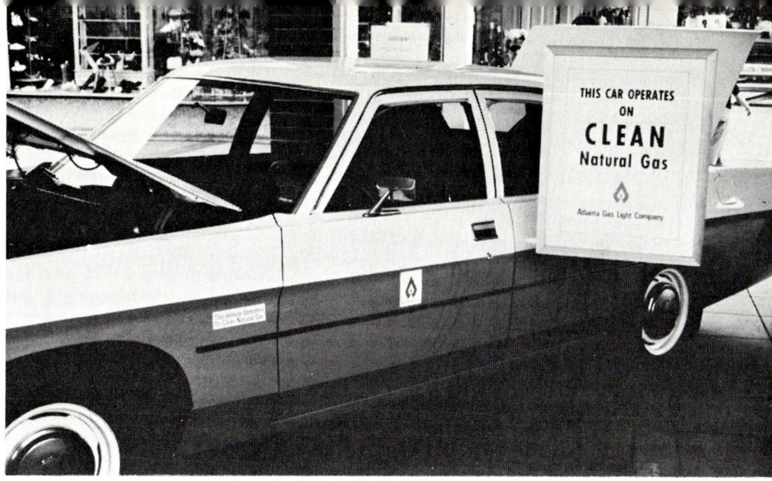

Figure 17.6 Atlanta Gas Light Company converted company cars to operate on natural gas to reduce air pollution.

"tank" of compressed natural gas is enough for 100 miles. (But natural gas is also in short supply.)

Because gas is cleaner than coal, it would be desirable to convert coal into gas for fuel use. This *gasification* process is now under development. The energy content of the carbon atoms in coal can be transferred to the gas, and the fuel value is preserved. The development of the gasification process is associated with the trend toward increased strip mining of coal in the western United States, which will be discussed in Sec. 19.3. Although our supply of natural gas is limited, the large reserve of coal may replenish the supply through gasification. In this way we could make better use of the solid fossil fuel resource of the earth.

17.3 Radiation Pollution by Nuclear Fuels

In perhaps 300 years, all fossil fuels will be depleted. The air pollution problem generated by fossil fuels will be over by that time. Then we will be faced with the more serious problem of finding substitute fuels. These in turn will have their own pollution problems.

Nuclear Fission Fuel. The substitute fuel most likely to be used first is fission fuel. Fission energy is released when a heavy atomic nucleus such as ^{235}U splits into two fragments. These fission products, atoms of medium weight, are mostly radioactive. Because the uranium nucleus may be divided in various proportions, a large variety of radioactive fission products are formed. Some of these have short lifetimes, and their radioactivity dies out quickly; but others have lifetimes of many years. Disposal of nuclear industrial wastes composed of these long-lived radioactive fission products presents special problems. Unlike sewage, they cannot be treated in any way to reduce the harmful radioactivity, because this property is one associated with the nucleus of the atom, which remains unchanged in chemical processes like those used in sewage treatment. If improper disposal methods are used, they may enter the air, surface water, and soil, causing hazardous contamination. Currently there are three ways to dispose of radioactive wastes according to the degree of activity: dilute and disperse for low-activity wastes, delay and decay for medium-activity wastes, and concentrate and contain for high-activity wastes.

The *routine emission* of a reactor—the release of very-low-activity radioactive fission products into the air during normal operation of a reactor—can be controlled and reduced to a very low level, much lower than that of natural background as mentioned in Sec. 2.6. As far as the hazard of routine emission is concerned, living near a reactor is more than 20 times safer than living in Colorado (Sec. 2.6).

Fear of the nuclear reactor comes from the possibility of mishap. The most common fear is that it will explode. Laymen frequently think that a run-away reactor will explode like an atomic bomb. This is a popular misconception, but a scientific impossibility. During the recent hot debate on nuclear safety this possibility has never been raised as a legitimate point of contention. In an atomic bomb the time between two successive generations of fission in the chain reaction is 10^{-8} sec. In a reactor this time is lengthened to 10^{-3} sec because the neutrons have to be slowed down first by the moderator. The explosion is thus slower than the atomic bomb explosion by 100,000 times and therefore is not of the uncontrollable, run-away type. Once the reactor core is overheated and starts to expand, the change of the geometry causes the chain reaction to become convergent and to stop. If a reactor should run out of control, it will "explode" merely with a force comparable to that of a few pounds of conventional explosives, which is not a major hazard.

Explosive power is not a point of concern, but the dispersal of radioactive material during such a mishap is. Because of the small power, the explosion can be contained. The dispersed radioactivity can be confined and cleaned up later without jeopardizing the general public.

Nevertheless all precautions are taken to insure against accidents. A separate conventional power system, which is kept on stand-by, can be activated instantly to supply the necessary power to shut down the reactor during a mishap when the reactor is "powerless" to help itself. The control room (Fig. 17.7) is manned by *two* licensed operators all the time; thus even the possibility of the operator having a heart attack is taken into consideration. However, every technological innovation has its risks, no matter how safely designed. In this connection we have to consider even the slight possibility of enemy sabotage, terrorist attack, and extortionist blackmail, all because of the hazard of dispersing radioactive material. Safeguards must be taken against these possibilities.

As far as technical accidents are concerned, the

Figure 17.7 A typical highly automated control room of a nuclear reactor. The dials, gauges, and lights notify the operator of the operating conditions. Any of 60 types of malfunction will cause an automatic shutdown of the plant. [U.S. Environmental Protection Agency]

point of contention on reactor safety concerns largely the possibility of a *melt-down* (loss-of-coolant accident). When a cooling tube bursts and heat is not removed from the reactor core, then, even though the chain reaction is stopped, heat is still generated from the radioactivity of the fission products in the core and the heat will cause the core to melt. The hot molten liquid descends and will melt anything in its way. Thus it cannot be contained. The radioactivity of the reactor core may thus be turned loose. The liquid core material, after sinking deep into the ground, will become dispersed. To prevent the melt-down there is an emergency core-cooling system (ECCS), which injects water into the core when the regular cooling system breaks down. A central point of argument is whether the ECCS will work as planned. Although separate components of the ECCS have been tested, the whole system has not been under the melt-down condition. An experiment is now planned to let a reactor proceed to melt-down and see if the ECCS actually works to stop it.

The frequency of melt-down, according to one study (Rasmussen) is one per 17,000 reactor-years. This estimate should not be too far off the mark, certainly not more than 30 times too low, because the world nuclear industry has accumulated 2,000 reactor-years of operation without melt-down, an excellent record of safety. According to the Rasmussen study (a project of 60 man-years and $3 million), the average fatality per year resulting from such reactor accidents if all current U.S. electric power were produced by reactors, is 1.4 deaths per year, which is smaller than that of routine emission, 8.7 deaths per year. If the estimated frequency were 30 times too low, the fatality rate would be 42 deaths per year, five times that of routine emission but at least four times safer than living in Colorado. In this connection, one should note that the fatalities that result from using coal to generate electricity in the United States number 5,000 to 100,000 per year (through air pollution). Thus nuclear energy is vastly safer than fossil energy (coal) in this respect.

In the worst possible accident that could happen to a reactor, in which there would be a steam explosion, breachment of containment, scattering of half of the reactor core and the release of fission products right into a population center because of adverse weather conditions, the fatalities might include 2,300 acute deaths and 3,200 latent cancer deaths. If evacuated in time deaths could be reduced; the loss would be $5 billion for evacuation and clean-up costs. This is the kind of possibility that frightens people and dominates the headlines. However, such an accident, like the death of five members of a family in five separate auto accidents on the same day, is improbable, even if remotely possible. The study shows the frequency of such an accident to be once in a million years. In considering such a major catastrophe, one must compare other possible accidents. The study shows that an accident involving a fatality of 1,000 or more would happen once in eight years from natural disasters (earthquakes, hurricanes, and tornadoes), once in 25 years from other man-made causes (dam failures, air crashes, fires, explosions), but only once in 120,000 years from an all-nuclear electric industry. Thus nuclear energy does not make an imperfect world any more imperfect.

The above discussion is based on the Rasmussen study. This major work on reactor safety, being sponsored by the Atomic Energy Commission, has been criticized as being an in-house study defending the nuclear industry. Recently an independent study has been carried out by the American Physical Society. In the report, issued in the summer of 1975, it is stated that the study group has not uncovered reasons to believe that the reactor is unsafe. Although it expresses reservations about the Rasmussen report (because the major accident is too improbable to be properly analysed quantitatively), it does not challenge the general conclusions of the latter.

When nuclear power is used in a large scale, the fission products will be generated in large volumes. Even without accidents to set them loose there is still the problem of proper disposal. The radioactivity will not completely disappear for thousands of years; the nuclear waste must be safely stored and guarded for a long time. A thousand years' vigilance! There is the question of whether we are justified in polluting the earth with that much radioactivity. Even though our present safety may be assured, what is the guarantee thousands of years later? On the issue of contaminating the planet it may be noted that the radioactivity in a reactor is 2 per cent of that of ^{40}K in the oceans and the total radioactivity of all nuclear wastes in an all-nuclear world electric industry in the steady state (10^{12} curies; 1 curie is about 4×10^{10} radioactive decays/sec) is $\frac{1}{20}$ of natural radioactivity in the top 500 m of the earth crust (10^{13} curies from ^{40}K and 10^{13} curies from U, Th, and their descendants). The radioactivity of a reactor amounts to 10^{10} curies. After shutdown it reduces to 10^9 curies in one day and to 10^7 curies in 10 years. Thereafter it has a half-life of about 30 years. A volume of stored waste, after 600 years, will have its radioactivity reduced 10^9 times to a harmless, negligible level, comparable to that of a 4-m-thick layer of uranium ore. The radiation hazard and genetic hazard will be negligible. The worst that could happen would be to ingest it: the effect would be as small as ingesting 10 times as much pitchblende, an ore. Thus the death-watch of the nuclear waste is not an excessive burden for future generations. Various ways of nuclear waste disposal have been studied, including sending it to the sun by space vehicles (it will leave earth for good). The current practice of burying it in deep salt mines is quite safe.

Finally there is the problem of plutonium as a nuclear fuel produced by breeders. Plutonium-239 is radioactive (with a half-life of 2.4×10^4 years). It may be lodged in the lung and cause cancer. It has been said that one teaspoon can kill a million people. This would be true if the teaspoon were divided into a million parts and administered to the lungs of a million people, a feat even the most skilled terrorist-extortionist could not accomplish. As a matter of fact, the nuclear tests have already distributed 15,000 lb of ^{239}Pu equivalent on the earth without killing everybody. Under realistic dispersal conditions the plutonium hazard is about one cancer death for each 5 to 10 g of plutonium dispersed in a city without warning. The expected routine release of plutonium in a breeder economy is 1 g/year, and not in a city. On the other hand, if ingested, the toxicity of plutonium is comparable to that of an equal amount of morphine, which is not a hair-raising poison. Other effects have been studied and have also been found to be small.

Nuclear Fusion Fuel. Finally, we consider the environmental effects of fusion energy utilization. Unlike the fission process, fusion based on deuterium–deuterium reaction does not lead to the production of hazardous radioactive waste material. Thus fusion energy is more desirable from the environmental point of view, for there is no problem of radioactive waste disposal. On the other hand, the reaction releases free neutrons that, when absorbed by the machine casing (possibly made of niobium), will produce radioactivity. Thus some radiation hazard will exist. Perhaps in the early stages of fusion technology development the deuterium–lithium reaction will be used; it does not produce neutrons as an end product, which is desirable. But in this reaction radioactive tritium is produced as an intermediate product, and as much as 13 per cent of it might leak from the reactor. The radiation pollution would be vastly more serious than that of fission reactors. Incidentally, neutrons are also involved in this reaction; in the first reaction tritium is also involved.

Before we leave this subject we would like to touch upon the problem of clean-up costs and their social consequences. Some pollutants can be reduced to an acceptable level, but it is likely that no reduction can be accomplished without incurring increased costs. Rational decisions will have to be made on the basis of a careful balance of benefits and risks. In the past we accepted the use of coal with 3 per cent sulfur content. To reduce SO_2 pollution, we now have set a new standard of less than 1 per cent. As a result, power plants have to import low-sulfur coal from distant places, incurring higher transportation costs. The coal cost at TVA's Widows Creek plant was \$4 a ton until 1969. In 1970 it increased to \$9 a ton. The installation of environment-related facilities increased the cost of steam as well as that of nuclear power plants. In 1965 a 1,000-Mw steam plant cost \$100/kw, and a comparable nuclear plant cost \$120/kw. Now the figures have increased to \$200/kw and \$250/kw, respectively.[2] Nuclear plants, previously constructed in four or five years, now require seven years for completion. The delay is one reason for power shortages that have hit some regions in recent years. The increased cost of power inevitably leads to increases in utility rates and consumer prices, curtailment of production, slowdown of economic progress, and further unemployment, all affecting the disadvantaged members of the society more adversely than the affluent.

[2] Alvin M. Weinberg, "The National Energy Problem," invited paper, American Physical Society, November 23, 1970.

Problems

17.1 We use electricity in the house for lighting, heating, cooking, air conditioning, operating the radio, television, washing machine, dryer, refrigerator, electric typewriter, and so on. What becomes of the energy consumed in each of these applications?

17.2 Wood alcohol (CH_3OH) and grain alcohol (C_2H_5OH) can be used as fuels for the automobile engine. Discuss the effect on air pollution of using these fuels.

17.3 The heat of combustion of coal and gasoline is 7 kcal/g and 11.5 kcal/g, respectively. Using figures in this chapter find out the per calorie costs of coal, gasoline, and natural gas. Explain why the energy cost of the electric car or the wind-up car is much lower than that of the gasoline car. Assume that gasoline costs \$0.50/gal.

17.4 In addition to the lower fuel cost and the reduced air pollution, the natural-gas-fueled car also has the advantage of saving maintenance costs and providing greater safety in a crash. Explain. (In municipal fleet operations only the police complain, for the slow acceleration allows criminals to get away faster.)

17.5 If the consequences of a major reactor accident, albeit improbable, are so grave, why not divide a reactor into a number of small reactors so small that the maximum effect of a single reactor accident would be reduced to an acceptable level or perhaps even be made controllable? In other words, why put all eggs in one basket?

18

Strategies to Meet the Energy Crisis

18.1 Elimination of Waste of Energy

The energy crisis may be dealt with from a number of angles: exploring new energy sources, improving the use of existing energy resources, and eliminating waste of energy. Development of known energy sources, such as fission, fusion, offshore oil, shale oil, and so on, has been discussed earlier and will not be repeated here.

Elimination of waste is perhaps the simplest way to begin a program of self-dependence in energy. In the past when energy costs were relatively low, we tended to use more energy than was necessary. Generally it paid to do so. For example, in building a house, rather than trying to perfect the insulation, it was possible to cut construction costs and keep the house warm in the winter by burning more heating fuel. When fuel costs are low, we can afford to do this and it may even be the cheaper way to warm a house. However, it is an unnecessary waste of energy. In a time of shortages and high prices, energy waste of this kind should be reduced. Also many trivialities introduced in the age of cheap energy, such as the electric toothbrush and the electric blanket, can be eliminated.

A large part of the energy consumed in the nation is used to heat or cool buildings. Besides the improvement of insulation other measures can be taken to reduce waste. Unused rooms can be closed and the heating or air conditioning turned off. Air flow around door edges accounts for much heat loss, and this may be eliminated by tighter fitting. Above all, heating energy consumption can be reduced greatly by the use of the heat pump, which works as a refrigerator to take heat out of the earth and put it into the house. The heat energy of the earth is free; only the electrical energy running the heat pump has to be paid for, and this part of the energy also ends up as heat in the house. The heat pump has not been used widely because of high initial cost; in a cheap energy economy, it is more economical to pay for more energy than to install the heat pump for heating and cooling.

Transportation is another major item of energy consumption and here again opportunities for saving abound. By reducing jet speed and automobile speed a sizable amount of fuel may be saved. Before the energy crisis the airplane occupancy rate was about 50 per cent—one half of the seats were vacant and thus one half of the energy consumed was wasted. This kind of waste can be eliminated by improved scheduling. The efficiency of the automobile engine can be improved. Gasoline mileage of 20 miles per gallon is well attainable. In the past, because of low gasoline cost, American auto makers did not pay special attention to gas economy; thus most American cars are gas guzzlers. Air freight has an energy cost of more than 50 times that of rail freight. Private passenger cars consume much more energy per passenger mile than the public transportation system. Much energy can be saved by making more use of the less-energy-intensive facilities.

Industrial use of power may be saved by improving the efficiency of the heat engine and by eliminating unnecessary uses of power in production. Industrial use of heat may be reduced by better insulation. In the steel industry the water-cooled skid rail system removes heat from the furnace (through the steel slabs to be heat-treated in the furnace) very rapidly. This heat loss can be greatly reduced by applying insulation to these rails. Estimates are that an expenditure of $100,000 on insulation can save approximately $234,000 worth of natural gas per year. Another loss in industrial use of heat is the loss of heat in furnaces through the exhaust. In many furnaces as much as 50 per cent of the heat generated is lost through the chimney. Part of this loss can be recovered through a heat recuperator, which uses this heat to preheat the air for combustion.

Besides using heat and power, industrial and commercial plants use energy for lighting and space heating, just as do private residences. Here possibilities for saving are also numerous. Simply adjusting combustion equipment and controlling plant ventilation have yielded fuel savings of 10 per cent or more in many industrial plants. Often, broken windows through which heated or refrigerated air leak are left unattended in plants. In one plant the energy-saving program began by replacing the broken windows, which totaled, surprisingly, several hundred. Another plant found a main source of heat loss in keeping the large loading door open while loading trucks. By changing this practice, which does not cost anything, the waste can be eliminated. In most industrial and commercial plants lighting can be selectively reduced. Hall lights can be reduced by 50 per cent. Parking lot lights often unnecessarily burn all night. Heat in the halls can be turned off. In chemical laboratories and plants, for health reasons fresh air is heated and then used once instead of being recirculated, thus consuming more heat than usual. A considerable amount of heating fuel can be saved by turning off the heating system after work and over the week end and turning it on an hour before work starts in the morning. By reducing the lunch time from one hour to one-half hour, heating hours are reduced and heating expenses can be reduced proportionately. With all these measures one industrial unit found that fuel oil savings ranged from 31 to 33 per cent and electricity savings from 23 to 30 per cent.

18.2 Gasification and Liquefaction of Coal

The current energy shortage is largely a shortage of the liquid fuel derived from petroleum. We have plenty of solid fuel coal but coal is not a desirable and convenient fuel; in particular, it cannot be used for automotive purposes. If coal can be changed to fuel materials in liquid and gaseous forms, it can be used as gasoline and natural gas. The usefulness of coal will thus be greatly enhanced and it will go a long way in alleviating the energy crisis. The processes to convert coal to liquid and gaseous fuels are called liquefaction and gasification.

Research on these processes has been pursued in many nations at different times. First we consider liquefaction.[1] The Germans developed successful coal liquefaction methods that supplied synthetic liquid fuel for cars, tanks, and airplanes during World War II. The German success aroused American interest. From 1944 to 1955, \$60 million was spent on coal conversion (also \$22 million on shale oil and \$0.5 million on agricultural waste conversion). The program stopped in the 1950's after the discovery of the low-cost Middle East oil, which could be produced at 10¢ per barrel. Under such circumstances coal conversion could never compete economically. Now that Middle East oil is neither cheap nor reliable, interest in coal conversion has revived.

There are two methods for converting coal to a state where it can be processed like crude oil to yield various petroleum products, including synthetic gasoline. The *Bergius process* accounted for 85 per cent of liquid synthetic fuel produced in Germany during World War II. Coal is ground to a fine powder and mixed with a liquid hydrocarbon that is produced in the process itself. With the help of a catalyst the mixture reacts with hydrogen at a high pressure (10,000 lb/in.^{2}) and a high temperature (850°F). The carbon in the coal is thus converted into hydrocarbons. A conversion ratio of 55 per cent is achieved. Cost estimates in American projects ranged from 20¢ to 35¢/gal at a time when the gasoline price was 10¢/gal (all prices are the refinery price, not the retail pump price). In terms of the 1974 dollar the price is perhaps 42¢/gal compared with a gasoline refinery price of 20¢/gal. It is an expensive process. It calls for a \$.33 billion capital investment and a construction time of 3 to 4 years. No Bergius plant is currently in operation.

The other method, the *Fischer–Tropsch process,* was also developed by the Germans during the war. Coal is first completely gasified (see later discussion) into a mixture of carbon monoxide and hydrogen (synthesis gas). Under the action of a catalyst at a high pressure (450 lb/in.^{2}) and a high temperature (570°F to 640°F), the mixture reacts to form an assortment of straight-chain hydrocarbons. Conversion efficiency is only 38 per cent. The cost of liquid fuel produced by this method is higher than that made by the Bergius method. Currently only one F–T plant, in South Africa, is operating, and not entirely for reasons of economy.

Because the cost of the product derived from these two industrial methods is too high, new experimental methods are being developed in an effort to bring the cost down. There are four of these: the COED process, Project Gasoline, H-coal process, and the Synthoil process. These are still in the experimental and pilot plant stage, far from industrial production. There is hope that the cost of coal liquefaction can be brought down.

The process of coal gasification[2] has been mentioned several times previously. Before 1972 the United States could produce more natural gas than it could consume, but not anymore. Gasification of coal can make up this shortage. Gasification is not new. The old-fashioned household gas used before the advent of natural gas is such a product. Such processes usually involve the reaction of coal with air or with a mixture of oxygen and steam. In the former case the product, called *power gas,* is a mixture of carbon monoxide, hydrogen, and nitrogen. In the latter case the product, called *blue water gas* or *synthesis gas,* is a mixture of carbon monoxide and hydrogen. In the 1920's there were about 12,000 air-blown gas producers in the United States, consuming perhaps 25×10^6 tons of coal annually. By the 1960's only a few in the Pennsylvania coal-producing regions were left, the others having been eliminated by the use of natural gas, which previously was a petroleum industry waste product to be burned away. Natural gas is mostly methane. The heat content of natural gas is six times higher than that of power gas and three times that of synthesis gas. Besides methane is not poisonous, whereas carbon monoxide is.

[1] Harry Perry, *Bull. Am. Phys. Soc.*, **19** (1974), 26.

[2] A. M. Squires, *Science,* **184** (1974), 340.

In a typical gravitating-bed gas producer, fresh coal is added on the top of a column of coal and ashes are collected at the bottom. Above the ashes is a narrow combustion layer into which humidified air is blown. In this layer coal is burned to form carbon dioxide and to release heat. Moisture in air reacts with coal to form hydrogen and carbon monoxide. As the coal burns, fresh coal gravitates to the combustion zone. Beyond this combustion region oxygen in the air disappears. Under the high temperature produced, carbon dioxide is reduced by carbon to form carbon monoxide. This process and the reaction of coal with water are both endothermic (heat absorbing)—this is how the heat of combustion produced in burning carbon to form carbon dioxide in the combustion layer is absorbed and stored in the product gases, hydrogen and carbon monoxide, which rise to the top and can be collected and pumped away for consumer use.

The historic gas producer, which works satisfactorily, has too small a capacity to meet current coal gasification needs and it is not easy to increase the scale for large-quantity production. One way to increase capacity is to develop facilities in which ash is removed as a molten slag. As a matter of fact, the iron blast furnace (Sec. 19.3) is a slagging gasifier. In experimental gasifiers, which resemble the blast furnace with oxygen and steam blown in, as much as 635 tons a day of hard coke can be gasified (in Germany). Although many new developments are in progress the historic producer is still a quick way to supply energy needs for small industries as a first step in coping with the energy crisis.

If coal is being mined for gasification, why not gasify it underground in the coal bed? The expenses and hazards of tunnel mining and the environmental destruction of strip mining can be avoided. Transportation costs, which are a large part of the coal price, can be greatly reduced. Theoretically this can be done by pumping an oxidizing agent (air or oxygen) down a hole to maintain incomplete combustion of the coal and collecting the fuel gas produced, comprised of H_2 and CO, from another hole. Work on underground gasification has been going on for a century and peaked between 1945 and 1960. The process competes well with other fuels except the cheap Middle East petroleum. In a U.S. Bureau of Mines experimental installation currently in operation in Hanna, Wyoming, air is used as the oxidizing agent and the product gas consists of 21.4 per cent H_2, 4.5 per cent CH_4, 9 per cent CO, 17.1 per cent CO_2, 46.4 per cent N_2, 0.11 per cent H_2S, and 1.5 per cent others by volume. Oxygen can now be produced more cheaply and may be used as a better oxidizing agent. The mixture of H_2 and CO can also be used not as a gaseous fuel, but as a chemical raw material to make methane, methanol, or straight-chain hydrocarbons (the F–T process). There are environmental hazards—contamination of underground water and subsidence of ground surface. The process is ideal for coal seams deep underground for which conventional mining is not feasible.

18.3 The Hydrogen Economy[3]

The coal gasification and liquefaction processes discussed earlier represent ways to make better use of conventional energy resources. In this section we are concerned with better ways to transport and store energy. Even though no new energy sources are involved, they may greatly help alleviate the energy shortage.

Hydrogen is a combustible gas. In the combustion process it combines with oxygen to form water and releases a large amount of energy. The amount of energy released per gram is highest (1.25×10^5 joules/g) among all fuels (because hydrogen is lightest in weight). It is thus a potent fuel. The combustion product, water, is harmless. It is thus a clean, pollution-free fuel.

But hydrogen is not a primary energy source. It is produced, say, by electrolysis of water, which con-

[3] D. P. Gregory, *Bull. Am. Phys. Soc.*, **19** (1974), 57.

sumes energy—exactly the same amount of energy that is released in combustion. Thus the energy delivered by the hydrogen fuel is actually the electrical energy consumed in electrolysis and stored in the hydrogen. Like many other chemical energy sources, hydrogen is an instrument to utilize primary energy sources in a more useful and convenient way.

What are the major advantages of using energy through hydrogen besides its high energy content and pollution-free nature? It is here that the issues of transportation and storage of energy make themselves manifest. A hydrogen-based economy is considered to be best suited to deal with these problems.

It is not generally realized that transportation of energy is costly. A large fraction of the cost of coal is transportation, solid fuel being the most costly to transport. Liquid and gaseous fuels can be transported by pipelines, a much less expensive method. Even the transport of electrical energy costs dearly. From the generating plant to the consumer, the energy is transported through high-tension transmission lines. Electric energy is partly lost through the resistance R of the transmission lines at the rate of I^2R, where I is the current. To reduce this energy loss the voltage V is raised high (thus "high-tension") to keep the current I low (the product VI is a constant because of conservation of energy). However, the longer the distance, the larger the resistance R (Ohm's law) and the larger the energy loss in transit. About 12 per cent of electric energy is lost through transmission; this amounts to a waste of 1 million barrels of crude oil a day

One way to reduce the energy loss is to reduce the resistance R. Resistance decreases with the temperature. One approach is to cool the transmission lines by using liquid nitrogen to reduce the resistance. Another approach is to cool the transmission lines by using liquid helium to make them superconducting. (The resistance is practically zero.) The transmission energy loss can be eliminated, but energy has to be spent to maintain the low temperature by refrigeration. Research is under way, and preliminary results show that refrigerated transmission lines are feasible and economically competitive. These lines are buried underground, which has the environmental advantage of eliminating the landscape-spoiling high-tension lines. In the future, with increased use of electric energy there may be problems in finding room for additional aboveground transmission lines.

Hydrogen is a gaseous fuel and can be transported by pipelines. The energy transport capacity of a 3-ft-diameter pipeline is 11 times higher than that of a 500-Kv overhead transmission line and the cost is a fraction of that of the latter. The same amount of energy is thus delivered by hydrogen at a cost that is more than 10 times cheaper than that of electricity. Even including the cost of conversion of electric energy to hydrogen energy at the generating plant, there is still a saving from transportation. The lower transmission cost removes a serious restriction in choosing the site of the generating plant, which is often a troublesome problem.

Hydrogen can be stored in tanks to be used anytime, whereas electric energy cannot. Electric generating capacity must be designed to meet the peak load. In off-peak hours the capacity stands idle and the average rate of use is only 50 per cent, which is a great waste. Hydrogen energy can be stored to even out daily variations in energy needs and the power plant may be utilized 100 per cent. The storage cost of hydrogen gas is two to three times that of natural gas. The pipeline itself is a storage unit and may be tapped in time of emergency. With these advantages of transportation and storage, it seems natural that a hydrogen-oriented economy may replace the present electricity-oriented economy for energy delivery in the future.

If hydrogen is the principal form of energy delivery in the future economy, how will electric appliances be run? Electricity will be generated at the site of consumption from the energy of hydrogen (transported from a primary power station by pipeline) by *fuel cells*. A fuel cell is a device that converts the chemical energy of a fuel directly into electric energy, and hydrogen can be used as such a fuel. The fuel cell is similar to electric cells and batteries, but the latter cannot use conventional fuels. Fuel cells have been developed and used for military and space applications where conventional electric power is not available. They use expensive rare metals and thus are still costly. Current research aims to bring the cost of the fuel cell down for wider uses. Research on batteries is also in progress. In general, fuel cells are for electric energy generation and batteries are for electric energy storage. The consumption of hydrogen in fuel cells, of course, does not generate air pollutants as conventional generating plants do.

Electric energy accounts for 25 per cent of our total energy consumption. The other 75 per cent is for heat production, either for various heating purposes or for powering heat engines. For these purposes hydrogen can be used readily. It can replace natural gas for all heating purposes. It can be used in external and internal combustion engines (with NO_x emissions much lower than those of gasoline engines). The difficulty in using hydrogen for the automobile is not that of powering the engine (unmodified auto engines can be run on hydrogen; better performance may be obtained by adjusting carburetor and ignition), but rather that of storage in the automobile. With several hundred pounds of hydrogen storage tanks filling up the trunk space, the fuel is enough for only a few dozen miles, comparable to the range of an electric car. Hydrogen has a very low critical temperature and thus cannot be liquefied without low-temperature facilities, which is impractical for a car. It seems more feasible that airplanes may be equipped with low-temperature facilities for carrying liquefied hydrogen as the engine fuel.

The use of hydrogen is not without hazards. Since the Hindenburg airship disaster in 1937 hydrogen has been looked upon with suspicion as far as safety is concerned. However, hydrogen has been widely used in industry. Leaks may be detected by odorization, a practice now applied to natural gas.

Hydrogen may be produced from water in three different ways: by electrolysis, by thermal decomposition, and by thermal chemical processes, for example,

$$H_2O + X \longrightarrow XO + H_2$$

$$XO \xrightarrow{>1{,}100^\circ C} X + \tfrac{1}{2}O_2$$

all with oxygen as the by-product. (In the above reaction X is a reducing agent.) There are many other ways of producing hydrogen. A hydrogen economy thus has versatility in meeting shifting primary energy sources in the next half century. In the distant future when the primary energy source will be essentially nuclear, hydrogen may be produced from the electric energy generated or even from the thermal energy of the reactor directly by thermal chemical reactions. At present, industrial hydrogen is produced from natural gas. Oil shale and low-grade fuels that cannot otherwise be utilized may be used to produce hydrogen. Solar energy may also be tapped through the thermal chemical processes for hydrogen production.

An additional advantage of hydrogen as a heating fuel is that because its combustion product is harmless, there is no need for a chimney. Not only can the cost of the chimney be eliminated, but 30 per cent of the heat lost through chimney exhaust can be saved.

18.4 Solar Energy

Solar Heat. Solar energy is the largest and the truly inexhaustible energy source on the earth. Naturally, various ways of tapping this energy have been contemplated. Direct conversion to electric energy by photovoltaic cells, though most desirable, is still too costly. Single-crystal silicon solar cells have at least 10 per cent efficiency in converting sunlight to electric energy. Installed solar-cell facilities cost more than $30,000/kw of electricity, which is much more expensive than the $200/kw for a coal electric plant. However, using solar energy to heat a house can be done with relatively simple technology. Some 20 experimental solar-heating houses have been built in the last two decades and some half dozen are being built now.

The principle of solar heating is simple enough.

Use a heat-absorbing material to make a solar collector to gather solar heat on the roof. Let water pass by it to remove the heat and transport it to a heat storage tank (e.g., a 1,600-gal water tank) in the basement. Blow air over the tank or the surface of a heat transfer material (e.g., 50 tons of stone surrounding the hot-water tank, which also acts as the heat storage tank) and circulate it to heat the house. More sophisticated designs are under development. The special feature of the system is the use of large volumes of low-temperature (75° to 80°F) and low-velocity air. Conventional heating systems use a small volume of high-velocity, high-temperature (140° to 160°F) air.

In general, a solar collector to be used on roofs is a flat metal surface coated black to absorb solar heat. The absorbing surface is in contact with a circulating heat transfer fluid (such as water) to transport the heat to a heat storage unit. The collector is insulated on the roof side to prevent heat loss. It is also insulated on the sun side by one to three layers of infrared opaque glass (using the greenhouse effect mentioned earlier).

To use solar heat for generating electricity, more elaborate collectors are required. One kind is a parabolic reflector used to concentrate the sunlight on the focal line. The reflector must be rotating at the same rate as the sun to keep focusing all the time. The temperature of the heat transfer fluid may reach 150° to 315°C. The collector efficiency is about 50 per cent. In another method many two-axis mirrors are used to track the sun and reflect the light on an absorber-boiler on a high tower. A temperature of 500°C may be reached. High-temperature heat thus collected may then be used in a heat engine to generate electricity. Flat-surface collectors may also be used for this purpose, but the temperature reached is only about 150°C. Remember from the second law of thermodynamics that high-temperature heat is more useful for conversion to mechanical work. From solar energy both high-temperature and low-temperature heat may be collected, the former more costly and more useful. The latter is used for space-heating purposes, not for energy conversion.

Solar-energy-operated air conditioners are under development by a number of companies, and commercial units may be available soon. In both heating and cooling the use of solar energy is restricted by weather conditions, a major drawback of solar energy. In most cases a back-up system for heating and air conditioning is necessary. Nevertheless the use of solar energy definitely reduces fuel consumption for indoor climate control purposes.

The Autonomous House.[4] The solar-heating idea is a part of the concept of an autonomous house that is free from dependence on centralized utility systems and thus is not vulnerable to fuel shortage, strikes, blackouts, brownouts, and so on. It is the fulfillment of a desire of man to be his own master. From the point of view of resources and environment it also provides the promise of regenerating energy and recycling material. Although heating by solar energy is not a great problem and even solar air conditioning is realizable, it is more difficult to be self-dependent for electric energy generation.

One way to be self-reliant in the area of electricity is to make use of wind energy. Windmills have been in use for a long time. Wind-power-operated electric generators are not new. Research on large-scale utilization of wind energy is under way. In meeting the needs of a household the problem is how to cope with unpredictable weather conditions. Ordinarily a set of storage batteries is needed to store electric energy when the wind is blowing and to supply electric energy when there is no wind. A back-up system may still be needed. It is possible to hook up with the utility network to sell extra electric energy to the utility when the wind is blowing (meter running in reverse) and to draw power from the utility for back-up and for emergency. In this manner the utility network acts as the storage unit and the expense of batteries can be eliminated.

From the environmental point of view an autono-

[4] Allen L. Hammond, "Individual Self-sufficiency in Energy," *Science*, **184** (1974), 278.

mous house is supposed to recycle its waste products; from these methane gas may be produced that can be used to generate electricity. Thus the autonomous house is an integrated system encompassing several interlocking components. Household wastes are of two kinds: kitchen waste and human waste. Both can be composted, say, with a Swedish clivus system, which converts them to a dry mulch that is an excellent fertilizer. This system enables a house to be completely independent of the centralized sewage system. Wastes are digested aerobically by the circulation of air. The turnover period is two to three years. Another way to recycle the organic waste is to build a *methane digester,* which converts waste anaerobically to methane gas, which can be used as a fuel. The gas produced from the organic waste of a house is not enough to heat the house, but may be enough for cooking purposes. Currently no commercial methane digesters are available in the United States, but it is possible to buy complete plans to build one.

Solar Energy by Photosynthesis.[5] As mentioned previously, plants make their own fuel molecules by photosynthesis, and man has been using solar energy indirectly through plants in many ways. Could this avenue of solar energy utilization be expanded? Considering the large amount of inexhaustible solar energy available, this may be one effective way to deal with the energy problem.

On the earth the area with the largest rate of solar energy input is the Sahara Desert, about 280 w/m^2. (The solar constant is about 1 kw/m^2 but that is the energy input before entering the atmosphere.) Northern New Mexico is not far from this rate (about 260 w/m^2). However, it is in the equatorial area that solar energy is converted by plants at the highest rate (producing 1 kg of organic carbon per square meter per year, mostly carbohydrates, and essentially sugar) because of an abundance of water.

[5]Melvin Calvin, "Solar Energy by Photosynthesis," *Science,* **184** (1974), 375.

The natural efficiency of solar energy conversion by plants is only a few hundredths of 1 per cent. However, some plants have high efficiencies. The most efficient is sugarcane, with an overall efficiency of 0.6 per cent.

Sugar can be used as an industrial raw material. About 6 per cent of the total energy consumption in the United States is for nonenergy purposes—as industrial raw material—which cannot be supplied by other energy sources such as nuclear energy but may be supplied from agriculturally produced organic carbon. Of course sugar may be burned as a fuel—a rather extravagant way of consuming a precious commodity that can at least be used as food. A better way of using it for energy is to convert it into ethyl alcohol to be used as a motor fuel (more later) to alleviate the gasoline shortage. For this purpose cane cellulose can also be converted to alcohol and the per acre fuel yield can be doubled. It takes 12.9 lb (64¢ in 1973) of sugar to make 1 gal of alcohol. The chemical conversion costs 20¢. Thus the production cost of alcohol by fermentation is about 84¢/gal. This cost is nearly competitive with the current retail price (about $1/gal) of alcohol produced from ethylene as a petrochemical product.

Sugarcane yields 4 tons of sugar per acre per year from which we can produce 2 tons of alcohol and 1.2 tons of ethylene. The overall solar energy conversion rate is about 0.23 per cent. Including the products from the cellulose, which is present in the cane in about the same amount as the sugar, the total efficiency of conversion can be raised to about 0.5 per cent.

Sugar and cellulose are carbohydrates (carbon, hydrogen, and oxygen). The most desirable fuels are hydrocarbons (carbon and hydrogen). For fuel purposes it would seem more desirable to grow hydrocarbons directly. The rubber plant does just this, for rubber is already a hydrocarbon. The yield of rubber is about 1 ton per acre per year, and there is hope this may be raised to 3 tons per acre per year.

If the equivalent of 2 million barrels per day of oil consumed for nonenergy purposes (for chemicals and materials) were to be supplied totally by sugarcane, making use of both the sugar and the cellulose, we would require 30 million acres of plantation. The actual sugarcane and sugar beet acreage is about 0.7 and 1.4 million acres, respectively, in the United States.

Alcohol as a Motor Fuel.[6] Ethyl alcohol (grain alcohol, also called ethanol) can be used as a motor fuel. From 1940 to 1945 ethyl alcohol was mostly made by grain fermentation, the age-old method of making liquors. Around 1950, when cheap petroleum products became available, industrial alcohol manufacturing was converted entirely from fermentation to petrochemical processes (from ethylene).

With the petroleum shortage the fermentation method may again become economical. In addition, spoiled grain can be utilized by making it into alcohol. Nebraska has 7 million bushels of spoiled grain per year, which may yield 20 million gal of alcohol. Nebraska has a law giving 3¢/gal tax credit to "gasohol," a gasoline with a 10 per cent blend of alcohol, which performs just as well as or even better than regular gasoline. This amounts to a subsidy of 30¢/gal for alcohol as a motor fuel. At the current price alcohol is still not economically competitive with gasoline. However, as gasoline prices increase further, alcohol will soon become competitive.

Not only grain alcohol C_2H_5OH, but also wood alcohol CH_3OH (methyl alcohol or methanol) can be used as a motor fuel. Methanol was originally produced by distilling wood. In the nineteenth century it was used instead of wood in Paris as a fuel for heating, lighting, and cooking to save the trouble of transporting the wood to and removing the ashes from the city. Now very little methanol is produced by distillation. But the method can easily be revived in an energy crisis and much forest waste can be converted to desirable fuel in this way.

Both methanol and ethanol by themselves can be used to run automobile engines without any adjustment of the automobile. This knowledge may be helpful in meeting local emergencies of gasoline shortage. However to use them as a regular motor fuel, some minor adjustment (cost about $100) of the engine is desirable. These compounds already have an oxygen atom combined in a hydrocarbon. Their combustion needs less oxygen and generates less heat than that of gasoline. Their "gas mileages" are reduced by one half for methanol and by one third for ethanol. A richer air-to-fuel mixing ratio (6 instead of 14 for gasoline) is needed to generate the same amount of energy. In some cases this can be done simply by applying the choke. Because more oxygen is available than in a gasoline engine, the combustion is more complete and air pollution is reduced.

A better way of using them as motor fuels without engine modification is to mix them with gasoline. Methanol of 5 to 30 per cent may be blended into gasoline. When used in unmodified cars, the performance of the blended gasoline appears to be better than that of regular gasoline: fuel economy is increased by 5 to 13 per cent, with the maximum increase at the 15 per cent blend, exhaust temperature is decreased by 1 to 9 per cent, and acceleration is increased by as much as 7 per cent. Moreover, CO emission is decreased by 14 to 72 per cent. Thus the use of blended gasoline helps reduce air pollution. The same is true in burning pure methanol; among the air pollutants produced the unburned fuel is reduced to one twentieth, the CO amount is reduced to one tenth, and the oxides of nitrogen remain about the same. Thus the use of alcohol as a motor fuel enables us to solve the energy and the environmental crises at the same time. The results with ethanol are generally the same—blends of up to 33 per cent have been used, with the optimum at 10 to 15 per cent. Pollution is greatly reduced.

Because alcohol from agricultural sources can be regenerated, it represents a source of motor fuel that is not threatened by depletion. In the not too distant future when petroleum resources are depleted, alcohol will be the most visible substitute, and it can be renewed indefinitely. Another important advantage is that it can be used in present automobiles; other forms of energy, such as nuclear energy, hydrogen energy, electric energy, and so on, would need

[6]T. B. Reed and R. M. Lerner, "Methanol: A Versatile Fuel for Immediate Use," *Science,* **182** (1973), 1299.

completely different automobiles (either new engines or new storage facilities), and the current investment in automobiles (over $100 billion) would be largely wasted. However, the agricultural source of alcohol is limited. If all the present corn acreage in the United States were used for grain alcohol production, it would meet only 14 per cent of current gasoline needs. However, by converting cellulose to ethanol the potentiality of this energy source can be greatly enhanced. Moreover, methanol produced from methane generated from waste material may provide 8 per cent of gasoline needs. Thus in the future the auto culture can be perpetuated indefinitely, even though on a reduced scale. It will not disappear together with the petroleum sources.

In 1972 methanol production in the United States was about 1 per cent of the volume of gasoline production, with an average price of 18 cents per gallon. Because it takes 2 gal of methanol to generate the energy of 1 gal of gasoline, the fuel cost is 36¢ (1972 prices) for the equivalent of 1 gal of gasoline, which is economically competitive. The octane number of methanol is 106 and that of ethanol is 99. Ethanol in this country costs about three times more than methanol but its heat of combustion is about 50 per cent higher than that of methanol.

About 50 per cent of photosynthetic products is carbohydrate, much of which is in the form of cellulose. Cellulose is inedible; most of it is burned either in its original form (leaves, branches, sawdust) or after being used for specific purposes (used furniture, worn-out fabric, used newspaper, discarded containers). If cellulose can be better utilized, then a large part of the solar energy captured by photosynthesis can be made more useful. The molecular formula of cellulose, like that of starch, is $(C_6H_{10}O_5)_x$. It is conceivable that the age-old chemical process of converting starch to glucose $C_6H_{10}O_5$ to ethanol can be reproduced for converting cellulose to glucose to ethanol. Indeed, this can be done. Recent experimental work[7] at the University of California shows that alcohol can be produced cheaply from waste material. The process makes use of an enzyme, cellulase, derived from a common fungus. This strain of fungus was bred by the U.S. Army in experiments related to the rotting of tents. The enzyme breaks down most fibrous material into glucose. Assuming the raw material, waste cellulose, to be free, it is estimated that glucose could be produced at a cost of 1¢/lb compared with the sugar price of early 1974 of 13¢/lb. The ethyl alcohol so produced would cost 30¢/gal. Used as a motor fuel in pure form the energy cost equivalent to 1 gal of gasoline would be 45¢, which is already comparable to the current gasoline price (used as a 15 per cent blend, the equivalent cost is less than 30¢/gal of gasoline). (Note that price is cost plus retail mark-up and taxes.) Besides, air pollution would be greatly reduced.

[7]G. Mitra and C. R. Wilke, Lawrence Berkeley Laboratory Report LBL-2334 (Berkeley, Calif.: University of California, February, 1974).

The waste paper generation rate is about 1 lb per person per day. From the cellulose in the readily available solid wastes, 13 billion gal of alcohol can be produced annually, equivalent to 10 per cent of the annual U.S. gasoline production. In the future fast-growing plants may be raised and the cellulose harvested may be used for making fuel alcohol. This way the cost of alcohol would be increased by 12¢/gal. When petroleum is depleted, alcohol may be used as the raw material to make ethylene, and then plastics, and so on. From glucose we may also make synthetic protein for animal feed to increase the world food supply.

Both methanol and ethanol were used as motor fuels during World War II and earlier in Germany, France, China, the Phillipines, and other nations. In 1938 there were 9,000 wood-burning automobiles used in Europe. Wood chips were distilled to make methyl alcohol vapor mixed with carbon monoxide and hydrogen, which is barely powerful enough to propel the vehicle. In wartime China "charcoal cars" were used with a cylindrical charcoal furnace, generating "gas" by incomplete combustion to power the auto engine. The charcoal car took a snail's pace in negotiating a moderate slope. When the engine was about to fail, an assistant jumped out of the driver's cab and quickly inserted a chock under the front wheel. Then came a time to build up a head of gas, marked by a rush of bellowing noise, which energized the vehicle to dash for several feet, only to stop and rally again. Nevertheless, it was the first automotive application of gasification of charcoal. In contrast, the ethanol-powered trucks (mostly American Dodges) could roam over hills unassisted and were the envy of all wayfarers on the highways.

Since 1925 methanol has been produced by synthesis of CO and H_2, which are obtained by partial oxidation of methane with water. Methane gas at the wellhead in the Middle East is often discarded. Plans are under way to use the methane at the wellhead to make methanol to be used in this country (6 cents per gallon). This is more economical than transporting the methane gas by refrigerated tankers in liquefied form (46 per cent more expensive per calorie of energy). Methane can be generated from solid waste and organic waste. The use of methane from such sources to make methanol helps with not only the energy problem, but also the environmental problem of waste disposal. Because methanol can be made from wood and forest waste material, it also provides a way to utilize solar energy by way of the forest as an "energy plantation." Commercial forests now cover 23 per cent of the area of the United States. It has been estimated that 5 to 20 per cent of our commercial forest used as energy plantations can supply all our energy needs for electric power generation.

18.5 Exotic Energy Sources

Geothermal Energy.[8] Ordinarily when we mention geothermal energy we mean the heat energy in *hot springs* and in *natural steam* that can be utilized in an obvious way. As such the amount is insignificantly small. However, there is another aspect of it, and with an eye on the future, geothermal energy presents a different picture. The interior of the earth is very hot, with a temperature above the melting points of rocks and alloys (this will be discussed in Chapter 24). The size of the earth and thus the total amount of heat energy in it is very large from a human standpoint. In fact, the heat energy in 40 cubic kilometers (km^3) of earth material is equivalent to all the energy consumed in the United States yearly. And the volume of the earth is about 10^{12} km^3, containing the equivalent of 25 billion years of use of energy. If this energy could be utilized, even a small fraction of it, it would provide a sizable part of our energy needs. Moreover, it is clean energy—no combustion process is involved, and so no chemical and nuclear pollutants would be released. However, the heat energy readily available near the surface of the earth is not very useful because the temperature is too low (the Carnot theorem). On the other hand, the high-temperature heat in the interior of the earth is not readily accessible. Thus this energy source cannot be easily utilized and was not seriously considered previously.

Nevertheless, the use of natural steam is a simple, economical process. In regions where natural steam is present, a hole is drilled to reach the steam, which then issues and can be directed to drive a turbine to generate electric power. However, such regions are quite rare and this source at best can only fulfill local energy needs. There are also chemical and environmental problems involved in using underground water in this fashion.

Underground superheated water is much more abundant than natural steam and may satisfy a sizable fraction of our energy needs. When brought to the surface at a reduced pressure, part of the superheated water is converted into steam, which can then be used to drive a turbine. However, not all the hot water is converted into steam. The part that is not is usually discarded, wasting probably one half of the original heat energy of the superheated water. Further, the minerals in the hot water discarded cause chemical pollution and the heat

[8] Morton C. Smith, *Bulletin of the American Physical Society,* **19** (1974), 98.

causes thermal pollution. The withdrawal of a large amount of underground water may cause the subsidence of the earth surface, which is another environmental problem. It would seem that the best way of dealing with the unused hot water is to pump it back to the underground source. In addition to underground hot water, lava pools and magma chambers (Chapter 24) are also large reservoirs of geothermal energy that may be more easily utilized.

Hot spring water often contains corrosive minerals that are damaging to the equipment making use of the geothermal energy. During the flashing process when superheated water is converted into steam, the pressure and temperature are lowered. As a result the solubility of minerals is reduced and some minerals may precipitate, clogging the pipes and fixtures. Drilling to reach underground superheated water, natural steam, and magma chambers is a hazardous process. The hot material that issues in this process poses a danger to the personnel working on the drilling.

The energy of the underground hot water comes from the hot rocks deep in the earth. The occurrence of a hot spring on the surface, which is connected with the deep hot rocks through cracks and channels, is a geological accident. To utilize the heat energy of the deep hot rocks—the vast reserve of clean heat energy in the earth interior—we may artificially create a hydrothermal system simulating the natural system. In principle we can drill a hole to the hot rocks, pour in cold water, and collect the hot water and steam from another hole after the water is heated by the hot rocks. A hole 15,000 ft deep will reach a region with a temperature of 300°C. Oil wells have been drilled to such a depth at such a high temperature. The high temperature makes conventional drilling processes more difficult, but it may be a feature that can be taken advantage of. A new drilling process that melts hot rock is under development and may enable us to reach greater depths.

Unlike other mineral reserves, geothermal energy is accessible from any point on the earth and thus is available to all nations. However, many practical difficulties remain to be solved before this energy source can be economically utilized. Considering all the merits, this source probably will figure importantly in the energy picture in the future.

Solar Sea Power.[9] Solar energy is an extremely large energy source, but we are still far from being able to use it for electric energy generation in an economical way. The idea of solar sea power is to use the surface of the tropic oceans as a solar energy collector (albeit an inefficient one) to act as the "boiler" and the deep (several thousand feet) cold water brought to the surface as the "condenser" to run a heat engine. We learned from the second law of thermodynamics that there must be a temperature difference between two heat reservoirs before the heat energy in the high-temperature reservoir can be converted into mechanical energy by a heat engine. The concept of solar sea power exemplifies this principle. The temperature difference is about 20°C. According to the Carnot theorem the heat engine so operated will have a maximum efficiency of 6.6 per cent. The actual efficiency is probably about 2 per cent. Here the engine is run in a very inefficient way, a condition never actually seen in current industrial applications. However, by virtue of the Carnot theorem, there is no reason why an engine cannot be run in this way. Because solar energy is free and the total energy input is tremendously large, even a low-efficiency engine may yield a sizable energy output. The conversion of a small fraction of the solar energy absorbed is enough to supply all our energy needs. A 22-kw electric energy plant using solar sea power was operated in 1929.

It has been claimed that the capital cost of such a power plant, which floats in the ocean, might be reduced to an amount comparable to that of a fossil plant of the same output. The electric energy generated would be used on the site for the electrolysis of water to produce hydrogen (Sec. 18.3). Hydrogen can be piped to land to be used as a fuel or can be used on

[9] Clarence Zener, *Bulletin of the American Physical Society,* **19** (1974), 98.

site to make ammonia as a fertilizer to be shipped to all parts of the world. These uses take advantage of the ocean location of the solar sea power plant.

Another way of tapping solar energy is to use the kinetic energy of the ocean current, such as the Gulf Stream, which is also enormous. A research project is under way.

In conclusion we may say our problem seems to be that we have too many options available for dealing with the energy crisis. We are already in hot debate as to which is the best course to take. In addition to technical and economic considerations, environmental impacts complicate the issue greatly. This situation prompted Abelson in an editorial in *Science* to tell the following story: Once there was a jackass who found himself exactly between two haystacks. Unable to make up his mind which way to go he starved to death.

Problems

18.1 What percentage of electric energy can be saved if instead of electric heating we use the heat pump for heating? Assume that the heat pump works at maximum performance, the winter outdoor and indoor temperatures being 0°C and 20°C, respectively, for the sake of calculation.

18.2 In the process of coal gasification show that both mass and energy are conserved. Explain how the mass and energy are transformed so that the solid fuel coal is converted into a gaseous fuel. As far as fuel energy is concerned, is there any loss in the transformation? Where does the loss occur?

18.3 If electric power is not transmitted by a high voltage of 500 kv but by the voltage at which it is used, i.e., 110 v (over a short distance, say), how many times would the transmission energy loss be increased?

18.4 What is the rationale of using daylight saving time in January, as was done in 1974, when more lighting energy will be consumed in the early morning hours? (*Hint:* To even out peak load at around 5 P.M.) Would this be necessary in a hydrogen economy?

18.5 If 10 per cent of the solar energy intercepted by the roof of a house is converted into electrical energy, what is the average power in kilowatts? Assume a typical house of 50 ft by 30 ft and a solar energy input rate at the roof of 15 per cent of the solar constant. Is this enough to supply the household needs for electric energy, assumed to be 2 kw on the average? How can we even out the rate of energy production and rate of energy consumption in the 24-hour period?

18.6 Explain the working of the insulation of the solar collector on the sun side.

18.7 When underground superheated water is brought to the surface, why is part of it converted into steam? Why only part of it?

18.8 Why is it that a natural-gas-fueled car has a decent range of about 100 miles for a full "tank" of fuel but a hydrogen-fueled car has not, since natural gas (mostly methane with a critical temperature of -82°C), like hydrogen, cannot be liquefied in cars?

18.9 Would a butane-fueled car have a longer range than a natural-gas-fueled car? Why? Use all information available before looking up the critical temperature of butane.

19

Material Resources and the Materials Industry

19.1 Material Needs

Just about the only thing primitive man needed was food. Food supplies the material substance for the growth of body and repair of tissues; it also supplies energy for muscular and other physiological activities. After man learned to use his hands in a more sophisticated way, which led to the development and use of tools, human needs expanded greatly. Clothing, shelter, and so on, were added to the list. At the same time, energy needs multiplied. Additional energy was required for manufacturing articles to fill material needs and for newly introduced energy-oriented enterprises such as transportation, communication, and indoor climate control. Here we first analyze man's material needs and then discuss the industries that fulfill these needs.

Next to food, clothing and shelter are the most obvious things that man cannot do without. However, there are aspects of clothing and shelter—such as decorations—that man can do without. Thus a distinction between necessities and luxuries has been made among the material needs. The material goods are not free; their acquisition requires effort and costs money, sometimes even to the extent that man may engage in combat, either individually or collectively. This often leads to disappointment, frustration, or disaster. Therefore from the inception of civilization wise men and philosophers, typified by Diogenes, have advised austerity: reducing material needs to the bare necessities. The distinction between a necessity and a luxury thus became important. The former was regarded as a legitimate claim, and the latter was considered an undesirable demand. In spite of this, the list of luxuries kept on growing as civilization progressed.

The development of modern capitalistic society changed this attitude. It was recognized that increased consumption was necessary for economic development and that austerity hindered expansion. In a period of lagging consumption, it was deemed desirable for the government to expand its activity to create the necessary demand. Because man can only eat three meals a day and wear one suit at a time, the increased consumption had to come in the area of luxuries. For the first time in history, luxury items were given a more respectable place and liberal spending was advocated timidly from an economic, if not an ethical, point of view. (Commercial advertising has advocated it all along, of course.)

This reversal in attitude had profound consequences, not only in determining personal conduct but also in formulating social standards. The age-old virtues of thrift and saving gave way. Making more and spending more became the prescription for prosperity. In the past, goods were made to last. Artisans were proud of the workmanship and durability of their products. Now durability became an evil because it entailed less spending. Appliances and automobiles were not built to last; they even had limited lifespans deliberately built into them. In this way the total volume of consumption and thus the gross national product could be increased.

Although this modern attitude is no more than 40 years old, another reversal is in the offing. Increased consumption means accelerated exploitation of natural resources and increased degradation of the environment. Shocked by and suffering from the pollution of air, water, and land, environmental advocates have denounced technology as an evil. The hero of yesterday now becomes a villain. Simplicity of life is again advocated. The distinction between necessities and luxuries is emphasized again. The automobile, once an idol, is considered a demonic scourge of urban life.

But where is one to draw the line separating necessities from luxuries? To avoid confusion, let us define a luxury as *anything that is not necessary for subsistence*. For Diogenes the necessities consisted of a barrel to live in and a bowl to drink from. Even the bowl became a luxury (unnecessary) when one day he saw a boy fetching water with the palms of his hands. The modern list of totally undisputed necessities is much longer. Prisoners are demanding air conditioning as a necessity, and their demands are falling on sympathetic ears. From a scientific point of view, man did not have even clothing and shelter in the beginning. It was not too long ago that they entered human life. Before that they were by definition unnecessary and therefore luxuries. How did they become necessities? Anthropologists suspect that primitive clothing began as a kind of ritual dressing of the body, which is a type of luxury. Once clothing was in use, its function as a means of protection against weather became recognized, and with it man was able to move into areas of harsher climate. Once he was there, clothing became a necessity. Its conversion from luxury to necessity is associated with its changing role in human life accompanied with social advancement. The transformation of luxury to necessity may thus be regarded as a general law of evolution. If evolution means progress—advancement from the simple and primitive to the complex and sophisticated—then, among other factors, the transformation of luxury to necessity plays an important part in this process.

In fact, many necessary items of the present day, such as watches and fountain pens, were unnecessary in agricultural societies. Thus they were luxury items that were used largely for decoration or as status symbols. When appointments were made by the year and season ("I will see you next spring"), not by the hour and minute ("Meet you at 9:45 A.M. tomorrow"), a watch had no practical significance; it was a mere novelty. It is now a necessary instrument; a difference of a few minutes often means the difference of a few dollars. A peasant engaged in farming may look upon the watch and the fountain pen as trivialities to be despised. But when he becomes a guerilla fighter, he finds them mortally useful, and they become treasured necessities. When first invented, such "luxury items" may be looked upon with suspicion by the moralist as being nonessential and only contributing to human woes. Yet when the moment comes, they can play a significant role in bringing about epoch-making changes.

As a general law of evolution, the transition from luxury to necessity has taken place from the beginning of life to this day. Once a bacterium has acquired the ability to use glucose, a kind of sugar, for energy, there is no absolute necessity for it to acquire the additional ability to digest lactose, another sugar. Thus the genes that make the enzyme β-galactosidase, used to break down lactose, are a luxury. Yet the environment may change: if the glucose supply becomes depleted, organisms with a more versatile sugar metabolism are better able to survive. It is this evolutionary process that causes genetic operon after genetic operon to be incorporated into the genome (genetic information system), endowing the organism with more various, more complex, more sophisticated abilities to deal with basically the same—and sometimes redundant—biological needs. The number of genes increases from 10^3 for a bacterium to 10^6 for a man. The increased complexity may have the undesirable effect of making the organism more fragile—there are more ways the whole system can go wrong. Nevertheless, a complex organism usually contains repair and immunological systems that tend to compensate for this shortcoming. Above all, the complex system is more versatile, and history proves it to be superior. The same may be said of social and economic systems. Not all innovations are worth preserving. But given a chance, the evolutionary process will select the useful ones and incorporate them into the social inheritance. The direction of social evolution is still from the simple to the complex. Underdeveloped nations are all eager to catch up; spears and arrows are being replaced by guided missiles. As of now, the increased complexity of society is still small compared with what has happened in biological evolution.

One case in which the evolutionary law manifests itself most obviously is the use of the automobile. In a remote Shangri-la the automobile is merely a novelty. There are no serviceable roads, motels, drive-ins, or shopping centers to make it useful. Once a rich Iranian bought a Cadillac. But he and his family still walked, and the expensive car was kept in a "house" covered with satin blankets. It was treated as a piece of jewelry, not as a tool of transportation. In developed nations just about everyone is likely to come to work in his own car: the automobile has become a necessary tool of transportation. Once the automobile becomes available, other economic activities sprout from it. Thus arise the drive-ins, motels, shopping centers, and, most important, suburbs. After the new institutions have taken over the functions of the old system, the automobile becomes an absolute necessity. Public transportation degenerates into such an insignificant and deplorable state that one is forced to keep an automobile for his own transportation needs, like it or not. After the family car becomes a necessity, one can still indulge in the luxury of the auto culture by keeping a second car. The development of suburbia is so geared to the use of automobiles that even the second car becomes a necessity for suburbanites, because there is no more corner drugstore or grocery nearby, and a housewife has to drive to a shopping center that is so distant that it would have taken her a day's walk just to buy a book of matches. The increase in the complexity of life is obvious. Whether these changes are desirable is a separate question. For one thing, these changes are associated with the dramatic postwar increase in gross national product that was generally considered a sign of progress and prosperity.

With the benefit of hindsight, it is easy to see how a novelty like the private automobile could become a necessity. Foresight of this kind is more difficult to come by. In the early years of nuclear research, Rutherford said that nuclear energy had no practical use. From an economic point of view its pursuit was considered a pure luxury (pure knowledge). But the atomic bomb changed that picture. Atomic energy is now an important industry, and may become the lifeline of the future. The same might happen to other novel scientific and technological developments. Space exploration is now considered a pure luxury. But who can say that it might not lead to significant economic developments in the future? The current environmental movement is characterized by antiscientific and antitechnological attitudes, because science and technology are blamed as the immediate causes of environmental degradation. If accepted uncritically, these antiscientific and antitechnological attitudes might affect adversely activities that are unrelated to or may even help solve environmental problems. Science and technology that are not useful now may become useful later. The potentials of technology have not been exhausted yet. The modern Diogenes, like the ancient one, cannot turn history around. The list of necessities will keep on lengthening, but the total material needs of a modern society may be kept under control by limiting the population and recycling the materials. The historical direction of evolution is still toward a more complex and more sophisticated technology that, although it gives rise to undesirable side effects, may contain a self-correcting system to cope with them. Science and technology may cause depletion of resources and degradation of the environment; science and technology may rehabilitate them.)

19.2 The Materials and Their Natural Reserves

The materials used by primitive man were taken mostly from the biological world. The American Indians used the buffalo in many ways; they used its meat for food, its hide for coverings (on the body or on shelters), and its bones for making household articles. The usage determined the ecological balance (how many buffaloes to an Indian), and the Indian way of life could be perpetuated with a stable relationship between the Indian and the animal populations. The white man killed the buffalo in mass quantities for its hide, not for personal use, but as merchandise for sale. The reduction of the buffalo population had a disastrous effect on Indians. In our more complex modern society the materials come from a wide variety of sources, and the balance of nature assumes a more subtle form. Yet the basic principle is the same. An imbalance between the source and consumption of materials will lead to disastrous consequences.

Let us consider the material resources required to fulfill basic needs besides food, such as clothing, shelter, and transportation. To fulfill these needs, we require raw material, and the raw material has to be processed with the help of tools to be made suitable for various purposes, such as supporting a structure and covering a body. Thus three kinds of materials are needed generally: the supporting material used in shelters, vehicles, and so on; the covering material used in clothing, containers, and so on; and the working material used in tools and instruments. As already mentioned, in early times most materials came from the biological world: lumber for construction and cotton for clothing, for example. Besides, stone and clay were used as housing materials. The materials industry was relatively simple and the ecological balance uncomplicated. The situation began to change when the use of metal was introduced, a milestone in the development of civilization. First copper and then iron were used. In a variety of ways steel is used for supporting, as in construction; for covering, as in appliances and containers; and, most importantly, for making tools. It was the higher efficiency of the metal tools that spurred the rapid advance of civilization. The consumption of metal resources in early times was largely limited to tool making, the volume of which was relatively small. Only in recent years has metal been used as a material for construction and surfacing. The rate of consumption has increased rapidly, and the problem of resource depletion has become a realistic consideration.

Iron is the most important metal in human history. It is still the most widely used one today. Besides copper and iron, tin has been in use for a long time. The smelting processes for these metals are relatively simple, and thus these metals were discovered first. In modern times, aluminum was added to the metals. It has the advantage of being lightweight. An economical way to smelt it was not developed until 1886; this explains its late entry. Other metals such as magnesium and beryllium were brought into use recently for special purposes. Beryllium is a toxic substance; the environmental effect of its use has been brought to attention recently.

Besides these metals, there are a number of other chemical elements that chemically are metals but are used for other purposes. Gold and silver have been in use from ancient times. Mercury is a liquid metal and it has also been in use for a long time. Antimony, chromium, nickel, zinc, tungsten, and lead each has its special uses. Newly developed technology employs more and more unfamiliar metals for special purposes.

From the amounts of world mineral reserves and the current consumption rates, we can calculate roughly the time periods for which the reserves will last. The times may be considered as the lifetimes of the minerals. Improved exploration and mining techniques will lengthen the lifetimes, but increased use by increased population and by technological expansion may shorten the lifetimes. Nevertheless, the following lifetimes[1] may be taken as indicators for assessing the crisis of the impending depletion of resources. The world's iron reserve has a fairly long lifetime that will take us to 2500 A.D. The other iron-alloy metals are not quite so plentiful. Although chromium may last as long as 2500 A.D., manganese, nickel, molybdenum, tungsten, and cobalt will last until about 2100 A.D., with small variations. Among the nonferrous industrial metals, copper may last until 2000 A.D., and aluminum may last until 2150 A.D., but lead, zinc, and tin will be depleted by 1990 A.D. The precious metals—gold, silver, and platinum—are also expected to be depleted by 1990 A.D. New mineral resources, such as those under the ocean, may modify the picture to a certain extent.

19.3 The Materials Industries and Their Environmental Impact

Some metals are used not physically, but chemically as raw materials for making new chemical compounds. A large quantity of lead is now used in making the antiknock gasoline additive. Also in this category are the uses of many of the nonmetallic minerals, such as sulfur and phosphorus. These minerals are put to use in the chemical industry, a materials industry producing substances that are used either directly, such as drugs and dyes, or indirectly in other manufacturing processes. The chemical industry is a primary culprit in causing environmental pollution. Waste products of manufacturing processes frequently are discharged into river systems and cause water pollution.

Two other materials industries were brought into existence by the use of metal and nonmetal mineral resources: mining and metallurgy. Both figure importantly in environmental considerations.

Mining. Mining is essentially a physical process. Ores may be collected by the traditional tunnel-mining method or by the more recent strip-mining method. The mining industry is now deeply involved in environmental problems. Take coal mining as an example. In the past the tunnel-mining method was used and environmental effects were limited. To be sure, the environmental problems inside the tunnel itself are quite serious: poisonous gas, inflammable gas, dust particles, and so on. But these are looked upon as occupational hazards, not as environmental problems for the general public. Nowadays, the strip-mining method is used more widely, and the environmental problems have become more serious. In strip mining, coal is collected by mechanical removal of the overburden (the earth above the coal bed), thus exposing the coal deposit (Fig. 19.1). Because coal is formed by buried vegetation, it exists beneath the earth's surface in horizontal layers; the earth above the coal bed consists of sedimentations laid down during the geological age that followed the interment of the forest. Development of powerful earth-removing machines has made it possible to remove the layer of earth above the coal bed and expose the deposit for mining.

Strip mining of coal is now developing rapidly in the American west, partly stimulated by environment-protecting action. The air pollution in the eastern cities became so severe that municipal authorities began to impose restrictions on the quality of

[1] Preston Cloud, in *Environment,* William W. Murdoch, ed. (Stamford, Conn.: Sinauer Associates, Inc., 1971), p. 71.

Figure 19.1 Strip mining. [U.S. Environmental Protection Agency, Project DOCUMERICA. Photo by Lyntha Scott Eiler]

coal to be used. As already mentioned, coal of 3 per cent sulfur content was acceptable in the past; now the acceptable sulfur content has been lowered to 1 per cent. Thus industrial cities in the traditional coal-producing regions in the east are now turning to the west for low-sulfur coal, usually more than a thousand miles away. The deposits in the west were not exploited in the past because of the high cost of transportation to the industrial cities in the east.

The most obvious environmental effect of strip mining is the ruining of the natural landscape. A whole area of the earth is ripped open (Fig. 19.2). The overburden that has been removed is piled up in unsightly heaps. In mountain coal fields, trees, rocks, and earth are removed and cast down mountainsides to expose the coal to be mined. Silt fills streams for many miles. The sulfur in the coal, exposed to the air, undergoes chemical reactions that produce sulfuric acid, which pollutes water and kills fish downstream (Figs. 19.3 and 19.4). A land surface equal to half the size of New Jersey has been peeled off for coal mining in Arizona, Colorado, Montana, New Mexico, North Dakota, and Wyoming.

The problem will be intensified in the next decade or two because strip mining is expected to increase markedly. Coal is an old, unglamorous industry, but a renaissance seems to be in the making. Depletion of the natural gas reserve is expected within 15 years. How are the needs of gas-dependent facilities to be fulfilled? Gas is a desirable fuel, for it is pollution-free and can be transported conveniently. Therefore attention is being focused on the gasification of coal, the conversion of coal into gas. For western coal mines, gasification has the added advantage of saving transportation cost. Coal must be hauled by rail, and transportation cost is the major component of the cost of coal. Gas can be transported through pipelines, which costs much less. It is estimated that in 1990, probably 300 million tons of coal—half the 1970

Figure 19.2 Twin-Buttes Mine, Tuscon, Arizona. This open-pit mine is 1,000 ft deep. [U.S. Environmental Protection Agency, Project DOCUMERICA. Photo by Cornelius Keyes]

Figure 19.3 A pool of acid water lies at the bottom of a mined-out area near Elkins, West Virginia. The water drains into the Tygart River and pollutes it with acid. [U.S. Environmental Protection Agency]

total production—will be processed by gasification plants built around strip-mining fields in the western states. Strip-mining output has increased rapidly from three tenths of the annual tonnage in 1968 to four tenths in 1970. The cost advantage over tunnel mining is about three to one. Productivity per worker has an advantage of five to one.[2]

Metallurgy. The other industry, metallurgy, is essentially chemical. Most metals, except the precious metals gold and silver, exist in the earth's crust in the form of ores of oxides. The smelting process involves a chemical reaction in which the metallic form is recovered by a reduction reaction. The smelting of iron in the blast furnace (Fig. 19.5) will be discussed as an example. Iron ores containing iron oxides (such as hematite Fe_2O_3 and magnetite Fe_3O_4) are mixed with coke (made of coal, containing essentially carbon) and a flux (such as silica SiO_2 from sand or clay to remove limestone $CaCO_3$ or other contaminants in the ore). The mixture is poured into the opening in the top of the blast furnace, and preheated air is blown in from the bottom. The coke burns incompletely to form carbon monoxide:

$$2C + O_2 \longrightarrow 2CO$$

The iron oxide is then reduced by carbon monoxide:

$$3CO + Fe_2O_3 \longrightarrow 2Fe + 3CO_2$$

The flux combines with the impurities to form a liquid slag for easy removal:

$$CaCO_3 \longrightarrow CaO + CO_2$$

$$CaO + SiO_2 \longrightarrow CaSiO_3 \text{ (slag)}$$

Sometimes the ore contains silica, so limestone is used as flux; the reactions remain the same. The gaseous products escape from the top, while both liquid slag and liquid iron descend to the bottom and are then separated and tapped off.

The iron so obtained, containing a large percentage of carbon (3 per cent or 4 per cent) because of the contact between molten iron and coke, is called cast iron. It is not very useful because it is too brittle. Most cast iron is made into steel, with carbon content reduced to amounts ranging from 0.1 to 1.5 per cent and with other impurities eliminated. Many kinds of steel can be made for various uses by varying the carbon content and by adding other iron-alloy metals.

Because coal contains sulfur, the burning of coke produces sulfur dioxide, which escapes the blast furnace and pollutes the air (Fig. 19.6). Other smelting processes lead to air pollution for the same reason.

After being used for a time, iron may become rusted. Rust is an oxide of iron, so iron scrap can be reprocessed according to the same principle as is used in smelting. Thus iron is not entirely destroyed through use and can be recovered afterward. Most iron used in the auto, rail, construction, and manu-

[2]Ben A. Franklin, *The New York Times,* August 22, 1971, p. 1.

facturing industries can be recovered. Yet iron used in household and commercial products, such as hairpins and beer cans, is lost in the ocean of waste heap and cannot be easily recovered under present conditions. These uses represent a drain on material resources which, if not averted, could drain dry all material resources.

Manufacturing. The manufacturing industry is another one concerned with materials. It reshapes the raw material into useful forms. The reshaping is a physical process and requires the use of energy. As far as the environment is concerned, the manufacturing industry is involved because of the consumption of energy. Besides, during the reshaping process some of the material becomes dust particles or waste material, which may cause pollution (Fig. 19.7). Special hazardous substances such as asbestos and beryllium may become air pollutants through the manufacturing process as already mentioned.

Synthetic Materials. A significant recent development concerning material resources is the use of synthetic material. Man-made fibers such as rayon and nylon have been in use for several decades. In large measure, natural materials once used for clothing have been supplanted by synthetic materials. The dependence on biological resources for clothing material has been reduced. Synthetic material to replace leather has been made and used, but unlike nylon and rayon, the substitute (Corfam) turned out to be a $100 million commercial failure. (Still, the DuPont company can count on recovering its investment if the human population explosion squeezes cattle off this planet and no more natural leather is available.)

Another important advance is the development of plastics. Plastics are synthetic materials that can be used for making containers and as the raw material for making household and commercial products, thus acting as a substitute for metals and organic mate-

Figure 19.4 Acid mine drainage in a Pennsylvania stream. [U.S. Environmental Protection Agency]

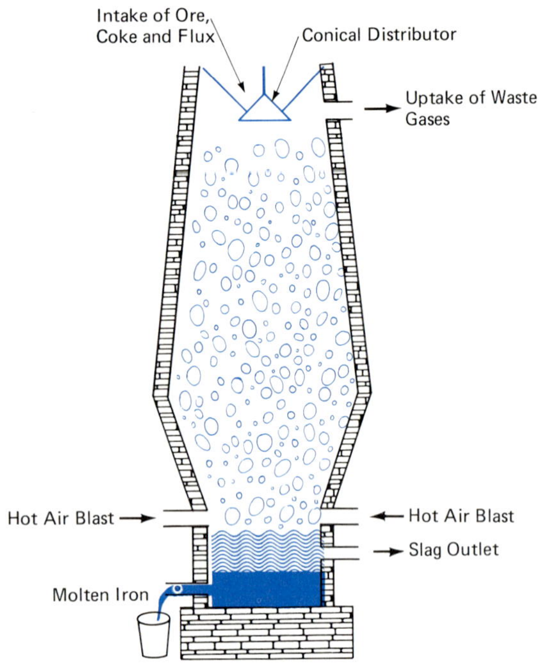

Figure 19.5 The blast furnace.

rials in many cases. Metal reserves are limited and may be depleted. Plastics are made of common elements such as carbon, hydrogen, oxygen, and nitrogen that will not be depleted, even though currently the plastics are produced by syntheses of coal, limestone, petroleum, salt, and water, some of which have a limited supply. Coal and petroleum are now largely used as fuels; their present nonenergy uses account for only 6 per cent of total consumption. It is obvious that once another energy source is available, petroleum should be saved for petrochemical use to help conserve material resources, instead of being wasted for mere energy production.

Synthetic materials are created through chemical reactions. Nylon, which is a plastic, will be discussed as an example. Silk and wool are constituted of protein molecules, each of which is a long-chain molecule made of many amino acid units. Each such unit has an amino group (NH_2) and a carboxyl group (COOH). The amino group of one combines with the carboxyl group of the other to form a *peptide bond* connecting the two units as discussed in Sec. 10.7. The process may be repeated many times to form a long-chain molecule made of many units. Nylon is a long-chain synthetic molecule formed by a similar process. It is made of adipic acid

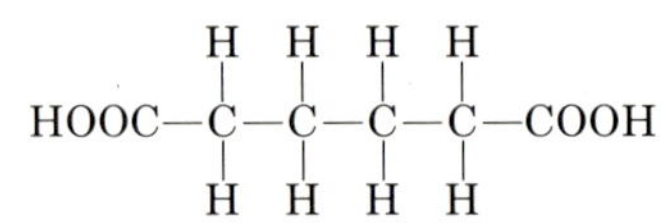

and diaminohexane

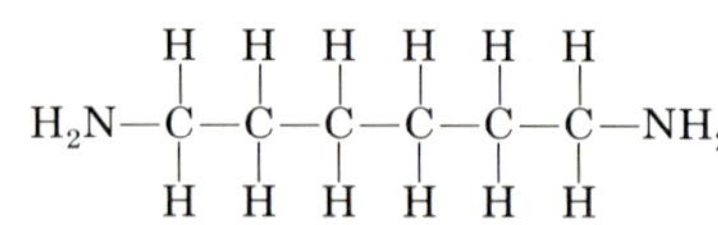

The amino group of one can react with the carboxyl group of the other to form a bond linking the two in the following way:

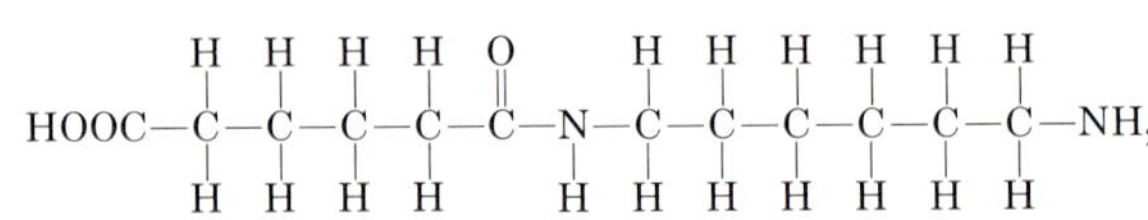

The process can be repeated to form a long-chain molecule with adipic acid residues alternating with the diaminohexane residues. A number of these long-chain molecules in parallel orientation bundled together form a nylon fiber.

Other artificial fibers and plastics in general are made by similar *condensation reactions* in which long-chain molecules (*polymers*) are formed from small units (*monomers*). Actually, many man-made fibers are plastic materials made in fibrous form, and their production may be considered a part of the plastics industry. Because many of the materials for which we want to find substitutes are organic polymers, the problem largely centers around *polymer chemistry,* of which plastics chemistry is a part. As the properties of a plastics material are determined by the polymer constitution and structure, they can be controlled by manipulating the latter. Thus man-

Figure 19.6 Double-barreled pollution of air and water by a steel mill, Hammond, Indiana. [U.S. Environmental Protection Agency]

Figure 19.7 Some industries are building their own waste treatment plants. Shown here is a waste treatment plant built by a papermill on the Delaware River. [U.S. Environmental Protection Agency]

ufacturers can create an artificial material that has properties that are close to, or even better than, the original natural material.

The plastics industry in the United States began about 1845. Natural "plastics" materials were molded to form useful articles such as brush handles. The first synthetic plastics material of commercial significance was celluloid, invented by Hyatt in 1869. It was made by combining camphor with pyroxylin (cellulose nitrate). Because it can be sawed and carved, it can be made into various articles. But it is hard to mold and catches fire easily. The first man-made fiber, viscose rayon, was invented in 1884. Cellophane followed in 1908. The first synthetic resin, Bakelite, was invented in 1909. It can be molded and is widely used to make such things as telephone casing and handles for household items. In the 1920's and 1930's the plastics industry expanded to large-scale production and use of cellulose acetate, acrylics, and polystyrene. The 1940's saw the introduction of polyethylene (for food packaging), silicones (lubricants), and epoxy (adhesives). New materials and wider uses of plastics were developed rapidly.

Polystyrene, first used in 1937 for such things as radio cases, will be discussed here to show how a plastic is made. The basic substances from which plastic articles are made are polymers called synthetic resins. The polystyrene resin is made from the liquid benzene and the gas ethylene, both of which are derived from petroleum:

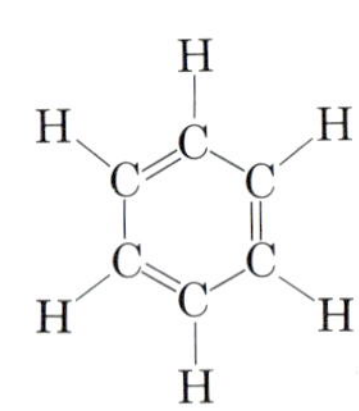

Benzene

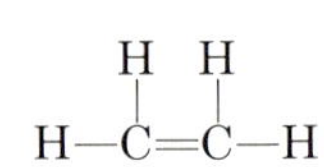

Ethylene

When the gas bubbles through the liquid, the two combine to form the liquid ethylbenzene:

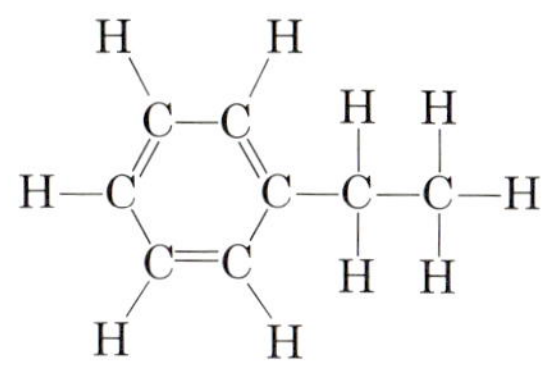

When ethylbenzene is heated with metal oxides, a hydrogen molecule is removed and the substance changes into the liquid styrene.

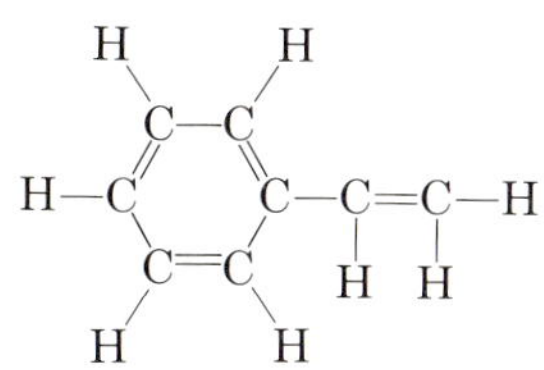

Many styrene molecules are then linked together to form a long-chain polymer, and the substance becomes the solid polystyrene. The polymerization is accomplished by adding heat, pressure, and chemicals. The solid is then ground into grainlike particles, to be used as the raw material that can be molded to make various polystyrene products.

Plastics can be molded into various shapes and therefore are convenient materials for making a large variety of products. Some plastics are so hard that they can be used to make gears and bearings as metal substitutes. Nylon is a hard plastic; yet when made into thin fibers, it can be woven into soft, spongy fabric. Other synthetic plastic fabrics include Orlon and Dacron. Some plastics are soft and can be used to make squeeze bottles, toys, and so on. Special uses are being added day after day. Because plastics do not harm the human body and do not react with body fluids, they are the ideal material for making articles for surgical implantation inside the human body. Special plastics have been developed to replace glass as a material for lenses. One example is the recent development of hydrophilic plastics for contact lenses. Special plastics resist heat and chemicals; asbestos-filled phenolics have been used to make rocket nose cones capable of withstanding heat generated by high speeds. Other exotic uses include those of scintillation plastics, employed in nuclear radiation counters based on their special optical properties, and of plastics counters, which are used to register nuclear particles based on the mechanical properties of the plastics material.

We can foresee increased material needs and depletion of mineral resources in the future. The problem will be dealt with by (1) conservation and recycling of mineral resources and (2) increased use of synthetic material. The chemical elements of synthetic materials are readily available. If we have an abundant supply of energy, more synthetic methods will be made economically feasible, and more synthetic materials will be made available to fulfill our material needs. Recycling mineral resources also requires energy. Thus we conclude that material resource depletion can be solved by abundant energy. This point will be elaborated further in the next chapter. The materials problem is thus tied to the energy problem. This lends additional importance to the energy resource problem we have already discussed.

Problems

19.1 Explain that petroleum is not only an energy resource but also a material resource. From a global point of view what is the best way of utilizing this resource?

19.2 In which uses can iron be replaced by synthetic material or material that can be regenerated so as to reduce the rate of depletion of iron?

19.3 In the history of metallurgy, copper was used before iron and aluminum was the last. Why?

20
Material Utilization and Waste Disposal

20.1 The Solid Waste Problem

In the preceding chapter we discussed some of the environmental effects of the utilization of materials and some instances of environmental pollution by the various materials industries. Now we will consider the main issue, the general effect of material utilization, that is, the depletion of material resources and the problem of waste production and disposal. Material goods, after being consumed, leave a useless residue, which we call *waste*. In the short run, the disposal of wastes gives rise to environmental problems. In the long run, the accumulation of waste reflects the diminishing of resources; therefore the problem of handling waste must be considered together with the problem of depletion of resources.

The waste disposal problem may not be the most romantic subject for an intellectual discourse. It concerns many unsightly things that are not fit to mention before a lady or a gentleman. We would not be interested in it except for the fact that we are forced to face the dire consequences that will ensue if the problem gets out of hand.

In the year 1968 the United States generated 360 million tons of solid waste,[1] corresponding to 1.8 tons per person per year, or 10 lb per person per day. The rate is expected to double in two decades. At this rate of growth, the accumulated trash could easily cover the whole earth and bury its entire human population in a short time.

Consider garbage. It used to be disposed of in a city dump somewhere out in the country, or used for landfill material, or dumped into the ocean. Garbage dumps breed flies and rodents and become health hazards. Burning the trash may eliminate the problem, but it may increase air pollution. Rain water washes the remains into the river system, causing water pollution. Landfill sites (12,000 of them, such as the sites of the New York World's Fair

[1]U.S. Bureau of Solid Waste Management.

and the Montreal Expo) are being used up. Ocean-dumped garbage frequently washes ashore; it sometimes produces a region of "dead sea," which contains no aquatic life at all.

In 1970 we discarded 28 billion bottles, 48 billion cans, 30 million tons of paper, and 4 million tons of plastic containers. In addition, we junked 100 million tires (Fig. 20.1) and 6.3 million automobiles (Fig. 20.2).[2] The rate of increase of solid waste is greater than the rate of increase of the production of goods, because of our "prepackaged" civilization. Instant food and drinks must be prepackaged. Division of labor and transportation further increase the need for packaging. The container industry is literally a waste-producing industry that never existed in a less sophisticated society.

With the rapid increase in the generation of waste, methods are being developed to dispose of waste products or even to make some use of them in the disposing process. The easiest way to get rid of them is to burn them. Incinerators are developed. As the waste problem intensifies, incinerator technology also advances. For one thing, the heat generated during incineration can be used for generating electricity. In fact, municipal waste is a reasonably good fuel; it has a heat-generating power about one third of that of coal and can provide about 10 per cent of a community's electrical needs. In Montreal a sophisticated incinerator has been developed to use the heat it produces to generate steam for sale to industries. Some incinerators are designed to reach such high temperatures that all metals are melted and thus can be recovered from the trash. It is difficult to separate one metal from the others, but a jumbled mixture may be used for making such things as doorknobs.

Another way of making use of solid wastes is to compress them under high pressure into a small, compact volume and use them as landfill (Figs. 20.3 and 20.4) or, after proper coating, as construction material. Outside Chicago, in the flat plains, compressed wastes are used to build man-made hills that may be used for recreational purposes, such as skiing.

A large proportion of solid wastes consists of paper, glass, and metal containers, which, if properly separated, can be processed for reuse. The difficulty lies in the separation. High wages make manual separation too costly. Some mechanical devices have been developed to do the separating. Iron cans and other iron products can be separated from other solid wastes by magnetic devices. Light materials may be separated from heavier ones by blowing air through masses of waste.

[2] Peter Gwynne, *The World Book Encyclopedia,* 1971 Yearbook (Chicago: Field Enterprises Educational Corporation, 1971), p. 78.

Figure 20.1 Discarded tires in city dump, Redding, California. [U.S. Environmental Protection Agency]

Figure 20.2 Cars stacked in junkyard, Escondido, California. [U.S. Environmental Protection Agency, Project DOCUMERICA. Photo by Gene Daniels]

Because the container is a major source of solid waste, improvement in container technology can help solve the waste disposal problem. If the food-packaging material could be made edible, like ice-cream cones, then it could be disposed of right in the stomach without affecting the environment. Soup packaging of this type is under development. Another solution is development of a container material (a plastic bottle) that will decay and thus disappear under exposure to natural conditions. Various attempts have been made along this line.

In the United States the municipal wastes produced annually contain 10 million tons of iron, 1 million tons of nonferrous metals, and 15 million tons of glass. About 50 per cent of the iron, copper, and lead; 30 per cent of the aluminum; and 20 per cent of the zinc are recovered and reused. Less than 10 per cent of the glass is recycled. Although glass accounts for only 6 per cent of the solid waste, glass containers are troublesome because they do not rot and break down on trash heaps (even metal cans rust and break down). Paper is the major component of solid waste, but only 20 per cent of it is recycled.

One wild idea for disposing of solid wastes is to burn them up by exploding hydrogen bombs underground. It has been estimated that 50 hydrogen bombs would be sufficient for all solid wastes and the burning could be repeated a few years later. The melted metal could be separated (as in the natural formation of mineral ores) and reused. (This may not be such a wild idea for disposing of hydrogen bombs.)

20.2 The Nature of Waste Production

From the commercial point of view, when we say consumption we imply that the goods, after being used, vanish and that any residue that does not vanish is useless waste. The residue is a minute fraction of the original amount, and its disposal is a minor side issue of consumption. In fact, economists have never considered waste disposal a main issue in economic processes.

From the point of view of physical science, the situation is quite different. The essential point may be brought out by a fundamental theorem on waste production in the following. From this theorem, we see that waste disposal is not a trivial side issue but a major economic problem.

The Fundamental Theorem on Waste Production. *The amount of wastes in weight after consumption is exactly the same as the amount of goods before consumption, provided that the terms* goods *and* wastes *are properly defined and that the wastes disposed of in all forms and at all times are included.*

Figure 20.3 Mount Trashmore near Virginia Beach, Virginia, as it reached a height of 65 ft and covered 7 acres of land. An above-grade sanitary landfill, Mount Trashmore is built up with municipal refuse compacted and covered each day with a 6 in. layer of earth. When finished it will be covered with topsoil and vegetation. [U.S. Environmental Protection Agency]

Figure 20.4 Municipal golf course built on sanitary landfill, Riverside Park, Jackson, Mississippi. [U.S. Environmental Protection Agency, Project DOCUMERICA. Photo by Bill Shrout]

The theorem follows directly from the law of conservation of mass in physical science. All goods are material substances. Immaterial economic products are called *services* by economists and are treated as a separate category. The weight (or more correctly, the mass) of material substance is conserved, always remaining exactly the same no matter what physical and chemical changes take place. The process of consumption is largely composed of physical and chemical changes, and therefore the total weight of all substances after consumption (wastes) is exactly the same as that before consumption (goods), as stated in the theorem. The following examples are discussed to illustrate the point.

A copy of the daily newspaper on the day of issue, before it has been read, is a piece of goods. After being read it becomes waste, and goes into the trash can. Even without being read, it becomes waste the next day from the seller's point of view. Yet the weight of the paper in the trash can (waste) is exactly the same as its weight on the newsstand (goods). The consumption process (reading) involves shining light on the paper and reflecting it into the reader's eyes. It is a purely physical process, and the weight of the paper is not changed by the process. This example is rather trivial, but more complicated examples can be analyzed in exactly the same way.

Let us consider the class of commodities called durable goods, such as furniture. When economists make this classification, it is implied that these commodities will not be consumed, that is, they will not vanish after being used. However, furniture eventually will be worn out or broken down. Even when used as an antique, a piece of furniture eventually (many centuries later) will wear out or break down. The weight of the junk furniture is the same as when it was brand-new. It may be slightly smaller if some of the fabric has worn out. Yet the worn-out fabric becomes dust and is eventually collected in the dustbag of a vacuum cleaner and disposed of as waste. Thus the wastes disposed of in all forms (junk and dust) have the same weight altogether as the original goods. After all, the consumption process for furniture, i.e., the continual seating of human bodies, is merely a physical change and does not alter the total weight of the substance.

The conventional concept of consumption is derived from the use of nondurable goods, or consumables. Take the candle as an example. After being used, it burns out and disappears nearly completely, leaving perhaps a bit of residual wax to be considered as waste. It is from this kind of experience that one gets the idea that goods are consumed and vanish after being used, and that the resulting waste is a minute residue. But from the view of physical science, the weight of the paraffin of which the candle is made never vanishes. Paraffin is a compound of hydrogen and carbon. The burning process involves a chemical reaction in which the hydrogen and carbon each combine with the oxygen in the air to form water vapor (H_2O) and carbon dioxide (CO_2), two gaseous substances released into the air. The weight of the candle does not disappear; it reappears in the weight of the combustion products H_2O and CO_2. These products are truly waste of the consumption process, because they appear after the process and are not useful in an economic sense. In this case, the consumption involves a chemical change, and the total weight of the products is exactly the same as that of the reactants according to the law of conservation of mass. Not only does all the weight of the candle reappear in the wastes, but also the total weight of the wastes, H_2O and CO_2, is greater than the weight of the candle because of the addition of oxygen from the air. If we burn a candle inside a sealed jar and balance it on a scale, we will see the scale remaining balanced as the candle burns, indicating that the total weight of the candle and the air is unchanged. If the jar is not sealed and we place materials inside that will absorb H_2O and CO_2, we will see that the total weight actually increases as the candle burns, because an extra amount of oxygen is added to it.

Candles are considered useful goods for which we must pay, but we may also consider oxygen in the air to be a useful good in the sense that without it, a candle will not burn. Although the visible solid wax residue of a candle is obviously a waste, we have every reason to regard the invisible gaseous combustion products H_2O and CO_2 as wastes. When goods and wastes are so defined, the total weight of all wastes after consumption is exactly the same as that of all the goods consumed, as the theorem stated.

A large proportion of consumables is composed of foodstuff. The consumption of food encompasses a series of complicated physical and chemical changes, and again the law of conservation of mass holds true. Therefore we expect the total weight of the wastes to be exactly the same as the foodstuff consumed. The fact that food disappears through the mouth does not mean that the substance has disappeared. The human body, like most animal bodies, may be regarded as a tube from anatomical and embryological points of view. The digestive tract represents the inner lining of the tube. Foodstuff passing through the human body is thus like material moving through a tube, which is a physical change that does not change its weight. As foodstuff passes through the digestive tract, it also reacts chemically with the secretions of the digestive glands. Some of the reaction products are absorbed by the membrane of the digestive tract and transported to the other parts of the body as nutrients. The residue moves out of the tract as excreta. The total weight of the nutrients and excretions is exactly the same as that of the foodstuff ingested, the processes involved all being physical and chemical ones obeying the law of conservation of mass.

The nutrients transported to the other parts of the body are used in two ways: either as building material to be incorporated into the body tissue (gaining weight by eating) or as fuel to supply energy to maintain body functions (as in an engine). The use of nutrients as fuel involves essentially a combustion process somewhat similar to the burning of a candle. Oxygen inhaled into the lungs reacts with nutrients in a series of complicated biochemical processes, and the final waste products are again largely H_2O and CO_2 exhaled from the lungs into the atmosphere. If we count oxygen as a "free" foodstuff, and H_2O and CO_2 as part of the waste, we again reach the conclusion that the total weight of the wastes equals that of the foodstuff.

We still have to consider that portion of the nutrients that eventually becomes body tissue. In human bodies, old cells break down and are replaced by new cells assembled from materials supplied by the nutrients. The broken-down products are excreted in the urine as wastes. Eventually an organism must die, and the body decomposes largely into gases, with some solid residue. Again the changes are physical and chemical, and the total weight of decomposition products equals that of the nutrients that become the body tissue. Thus, considering all the processes involved, the total amount of wastes—including feces, urine, sweat (water and salt), exhalants (H_2O and CO_2), and eventually decomposition gases and the skeleton—is exactly the same in weight as the foodstuff and free oxygen consumed.

20.3 The Disposal of Wastes

The consequence of the fundamental theorem is obvious. The wastes amount to just as much as the goods themselves. Waste disposal is thus not a trivial side issue. It is a part of the economic process as important as the production itself. As production increases, the magnitude of the waste problem increases proportionately. So far, economics has considered only the problems of production. The waste disposal problem remained hidden until recent years.

Consider the transportation aspect of the problem. We have built a large transportation system mainly for the moving of goods. Because the total amount of wastes is the same as that of the goods, we would need a transportation system of comparable size to move the wastes. But we see only a few garbage trucks and an underground sewer system. How is the large volume of wastes removed?

For one thing, a large portion of the waste products is in the form of gases, such as the combustion products of coal, gasoline, and natural gas. They are dumped into the atmosphere, and so no special transportation facilities are required. Burning of rubbish is a quick way of getting rid of solid waste. By conversion from solid to gas, it is disposed of in the atmosphere without having to be hauled away. Next to gas, liquid is easiest to transport. The sewer system is built to remove waste water and human excreta. In analogy to conversion to gas through the burning of rubbish, the garbage disposal unit in a kitchen sink is a device that converts solid waste to "liquid" that can be easily moved away. The hard-core, unconvertible solid wastes have to be hauled away by land transportation. Yet the total amount of wastes dumped in the atmosphere, in the water system, and in trash dumps remains the same as that of the goods consumed. Any device lightening the burden of solid waste disposal increases the burden of the atmosphere and the water system.

The consequence of burning and flushing the wastes away is the pollution of the atmosphere and the water system. The amount of goods produced every year is vast. Except for a relatively small amount converted into hard-core solid waste, most of it is disposed of in the atmosphere or in the water system after being consumed. Once a large amount of goods is produced, the same amount must be disposed of in the environment one way or another. Pollution is thus a necessary consequence of production and issues in an amount equal to that of production. No moralizing and social consciousness can help. Pollution and production are as inseparable as the two sides of a coin.

In our everyday life, when we wash some unsightly substance down the drain, we think the task of waste disposal has been accomplished. The underlying idea is that there exists a place such as the city dump or the sewer, which can be safely segregated from the rest of the world, and that undesirable objects may be discarded in it and forgotten. In earlier times, when small amounts of biological wastes were dumped in rivers, they decomposed while flowing downstream, reacting with oxygen in the atmosphere. However, as technology advanced, there was an increase not only in the amount of waste dumped into the water system, but also in the variety of its constituents. After World War II, detergents were used in large quantities. Unlike soap, some of the detergent ingredients resist decomposition when released into the environment. If it does not decompose, any liquid waste that is disposed of eventually goes into the river system. Detergents flushed into a septic tank are not safely segregated from the rest of the world. The undecomposed ingredients eventually seep out through the soil, and enter first the underground water and then the river system. Of course, in the process, many other organic compounds are broken down by bacteria in the soil, and the water is cleansed of them. The resistant detergents show up as foaming masses of suds on lakes and rivers, a common phenomenon in the 1950's. The detergent industry has responded by changing ingredients, and this particular environmental problem has now largely disappeared.

Thus the drain is not a safe place for waste disposal. In the past we expected wastes in the sewer to disappear when they decomposed chemically. Now that more sophisticated chemicals are present in the sewer this is no longer true. Because all the waters on the surface of the earth are connected through the surface and underground water systems, the pollution of any body of water—a stream, a lake, or even a septic tank—eventually leads to the pollution of a large part of the water system. The widespread nature of the pollution makes it difficult to clean up. There is really no safe place to dump liquid waste.

Like water pollution, air pollution also spreads widely and is difficult to clean up or to quarantine. Even though solid waste can be localized in a limited region, this does not mean that a solid waste dump is perfectly safe. The problem is brought out by two examples, the disposal of radioactive wastes, which we have already discussed, and the disposal of abandoned poison gas weapons.

Poison gas weapons (shells, for example) are encased in concrete "coffins" for disposal. Even that is not a guarantee against leakage of the lethal gas, which sooner or later will happen. Discarding these hazardous tools of war in a remote wilderness is obviously no solution. The weathering process—not to mention lightning, earthquakes, and so on—will eventually break the concrete casings open and release the lethal substance. Burying them underground is also unacceptable. During future construction, perhaps hundreds of years later, the "coffins" may be dug up and broken, with fatal consequences. The disposal measure that actually has been taken is that the weapons are dumped into the sea to rest deep down on the ocean floor. Not that the bottom of the ocean is a safe dump; rather it is hoped that when the cases break up eventually, the lethal substance will react chemically with the ocean water and become harmless before bubbling to the surface. Even the transportation of the "coffins" to the seacoast has posed a problem. Citizens protested the routing of the "funeral trains" through their communities. All sorts of hazards are posed; e.g., an automobile might accidentally crash into the train and break open lethal containers.

Thus there is really no absolutely safe dumping place on earth. To avoid the problem wastes must be recycled and must be capable of being recycled.

20.4 Recycling of Waste Materials

What happens to waste that has been disposed of? To throw a beer can into the trash pile (Fig. 20.5) is not the end of the affair. From a global and historical point of view, trivial acts like this may accumulate to bring about serious problems. As beer cans and other iron scrap keep on piling up and iron ores keep on draining away, eventually a time will come when the iron ores are completely depleted and all the iron previously in the ores turns up in the trash heap. Not only will we need a large space comparable to the ore fields to store all the waste iron that was originally in the ores, but also, more seriously, we will have no further supply of iron to meet the vital needs of an industrial society, not to mention the manufacture of beer cans. If the flow of material substance from goods to wastes is a one-way traffic, the inevitable consequence is the conversion of all resources to wastes. In order to avoid the catastrophe, it is necessary to convert the flow to two-way traffic—that is, the waste must be reprocessed to replenish the depleted resources. This will solve the problem of waste disposal at the same time.

Indeed, there exists now a scrap metal industry for reprocessing junk cars and the like to yield useful metal. It exists because it is profitable. It does not reprocess used beer cans because it cannot make money this way. The human labor expended on collecting the beer cans would be much more expensive than the metal recovered. Yet in low-wage, underdeveloped regions of the world, a little beer-can-supported industry has sprung up which uses beer cans as raw material. The products of this industry include small articles, pots, cans, house roofing and siding, and even arrowheads. In these regions it is a profitable enterprise.

From a historical point of view, the question of whether a recycling industry will make money or not is secondary. In the long run, all materials will have to be reprocessed and reused to avoid a depletion crisis. The crisis of depletion of iron may happen in less than several hundred years.

Although civilized history extends back for only a few thousand years, the human race has been in existence on earth for 1 million years or more. Life has been in existence for 2 billion years or so. Over such a long period of time, no crisis of this kind seems to have occurred. What has happened?

The answer is that in the past all wastes returned to the natural environment and eventually became useful material that replenished the depleted resources. This is what happens in a natural ecological system, and it also happened in the farming ecological system before industrialization. In the natural ecological system, the material that goes into an animal body eventually returns to nature, then to the vegetable kingdom, and finally to the animal kingdom again, forming a cycle. Biologists have studied thoroughly the cycles of carbon and nitrogen in nature. The material keeps on circulating, without leaking out of the cycle to be collected and accumulated in a trap like the waste heap. As long as there is no leakage in the cycles, the ecological system can be maintained in a steady state and perpetuated indefinitely.

The cycles in a primitive farming ecological system were more complicated but worked just the same way. Everything taken out of the field went back to the field; materials circulated in closed cycles, and

Figure 20.5 Anza-Barego State Park, California. [U.S. Environmental Protection Agency, Project DOCUMERICA. Photo by Gene Daniels]

the ecological system could be perpetuated in a steady state. For one thing, human and animal manure and urine were used as fertilizer, guaranteeing that things coming out of the earth returned to the earth. Those things taken out of the field that were not eaten, such as the straw, were burned as fuel, and the ashes went back to the field as fertilizer. The primitive farming community was self-sufficient because the circulation of materials was in closed cycles. In this way there was no crisis of resource depletion and waste accumulation.

Yet the advent of civilization has caused great distortions of the natural cycles. Suddenly it was deemed that human excrement, in spite of its very human origin, was unsightly and should be treated as waste to be segregated and destroyed. The invention of the flush toilet was looked upon as the watershed separating civilization from barbarism. World travelers use it as a yardstick to gauge a nation's level of culture. It was decided that fertilizers should be manufactured by the chemical industry and should come in neat packages. In spite of the superficial improvement in the style of farming, these changes broke the balance of natural cycles. The ecological system of farming is no longer a closed, circulating one, but an open system with a source (the chemical fertilizer) and a sink (the drainage system for human excrement). The civilizing revolution in agriculture has disastrous effects. Human waste eventually goes into the earth's water system; the organic nutrients in it cause uncontrolled growth of algae in lakes and rivers, giving rise to environmental problems (the same as those caused by phosphates from detergents). Because the organic nutrients are not being returned to the soil, its organic content is being depleted. The long-range effect cannot be predicted. Many old agricultural lands remain fertile after thousands of years of farming because of well-balanced ecological cycles. Newly opened virgin lands are more likely to be beset with productive problems. We do not know whether modern scientific farming will lead to problems in the future. Natural food (grown without chemical fertilizer) has become a current fad, and perhaps it is time to give more thought to natural farming.

The cycling of material in the man–nature system becomes more complex in a modern industrial society. Foodstuff is no longer the main item of consumption. A variety of industrial products appears, together with a large number of intermediate products used not directly by the consumer but by the manufacturers and farmers. These intermediate products include farm machinery, transportation facilities, fertilizers, insecticides, and so on. The manufacture and consumption of these additional material goods complicate the cycling of materials between man and nature.

Because historically the development of industries was spurred mainly by the pursuit of profit, the early stage of industrialization was characterized by a ruthless exploitation of natural resources. Many of the natural resources were "worthless" or quite cheap before technology found some use for them, and therefore their exploitation yielded great profits. The industrial society is only a couple of centuries old, a very short time even compared with the recorded history of man. At present, we can still witness some of the most senselessly wasteful uses of our natural resources, such as the use of iron for beer cans. Humans are going to live on the earth for a long time—they have lived here for at least a million years already. If we use up all the iron resources as fast as we can in several centuries, what will happen to the many future generations thus deprived of a supply of iron? Eventually man will have to return to a state in which materials circulate in cycles rather than going one way to the waste heap.

New technology causes the disturbance and dislocation of the balance of nature in many ways. The use of insecticides is a good example. It has increased crop harvest and improved the control of some communicable diseases such as malaria. Yet by lavish and extensive use, sometimes more for convenience than by necessity, the environment is polluted by insecticides, with damaging effects on wildlife and humans. The insecticide DDT, once regarded as a scientific wonder, is now banned from use in many areas.

In the future all materials will have to be recycled. The wastes will be the main source of raw materials;

there *cannot* be any waste! This runs counter to common sense to an extent, even though it has been the case in primitive agricultural societies for thousands of years. Actually, we have a good example of this style of living in space exploration. The total amount of water in a moonship, including that contained in the astronaut's body, in containers, in the air, and so on, is fixed and conserved. The amount cannot be very great because of the weight limitations of the spacecraft. When we consume water, fluids pass through the body and come out in excreta, perspiration, and exhalation. On the earth, we can afford to throw them away as wastes and drink only "fresh" water. In a spaceship we do not have an unlimited supply of fresh water. The same amount of water must be used over and over again. This is possible in principle because of conservation of matter. With proper technology, the reuse of water can be realized. In fact, in a spaceship provisions are made to collect the moisture condensed on the windows and put it back into the water system to be used again.

The earth itself may be regarded as a spaceship with a limited and fixed amount of resources on board; after all, it is a globe floating in endless space with no supplies coming in from the outside. In the future, all materials and all resources on earth must be used in a way comparable to the use of water in Apollo moonships. Actually, water on the earth has been used and reused in cyclic circulation. It is the sun that provides the energy to recycle water after use. The heat of the sun causes evaporation of water from lakes, rivers, and oceans. The moisture condenses and returns to earth as "fresh water." In this sense there is no "virgin water"; all water is secondhand, as in the Apollo moonships.

The recycling of water is a simple matter. The recycling of other industrial wastes is a more difficult task. It is possible in principle because of conservation of matter. The total amount of each of the 92 elements on earth always remains the same (disregarding the minute number of nuclear transformations taking place on earth). Any use of materials involves the rearrangement of atoms in molecules (disregarding nuclear applications), and it is always possible to reverse such rearrangements. Yet there are two difficulties.

First, although it is possible to return atoms to their initial configuration to retrieve the original material, such changes usually cost energy (and thus money). The source of future energy supplies is already a problem. The retrieval of useful materials from wastes would cost a sizable amount of energy, thus adding to the burden of energy demand in the future.

Another problem is related to the principle of increase of entropy. Whereas mineral resources are usually pure (or nearly pure) concentrated substances and therefore easily processed and utilized, waste material is usually a mixture of a large variety of compounds widely scattered and dispersed. To retrieve a useful substance, the compounds will have to be collected and separated from the mixture first, which will involve technical and economic difficulties. This point is intrinsic to the waste problem and is a consequence of the principle of increase of entropy. Among other things, this law implies that in all natural changes the distribution of matter tends to change from nonuniform to uniform. It is just a matter of probability. Uniform distribution is more probable than nonuniform distribution. In other words, things tend to become randomized (increase of entropy). Human use of materials is no exception. Each raw material is generally composed of pure substances of a few kinds. The waste material is a jumble of all kinds. This mixing of various kinds of material in waste makes the task of recycling material much more difficult. Besides, waste material is not concentrated in waste dumps; it is scattered all over the world, including the oceans. In fact, at the present time reprocessing is still very limited in scope and encompasses only such activities as recycling bottles and salvaging junk cars. As to cans, buttons, pins, and so on, they are "lost" in the waste heap.

Is industrial civilization doomed by the waste problem? Because the trend of the increase of entropy may be reversed by supplying energy, the material resource problem may be solved accordingly. When metals rust in waste heaps, the products are eventually flushed by rainwater into the ocean. Ultimately the ocean collects a large amount of mineral compounds, and if we have vast quantities of energy available, we can extract minerals from the ocean. Thus the problem of resource depletion is tied to the energy problem. Once we find an inexhaustible source of energy, the problem of materials depletion can be solved without fundamental difficulty.

Although no simple solution is now available, we do see the possibility of a solution to both the energy and material resource problems in the future in the development of a nuclear fusion industry. In several centuries, when the fossil fuels and even cheap fission fuels become depleted, we may need the fusion fuels as the main source of energy supply. The fusion energy reserve is so vast that we can afford to divert a part of it for the purpose of recycling materials. In fact, in the fusion process the temperature reaches a point so high that any chemical compound is decomposed to pure elements. Thus we can foresee a day, as has been suggested by some scientists, when a "fusion torch" is used to "burn up" all types of trash, beer cans included, and to reduce them to the 92 basic elements. The elements will still be mixed together, but they may be separable by fractional cooling in the same way that the minerals of the earth are fractionated during the cooling process of the earth's crust. Different materials have different freezing points and thus solidify separately, one after another, as the temperature of the molten mixture is lowered. The solid parts can thus be removed and separated one at a time as pure substances. As long as we can count on a large energy supply from the fusion source, the recycling problem can be solved. From the point of view of the human time scale we still have a long time (several hundred years) to work out the necessary technology to solve the problems of depletion of energy and materials. Even though many societal problems have been created by technology, they will eventually be solved by technology.

Before fusion energy and the fusion torch become available, we can at least take rational measures to correct the abuses of natural resources. It first requires a reorientation of thought away from worship of the flush toilet and from profit-seeking motivations and toward conservation of natural resources and recycling of waste materials. Before any material is used for an economic purpose, consideration must be given to the depletion of available resources and the recovery of the material after use. For instance, plastic containers, because they can be burned after use and their chemical elements returned to the natural environment to be used again, are obviously more desirable than disposable metal containers. If recycling cannot be accomplished readily, at least the waste products should be harmless or rendered harmless by proper treatment. The present exploitation of natural resources spurred by the pursuit of profit should go down in history as an inexcusable and deplorable act, just as was the destruction of the American Indian culture by the Spanish conquistadores for mere plundering. Future generations will view the beer can as an oddity of the past, reminding them of human self-indulgence, egotism, frivolity, and absurdity.

Problems

20.1 In conventional terms the total amount of waste production is always less than the total amount of production of goods. Yet the current rate of increase of waste production is greater than the rate of increase of production of goods. Why?

20.2 When all mines on land become depleted, where will all the minerals go? Explain why? Can they be retrieved?

20.3 What is the best use of junk auto tires? Remember rubber is a hydrocarbon.

20.4 A typical grocery basket may contain, besides food and drink, paper products, detergents, cooking utensils, china, plastic articles, and cotton products. These are all material substances and their mass is conserved. Where does each of them go after being consumed?

20.5 Consider the following scheme: Use human wastes as fertilizer to raise corn; use cornstarch and cellulose to make alcohol for automobile fuel. Discuss how energy and material resources are regenerated or recycled. Discuss how water pollution and air pollution are alleviated. Discuss the economic impact on employment and income. Is this the panacea for all modern woes?

21 The Origin of the Universe

21.1 The Physical Universe

What is the origin of the universe? How did it come into being? How did it evolve into its present state? These are questions the insatiable curiosity of mankind has posed for centuries. Various answers have been provided in myths, religions, and folklore. Indeed, in many of these media an origin for the universe has been hypothesized. There was "nothing" before this origin; then came a catastrophic event, a grand opening, in which the world came into being, and the world developed gradually to its present state. Invariably, creation is envisioned in a manner that provides answers to present-day problems and justification for current actions. (Thus the creation myth of one tribe may give its members justification for stealing cattle from another tribe, which allegedly had stolen theirs at the time of creation.)

From the scientific point of view, we are not obliged to create a theory to meet special human needs. In fact, there is a logical possibility that the universe may have no origin at all; it may always have been the same. But there is unmistakable scientific evidence that our present universe does have an origin, and it is on the basis of this evidence that scientific study of the origin of the universe begins. Nevertheless, the scientific theory of creation and origin serves the same purpose of explaining the present world in terms of its past.

What Is the Physical Universe? Before we ask about its origin, let us ascertain what the physical universe is. The physical world around us in our everyday life is made up of a great variety of things: fields, mountains, rivers, oceans, atmosphere, and the animal and vegetable life inhabiting them. Yet physically all of these are material substance made of the 92 chemical elements. Our problem thus concerns the origin of the elements and the transformation of the aggregation and organization of the elements into the present physical world. Moreover, the substances found on earth constitute a very small part of the whole of material substance in the universe, and their origin must be considered as a part of the larger problem of the whole universe.

Looking beyond our earth, we find that the earth is but a small part of a larger entity, the solar system. The solar system consists of the sun, nine planets, and other minor entities. The planets' sizes and distances from the sun are shown in Fig. 21.1. Our earth is but one of the planets. The sun takes up more than 99 per cent of the mass of the whole system. The sun is one of the stars. Most problems of the universe concern the stars. The earth as one of the planets is just a speck of dirt, and our interest in it is due to the fact that we live on it.

The earth has a mass of 6.0×10^{24} kg with a mean density of 5.5 g/cc. The radius may be rounded to the convenient figure of 4,000 miles. The distance from the earth to the sun is 1.5×10^{11} m; this distance is defined as an astronomical unit (AU). For later comparison, it is well to remember that light takes 8 minutes to travel 1 AU. The distance from the outermost planet, Pluto, to the sun is 40 AU; this gives us an idea of the size of the solar system. Our closest neighbor is the moon, a satellite of the earth. Its distance from earth is 60 times the earth's radius (240,000 miles), and its mass is 1/81 of the earth's mass.

The sun is but one of the 100 billion stars forming the assembly called the *Milky Way galaxy*. It is a typical star. Its mass is 2.0×10^{30} kg, with a mean density of 1.4 g/cc (density increases to 158 g/cc in the center). The stars nearest to the sun are about 10 *light-years* away. A light-year is the distance that light travels in a year. Thus the distances from the sun to the earth and to a neighboring star are in the ratio of roughly a minute to a year in terms of the traveling time of light. This provides an order of magnitude of the immense expanse of the stellar system. Because the stars are far away, they appear as a speck of light even though they are as big and as bright as the sun.

The Milky Way galaxy is a disklike entity with a diameter of about 80,000 light-years and a thickness of about 16,000 light-years (Fig. 21.2). Its contents of 100 billion stars have a total mass of 1.6×10^{11} times the sun's mass. Stars in the galaxy are not uniformly distributed; many stars form clusters (Figs. 21.3 and 21.2).

Outside the Milky Way galaxy at great distances are other galaxies, each containing 10^9 to 10^{11} stars. The distance between galaxies is of the order of 10^6 light-years. The nearest one is Andromeda (Fig. 21.4). The greatest distance at which we can make observations is about 3×10^9 light-years. In the observable universe (a sphere of radius 3×10^9 light-years), there are a total of 10^{10} galaxies, with a total mass exceeding 10^{51} kg. This provides an idea of the size of the whole universe. Because the galaxies are far away, they look like tiny faint clouds in the night sky and thus are also called nebulae. The near nebulae can be resolved into an assembly of stars in large telescopes. Nebulae also aggregate in clusters (Fig. 21.5).

Most galaxies are either *spiral* (Fig. 21.6) or *elliptical* (Fig. 21.7). They are further classified into several subclasses (Figs. 21.8 and 21.9). A third rare class is the *irregular,* e.g., the Magellanic Clouds (Fig. 21.4).

Evolution of the Physical Universe. The physical universe has not always been the same. From geological studies (or merely a look at a glacier gorge or a river canyon), we know there is a long history of changes of the earth's surface. Changes took place in stars too, we learned, and there were gradual, steady changes as well as catastrophic and explosive ones. Thus we can speak of a natural history of the evolution of the physical universe. These changes are manifestations of the transformations of matter and energy; the transformations are dictated by the dynamical and physical laws of nature and are governed by the three general principles of transformation. We learned in earlier chapters that an important application of the basic laws of physical science is the elucidation of natural history. This study will give us a better understanding of how our present world came into being. Furthermore, because current environmental changes are actually an ex-

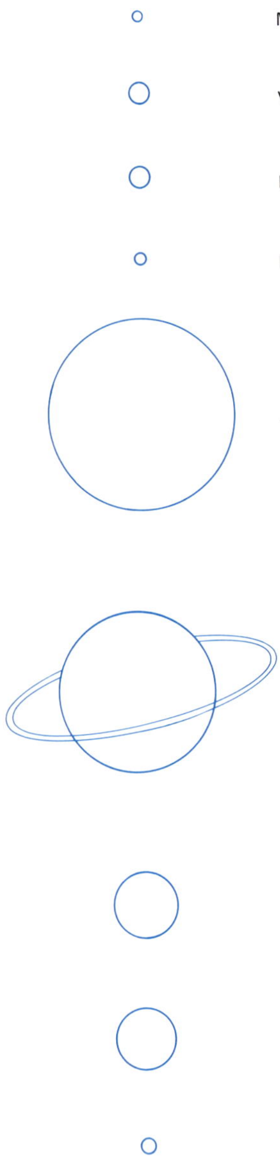

	Diameter at Equator (in miles)	Average Distance from the Sun (in millions of miles)	Period (in years)
Mercury	3,100	36	0.24
Venus	7,600	67	0.62
Earth	7,913	93	1.00
Mars	4,140	142	1.9
Jupiter	86,800	483	12.
Saturn	71,500	886	29.
Uranus	29, 400	1,782	84.
Neptune	27,000	2,793	165.
Pluto	3,600	3,670	248

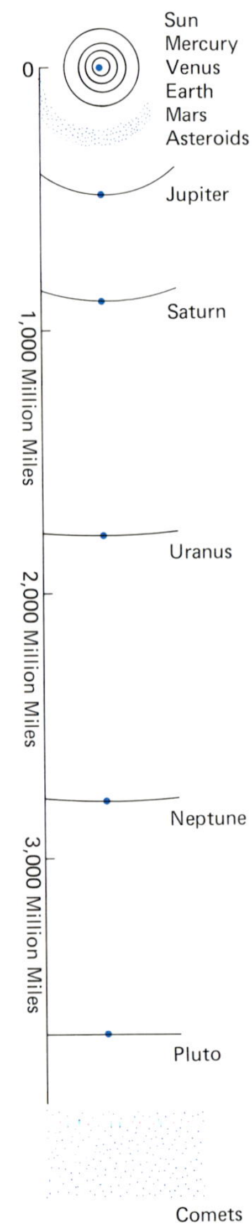

Figure 21.1 Planets of the solar system.

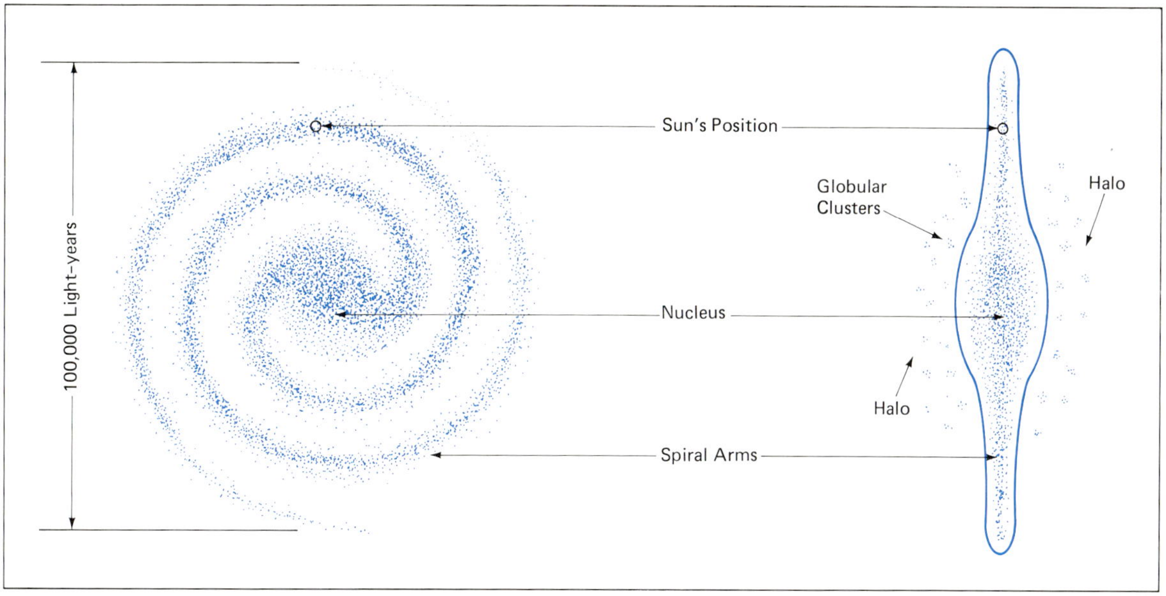

Figure 21.2 Schematic representation of the Milky Way galaxy. [From J. Allen Hynek and Necia Apfel, *Astronomy One,* copyright © 1972 by W. A. Benjamin, Inc., Menlo Park, California]

Figure 21.3 Globular star cluster in Canes Venatici. [Hale Observatories photo]

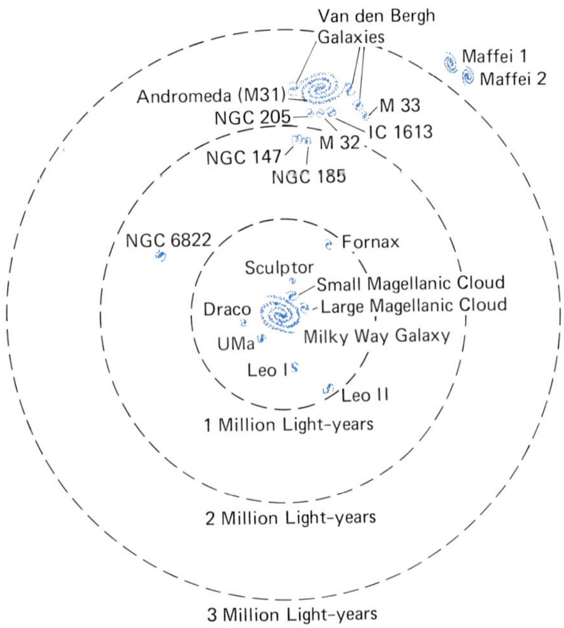

Figure 21.4 Galaxies in our neighborhood. Among the two dozen mostly small or satellite galaxies within 3.6 million light years (one megaparsec) the Andromeda (M31) is the greatest and is comparable to the Milky Way. [From J. Allen Hynek and Necia Apfel, *Astronomy One,* copyright © 1972 by W. A. Benjamin, Inc., Menlo Park, California]

tension of the evolutionary process of the physical world spurred on by human activities, the same methodology may be applied to study current problems related to them.

Inasmuch as the present universe is made up of compounds of the 92 elements, the same elements or the same elementary particles they are made of must have existed at earlier times. Changes in natural history take place when atoms or elementary particles change associations through physical, chemical, and nuclear reactions. Although the laws governing these changes are generally known, a complete prediction of the outcome of these changes requires a knowledge of the initial condition, which must be given as a fact and cannot be derived by reason. In studying natural evolution, we look for an initial condition of the physical universe beginning with which physical, chemical, and nuclear changes took place, transforming the material system from one stage to another until it reached its present state. The initial condition, like the axioms of a system of deductive logic, cannot be explained. Its validity rests on its ability to give a satisfactory account of the present state of the universe. We try out various assumptions for the best fit. No matter how successful we are, the initial condition at the "beginning" of the universe remains an irrational element of the theory which cannot be eliminated.

21.2 Evidence of the Beginning of the Universe

We now will consider the empirical evidence showing that the universe had a beginning. Simple and obvious evidence is the existence of natural radioactivity. The very existence of uranium on earth and in stars is a sign that our present universe could not always have been the same for an infinitely long time. If uranium had been in existence for an indefinitely long time, it would all have been converted to lead by radioactive transformation. Yet only a fraction of uranium has been so converted, indicating that the age of the universe—or at least the age of the presently existent uranium—must not be much longer than the half-life of uranium, which is 4.5 billion years. At least the uranium (and thorium) must have originated or been created from other things, somehow, somewhere, at a finite time several billion years ago. If such an origination occurred in two of the elements, it may well be that it happened with all the other elements. The physical world, made of the elements, thus may have an origin of some kind.

The origin of the elements is not necessarily the origin of time itself. There may have been things

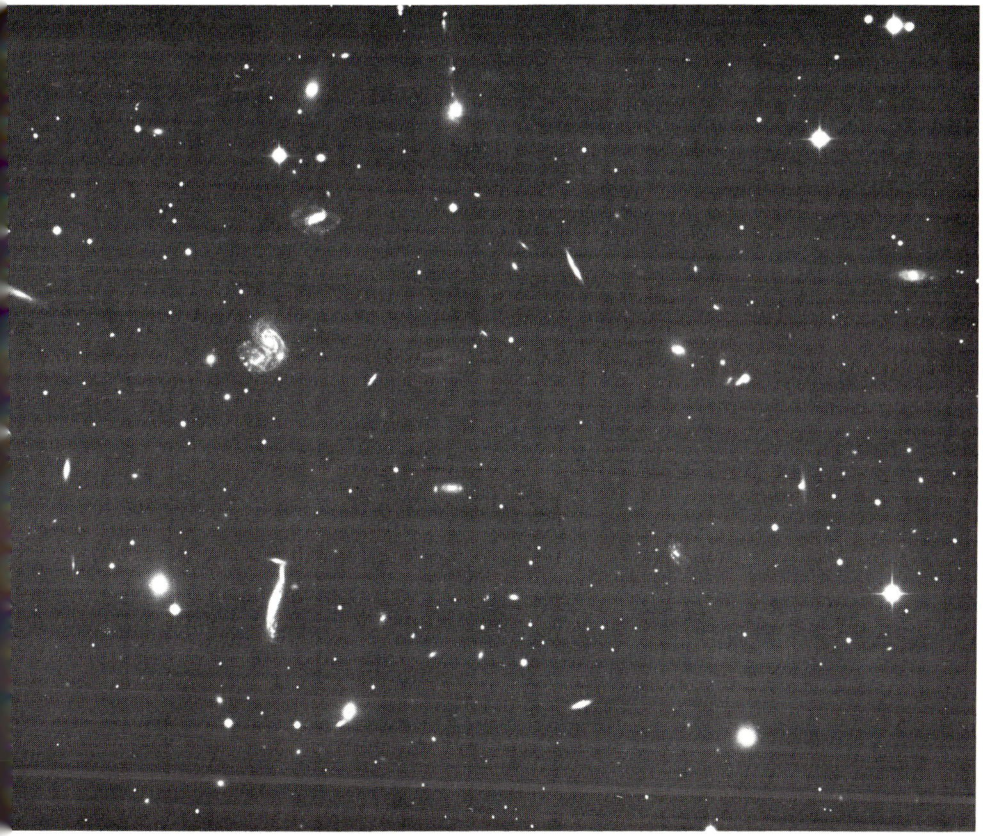

Figure 21.5 Cluster of galaxies in Hercules. [Hale Observatories photograph]

going on before the origination of the present elements. After all, the elements are made of things that existed before their creation. Therefore the term *creation* is not used in an absolute sense. However, what went on prior to origination of the elements is not very relevant to present-day problems and thus is not the most pressing scientific question. In other words, no matter what happened beforehand, occurrences in the present world are not dependent on it. Things that happen today can be traced as far back as the creation of the elements or some similar events, but not much further. Therefore from a practical point of view we are justified in talking about the creation or origin of the universe in a scientific sense.

Another important evidence of the origination is the expansion of the present physical universe, which is composed of numerous galaxies. It is a fact established by astronomical observations (by the Doppler effect on the spectral lines; see Fig. 21.10) that the galaxies are flying apart as if in a state of expansion (like an expanding bubble). The speed of expansion can be determined by observation. From this speed, we can trace back to a time when the expanding bubble was a point and all the galaxies were packed together (by dividing the distance between two galaxies by the relative speed at which they are flying apart). This time turns out to be several billion years ago. That must be the time the expansion started; the physical world has been expanding since and is

Figure 21.6 Spiral galaxy NGC 5194 in Canes Venatici. Satellite galaxy is NGC 5195. [Hale Observatories photograph]

still expanding now. That must be the starting point for the present world; the expansion cannot be traced back any further. Interestingly, the expansion time of several billion years roughly coincides with the age of the radioactive elements.

It thus appears that something catastrophic happened several billion years ago in which not only were the elements formed, but also the galaxies began their expansion. In a practical sense, that is the origin of our present universe.

We still may ask and even speculate on what went on before that catastrophic event. One possibility is that the universe previously may have been contracting from a large volume. After it reached its smallest volume and thus generated a tremendously large amount of heat, it exploded and started to expand. Moreover, the current expansion may eventually end in contraction, and the cycle may be repeated over and over again. From this point of view, the universe really has no beginning and no end. Yet such speculation has little scientific relevance because, as already mentioned, what happened before origination can hardly be tested and verified experimentally from the present world, all the vestiges of the previous world having been wiped out completely during the initial catastrophic explosion.

Figure 21.7 Elliptical galaxy M87 in the Constellation Virgo. [Lick Observatory photograph]

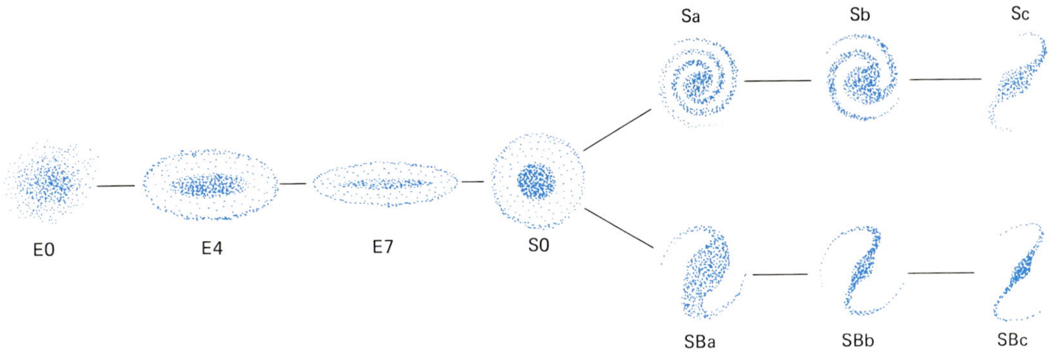

Figure 21.8 An early classification of galaxies, by Hubble, indicating the possible relationship between the various types. E = elliptical, S = spiral, SB = spiral barred. [From J. Allen Hynek and Necia Apfel, *Astronomy One,* copyright © 1972 by W. A. Benjamin, Inc., Menlo Park, California]

21.3 Theories of the Origin of the Universe

Apart from empirical evidence, the problem of the origin of the universe may be contemplated from a theoretical point of view. As discussed earlier, we have gained a body of knowledge in physical science expressed in terms of a set of dynamical and physical laws through which most natural phenomena can be accounted for and future development predicted from a given initial condition. These laws can be used in reversed order in time to reconstruct the past condition from the present one. In principle, this can be done for any time period all the way back to the distant past. If this is worked out for a sufficiently long time back, we find ourselves confronted by one of three possibilities: either the past was much simpler than the present, or it was about the same, or it was much more complicated. In the first case, the study of the origin of the universe is a logically and physically interesting and significant problem because based on some simple beginnings we can explain the complexity of the present situation. In the last case, the problem is uninteresting and unprofitable. In the second case, the problem is interesting because the situation calls for a stringent correlation between the past and the future (that they are always the same) that is scientifically significant. Thus scientific theories of creation are usually related to either the first or the second alternative.

In the first case, we may trace back to a time when the situation was so simple that it could not be reduced to any further degree of simplicity. That situation may be considered as *the* beginning of the universe. The theories developed along this line are the *catastrophic theories,* according to which the universe started in a catastrophic event from which everything else followed and developed from the simple to the complex. Catastrophic theory has a strong intellectual appeal, for it explains the complex by the simple. The expansion of the universe suggests that it may actually be the case. Because the hypothesized origin is an explosion, the theory is usually dubbed the "big-bang" theory.

We can also explore the second possibility. There is an appeal in regarding the world as always having been the same. If this should be the case, then the world has no beginning and no end. The theory developed along this line is the *steady-state theory,* which is the principal rival of the catastrophic theory. The fact that the universe is expanding seems

NGC 1201 Type SO

NGC 2841 Type Sb

NGC 2811 Type Sa

NGC 3031 M81 Type Sb

NGC 488 Type Sab

NGC 628 M74 Type Sc

Figure 21.9 Classification of normal galaxies. [Hale Observatories photograph]

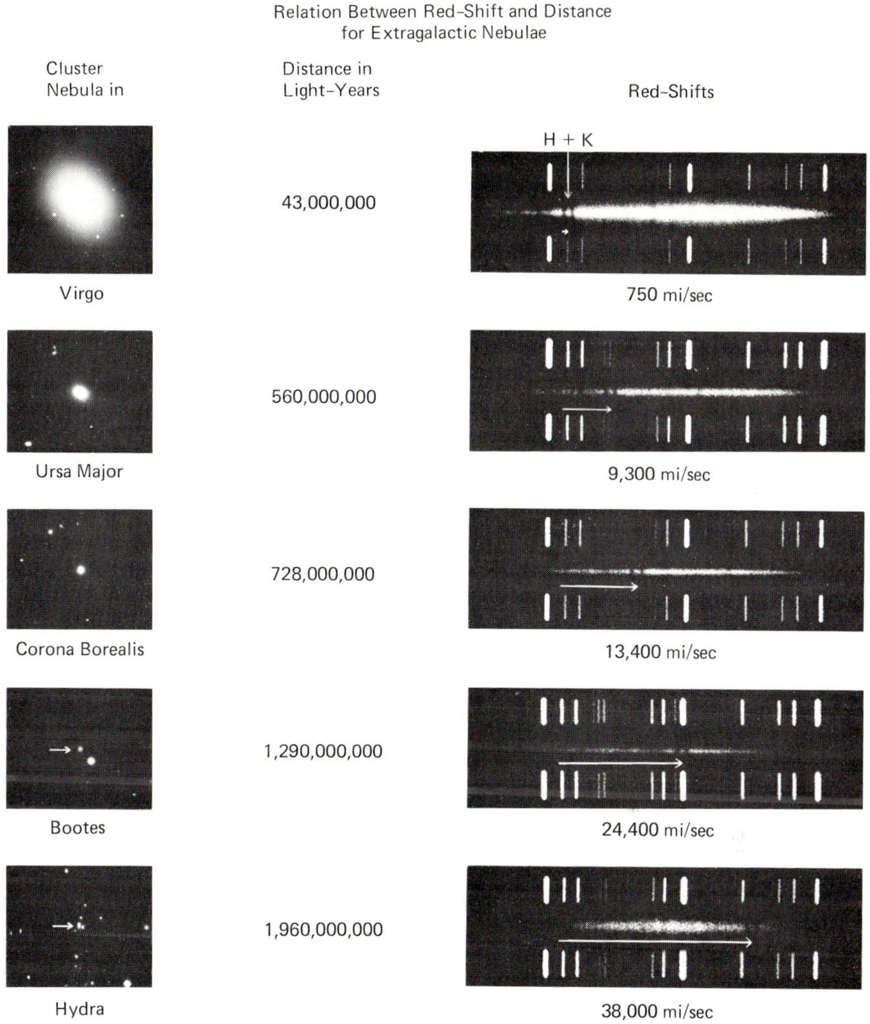

Red-shifts are expressed as velocities, c dλ/λ.
Arrows indicate shift for calcium lines H and K.
One light-year equals about 6 trillion miles, or 6×10^{12} miles.

Figure 21.10 Relation between spectral red shift (velocity away from us) and distance from us for galaxies in the universe. The two are proportional to each other, an indication the universe is expanding. [Hale Observatories photograph]

to contradict this view. Yet advocates of this view have been able to present a reasonably good case for this theory, which is more demanding. From the experimental point of view, this theory is easy to disprove but difficult to establish. Thus the proponents of the theory are constantly threatened by new experimental evidence to the contrary.

A recent experimental discovery seems to lend strong support to the catastrophic theory. If the world began with a gigantic explosion, then there should have been a *primeval fireball.* Such a fireball would emit electromagnetic waves the characteristics (spectral distribution) of which would depend on the temperature of the fireball (Sec. 13.2,

Fig. 13.5). As the fireball cooled down, the spectral distribution would change, with the dominant waves emitted shifting to lower frequencies. Currently, after several billion years, the fireball would have cooled down almost to absolute zero. We would hardly see any fire in the sky. But the fire would not be completely extinct. Based on the law of thermal radiation (Sec. 13.2), we can calculate the temperature of the remnant of the primeval fireball after cooling over the supposed lifetime of the universe (several billion years). This calculation was made in the formulation of the "big-bang" theory of the 1940's; the present temperature of the remnant was estimated to be 5°K, slightly above absolute zero. As long as the temperature remained above absolute zero, the remnant fireball would emit electromagnetic waves with a spectral distribution characteristic of this temperature. In 1965 radio astronomers discovered a radio "noise" from the sky that is ubiquitous and cannot be identified as coming from any definite source. The suggestion was made that it be identified with the remnant of the primeval fireball. Efforts were made to determine the spectral distribution of this radiation and to compare it with that of the presumed radiation of the remnant of the primeval fireball. The evidence to date seems to indicate that the radiation indeed has the characteristics expected of a fireball remnant, with a present temperature of about 3°K. Thus the idea of an explosion at the beginning of the universe received direct experimental support.

Development of the Theory of Creation. The development of the catastrophic theory is greatly helped by the *general theory of relativity,* which predicts the possibility of an expanding universe. In fact, this feature emerged in the theory before actual astronomical observation of the expansion of the universe, but Einstein at first disregarded it because empirical evidence was lacking. Later, after evidence had been found, he regarded this slip as the great mistake of his scientific career. Since then, the general theory of relativity has been the scientific basis for cosmological investigation.

The development of nuclear physics made it clear that the 92 chemical elements are not immutable and that all atoms are made of three elementary particles. The possibility arises that perhaps the universe began with a simpler primordial substance (e.g., elementary particle) out of which the 92 elements of the present physical world were created according to the laws of nuclear physics. As mentioned before, at least the radioactive elements uranium and thorium must have been created a finite time ago. When we consider the chemical composition of the universe we find, based on evidence coming largely from the solar system, that most of the substance is in the form of hydrogen. About one-fourth of the total is helium, and all other 90 elements together account for only 1 per cent of the total. This distribution suggests strongly that perhaps the material universe began with nothing but hydrogen (or neutrons that decay readily into protons and electrons), and other elements were formed from it, with only a small fraction having been converted into helium and a minute amount into all other heavier elements. In other words, the syntheses have barely started, and the original composition of pure hydrogen has not changed much.

The relative abundances of the elements in the universe (Fig. 21.11) in terms of percentages of weight are hydrogen, 76 per cent; helium, 23 per cent; carbon, nitrogen, oxygen, neon, 1 per cent; sodium to scandium, 0.1 per cent; the iron group, $50 < A < 62$, 0.1 per cent; medium-weight elements, $63 < A < 100$, 10^{-4} per cent, and heavy elements, $A > 100$, 10^{-5} per cent. The list is in the order of increasing atomic weight. The remarkable feature is that the abundance decreases rapidly as atomic weight increases, suggesting that heavier atoms are synthesized from lighter ones. The preceding amounts are the percentages by weight. The percentages by number of atoms are different, and show the trend even more drastically: 93 per cent of the atoms is hydrogen, 7 per cent is helium, and all others less than 1 per cent.

The two lines of thought merged together, and in the 1940's early catastrophic theory was developed

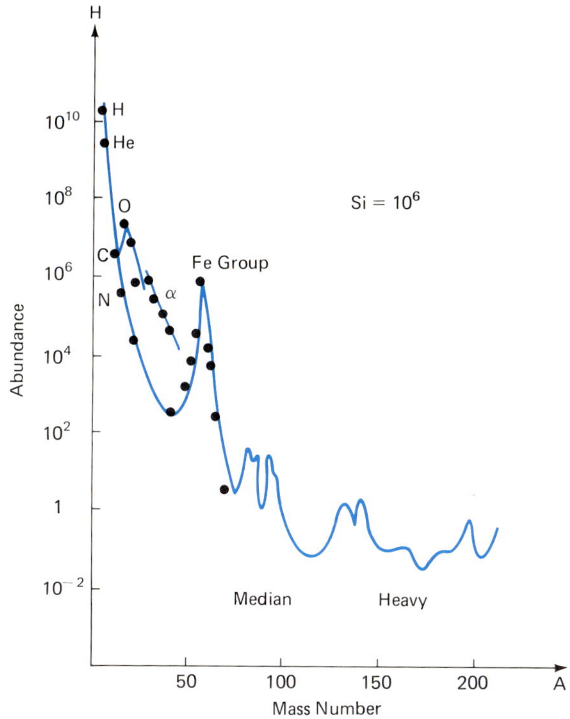

Figure 21.11 Relative abundances of elements. Most abundant is the gaseous group (H, He). Next is the icy group (C, N, O). Last is the earthy group (α-particle nuclides and iron group elements with traces of median and heavy elements).

in an attempt to explain the present world by a very simple initial condition. The initial condition was the assumed "big bang," following which the laws of physics would dictate everything that happens, thus explaining the evolution of the universe to its present state. It was thought at that time that the whole material universe began with a primordial substance packed in a small space at a very high temperature. As the simplest possible choice, it had been assumed that the substance was all neutrons; other variations (combinations of neutrons and protons) had been tried also. Once such an initial condition is given, the laws of physics demand the following to take place: First of all, the high temperature caused the primordial substance to expand and thus accounted for the expansion of the universe. Then the β-decay of neutrons proceeded and produced protons and electrons, the other required building blocks. Protons and neutrons combined to form deuterons ${}_1^2\mathrm{H}$, which in turn absorbed neutrons to form heavier nuclei. Starting with a small "seed" nucleus, all other heavier nuclei were formed by the addition of neutrons. Successive addition of neutrons to a nucleus ${}_Z^AX_N$ changed it to the neutron-rich isotopes of higher N and caused it to move away from the β-stability line and become negative β-active. The ensuing β-decay then brought it back to the stability line and changed the nucleus to one with a higher value of Z. Thus not only N but also Z was increased by adding neutrons (without adding protons). And the neutron is easier to add than the proton, this being a great advantage. By the laws of nuclear reaction, we can calculate the rate of successive formation of heavy elements. The result shows that it takes less than an hour to form all the heavy elements. The theory also enables us to calculate the relative abundances of the elements formed—the heavier elements formed at the later stages will be less abundant than their lighter precursors. This prediction may be compared with the experimental relative abundances, which indeed show a drastic decreasing trend as atomic weight increases. The agreement makes the theory more credible.

Thus a simple, neat origin of the universe was theorized. Within an hour after the start of the "big bang," all the existing elements would be formed, providing the material basis of the physical world. The explosion would send all the substances rushing out in all directions. In due course they would condense to form stars and galaxies. According to Newton's first law, once the substances started flying at a definite speed, they would keep flying at the same speed. Thus the galaxies are still flying apart.

It would be nice if such a simple theory were to represent the true story. Yet nature is always more subtle than we suppose. Now no one holds such a simplistic view. The simple theory has obvious difficulties. If the heavy elements are to be built up from light ones starting with mass number $A = 1$ by adding neutrons, then we run into the difficulty of the mass number gaps at $A = 5$ and $A = 8$. These are mass numbers for which there exist no stable isotopes. Once a nucleus of mass number A equal to 5 (or 8) is formed, it decays immediately and the mass number decreases below 5 (or 8). Thus the mass number cannot grow greater than 5 (or 8), and the elements heavier than these mass numbers cannot be formed by neutron addition. There is another gap before uranium and thorium corresponding to the natural radioactive elements. This gives rise to difficulty if uranium and thorium are to be formed from lighter elements by neutron addition.

At the present time, we believe that the elements were not created in the "big bang," but in the stellar interior over a long period of time by a complex set of nuclear reactions. The problem is worked out together with another important problem in astrophysics—the origin of stellar energy. In fact, creation of the elements and release of stellar energy go hand in hand because they are the principal effects of nuclear reactions in stars. Nuclear reactions largely determine the evolutionary history of the stars, so nuclear processes are the principal concern of stellar astrophysics. In fact, this branch of science has become the most successful proving ground of nuclear physics. Because of our well-developed knowledge of nuclear physics, we actually have learned more in this field than in any other field (such as about planet formation) in natural history. Therefore we will emphasize this aspect in our discussions in these chapters. The next chapter is devoted to the origin of the elements and the evolution of the stars based on the view that elements heavier than hydrogen were formed from hydrogen in the stellar interior. In this way, the process of element formation is separated from the more nebulous problems of cosmology and is confined to a more definite process of stellar evolution that can be investigated more conclusively by the known laws of classical physics and nuclear physics.

21.4 Formation of the Galaxies

As to the basic cosmological problems pertaining to the far distant past in natural history, before stellar evolution, we will satisfy ourselves by imagining a certain beginning—perhaps a "big bang"—that brought forth a large, exploding mass of hydrogen atoms. As volume expanded, the mass cooled, leaving a large body of cool hydrogen gas. This cool hydrogen gas is the initial condition we propose for stellar evolution to begin, and the more ascertainable part of our study of nature may commence. Indeed, this is the same initial condition the steady-state theory provides for stellar evolution, and therefore the uncertainty about the distant past

Figure 21.12 The break-up of a mass of hydrogen gas, resulting in the formation of galaxies and stars.

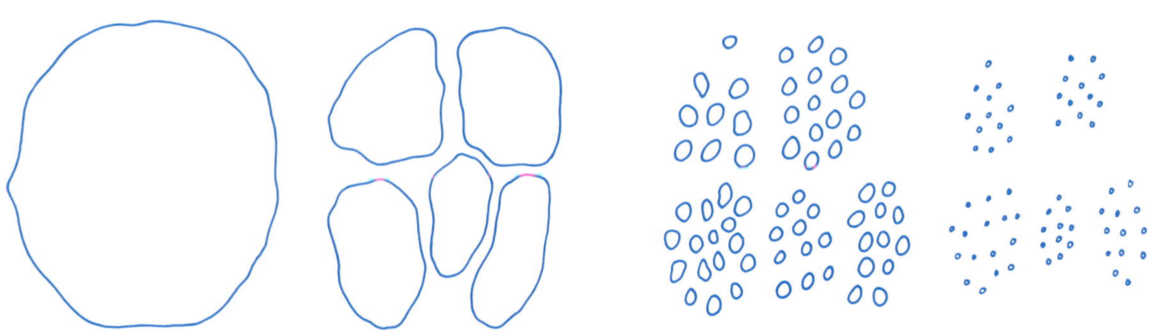

Figure 21.13 Stephan's quintet: a cluster of distant galaxies. [Lick Observatory photograph]

really is immaterial as far as the more significant and interesting problems of stellar evolution and elements synthesis are concerned.

The very large body of cool hydrogen gas, under the action of the gravitational forces among its various parts according to the laws of mechanics, will not remain uniform but will break up into a number of smaller bodies, each destined to become a galaxy (Fig. 21.12). Each will break up into a large number of even smaller bodies, each of these contracting under its own gravitational force to become smaller and smaller in volume and eventually becoming a very small but dense body that is the beginning of a star. The formation of the stars and galaxies from an initially uniform gas is not difficult to understand in the light of the fundamental laws of physics. The breaking-up process is random, so clusters of stars and galaxies (Figs. 21.13 and 21.14) may appear. If the initial body of hydrogen gas is expanding, the resulting collection of galaxies will be expanding according to Newton's first law; this accounts for the observed expanding universe. How a small, dense body of hydrogen gas becomes hot and bright (i.e., becomes a star) and how it yields heavier elements are the main problems of the next chapter. According to the best current estimate, the universe began about 20 billion years ago, the Milky Way galaxy originated about 12 billion years ago, the sun is 4.5 billion to 5 billion years old, and the earth has an age of 4.55 billion years.

Figure 21.14 Virgo cluster of galaxies. [U.S. Naval Observatory photograph]

This view is corroborated by observational evidence. Interstellar gas clouds, essentially hydrogen, do exist in our galaxy that may condense to form new stars in the future. Before a small, dense body of hydrogen gas heats up to become a shining star, it will pass through a stage when the temperature is not hot enough to emit light but is sufficient to emit radio waves (by the law of thermal radiation, Sec. 13.2). Such a body cannot be seen by the naked eye or through optical telescopes (Fig. 21.15), but can be detected by a *radio telescope* (Fig. 21.16). The rapid development of *radio astronomy* in recent years has led to the discovery of a large number of such radio sources in the sky (Fig. 21.17), many of which may be considered early stages of a star. This evidence not only corroborates our conception of stellar formation, but also shows that the process is continuing. In some of the "young" galaxies,

Figure 21.15 Moonlight view of 200-in. Hale telescope dome with shutter open. [Hale Observatories photograph]

Figure 21.17 Radio source in Cassiopeia. [Hale Observatories photograph]

Figure 21.16 The 300-ft radio telescope at Green Bank, West Virginia. It is made of 600 tons of steel, with a surface area of 1.8 acres. [National Radio Astronomy Observatory photograph]

Figure 21.18 The Orion Nebula, a giant luminous mass of gas inside our galaxy. [Lick Observatory photograph]

atomic hydrogen gas accounts for 17 per cent of the galaxy, and therefore the star formation process is still actively in progress. In our galaxy, hydrogen gas (Fig. 21.18) accounts for about 1 per cent of the total mass, and not much new condensation is expected.

Moreover, many stars spew stellar material containing newly synthesized elements back into the interstellar space to be mixed with the hydrogen gas there—so newer stars formed by condensation from this mixture are not formed from the "virgin" material of hydrogen but are "polluted" by the "waste" products of previous stars. The exchange of material between stellar interior and interstellar space proceeds on a large scale, carrying on a "recycling" process. In *solar flares* (Fig. 21.19), the sun spews out a small amount of material, mostly hydrogen, from its interior. It blows by the earth as the *solar wind.* Thus solar material goes into interstellar space. Moreover, there are stellar explosions (Figs. 21.20 and 21.21) of larger scale, such as the *supernova* explosion in which a large fraction of the stellar material is returned to interstellar space. In fact, in observational astronomy we find that "aged" stars (*white dwarfs*) are much less massive than youthful stars. This may be taken as evidence that in the course of evolution of a star, a large fraction of its mass is returned to interstellar space to be incorporated in the next generation of stars. Each generation contributes its pollutants (elements formed by nuclear reactions in the stellar interior). Our sun contains all 92 elements, most of which cannot be formed by nuclear reactions in the sun, indicating that some of the material has been in and out of previous stars several times. The sun is thus a later-generation star. Our earth is formed of the heavy elements in the primordial gas cloud that evolved into the sun and the planets. It is made almost exclusively of cosmic pollutants from the nuclear processes in the interior of many generations of stars.

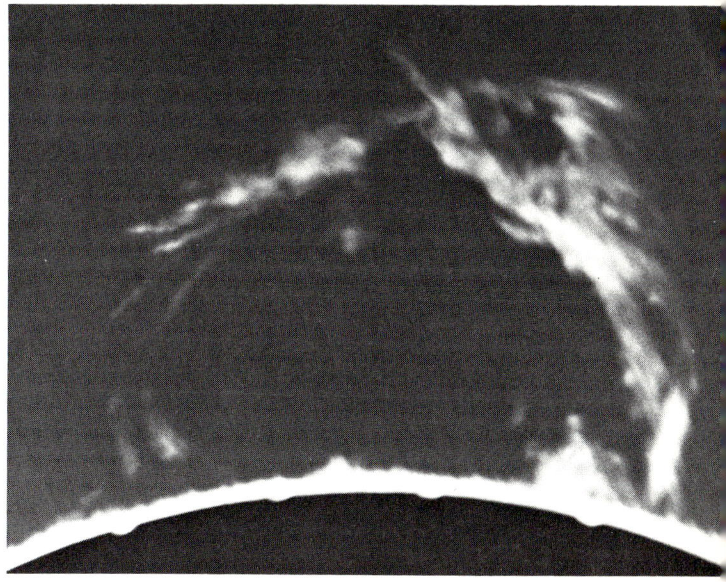

Figure 21.19 Solar prominence of July 2, 1957, rose 205,000 miles high. [Hale Observatories photograph]

Figure 21.20 Expanding nebulosity around Nova (new star) Persei 1901, taken November, 1949. [Hale Observatories photograph]

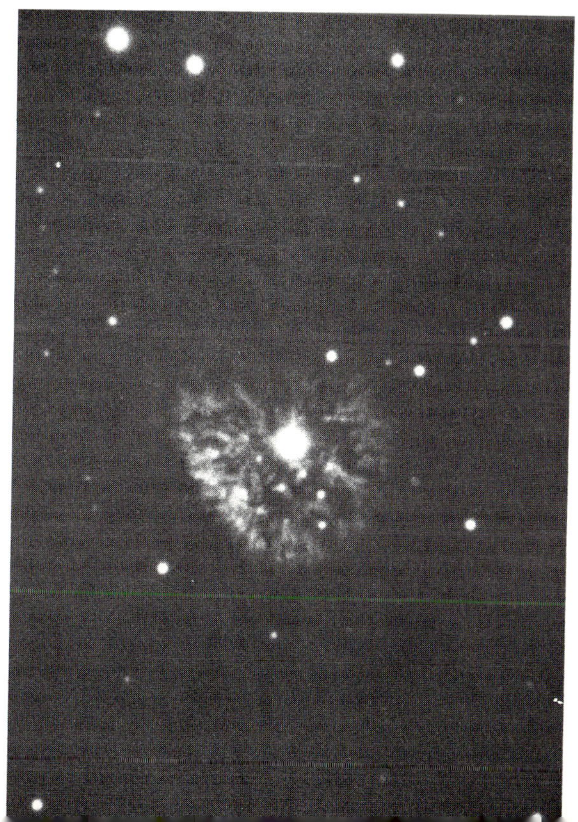

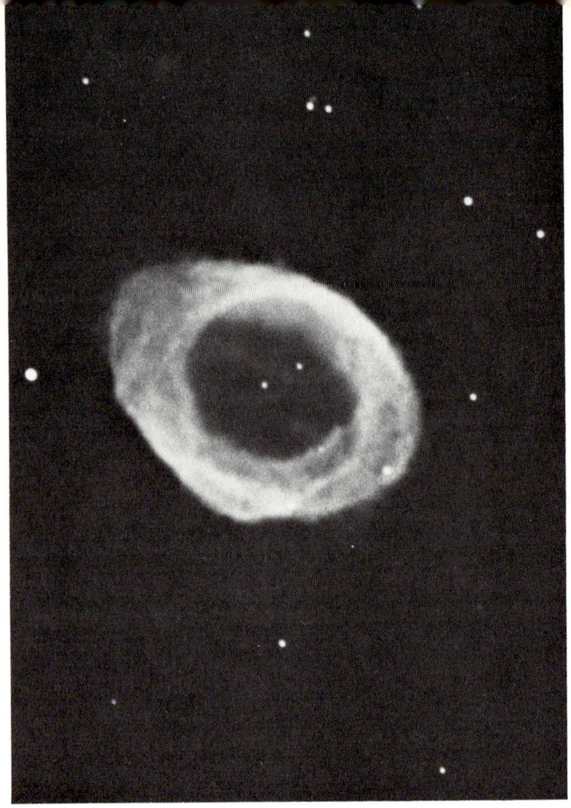

After we discuss stellar evolution in the next chapter, we will discuss the formation of the solar system, the geological evolution of the earth, and the origin of life, giving an account of the nature, origin, and evolution of the present physical universe.

Figure 21.21 Planetary nebula (M57) in Lyra. [Hale Observatories photograph]

Problems

21.1 In a uniformly expanding universe show that every point is a center of expansion. Thus the fact that we are a center of expansion does not mean that we are at the center of the universe.

21.2 In a uniformly expanding universe, show that the expansion velocity of a distant galaxy relative to us is directly proportional to the distance of the galaxy from us.

21.3 Assume the sun to have a surface temperature of 6,000°K and a wavelength of maximum intensity of 5,000A. Find out the temperature of a radio star that emits radio waves in the neighborhood of 10^6 megacycles/sec.

21.4 Why does the periodic table stop at uranium, not earlier or later?

22 Stellar Evolution: Origin of the Chemical Elements and Stellar Energy

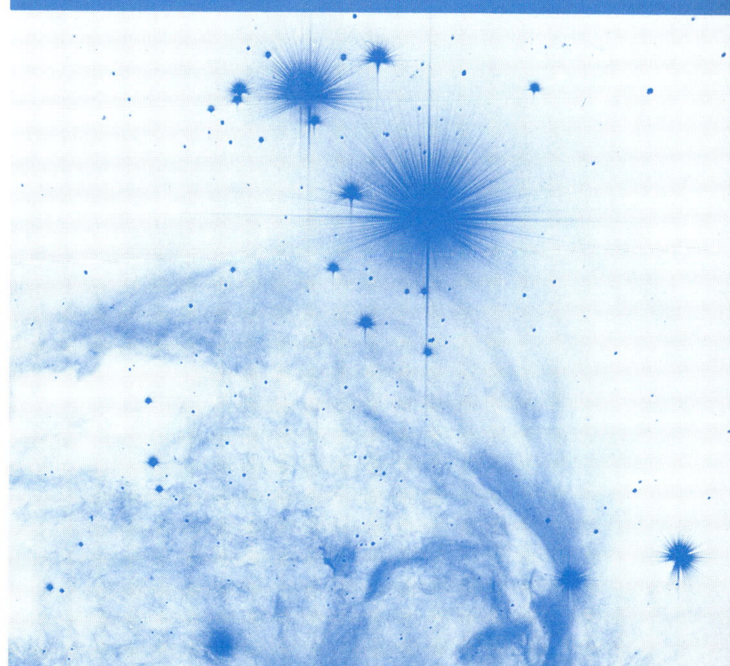

22.1 Hydrogen-Burning Stage

The problem of stellar evolution will be taken up here in a limited setting dissociated from general cosmology. We are concerned mainly with the formation of elements and release of energy by nuclear reactions in the stellar interior. As discussed earlier, the appearance of the element technetium in some stars indicates directly that nuclear reactions take place in the stellar interior. Moreover, the study of the source of stellar energy suggests strongly that nuclear rections take place. The important point to be made here is that we want to ascribe the origin of *all* elements, all substances heavier than hydrogen in the universe, to nuclear reactions in the stellar interior. The energy released in these reactions accounts for the energy source of stars. Thus two basic questions may be answered by studying stellar evolution in terms of nuclear processes.

Stellar Energy Source. The problem of the source of stellar energy will now be discussed. The sun, a typical star, pours out a large amount of energy at all times. The earth intercepts solar energy at the rate of 2 cal/cm^2/min. From this figure we can calculate the total amount of energy released per unit time, which is tremendously large. Where does this energy come from? A naive approach once adopted was to regard the sun as something like a giant burning charcoal ball. We know how many calories of heat 1 g of charcoal will produce when it burns through. We know the total mass of the sun and so can calculate the total amount of heat available by burning. From the rate of energy release, we can figure out how long the sun will last if it keeps on burning at its present rate. The answer obtained is about 6,000 years. For any other fuels (chemical energy source), this figure will be changed merely by a fraction. This calculation was made at a time when our information on the age of the universe was largely derived from the Bible, and biblical scholars had figured out that the creation had occurred about

6,000 years before. The sun's life expectancy so calculated turned out to be of the same order of magnitude, and at the time lent credence to the biblical version of the creation.

Yet geological evidence (salt in the oceans, erosion of the riverbeds, and so on) began to show that the earth is at least several hundred million years old. The age of the universe lengthened nearly a million times. The sun must have been burning for that long. So what kind of fuel is used in the sun? It must be a million times more potent than charcoal to last that long. Although there are better chemical fuels, a difference of a million times is unthinkable. The energy source of the sun thus became a great mystery.

After the development of nuclear physics, it became evident that nuclear energy is the likely source of solar energy because it is a million times more potent (compared atom for atom) than a chemical energy source. In the 1930's, successful attempts were made to figure out nuclear reactions in the stellar interior that would yield the observed rate of energy release. With a nuclear energy source, the sun can keep on burning for billions of years.

Nuclear reactions not only produce energy, but yield new elements. If stellar energy is of nuclear origin, then there must be element synthesis in stars. It may well be that all heavy elements originated this way. The current theory of the origin of elements is thus worked out together with the energy problem as a part of the theory of stellar evolution. The theory is briefly described in this chapter.

Hydrogen-Burning Stars. Let us imagine, following our cosmological initial condition, a large mass of cool hydrogen gas condensing into a smaller body through the gravitational attraction among its parts. As the masses "fall" together, gravitational potential energy is converted into kinetic energy. Some of the energy is used to increase the temperature of the hydrogen gas. As the volume becomes smaller and smaller, the temperature becomes higher and higher. When it becomes sufficiently hot, it emits electromagnetic radiation in the range of radio waves (Sec. 13.2) and can be detected as a radio star, though it cannot be seen as a shining star. As the contraction proceeds further, the volume becomes even smaller and temperatures rise even higher. When the outer surface temperature reaches a few thousand degrees, the electromagnetic radiation emitted is in the range of light waves, and the star begins to shine. The emission of light depletes the energy and causes the star to cool down. But contraction by gravity heats it up again and makes it even smaller. Finally, the contraction makes the volume so small and the temperature and pressure of the interior so high that thermonuclear reactions (Sec. 15.4) take place. This opens up a new page in stellar evolution. The early contraction stage is completed in a matter of a few million years, which is but a brief prelude to the billions of years of new life derived from nuclear reactions.

Because the star under consideration is assumed to be formed from hydrogen exclusively, the only thermonuclear reactions possible are those involving hydrogen and its reaction products. It has been calculated that when the temperature in the stellar interior reaches $10^7\,°K$ and density reaches 10^2 g/cc, a chain of nuclear reactions called *p-p chain reactions* will take place. This is a major mechanism of producing stellar energy in many stars. It begins by two protons reacting with each other when they collide with the prevalent thermal energy at that temperature. The two fuse to form a deuteron ${}^2_1H$ with the emission of a positron and a neutrino:

$$ {}^1_1H + {}^1_1H \longrightarrow {}^2_1H + e^+ + \nu $$

In turn the deuteron reacts with another proton to form ${}^3_2He$, with the emission of a γ-ray photon:

$$ {}^2_1H + {}^1_1H \longrightarrow {}^3_2He + \gamma $$

Two ${}^3_2He$ then react to form a ${}^4_2He$ and two protons:

$$ {}^3_2He + {}^3_2He \longrightarrow {}^4_2He + {}^1_1H + {}^1_1H $$

The net result of the three reactions together is to convert four hydrogen nuclei ${}^1_1H$ into one helium nucleus ${}^4_2$He. The total amount of energy released in this conversion may be figured out from the mass difference between the final helium nucleus and the initial four hydrogen nuclei, which corresponds to about 26 MeV or about 7 MeV per hydrogen. The reaction rate can be calculated, and thus the rate of energy production. It was found that the rate of energy release by this mechanism is sufficient to supply the energy radiated by a typical star.

The p–p chain reactions also include reactions of ${}^3_2$He with ${}^4_2$He, leading to the formation of Be and Li isotopes, which combine with protons and eventually end up with ${}^4_2$He. Thus in all routes the net effect is the conversion of four hydrogen to one helium.

The energy released in the reactions causes the interior of the star to be heated further, and thus the star tends to expand. This tendency is counteracted by gravitational contraction. The two eventually reach an equilibrium, and the contraction of the star stops. The star maintains the same size, the same temperature, and the same rate of energy production as the hydrogen "burns." It will remain so as long as the supply of hydrogen lasts. For a typical star of the size of the sun, the burning of hydrogen is sufficiently slow that the fuel may last for some 10 billion years. Many stars now are in the early stages of hydrogen burning. Thus most of the material substance of the universe remains in the original form of hydrogen, and only a small fraction has been converted into heavier elements.

Three Kinds of Stars. Stars are formed in various sizes. A star much more massive than the sun will contract with much greater force, and therefore the equilibrium temperature will be much higher in order to balance the force. At a higher temperature the nuclear reactions proceed much faster. As a result, the amount of energy released and so the luminosity are much greater and the lifespan much shorter. Therefore among the hydrogen-burning stars there is a definite correlation: the higher the temperature, the greater the luminosity.

In observational astronomy it was found that many stars do follow such a correlation—when the color (surface temperature) and luminosity of a star are plotted as the coordinates of a point on a chart (Fig. 22.1), the points of most stars fall into a narrow band called the *main sequence,* showing that the higher the temperature, the greater the luminosity. The behavior of the *main sequence stars* thus may be explained by the hydrogen-burning mechanism. There are two classes of exceptional stars besides the majority of main sequence stars: *red giants,* which are large, luminous stars with low surface temperatures, and *white dwarfs,* which are small, faint stars with high surface temperatures.

Red Giants. The massive stars burn up their hydrogen content at a faster rate and thus have a shorter hydrogen-burning life. Many massive stars have exhausted their hydrogen fuel long ago and have moved into advanced stages of stellar evolution in which elements heavier than helium are formed.

Figure 22.1 Luminosity–temperature correlation of stars. Stars may thus be classified into main sequence stars, red giants, and white dwarfs.

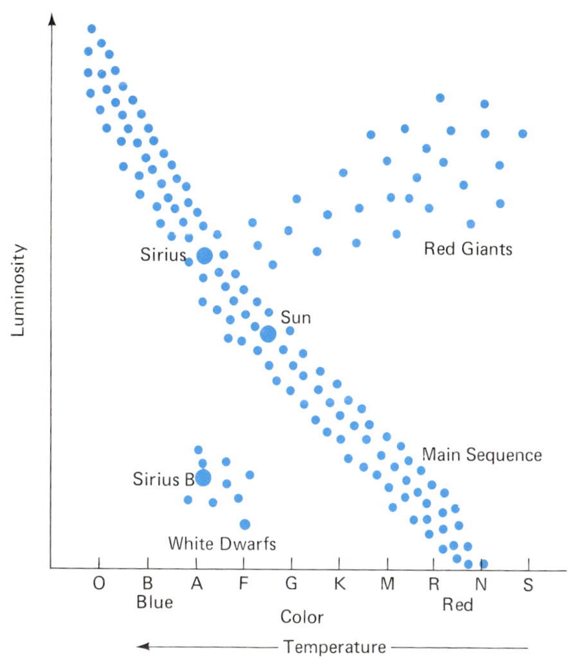

During the stage of hydrogen burning, the "ashes" of the nuclear fire—the helium formed in the center of the star—stay in the center and form a helium core. At the temperature of 10^7°K, the helium does not react and hydrogen burning takes place on a shell outside the helium core. As burning proceeds, the helium core increases in size. Because no nuclear reaction is taking place in the core, gravitational contraction commences again, raising the temperature of the core. The extra heat of the core raises the temperature of the outer layer, which burns faster and expands enormously. In spite of the great luminosity, the increased surface area makes it possible for the large energy output to be radiated at a lower surface temperature (Problem 22.2). The star is thus characterized by a larger size and a redder color (lower temperature) than main sequence stars have. In other words, it leaves the main sequence and becomes a red giant (see Fig. 22.1). One example of a red giant is Betelgeuse in the constellation Orion, one of the brightest stars in Northern Hemisphere's winter sky. It has a radius 800 times that of the sun.

White dwarfs will be discussed later.

22.2 Helium-Burning Stage

As the core contracts further, temperature and density increase. When temperature reaches 10^8°K and density reaches 10^5 g/cc, the helium begins to burn. Two helium nuclei may merge upon collision at this temperature to form ^{8}Be. But ^{8}Be is unstable and it immediately breaks up into two helium nuclei again. Here we return to the problem of the gap at mass number 8. There exist no stable isotopes with mass number 8, so any theory of synthesizing the heavy elements from the light ones must face the problem of overcoming this gap. On the other hand, when the reaction is sufficiently frequent there will be an equilibrium between the formation and break-up processes, and at any moment there will exist a small concentration of ^{8}Be. If another ^{4}He hits ^{8}Be, it may be absorbed to form the stable isotope ^{12}C, and the gap of mass number 8 may be jumped over. This is equivalent to saying that three ^{4}He nuclei collide at the same time and merge, which occurs much less frequently than two-particle collisions. By this mechanism the yield of ^{12}C would be extremely small, too small to be of any significance.

However, there is a phenomenon called *resonant reaction*—when the energy of the incoming particle is just right and corresponds to the formation of a compound nucleus in a definite excited state, the reaction proceeds very rapidly and a large yield may be obtained (like the mechanical resonance at a particular frequency). It has thus been speculated that if the preceding reaction is to be of any significance at all, it must be a resonant reaction. This would be the case if the ^{12}C nucleus has an excited state with an energy level in the neighborhood of 7.6 MeV corresponding to the merging of the three ^{4}He at the temperature of 10^8°K. This point was investigated in nuclear laboratories, and it was found that indeed ^{12}C has an excited state at 7.656 MeV. Therefore the reaction

$$3\,{}^4_2\mathrm{He} \longrightarrow {}^{12}_{6}\mathrm{C}$$

may be expected to proceed at a rate fast enough to account for the creation of a significant amount of ^{12}C. The ^{12}C so formed are at the excited state of 7.656 MeV; many of them go to the ground state of ^{12}C by γ-emission and become stable.

The significance of this reaction is twofold. First, a new source of energy is found. After the depletion of the hydrogen fuel, the star may obtain new life by burning helium by the preceding reaction. Second, elements heavier than helium can be created, and the difficulties of the gaps at masses 5 and 8 are overcome by a great leap from mass 4 to mass 12.

That the preceding reaction of helium burning is possible at all depends on the existence of the 7.656 MeV excited state. Where the energy of an excited state falls is determined by nuclear force and nuclear structure. That the nuclear force works in

^{12}C in such a way that the energy happens to fall on the desired value of 7.656 MeV is purely accidental, but this makes it possible for element synthesis to be carried out beyond helium. Otherwise the universe would be made of two elements, H and He only; there would be no earth, and needless to say, none of us would be here to meditate on the origin of the universe.

Although the helium-burning stage is extremely important for element synthesis because all the 90 elements heavier than helium have their beginning there, it is not very significant as far as the energy source of the star is concerned. The energy released per ^{12}C nucleus (7.656 MeV or 0.07 per cent of the mass) is much smaller than that released per ^{4}He nucleus (26 MeV) in hydrogen burning. Total energy available from helium burning is only one-tenth of that from hydrogen burning, and helium burns at a faster rate (red giant stage). The new lease on life is a short episode before the star relapses into other catastrophic convulsions preceding its eventual death.

As helium burns, the reaction product ^{12}C may react with a helium nucleus to form an oxygen nucleus:

$$^{12}_{6}C + ^{4}_{2}He \longrightarrow ^{16}_{8}O$$

The ^{16}O nucleus may be regarded as being formed of four helium nuclei. Thus the ashes of helium burning include both ^{12}C and ^{16}O.

Hydrogen Burning by CNO Cycle. Stars occasionally spew stellar material back into interstellar space. At the helium-burning stage, some stars may become unstable and eject a large amount of stellar material, including unburned hydrogen and helium as well as the reaction products carbon and oxygen. Once back in interstellar space, these elements mix with the virgin hydrogen not yet condensed previously. When this "polluted" interstellar gas condenses into a star, the star will contain the reaction products ^{4}He, ^{12}C, ^{16}O of the previous generation. It is a second-generation star, differing in composition and properties. A star of this kind can burn its hydrogen by an additional mechanism called the *CNO bi-cycle.* It involves a main cycle of nuclear reactions using C, N, and O as catalysts together with a branch cycle involving ^{16}O, ^{17}F, and ^{17}O. The main cycle starts from ^{12}C, proceeds to ^{13}N, ^{13}C, ^{14}N, ^{15}O, and ^{15}N, and returns to ^{12}C. The branch starts from ^{14}N, proceeds to ^{15}O, ^{15}N, ^{16}O, ^{17}F, and ^{17}O, and returns to ^{14}N. In these cycles, beginning with a ^{12}C or ^{14}N nucleus, four protons are added to it in succession, and finally a helium nucleus splits off, with the remains returning to ^{12}C or ^{14}N, which can be used over again. The net outcome is again the conversion of four hydrogen to one helium, but this time the conversion is carried out with ^{12}C or ^{14}N as a catalyst.

It was first thought that solar energy derives from the CNO cycle. Later studies show that the p–p chain is more important. Both take place in hydrogen-burning stars that contain ^{12}C and ^{16}O, with the relative proportion determined by temperature.

22.3 Burning of the Last Nuclear Fuels

The α-Process. We now go back to the helium-burning red giants. Eventually helium burns out, converting itself to carbon and oxygen. The core of the star made of carbon and oxygen is not reactive at the temperature, so gravitational contraction sets in once again, heating up the core to higher temperatures. When the temperature becomes sufficiently high, i.e., 10^{9}°K, ^{12}C and ^{16}O begin to burn. An α-particle is split off from one such nucleus by a photonuclear reaction prevalent at this temperature and is added to another to form a heavier nucleus. The final result is the production of many intermediate mass nuclides, of which the most abundant are ^{20}Ne, ^{24}Mg, ^{28}Si, and ^{32}S. These isotopes may be regarded as being made of a number of α-particles and therefore are very stable (they have large mass defects).

Isotopes that are more stable are formed in greater abundances, and these isotopes are indeed the more abundant ones. Because they are formed by the successive addition of α-particles, the process by which they are formed is called the α-process.

As the temperature rises even higher in the continuing contraction process, driven by the gravitational force and beaten back from time to time by nuclear burning, the reaction products may be ignited and burn, leading to products of higher mass numbers. But the mechanism is not direct fusion, because the Coulomb repulsion force between the heavier nuclei is much larger. Instead, one ^{28}Si nucleus, for example, breaks down into seven α-particles by photonuclear reactions prevalent at that high temperature, and the α-particles are absorbed by another nucleus—say, another ^{28}Si—to form a nucleus of ^{56}Ni. ^{56}Ni is β-active and decays into ^{56}Co, which again decays into ^{56}Fe. Other isotopes in the neighborhood of ^{56}Fe may be formed in this manner. These isotopes together with ^{56}Fe form the "iron peak" in the abundance curve (see Fig. 21.11).

The e-Process. Once we have reached the iron peak, the evolutionary process of burning lightweight nuclei to yield heavier ones cannot proceed any further. In all the previous reactions, light nuclei fuse to form heavier ones in a process that sees the release of energy. The iron peak isotopes are the most stable (they have the largest mass defects). They are at the rock bottom of the energy scale. We can get energy released by fusing lighter elements to form them or by breaking down heavy elements to form them. Once they are formed, no more energy can be extracted by nuclear reactions. The processes of nuclear burning have finally come to an end.

When the contraction of the core produces a very high temperature, the iron group isotopes will be formed in large abundance. The mechanism of element synthesis now changes from α-particle addition to a new one called the *equilibrium process* (*the e-process*) in which all possible nuclear reactions take place in direct and inverse directions at fast rates, and an equilibrium is reached among all reactants (similar to chemical equilibrium governed by the law of mass action). In the equilibrium distribution the most stable isotopes, i.e., the iron peak isotopes, are expected to appear in greatest abundance. The observed iron peak in the abundance curve corresponds to an equilibrium process of a core temperature of 3.8×10^9 °K and a density of 3×10^6 g/cc. The very existence of the iron peak thus suggests that many early stars did go through the evolutionary process all the way to the formation of the iron peak by the equilibrium process (the iron peak products were redistributed in interstellar space and carried over into the later-generation star systems including the solar system).

Once the stellar material has been converted into iron, there is no further nuclear fuel to sustain the life of the star. It shines in its last stage by energy from gravitational contraction. It becomes smaller and smaller. The rate of energy radiation also decreases because of a smaller surface, but the temperature remains at its original high level. The star thus becomes a white dwarf if it is not too massive. More massive stars will become neutron stars and black holes, to be discussed in the next section.

White Dwarfs. White dwarfs are a class of stars differing from main sequence stars in color–luminosity correlation (see Fig. 22.1). The average mass of the white dwarfs is a small fraction of that of a main sequence star, indicating that a large fraction of the stellar material has been ejected and returned to interstellar space to be incorporated into later-generation stars. But its volume is much much smaller in proportion to its mass. The white dwarfs, having no further nuclear fuel, are expected to cool down gradually and eventually become cool, dark celestial bodies of high density. This cooling process may last as long as 200 billion years because of the low rate of energy radiation (Fig. 22.2).

An outstanding example of a white dwarf is *Sirius B*. It has a high temperature and therefore a high rate of energy radiation per unit surface area. In order to account for the low rate of total energy radiation (low luminosity), the surface area and thus

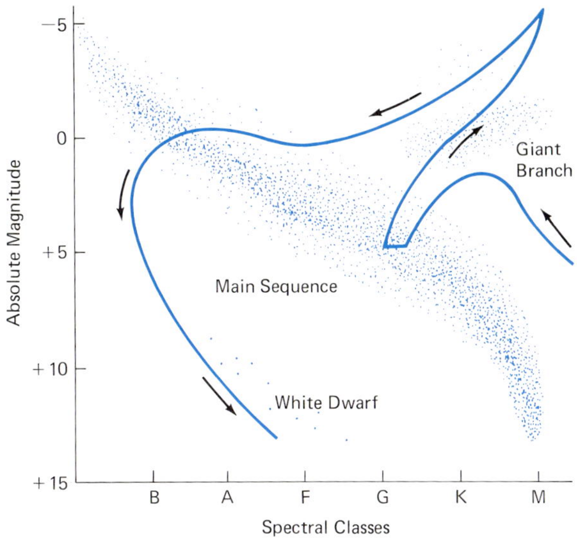

Figure 22.2 The possible evolutionary path of a star with a mass comparable to that of the sun from before the main sequence stage to main sequence to red giant and finally to white dwarf. [From J. Allen Hynek and Necia Apfel, *Astronomy One,* copyright © 1972 by W. A. Benjamin, Inc., Menlo Park, California]

the physical size of the star has to be very small. Sirius B has a radius 50 times smaller than that of the sun. The term *white dwarf* is thus as much scientific as it is poetic. But Sirius B has a mass about the same (95 per cent) as that of the sun. This means that the density of the stellar material is very high. The average density of Sirius B is 2×10^5 g/cc.

On the earth, one of the materials of greatest density is platinum, with a value of only 21.6. Solids are very difficult to compress—compression cannot increase their density very much, because the electron shells of neighboring atoms are mutually impenetrable. However, when the pressure is sufficiently great (more than 150 million lb/in.^{2}), the electron shells may be "crushed"—i.e., the electrons no longer circulate about a particular nucleus but can move freely from one atom to another. This crushed state of matter, consisting of a collection of freely moving nuclei and electrons, is more like a gas than a solid because it can be compressed to a smaller volume by pressure. Thus the density may be increased to a very great value. A heavenly body the size of Jupiter, which is much larger than the earth (317 times heavier) but still smaller than the sun, has a central pressure so close to the crushing point that it almost represents the limit at which a body can exist in the ordinary solid form. Any larger body will have its interior material crushed and will no longer remain a solid. The material in white dwarfs is in the crushed state, and therefore its density can be very high.

The crushed matter acts like a gas. It exerts a pressure and can be compressed. Once the atoms are crushed, the electrons emerge from the atomic shells to form a gas, and this gas exerts a pressure that increases as the volume is compressed. The gravitational contraction will proceed until the volume becomes so small that the electron gas pressure balances the gravitational pressure. The star thus again reaches an equilibrium, and the size contracts no more. White dwarfs are stars in such an equilibrium. Thus even the gravitational energy source is cut off and they can only cool down and die. As already mentioned, they must have a mass greater than that of Jupiter for the interior matter to be crushed. On the other hand, their mass cannot be greater than 1.2 times the sun's mass. For heavier stars, the gravitational pressure is so great that the electron gas pressure cannot balance it. The size of the star will not be stabilized at the white dwarf level, and contraction will continue. This will lead to densities much higher than the white dwarf density, until another support is reached to counterbalance the gravitational contraction, i.e., the formation of the neutron star, which will be discussed later.

Neutrino Astrophysics. A recent advance in the theory of the equilibrium process involves some elementary particle physics; thus this highly theoretical and esoteric branch of science may, after all, have some "practical" relevance. At the high temperature involved in the α-process and equilibrium process, the high-energy photon present in the thermal radiation may produce electron and positron pairs. The members of the pair may annihilate each other to become photons again. Yet under the action of the fourth fundamental force, the weak interaction, we expect that a very, very small fraction of the electron–positron pairs will annihilate to form not a photon pair, but a neutrino and antineutrino pair. The neutrinos do not interact with matter, and thus they escape the interior of the star to enter outer space with great ease, carrying the energy of the annihilated pair away with them. The neutrino reaction thus presents a significant energy drain that in turn speeds up the stellar evolution process. This shortens to one day the burning of nuclear fuel ^{16}O to form ^{56}Ni by the α-process, which would have taken 10^4 years by nuclear processes alone without the neutrino process. The time duration of the equilibrium process is also shortened greatly, to the range of 10^3 to 10^4 sec. Actually, the calculation of the relative abundances of the iron peak isotopes shows that agreement with observation will be obtained if the time for the equilibrium process is in the range of 10^4 sec (not zero, not infinity), which is just what is expected from the neutrino processes. Thus the abundance calculations strongly suggest that the neutrino process discussed on the basis of the universality of weak interaction is indeed correct, even though it has not been verified directly by laboratory experiment.

22.4 Supernovae, Neutron Stars, Pulsars, and Black Holes

Supernova: Collapse of the Core. The senile longevity of a white dwarf is not necessarily the only destiny of an aging star. For a heavier star, the last spasm of its dying years may lead to spectacular events in the heavens, resulting in nuclear reactions that we believe may account for the remaining part of the problem of element sysnthesis from iron up all the way to uranium in the periodic table. During the last stage of a red giant's life, when an iron core has been formed, the following sequence of events may occur. As the iron core has exhausted its nuclear energy, gravitational contraction sets in again. If the star is sufficiently massive, the pressure of the crushed matter cannot stop the gravitational contraction, so this contraction will proceed rapidly and the temperature of the core will be raised again. The increase in temperature shifts the nuclear equilibrium in the core in such a way that nuclear reactions that absorb heat will now proceed (Le Chatelier principle of chemical equilibrium, Sec. 11.4, equally applicable to nuclear equilibrium). In other words, some of the iron peak isotopes will now break up into α-particles, absorbing energy in the process. When $^{56}_{26}Fe$ breaks up, it becomes 13 α-particles and four neutrons. Thus a large flux of neutrons is produced all of a sudden, and new nuclear reactions may now take place.

This reaction absorbs heat, in contrast to the previous reactions that generate heat. The previous *exothermic reactions* (heat-releasing reactions) cause the star to expand, thus balancing the gravitational contraction. The *endothermic reactions* (heat-absorbing reactions) work the other way and therefore gravitational contraction proceeds unimpeded. The star collapses! Its parts fall into its center rapidly (nearly gravitational free fall). The star now becomes a *supernova,* the collapse of its core being a main feature. Other features will be discussed in the following paragraphs. One is the extreme brightness—as if a "new" superstar has suddenly appeared in the sky, thus its name.

Supernova: Neutron Reactions. Before the supernova stage is reached, all nuclear reactions are charged-particle reactions that are inefficient because of the Coulomb repulsion between the particles; the yield of reaction products is generally small. It took much effort and many ingenious mechanisms to explain the synthesis of elements up to $^{56}_{26}Fe$. Neutron reactions are different. They are more efficient because the neutron, carrying no electric charge, can easily enter a nucleus without experiencing the repulsive force that impedes the entry of the charged particles. In the presence of a large flux of neutrons in a supernova, the iron peak isotopes may react with them rapidly to synthesize heavier elements. First, neutrons are added to them one after another to produce neutron-rich isotopes. The more neutron excess, the more β-unstable, so they decay to form nuclides of higher atomic numbers. Neutron addition then takes place again in the newly formed elements, and the processes repeat themselves to form elements of higher and higher Z values. In fact, once the neutron flux is available, all heavy elements may be formed at the same time in one and the same mechanism. Calculations of relative abundances of heavy elements have been carried out on this basis with remarkable success, supporting the mechanism of element synthesis by this rapid neutron addition process (r-process). Incidentally, there also exists a slow neutron addition process (s-process) and a proton addition process (p-process); these complete the list of nuclear processes that explain the origin and abundance of all existing isotopes.

Supernova: Explosion of the Envelope. As the iron core collapses gravitationally, the outer envelope, having its rug pulled from under it, falls toward the center and gains huge amounts of energy suddenly. The envelope is thus heated, and the remaining nuclear fuel (hydrogen) burns rapidly, releasing vast amounts of energy instantly. This results in an explosion, blowing away the outer envelope, dispersing the envelope material back into interstellar space, and carrying with it some of the core material as well. The core material consists of the iron peak elements and all heavier elements formed in neutron reactions. Interstellar space thus contains an even larger variety of nuclear pollutants. The next generation of stars condensed from such a polluted interstellar gas will thus contain not only the iron group elements, but also all the heavier elements from iron to uranium. Our sun contains all the heavy elements and thus contains the material that has gone through such an explosion, according to the present theory.

The sudden release of a large amount of energy and the explosion provide an explanation of the observed phenomena of the supernova. What is observed is a star that suddenly increases its luminosity several billion times and then cools down and fades out in a matter of months (Fig. 22.3). After the initial explosive period, the luminosity of a supernova (type I) decreases with a half-life of 55 days. During the past thousand years, there were three recorded supernovae in our Milky Way galaxy. (Many in other galaxies have been studied recently.) The first one was recorded by the Chinese in 1054 A.D. That supernova was so bright that it was visible in the daytime. The royal astronomer recorded the luminosity changes for months, and modern reconstruction of the light intensity curve shows an exponential decay of a half-life of 52 days, rather close to the figure 55 days. In the position where the Chinese astronomer recorded the supernova we now find a nebula, an interstellar gas cloud (not a galaxy), called the *Crab Nebula*. This body of gas is actually expanding, and the expansion velocity can be measured by the Doppler effect. From the present dimensions of the nebula and its expansion velocity, we can work backward to figure out the time when the expansion began from a point source. This was the time of the initial explosion that produced the Crab Nebula. The point in time reached through this reconstruction turns out to be just about 1054 A.D.[1] Thus it is established that the Crab Nebula is the remnant of the exploding envelope of the Chinese supernova of 1054 A.D. Recently, even the remnant of the core of the supernova has been identified with a pulsar in the center of the Crab Nebula, which will be discussed later.

Neutron Stars. What happens to the core of a supernova after the envelope is blown away? To begin with, gravitational contraction after the exhaustion of all nuclear fuels has increased the density of the core material. The core material is already

[1] Because the Crab Nebula is about 5,000 light-years away, the whole time sequence should be moved back 5,000 years. The actual time of explosion is four millenniums before Christ.

SUPERNOVA IN IC 4182

a

b

c

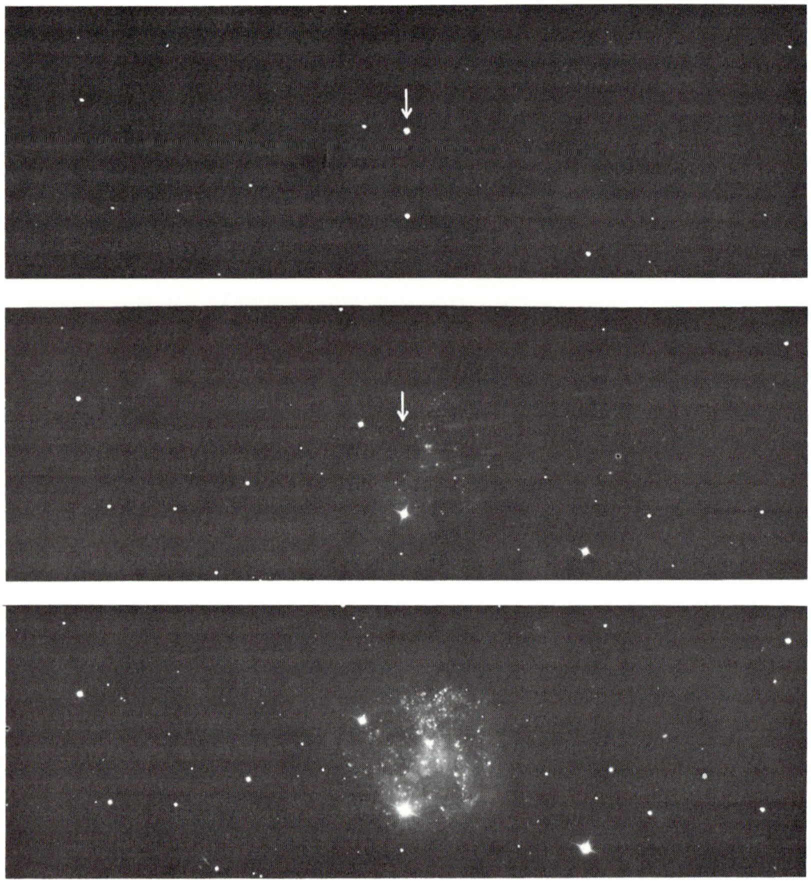

a) 1937 Aug. 23. Exposure 20^{m}. Maximum brightness.
b) 1938 Nov. 24. Exposure 45^{m}. Faint.
c) 1942 Jan. 19. Exposure 85^{m}. Too faint to observe.

Figure 22.3 Supernova in IC 4182. Three Views: 1937, 1938, and 1942. [Hale Observatories photograph]

in the crushed state as that inside a white dwarf, in which the electron shells break down but the nuclei are still intact and separate. In heavier stars, the electron gas pressure cannot counterbalance the gravity, and contraction continues. As density increases further, the nuclei begin to be packed together and merge into one giant nucleus—in other words, the whole star now becomes one single giant nucleus. The star density now becomes comparable to nuclear density, which is of the order of 2×10^{14} g/cc. Here a new support is reached against further contraction, and the size may be stabilized. Nuclear force now works inside this giant nucleus to maintain a nearly constant density under the pressure of contraction—a nucleus is like a liquid drop that has very small compressibility, not like a gas. If the gravitational force is not overly large, the contraction will stop at the formation of the giant nucleus.

This giant nucleus is peculiar in another way. It has to be one made of neutrons only; all the protons will have changed to neutrons. We have already mentioned that electrons cannot exist inside a nucleus. As the contraction proceeds and density becomes higher, the electrons of the star are being absorbed by the nuclei, converting protons to neutrons. When the whole matter of a star is compressed into one giant nucleus, all protons will already have been converted into neutrons, and the giant nucleus is made of neutrons exclusively. Such a star made of neutrons only is called a *neutron star*. Thus one possible outcome of the evolution of the core of a supernova is the formation of a neutron star.

Pulsars. The existence of the neutron star has been speculated on for a long time. In recent years the spectacular discovery of the *pulsars* has led scientists to believe that neutron stars may indeed exist in the form of pulsars. The latest entry of the exotic heavenly bodies, pulsars are stellar bodies (mostly invisible but some visible) that emit radio waves in pulses at precise intervals (from 4 to $\frac{1}{33}$ sec). As if they were telegraph messages, the pulses were first thought to be communication signals from intelligent beings in outer space. Radio astronomers, pointing their radio telescopes into space, quickly discovered a number of radio sources of this kind (about 20 to date) and named them pulsars. Later, pulsars that emit light were discovered and were named *optical pulsars*.

A reasonable explanation of the radio pulses is this: A stellar body may have a local region that emits radio waves in one direction. As this body rotates, the radio beam sweeps around like a beacon, so we receive the radio waves in pulses. The frequency of the pulses we observe is thus the frequency of rotation of the emitting body. This calls for a period of rotation of the emitting body on the order of seconds. The periods of rotation of the stars and planets are in the range of days and months. A body as large as a star cannot rotate as fast as a few seconds per revolution because the centrifugal force would tear the body apart. Only if a body, and so its radius, is extremely small can it rotate that fast. The neutron star, with a density comparable to that of a nucleus and thus a radius as small as several miles, is the ideal candidate for such a rotating body. The neutron-star theory of pulsars thus has gained wide acceptance. Moreover, as already mentioned, an optical pulsar has actually been discovered in the center of the Crab Nebula, the site of the 1054 A.D. Chinese supernova. It is expected that the core of the supernova left behind will turn out to be a neutron star. The discovery of the pulsar in the Crab Nebula, which is likely a neutron star, thus further supports the theory.

Black Holes. Can a neutron star withstand the pressure of gravitational contraction without limit? For very massive stars, the gravitational pressure is so strong that even the nucleus cannot withstand it and will be "crushed." The nucleus is the last support against contraction known to us. After it has been crushed there is no further defense. The gravitational contraction will go all the way unimpeded and the star body will shrink without limit. The whole mass of the massive star falls toward the center. The resulting body becomes so small (a point?), and the gravitational force (inversely proportional to the distance squared, with the distance approaching zero) so extremely large, that nothing can come out of it. Even light cannot come out of it, according to the general theory of relativity. If no light can come out of it, then it will be completely dark. Such a hypothetical entity is designated a *black hole*. Thus according to the current theory, the core of a supernova may end up in one of two possible states: the light ones as neutron stars and the massive ones as black holes.

Whereas the theoretical concept of the neutron star has gained some experimental support from pulsars, evidence for black holes is scanty. Although the black hole cannot be seen by telescopes, optical or radio, its existence may be revealed by its gravity effect. In a binary star system the two stars revolve about each other because of the force of gravity. Among the binary systems there are many in which one member cannot be seen. It could be a faint white dwarf—Sirius B was discovered this way. It could also be a black hole. If it is a black hole, it may suck in gaseous material from the other member. The in-falling material will emit intense X rays and allow the black hole to be detected. Recently it was found that a visible star, a supergiant known as HDE 226868, is revolving once every 5.6 days about an invisible object eight times more massive than the sun—too massive to be either a white dwarf or a neutron star—which also coincides with Cyg X-1, the strongest among 100 or so X-ray sources in the sky discovered by the satellite Uhuru. The X-ray intensity also varies in cycles of 5.6 days. This discovery is considered evidence supporting the existence of black holes.

22.5 Extended Radio Sources and Quasars

Radio Astronomy. In the course of investigating the origin of the chemical elements, we find it necessary to look into various aspects of stellar structure and evolution. Indeed, we have studied most of the significant stages of stellar evolution and discussed most of the stellar species, i.e., main sequence stars, red giants, white dwarfs, and supernovae, each having a proper role to play in element synthesis. These species have long been known to exist, through observations based on the optical telescope. Yet the optical telescope detects only the light-emitting bodies and misses many others. Advances during the last two decades were made possible by the introduction of the radio telescope, which revealed to us a large number of previously unknown heavenly bodies that emit radio waves. Two spectacular discoveries stand out above all others: first the quasar and later the pulsar. Pulsars have been discussed already. To complete our survey of the stellar universe, we now turn our attention to the quasars as well as some of the unusual radio sources.

Extended Radio Sources. During the past two decades, radio astronomers have found a large number of radio sources and have determined their positions accurately. The positions may be compared with positions of objects already identified optically, and frequently the radio source may be identified with an optically observable object. Many of the extragalactic radio sources are associated with certain galaxies. Some of them are very puzzling. Their rate of energy radiation is tremendously high, and it appears as though some explosion has taken place in the galaxy. In one case we see two chunks of material shooting away from each other; these also act as the centers of radio emission. The amount of radio energy emission in radio source 3C295 reaches a value 10 times the optical emission of our galaxy. Where the energy is coming from and how it is converted into radio waves are problems under intensive investigation. These *extended radio sources* represent one of the exotic recent discoveries that we still do not know very much about.

Quasars. There are other mysterious radio sources identified with a starlike object, a point of light seen in an optical telescope that cannot be resolved into a structure. These objects were puzzling because their spectral lines at first could not be identified with any known atoms. They are called *quasars* (Fig. 22.4). We have mentioned that by identifying the spectral lines in a star with those of the known elements, we learn the chemical composition of the star and we also conclude that the elements in the heavens are the same as those on earth. The spectroscopic identification method may be compared to identifying friends in the next room by their voices. This time there are strange voices that do not belong to known friends. On the other hand, a friend who has caught cold may have a hoarse voice that may seem strange, but a judicious listener may still reconstruct the original voice pattern from the distorted voice and make positive identification. The same thing may happen with light waves. We know that light wavelengths change by the Doppler effect: When the light source is moving away from us, its wavelengths become longer and the color shifts toward red, just as the siren of a police car moving away from us appears to have a lower pitch (longer wavelength). It has been found that the strange

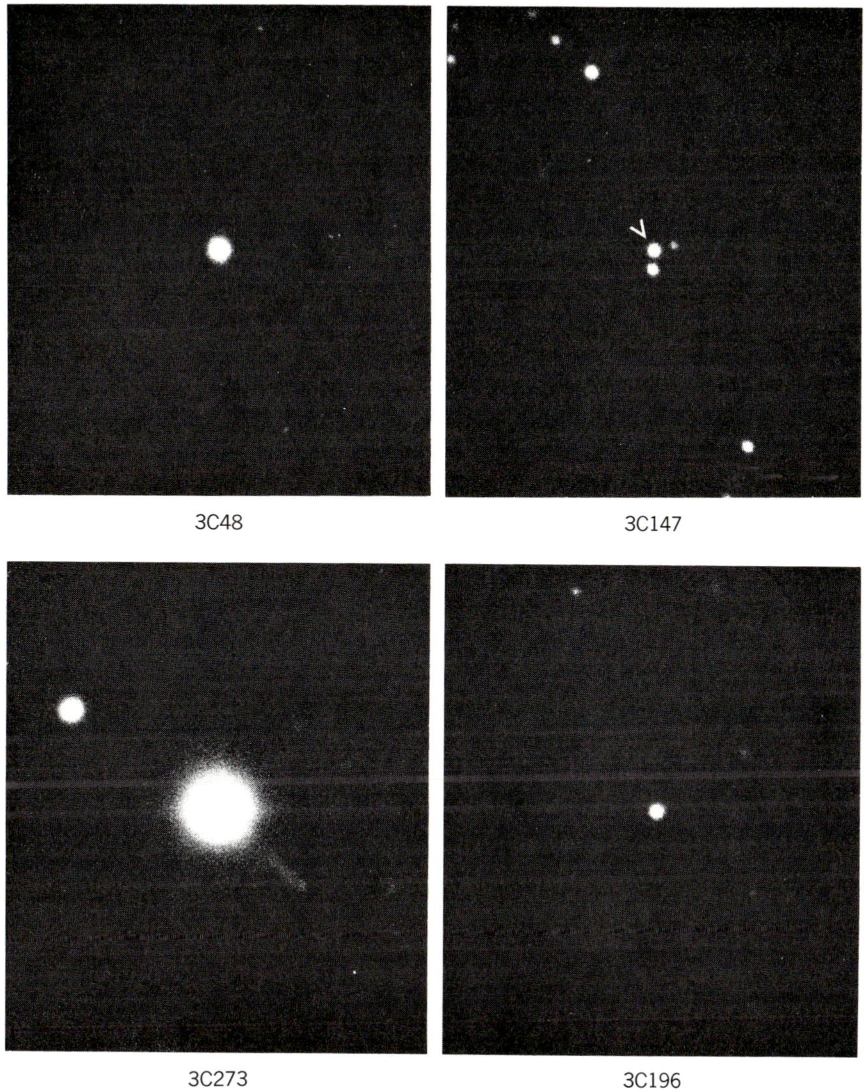

Figure 22.4 Quasars 3C48, 3C147, 3C273, and 3C196. [Hale Observatories photograph]

spectral lines of the quasars are actually lines of familiar atoms such as hydrogen, oxygen, neon, and magnesium suffering a very substantial Doppler red shift (37 per cent red shift in source 3C48). The "old friend" is suffering from such a severe cold that his voice has changed beyond recognition, but it can still be identified correctly by mathematical calculation. Thus these objects are moving away from us at a very high speed (37 per cent of the speed of light for 3C48).

A massive material object moving at a speed close to that of light has never been seen before. In the study of distant galaxies, it was found that the farther away a galaxy is, the larger the red shift; it is on this basis we conclude that the universe is expanding—the galaxies are running away from each other. In an expanding system (like an expanding balloon), every point will see all other points receding from it with a relative velocity (and thus the amount of red shift) proportional to the distance. It is natural to associate the red shift of the quasars to the expanding universe, and the large measure of their shift means that they are at a very great distance from us. In fact, the distances are on the order of several billion light-years. Because the age of the expanding universe is about 20 billion years, the farthest distance that could be reached in that time by a speed that is a fraction of the speed of light is a few billion light-years. This probably is the size (radius) of the present universe. The great distance of the quasars means that they are probably at the "edge" of the expanding universe. Previously, our astronomical observation was limited to our immediate neighborhood; myopic sight cannot give us an overall view of the whole universe. The quasars for the first time provide us with objects likely to be at the edge of the universe. Perhaps a more extensive survey of the quasars may bring us a more complete view of the stellar universe.

If the quasars are indeed extragalactic and at a great distance, then their luminosity must be tremendously high, because they still can be seen here as a starlike light source. Calculations show that they must be more than 100 times brighter than our galaxy.

How can a small starlike object release such a large amount of energy? Many theoretical speculations have been advanced. Perhaps the quasar is a very massive star, one with a mass 10^8 to 10^{10} times the mass of our sun. Then the nuclear energy would be sufficient to account for its luminosity. Perhaps it is a dense galaxy of ordinary-sized stars in which one supernova explosion ignites another in the manner of a string of firecrackers, thus releasing a tremendously large amount of energy.

If the energy problem posed by the quasar is too difficult to solve, perhaps we can get around it by considering the quasar a local object inside the galaxy. If this should be the case, the total energy emission rate would not be tremendously large and there would be no need to find exotic mechanisms of energy production. Yet how can we account for a local object with a speed close to that of light? This seems to be equally difficult to explain. Thus the controversy remains, and many questions are still unanswered.

We have discussed a variety of celestial objects known to us, conventional as well as exotic. Whereas the physical principles underlying the conventional objects are well understood, uncertainty increases when we deal with the more exotic objects, from the supernova to the neutron star, black hole, extended radio source, quasar, and pulsar. Our knowledge of their evolution and the nuclear processes taking place in them is tentative; in many instances controversies still remain, and drastic revisions may be made in the future. Undoubtedly new observational discoveries will lead the way. Already we are going beyond the optical telescope and radio telescope and embarking on the X-ray telescope and γ-ray telescope to study radiations of higher frequency to learn more about the electromagnetic and nuclear processes in galaxies and stars.

22.6 Conclusion

The preceding discussion accounts for the major events that have happened in natural history. It is interesting to note that our discussion revolves around two basic concepts: matter and energy. The evolution of the physical universe is a manifestation of the transformations of matter and energy, the principles of which we have discussed in the preceding chapters. Our present view is that the physical universe began with a fixed amount of matter in the form of a fixed number of hydrogen atoms. In the course of transformation, the number of nucleons (protons and neutrons) in the universe has remained the same even though some of the protons have been transformed into neutrons and have combined with other protons in various aggregates to form the other 91 elements. The transformation is carried out by a series of nuclear reactions. The transformation of matter is accompanied by the simultaneous transformation of energy. The heat and light emitted by stars come from the gravitational and nuclear energies. Each step of the transformations follows definite laws of physics that have been previously established in our laboratory investigations; these laws are summarized in earlier chapters. In applying these laws to the study of natural history, the only additional information necessary is the assumption of the initial condition that in the beginning only cool hydrogen gas existed. On this basis, we can give a reasonably good account of past evolution as well as the present status of the physical universe; the latter include the aggregation of matter in stellar systems, the relative abundances of elements, the radiation of stellar energy, and so on.

What is left to be said concerns the earth, which is of vital interest to us. In the next three chapters we will be concerned with the origin and evolution of the solar system, the earth, and life on earth. Again we begin with an initial condition of the solar system and apply physical laws to predict the evolution of the system, leading to an explanation of the present state of our earthly world. This time, the initial condition is not an arbitrary assumption but is a conclusion obtained through the study of the evolutionary history of the stellar system.

We will be less egocentric if we remember that from the cosmic point of view the earth is but a speck of dirt in the vastness of the universe. It is made of medium-weight and heavy elements formed as the products of stellar nuclear reactions. They came to the solar system as cosmic pollutants from explosions of earlier stars. And it is of these materials that all of us are made.

Problems

22.1 At least how many generations of stars must have preceded the sun? What was accomplished in each generation?

22.2 Explain in detail why the red giant has a lower surface temperature in spite of having a larger luminosity.

22.3 Write out the nuclear reactions one by one in the CNO main cycle.

22.4 If the sun rotated at the rate of a pulsar, compare the centrifugal force mv^2/r with the gravitational force GmM/r^2 at the surface of the sun (radius of the sun $r = 7 \times 10^8$ m; mass of the sun $M = 2 \times 10^{30}$ kg). What would happen to the sun under these conditions?

22.5 Similar to the mass number gaps at $A = 5$ and $A = 8$, there is a wide gap between lead ($Z = 82$) and thorium ($Z = 90$), in which no stable isotopes exist. How can this gap be overcome for the formation of thorium and uranium ($Z = 92$)?

22.6 In the process of formation of thorium and uranium would the transuranian elements be formed? They do not exist in nature now. What happened to them? Would the hypothetical superheavy elements be formed?

23

Origin and Evolution of the Solar System[1]

23.1 Earlier Theories: Nebula Theory and Collision Theory

The solar system consists of the sun, nine major planets (Figs. 23.1 and 23.2), and numerous minor bodies in the form of asteroids and comets.

Asteroids are very small planets, the largest of which, Ceres, is only 500 miles in diameter. Some 2,000 asteroids have been observed; they have a total mass of perhaps one-thousandth of the earth's mass and are congregated in a belt between the orbits of Mars and Jupiter.[2] We believe that *meteorites* that have landed on earth are fragments of asteroids broken off by collisions.

There is reason to believe that beyond the outermost planet, Pluto, there exists a comet belt consisting of 10^{11} comets (Fig. 23.3) with a total mass of one earth mass and extending from a distance of 50 AU to 100,000 AU, but still gravitationally attached to the sun and thus a part of the solar system. Some of them happen to wander into the inner reaches of the solar system and delight us with their characteristic sweeping tails. From spectroscopic analysis of their light we have learned that they are made of ice and dirt. *Meteors* (shooting stars), similarly investigated through analysis of their light, appear to be composed of cometary debris, fragile in nature, that usually burns up in the atmosphere and leaves no solid remains.

The sun is an average star. The origin and evolution of stars were discussed in the preceding chapter. The sun is a hydrogen-burning, main sequence star. Although hydrogen and the less abundant helium account for 99 per cent of the solar mass, the remaining 1 per cent includes *all* other stable elements. It is thus a later-generation star. The sun accounts for 99 per cent of the mass of the solar system and

[1] As a general reference, see F. L, Whipple, "The History of the Solar System," *Proc. Nat. Acad. Sci.,* **52** (1964), 565.

[2] In the position where the missing planet corresponding to $n = 5$ in the Bode–Titus law (an empirical law correlating the orbital radii of planets) is located.

Figure 23.1 Planets Mars, Jupiter, Saturn, and Pluto. [Hale Observatories photograph]

therefore is the main body of the system. The other bodies are minor components and are very close to the sun compared with other stars. The formation of the planetary system thus appears to be a minor episode in the main drama of the evolution of the sun.

The nine planets revolve about the sun in the same direction on nearly circular orbits (ellipses of very small eccentricity). All the orbits except that of Pluto are on nearly the same plane (Fig. 23.4). Moreover, the sun and most of the planets rotate in the same direction, which in turn is the same as the direction of orbital revolution. Most satellites of the planets revolve in this direction also, although there are exceptions. The sun rotates once every 25 days.

The Nebula Theory. Because the sun was formed by condensation of a primordial gas cloud, it is natural to suppose that the original mass of gaseous material did not condense completely into the sun but rather that small fractions of it were left out and remained in the immediate neighborhood, condensing into planets. The relative abundances of chemical elements on earth and in meteorites are very similar to those in the sun (with a few exceptions, mostly light elements), suggesting that all of them originated from the same primordial material. The exception that there is little hydrogen and helium on earth may be explained by the fact that they are both lightweight gases—earth's gravity is not strong enough to hold them.

The idea that the planetary system originated from condensation of a gas cloud or a nebula may be traced back to Kant and Laplace in the eighteenth century. This hypothesis explains the common orbital plane and the common direction of revolution and rotation. However, there is a major difficulty. If 99 per cent of the mass condenses into the sun, then most of the rotational motion (angular momentum) should be carried into the sun and it should rotate much faster—perhaps once every 10 hours. But the sun rotates once every 25 days, and 99 per cent of the angular momentum of the solar system

Figure 23.2 Mars' surface as shown by this Mariner 6 photograph. [NASA photograph]

is associated with the less than 1 per cent of the mass in the planets. This is difficult to explain by the condensation theory.

The Collision Theory. To overcome this difficulty another line of approach was advanced, typified by James Jeans's tidal theory. Another star may have happened to pass very close by the sun, inducing tidal waves on the sun. Some of the solar material may thus have been pulled out of the sun and set into rotation. It eventually condensed to form planets. This would explain the large amount of angular momentum associated with the planets. This view is supported by the fact that the earth's interior is hot—perhaps the earth was formed from material that originally was hot, presumably solar material. If this were the case, then the creation of a planetary system is a rare and exceptional event depending on the unlikely occurrence of the near collision of two stars. Most stars in the galaxy may then not have planets. The chance of existence of life beyond earth would be extremely small and we may consider ourselves as being specially blessed. According to the nebula theory, the formation of planetary systems may be considered a regular and frequent affair, and billions of stars may have planets similar to earth.

The probability of life beyond earth would be very great and we are among the dime-a-dozen. Thus the theory of planetary origin is highly relevant to the problem of life.

With the advance of our knowledge of stellar structure, the collision theory can be shown to be untenable. The stellar material of mostly hydrogen and helium at a temperature of millions of degrees, when taken out from the stellar interior and thus freed from the pressure of gravitational contraction, would explode immediately. It would disperse widely and could not condense to form planets.

Transfer of Angular Momentum. Thus we are forced to go back to the condensation theory. Von Weizsaecker proposed a turbulent eddy theory by which angular momentum may be transferred from the sun to the Laplace nebula disk, but the mecha-

Figure 23.3 Halley's comet, May 12 and 15, 1910. [Hale Observatories photograph]

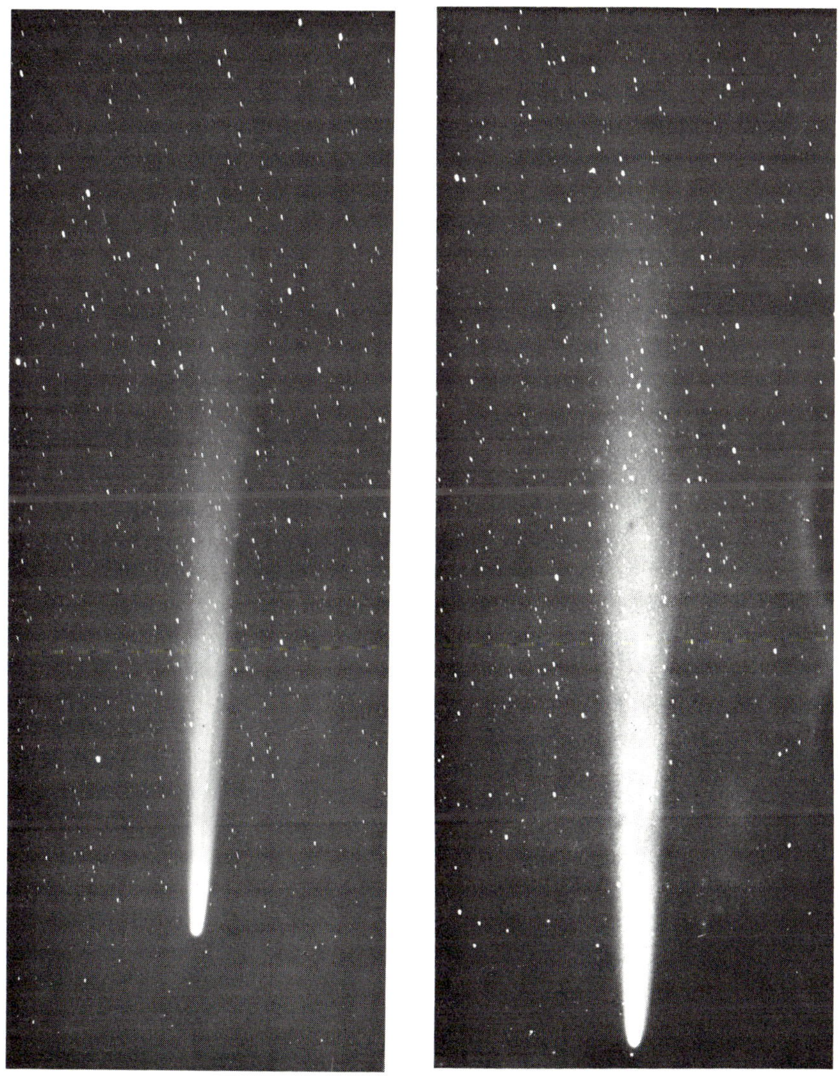

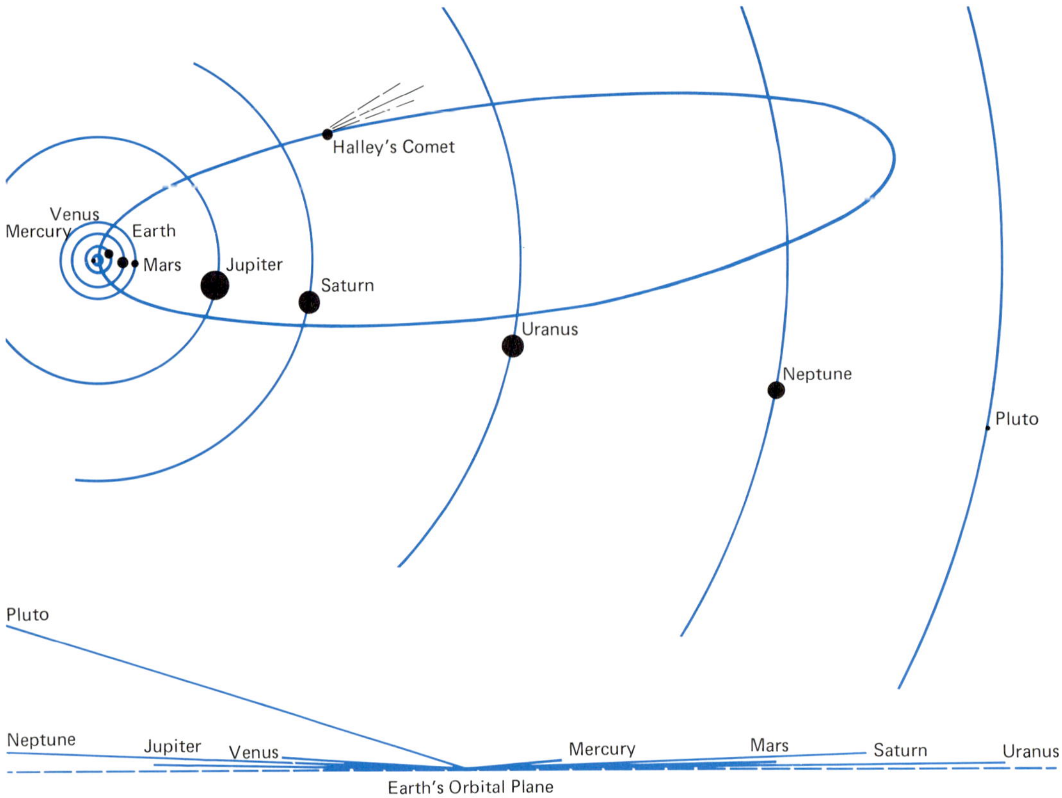

Figure 23.4 The solar system: sizes, orbit radii, and orbit plane inclinations of the planets.

nism is not efficient enough; the sun would still be rotating too fast. In recent years, attention has been directed to *magnetohydrodynamics,* by which a mechanism of transferring angular momentum from the sun to the planets may be worked out based on the magnetic field produced by the motion of charged particles. The mechanism seems promising, but the mathematical problem involved becomes exceedingly complicated and difficult. Young stars are indeed rotating rapidly and old stars like the sun are rotating slowly. This indicates that a certain mechanism is operating to transfer angular momentum away from the star in the later stage of its evolution.

Why Is the Interior of the Earth Hot? If the planets were formed by condensation of cold material, why is the earth's interior hot? This problem can now be answered by nuclear physics. The radioactive material in the earth's interior—U, Th, and particularly ^{40}K—emits radioactive rays and releases energy. The total amount of radioactive material is known, and the rate of energy release can be calculated. This energy becomes heat trapped inside the earth's interior. The heat is enough to have melted the earth in a time much shorter than its geological life (4.6 billion years). Therefore it is quite possible that the earth began through accretion of cold material and then went through a stage of melting. As the radioactive material was used up, the rate of heat production declined, the earth cooled down, and a solid crust was formed. When molten, the heavy metallic elements iron and nickel (the iron peak isotopes in the abundance curve, Fig. 21.11) sank to the center and the stony material (silicates) floated to the top. Geological evidence does show that the central part of the earth is a spherical *core* of molten iron and that wrapped around it is a *mantle* composed of iron and magnesium silicates. The earth's *crust,* made of rocks with which we are familiar, is a thin skin covering the mantle.

The meteorites we studied may be divided into two classes: the iron meteorites and the stony meteorites. Presumably they were formed by the breaking up of planetlike objects. The former are fragments from the central (core) region, and the latter are fragments from the outer (mantle) region.

The melting process complicates the history of the planets. The process greatly changed their chemical composition and physical structure. Here our study relies heavily on chemistry (astrochemistry) rather than nuclear physics, and many of the recent advances in planetary studies, pioneered by Urey, are based on chemical considerations.

23.2 Some Important Facts

Age of Planets. A few important facts must be stated before we consider the modern theory of the origin of the planetary system. The age of the earth may be determined in a number of ways. The most accurate calculation, based on radioactive decay (the lead-lead method of Patterson), determines the age of the earth since it first assembled, disregarding later melting, mixing, fractionating, and loss of gases; it is 4.55 billion years. The age of meteorites determined by this method is very close to this figure. These facts give us a definite idea on the time scale of the formation of the solar system.

Densities of Planets. The densities of the planets vary. The giant planets Jupiter and Saturn have a density close to that of water, 1 g/cc, whereas the small, inner, terrestrial planets (Mercury, Venus, Earth, and Mars) have densities in the neighborhood of 5 g/cc. From what we know of high-pressure physics we have learned that Jupiter must be made largely of hydrogen. As mentioned before, Jupiter's size is just about as great as a solid body can be. For a more massive body, the size would be smaller instead (electron shells crushed). The average density of the earth, to be more exact, is 5.52 g/cc. The density of the surface rocks and outer layer is smaller than this figure, and therefore the density of the core must be higher. On this basis, we conclude that the core is made largely of iron. The density of the moon, 3.34 g/cc, is much smaller than that of the earth. This is puzzling. If the moon is made of the same material as the earth, its density should not differ so greatly. We will return to this problem in Sec. 23.4.

Abundance of Elements. The relative abundances of chemical elements on the earth and in meteorites have been elaborately studied experimentally. Generally they are similar to those in the sun; exceptions are the absence of H, He, the noble gases, and other volatile material on the earth and in meteorites.

The material of which all planets are made may be divided into three groups:

1. The *gaseous group,* consisting of H and He, which are the most abundant of all the elements in the entire universe. Hydrogen is primordial and helium is the product of the hydrogen burning processes in main sequence stars.
2. The *icy group,* consisting of the next most abundant group of elements, C, N, and O, in the form of their hydrogen compounds methane CH_4, ammonia NH_3, and water H_2O. Carbon and oxygen are products of the helium burning processes in red giants.

3. The *earthy group,* consisting of the remaining median-weight and heavy elements, the most abundant being Si, Mg, and Fe, originated in the α-process and e-process in late-stage red giants. The heavy elements originated in neutron reactions (r-process in supernova and s-process in red giants).

The terrestrial planets are almost exclusively made of the earthy group. The giant planets Jupiter and Saturn are largely made of the gaseous group. The outer planets Uranus and Neptune and the comets are made largely of the icy group. It is reasonable to suppose that the planets were originally made of the same materials as the sun, of which 99 per cent belong to the gaseous group, about 1 per cent to the icy group, and a trace amount to the earthy group. Jupiter apparently was formed directly from such a primordial material. On the other hand, after formation, the outer planets somehow had the gaseous group removed, and the inner planets somehow had both the gaseous and icy groups removed. How these components were removed is a major problem in the theory of evolution of the planetary system.

This problem, together with the study of chemical composition of meteorites, led Urey to propose the following sequence of events in the formation of the inner planets. Chunks of the primordial material condensed into objects of a size several hundred kilometers in linear dimension at a temperature below 0°C. Thus the gaseous group was separated from them, but the icy group was frozen inside. These objects were heated by internal radioactivity to perhaps 2,000°C, and thus were ridded of the icy group, the noble gases, and other volatile material. They were then broken up by collisions, and the fragments reaccumulated into the present-day asteroids, the moon, and the terrestrial planets.

The present asteroid belt, according to one theory, was thought to have been formed by the collision and break-up of two earthlike planets. This view finds support in the recent discovery that stony meteorites consist of two classes of different compositions, presumably originating from the two different parent planets. Yet many meteorites have remained cool for 4 billion years. The parent body must have cooled in a relatively short time and therefore cannot have been as large as the earth, which has not finished cooling in 4.6 billion years.

23.3 The Modern View of the Origin of the Planetary System

The Outer Planets. Having stated the basic facts, we can now make a sketch of the processes that probably led to the formation of the planetary system. Our knowledge in this area is far less complete than that of the evolution of the stars. Many conclusions are tentative. The system very likely began with a cloud consisting essentially of hydrogen gas containing dusts made of heavier elements, the total mass of which was about 1.2 times the present mass of the sun. It extended over an area 100,000 times the present radius of the earth's orbit, with a temperature around 50°K. As the cloud contracted, the gravitational energy released was largely consumed in the dissociation of hydrogen molecules and ionization of the atoms in addition to heating and radiation. These mechanisms of removing kinetic energy produced by gravitational contraction were sufficiently efficient that the contraction process was completed in a time comparable to free fall. In a short period of a few million years, the process was completed.

As the cloud contracted, the total amount of angular momentum remained the same. Rotation tends to throw things outward, and thus opposes the tendency toward contraction. The balance of the two tendencies made the cloud condense into a thin, flat, rotating disk.

When the disk became sufficiently small, the two tendencies balanced exactly and the disk contracted no more. At this time the temperature had already risen to perhaps 10^3°K. The halting of contraction removed the source of energy input from gravity, and the cloud began to cool down. Before the cooling, the temperature was high enough to vaporize the icy and even some earthy materials. In the cooling process, the vapor condensed on the unvaporized earthy dusts to form what may be called *cometesimals,* small bodies made largely of icy material.

In the outer regions of the disk, the cometesimals collected to form the planets Uranus and Neptune. They have little gaseous group material. The cometesimals also contributed to the formation of Jupiter and Saturn. These giant planets are made largely of the gaseous group material; they probably were formed directly from the original gas cloud, so they kept the original composition, with the gaseous group dominating. It is likely that the material in the nebula disk moved in such a way that certain regions happened to have greater density because of fluctuation. When the density exceeded a certain limit (the Roche limit), the region became a new center of condensation. Jupiter and Saturn may have originated this way. They also may have formed their own disks of condensation, leading to the formation of their satellites (and the ring of Saturn).

When a solid body is formed by condensation from a medium, whether a gas or a collection of small particles, the rotation (angular momentum) of the latter is expected to be carried over and to appear as the revolution and rotation of the condensed body. Thus the planets and the satellites are expected to revolve and rotate on the same plane in the same direction. This is indeed the case, with some minor exceptions. One outstanding exception, the retrograde motion of the inner and massive satellite Triton of Neptune, may be explained by supposing that Triton was a captured foreign object and that in the capture process it destroyed the original inner satellite system, perhaps ejecting an original satellite that became the present planet Pluto. Beyond Pluto, the cometesimals led to the formation of the present comet belt.

The Inner Planets. We now consider the formation of the terrestrial planets. All evidence suggests a process of accumulation of earthy *planetesimals,* very small planetlike objects. Yet there must have been a mechanism that prevented the appearance of a giant planet like Jupiter. Perhaps the temperature of the inner reaches of the solar system was high enough to evaporate the icy material, which, together with the gaseous components, was prevented from condensing by a magnetohydrodynamic process involving giant solar flares. The mechanism by which the gases were expunged for the formation of the terrestrial planets is an unsettled problem.

Much of the direct information on the early stage of planet formation comes from study of the meteorites. The meteorites vary widely in composition and in structure, indicating that they have been formed in a variety of ways under a variety of conditions. So must their parent bodies have been. Planetesimals of different sizes were formed from the original nebula disk with the gases and ices removed. Because size determined the internal heating by radioactivity, the degree of melting of the interior and the extent of differentiation of the iron varied. Large planetesimals melted and differentiated more completely. The planetesimals then cooled and solidified. They collided and broke up. The fragments, having varying densities because of varying differentiation, reaccumulated to form the present celestial bodies; thus planets and satellites have different densities because they had different compositions in the process of reaccumulation. The bodies so formed went through the heating cycle again; thus the earth is differentiated in its interior. In the inner region of the solar system, the planetesimals eventually led to the formation of the terrestrial planets. In the region between Mars and Jupiter, they formed the asteroid belt. The asteroids are dissipating; they are not coagulating through collisions but rather are fragmenting. Small pieces land on earth in the form of meteorites. Their composition and structure provide clues to reconstruction of the early history of the solar system.

As mentioned earlier, we know less about the origin of the planetary system than we do about the origin of the stellar system. The preceding discussion brings out the problems involved and the complications of the process. Expert opinions vary, and the conclusions are subject to revision. Before we finish, we will consider the origin of the moon, which is an interesting problem further illustrating the difficulty and complexity of the problem concerning the origin of the planetary system.

23.4 The Origin of the Moon

The moon (Fig. 23.5) is special among all satellites in that it is unusually large (more than 1 per cent in mass of the parent planet earth), its density (3.34 g/cc) is much smaller than that of the earth (5.52 g/cc), and it is the only satellite of the earth. Jupiter and Saturn have more but smaller satellites. Many theories of the origin of the moon have been advanced. The theory by Ringwood and Wise is particularly interesting in that it can answer many related questions. This theory is outlined below.

Figure 23.5 The moon, age 7 days, first quarter. [Lick Observatory photograph]

As the earth's interior went through melting, iron separated and sank to the center. The mass of the earth was thus more concentrated in the center; as a result, the speed of the earth's rotation suddenly increased. This follows the law of conservation of angular momentum in mechanics (Sec. 5.7) and is similar to the phenomenon that a spinning ice skater suddenly spins faster as she drops her outstretched arms, concentrating the mass of her spinning body closer to the center. The increased speed of rotation corresponds to a greater centrifugal force, causing the ejection of a part of the equatorial material together with the primitive atmosphere to form a massive ring separated from the earth. Density fluctuation caused the formation of a *protomoon* near the earth. Tidal friction of the earth induced by the protomoon caused the latter to spiral outward, sweeping up outer material of the ring to form the moon. Thus the earth has just one large satellite, whereas Jupiter and Saturn have numerous small satellites. The craters on the surface of the moon (Fig. 23.6) bear the marks of this accretion process during the outward spiraling sweep. [Compare them with the meteorite craters on earth (Fig. 23.7).] The density of the moon is smaller than that of the earth because the moon was formed from the earth's outer layer, which has a smaller density than the average.

Figure 23.6 The moon, Apollo 15 landing area as seen by Command–Service Module in lunar orbit. Hadley Rille meanders through the center; the Apennine Mountains dominate the picture. [NASA photograph]

Figure 23.7 The Barringer meteorite crater near Winslow, Arizona, is 4,200 ft across and 600 ft deep. [Yerkes Observatory photograph]

Figure 23.8 Apollo 15 Command and Service Modules in lunar orbit as seen from Lunar Module. [NASA photograph]

Figure 23.9 The Apollo 11 Lunar Module ascent stage as seen from Command-Service Module during rendezvous in lunar orbit. The earth rises above the lunar horizon. [NASA photograph]

Moreover, the theory at the same time provides a mechanism that explains removal of the primitive atmosphere, methane (CH_4), ammonia (NH_3), and excess water (H_2O), i.e., the icy group material of the earth. The present atmosphere of nitrogen N_2 and oxygen O_2 probably is of secondary origin (Sec. 24.3). In addition, because the moon was formed after the melting of the earth, the radioactive heat-producing material may have been largely used up, and the moon may not have gone through the process of complete melting. Observational evidence suggests that this is indeed the case; the surface of the moon is quite rugged, which would not be so if the moon had melted completely at one time. Calculating by the energy loss due to tidal friction, we can figure backward to find the time that the moon material spiraled out of the earth. The resulting span is much shorter than the age of the earth, further corroborating the theory.

If the moon never has melted, then there never should have been any volcanic activity on it. One of the major missions of the Apollo lunar expeditions (Figs. 23.8 to 23.12) was to ascertain the extent of volcanic activities on the moon. Judging by the Apollo 16 astronauts' report, evidence of volcanic activities in the lunar highlands seems to exist. Many unanswered questions remain.

In Chapters 21 to 23 we have covered a field conventionally belonging to astronomy. Astronomy began as an empirical science. Thus we studied celestial objects by observation one after another and we compiled a large catalog of heavenly bodies, including the sun, the moon, the planets, the comets, the stars, red giants, white dwarfs, supernovae, nebulae, galaxies, and so on. Each has its own properties that are ascertainable by observation and experimentation. Some of the properties can be correlated, and we established empirical laws (such as the Kepler laws and the Bode–Titus law). As far as an empirical science is concerned, this is as far as we can go. Yet astronomy did not stop there. The way astronomy is presented here is a dynamical description of the evolution of a physical system over the course of time dictated by the known laws of physical science. These laws are already established outside astronomy and are applicable to all material systems. The only specific information we need for astronomy is the initial condition. Once this condition is given, the laws of physical science predict the evolution of the system and the emergence of new features in later times. Thus the appearance of the heavenly bodies—i.e., the sun, the moon, the planets, the comets, the stars, red giants, white dwarfs, supernovae, nebulae and galaxies, together with each one's characteristic properties and the empirical laws relating them—can be predicted theoretically. To be sure, our theory at present is not complete. Yet this is the ultimate aim of a fully developed science. We have come a long way in astronomy toward this goal. Fundamentally there is nothing preventing us from reaching this goal. Our discussion thus serves the

Figure 23.10 Apollo 15 astronaut Irwin at Rover parked near Lunar Module. Mount Hadley in the background. [NASA photograph]

Figure 23.11 Neil A. Armstrong, Commander of Apollo 11, first man setting foot on the moon. [NASA photograph]

Figure 23.12 SKYLAB, Manned Orbital Scientific Space Station, is the center of space activities after the Apollo program and before the Space Shuttle program. The second crew completed 56 days' stay in 1973. [NASA photograph]

purpose of demonstrating the ultimate state of a science by using astronomy as a model. This is the goal all other sciences eventually are expected to reach. Needless to say, many other sciences are far from reaching this goal. Biological, behavioral, and social sciences are still in the early empirical stage, like the stage of Galileo and Kepler in astronomy, and there is a long way to go. The important point is to realize that there is a long way to go and not to be content with what has been accomplished so far in these sciences.

Problems

23.1 If the process of dissociation of H_2 were not taking place, what would happen to the process of condensation in the formation of the solar system? Without any such processes can a mass of gas be condensed by gravitational force?

23.2 Potassium probably accounts for 0.2 per cent of the earth. The current relative abundance of the radioactive potassium ^{40}K is 0.0118 per cent of natural potassium. What is the current rate of radioactive heat generation inside the earth? The mass of the earth is 6×10^{27} g. The half-life of radioactive potassium ^{40}K is 1.3×10^9 years. The energy of heating per ^{40}K atom is about 0.7 MeV.

23.3 In the preceding problem what would be the rate of heat generation 4.6 billion years ago at the beginning of the earth?

23.4 Approximately how much heat is generated in geological time from the radioactive potassium inside the earth? How many calories per gram of earth material? Is this enough to melt the earth?

24

Evolution of the Earth

24.1 Composition, Structure, and Age of the Earth

The origin and early evolution of the earth are part of the story of the solar system we have just discussed. We have studied the earth's history through melting and differentiation followed by cooling and solidification of the earth's crust. There remain the problems of the evolution of the earth's surface leading to the formation of the present-day *atmosphere, hydrosphere,* and *lithosphere.* These are the problems related to *meteorology, oceanography,* and *geology.*

As already discussed, the earth is formed of the earthy group of the primordial solar material. The gaseous group, the icy group, the noble gases and other volatile materials were lost during the formation of the earth in various ways as already mentioned. If not, the icy group would have dominated the surface of the earth. The atmosphere would be largely methane and ammonia, and there would be an excess amount of water (all lands would be submerged in deep oceans). With these largely disposed of, the present atmosphere and hydrosphere, secondary in origin, were formed of material largely squeezed out of the lithosphere during the geological history. Free oxygen in the atmosphere, on the other hand, has its origin in the biosphere, being the by-product of the process of photosynthesis.

The Composition and Structure of the Earth. The solid component of the earth, a globe (Fig. 24.1) of a radius of 4,000 miles, has a core of iron (and nickel) with a radius about one half and a mass about one third that of the earth. The core is largely molten, but its center is solid. Outside the core is a solid mantle extending nearly to the surface of the earth (where it changes from solid to plastic), with a mass about two thirds of the total. Chemically, it is largely iron and magnesium silicates. Outside

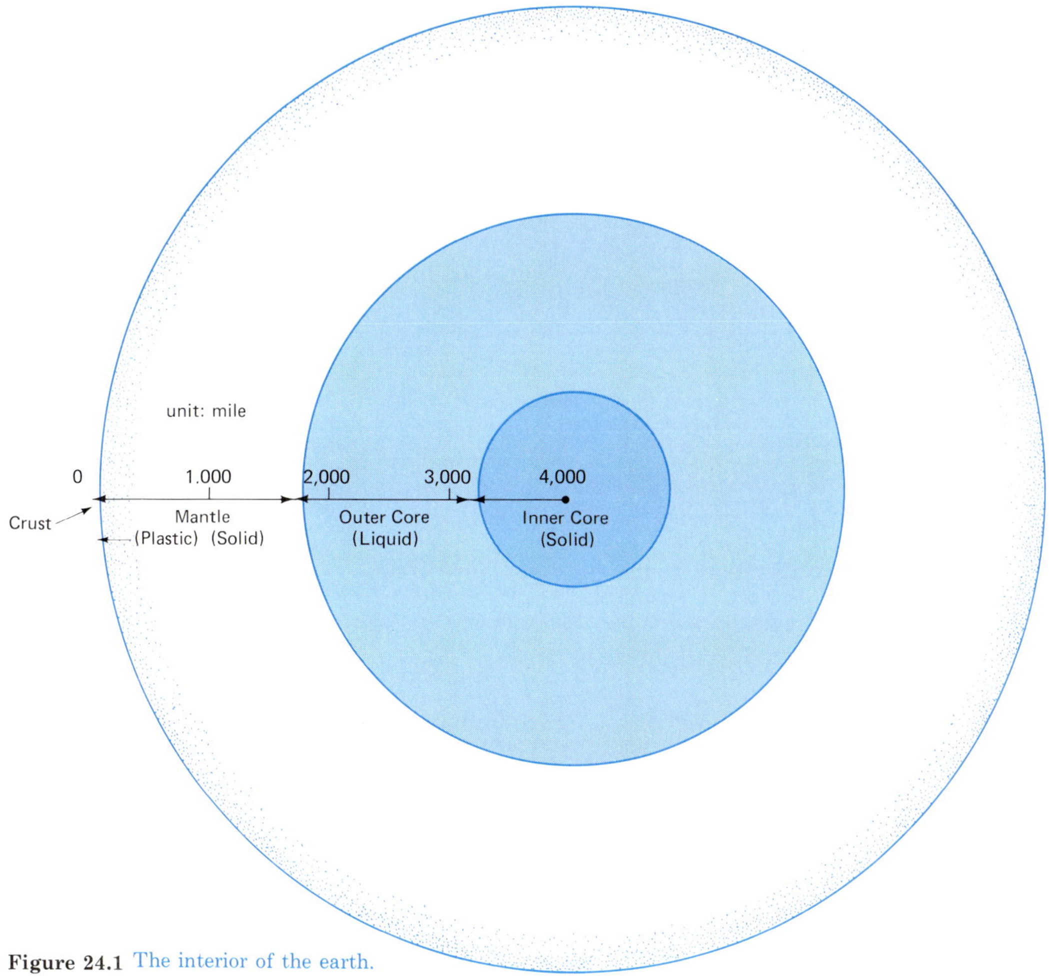

Figure 24.1 The interior of the earth.

the mantle is a thin, solid crust covering the surface of the earth, 4 to 22 miles thick, containing mostly silicon and oxygen with a mass 0.4 per cent of the earth's.

The solid crust (Fig. 24.2) is separated from the plastic upper mantle by a discontinuity called *Moho* (*Mohorovičič discontinuity*). The existence of the discontinuity and the core structure are established by seismic wave observations. The lower crust under the oceans and continents is composed of basalt, largely constituted of compounds of silicon and magnesium and thus is called *sima*. The upper crust, which is under the continents only, is made of lighter rock like granite, constituted of compounds of silicon and aluminum, and thus is called *sial*. On the surface of the continents is a thin veneer of sedimentary rocks. The continents and the oceans are bordered by the *shoreline*. Beyond the shoreline is the *continental shelf* and then the *continental slope* dropping to the *ocean floor*. The continent floats on a *crustal plate,* which may drift about on the plastic upper mantle.

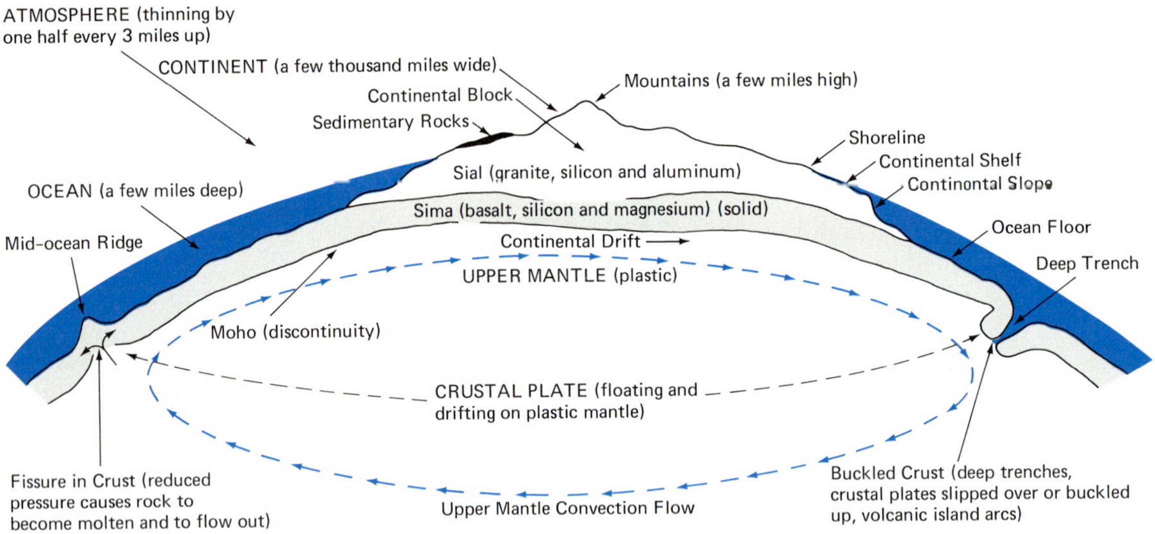

Figure 24.2 Cross section of the surface layer of the earth showing the atmosphere, hydrosphere, and lithosphere. The last includes the continent, the ocean basin, the Moho discontinuity, and the upper mantle. Convection flow in the upper mantle is assumed to explain many geological phenomena.

The hydrosphere, mostly water with 3.5 per cent of it composed of all other elements, accounts for 0.025 per cent of the earth's mass and 70 per cent of the earth's surface. The present atmosphere is of four-fifths nitrogen and one-fifth oxygen, with small amounts of water vapor, carbon dioxide, argon, and traces of other noble gases; its mass is about one millionth of the earth's. Its density thins out rapidly by an exponential law,[1] reducing by one half at every three miles' increase in altitude; very thin air extends to several earth diameters, where it becomes an ionized plasma.

Heat Generation in the Earth. In the stellar interior, heat is generated by nuclear reactions at the rate of 2 ergs/g/sec. In the earth's interior, heat is also generated by nuclear processes (radioactive decay), but at a very much smaller rate—4×10^{-8} ergs/g/sec. The rate of heat generation gives us important information about the earth's interior. In Apollo lunar expeditions, experiments were carried out to measure the rate of heat flow from the interior of the moon. Because the amount of radioactive material has been diminishing continually, the rate of heat generation in the earth was much higher at the start than it is now, perhaps by five to ten times.

[1]Similar to the law of radioactive decay in Sec. 14.5.

Age of Rocks. By the method of radioactive dating, we can determine not only the age of the earth (4.55 billion years) but also the ages of individual rocks from different parts of the earth. The oldest rocks have an age of less than 3 billion years. They occur at the center of the continents. As we reach the edge of a continent, the rocks become progressively younger. In coastal regions the age is down to a few hundred million years. It seems that the continent has grown outward in the course of time. In the ocean basin, no rocks are more than several million years old.

24.2 The Hydrosphere

As already mentioned, the present water on the surface of the earth is of secondary origin. Water recycles actively on the surface of the earth, carrying soluble material from the land to the ocean. One interesting topic of oceanography is the ocean basin, which will be briefly discussed here.[2]

A basic problem in explaining the history of the earth is the formation of the continents and the ocean basins. The earth's crust under the continents is 35 km thick, whereas under the ocean it is only 7 km thick. The ocean bed is uniform in composition and volcanic in origin, whereas the continental rocks are varied, reflecting complicated geological processes. Around the edges of the ocean basins are long, narrow, deep trenches (Fig. 24.2). In the center there exist undersea mountain ridges 1,000 to 2,000 m high. These ridges are very young, perhaps only tens of millions of years old. The *Mid-Atlantic Ridge,* running vertically down the Atlantic Ocean, actually extends into the Indian Ocean by bending around South Africa and then extends into the southeastern Pacific, finally turning up and reaching California, where it connects with the *San Andreas Fault.* The ridge at places is 1,600 km wide. Right in the middle of the width is a rift valley about 15 km wide running along the length of the ridge.

The production of heat by radioactive potassium deep in the mantle leads to the idea that perhaps convection heat transfer takes place. According to this view, the mantle rocks, not perfectly rigid but plastic, are slowly turning over in convection circulation. This enables us to explain many interesting geological phenomena. The heat flow rate is high near the mid-ocean ridges and low in the trenches. Perhaps the ridges are where the convection current rises to the surface (Fig. 24.2) and the trenches are where it returns to the interior. In between is the horizontal leg of the convection current. It is theorized that the ocean crust is continuously being formed along the ridges, dragged down the sides of the ridges and across the ocean basin into the trenches, broken up there and returned to the interior, with the whole movement being driven by the horizontal leg of the convection current in the upper mantle. Thus the ocean floor is continuously swept clean with a material of a uniform volcanic composition. All vestiges of past geological and oceanographical history are removed from the scene; this differs from the continents, where the "book of rocks" has

[2] H. H. Hess, *Proc. Nat. Acad. Sci.,* **52** (1964), 594.

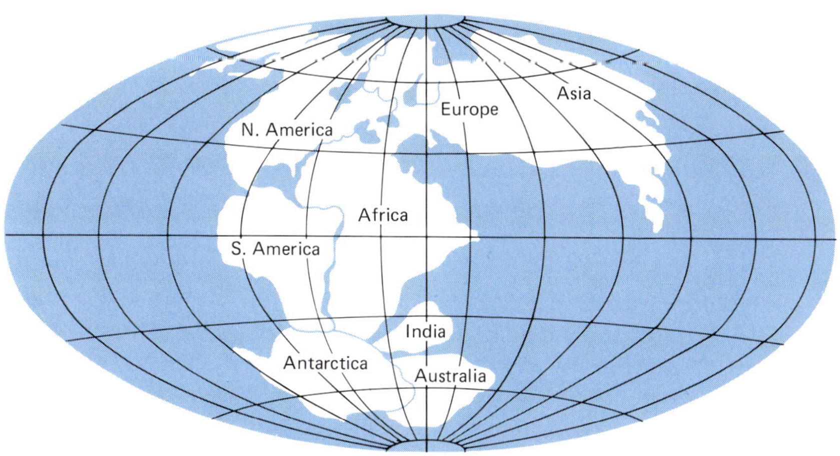

Figure 24.3 The continental drift. It is assumed that all continents were once connected in one land mass as shown in the figure, but later broke and drifted apart along the present coastlines.

preserved past history literally page by page. The ocean crust in the basins is sometimes broken up by long cracks; the crustal plates between these cracks sometimes have slipped past each other for distances up to 1,000 km, suggesting the drag by the mantle convection current.

The interesting old idea of continental drift (Fig. 24.3) may also be explained by the mantle convection current (Fig. 24.2). According to this view, the continents were originally unified, but they split apart about 200 million years ago along the Atlantic Ocean and have drifted away sidewise ever since, as evidenced by the complementary fitting of the coastline contours on the two sides of the Atlantic as well as by the geological and geomagnetic similarities of the two coasts.

24.3 The Atmosphere

At one stage in its early history the earth lost most of its primitive atmosphere and water. For a time the earth had little air and no oceans. The steady growth of the present secondary atmosphere is suggested by the relatively high content of ^{40}Ar (99.6 per cent of natural argon of three stable isotopes), which is the decay product of the radioactive ^{40}K. It appears that the original argon has largely disappeared and the present argon, essentially ^{40}Ar, originated from radioactive decay of ^{40}K (half-life 1.3×10^9 years) over geological history. The potassium content of the earth's crust is sufficient to account for one half the amount of ^{40}Ar. The other half must have been squeezed from the interior (mantle) of the earth. Much of the present air in the atmosphere and water in the oceans may have originated in the same way.

The evolution of the atmosphere is closely related to the evolution of life. If free oxygen were present in the primitive atmosphere, then life would not have originated from inorganic material because organic molecules cannot be formed and persist under oxidizing conditions. Urey and Miller turned to the idea that the primitive atmosphere may be a reducing atmosphere, perhaps a mixture of CH_4, NH_3, and H_2O, the icy group material the original planetary atmosphere was made of. They found through experimentation that a mixture of methane, ammonia, and water under the action of an electric spark or other agents can result in the formation of a large variety of organic compounds. Research along this line has progressed rapidly, and we have learned that most of the amino acids and nitrogenous bases that are, respectively, the building blocks of proteins and nucleic acids—the basic molecules of life—can be formed from inorganic material under simulation of the chemical composition and physical conditions of the primitive planetary atmosphere. The primitive atmosphere was the remnant of the gaseous and icy group material of hydrogen, methane, ammonia, and water after separation of the gaseous and icy groups from the earth. This composition is indeed that of the atmosphere of Jupiter, which did not go through the separation process. Organic compounds conducive to development of living organisms thus could be formed this way.

The earth's gravity is not strong enough to hold hydrogen in the atmosphere. The primeval hydrogen thus gradually escapes the earth. Without hydrogen NH_3 and CH_4 become unstable. Ammonia decomposes to form nitrogen. This is the origin of the nitrogen of the present atmosphere, which is secondary in origin. By this time no more organic synthesis can take place but the organic compounds formed already may proceed to form living organisms.

The first organisms were anaerobic ones, which did not use oxygen. In fact, oxygen was lethal to them. After the development of the photosynthesis process, the by-product free oxygen was released into the air. It was a lethal pollutant in every sense of the term. The increased amount of oxygen made it possible for organisms to develop oxidizing processes for energy utilization. Through a *tour de force* in the evolutionary process, organisms became aerobic, and the lethal pollutant became an indispensable nutrient. From then on, oxygen, carbon, and nitrogen have been recycling between the living species and the natural environment. The amount of oxygen in the air has increased gradually to its present amount. This is the origin of the oxygen of the present atmosphere.

Meteorology. A volume of air such as that in a cylinder enclosed by a piston, which we referred to in our earlier studies, is in an equilibrium state—it has a uniform temperature and a uniform pressure throughout the volume. If the pressure in one part of the gas is changed by some means, the gas will equalize the pressure by itself and a new equilibrium will be established in a tiny fraction of a second. However, the whole body of air on the earth, the atmosphere, is not in an equilibrium state, because of its enormous size. The temperature and pressure at one point (a city) may be different from those at another. The air still tends to equalize temperature and pressure by itself, but it takes hours and days to do it. In the meantime, external factors, such as solar heat, produce new high-temperature and high-pressure regions so that the equilibrium, before being established, is destroyed again. Atmospheric activities thus keep going on without end.

These atmospheric activities are what determine the weather. When two places have different pressures, air flows from the high-pressure area to the low. The air current is what we experience as wind, a major weather factor. All other weather features, such as clouds, rain, snow, thunder, can be traced to atmospheric activities. Besides pressure and temperature another important parameter is humidity, the percentage of water vapor in the air, which usually is measured on a relative scale (100 per cent for water vapor partial pressure in air equal to the saturated vapor pressure of water at the temperature). It is the interplay of humidity, temperature, and pressure that determines clouds, rainfall, snowfall, and so on.

Weather is an important consideration in everyday life. Weather also affects economic activities. Bad weather damages crops. A drought leads to a famine. In human history weather has been one major natural factor that determines the life and death of individuals and nations. Therefore prediction of weather and even the control of it, such as rain making, have been a major practical consideration of mankind. The science of *meteorology* developed to fulfill this need.

Once we know that the weather—that is, the wind, rain, snow, and so on—is determined by atmospheric conditions, the prediction of weather is reduced to a problem of predicting the changes in such atmospheric conditions as temperature, pressure, and humidity. Now the atmosphere is a physical system, a body of gas, characterized by these parameters. Such a system has been well studied in physics. It is a mechanical system and Newton's laws provide in principle a complete solution to the problems of the system. As we learned earlier, the dynamical laws of Newton are such that once we know the initial condition of the system we can predict its condition at any time in the future. Thus at least in principle the complete future can be predicted.

However, this system is much more complicated than those we are familiar with, such as a particle, a system of particles, a rigid body, and so on. The "conditions" of the system are not specified by a few numbers corresponding to a few coordinates but are specified by several three-dimensional functions representing temperature, pressure, humidity, precipitation, wind velocity, and so on. The temperature function comprises temperature values at every point in the three-dimensional space occupied by the atmosphere and thus comprises millions of temperature data. Yet the dynamical principle works just the same. Once we know the intial condition represented by the values of the functions at a particular initial time, the dynamical laws enable us to predict the values of these functions at any future time. The weather condition at any future time at any point (city) on earth can thus be predicted. The dynamical laws of a gas system were developed long ago in aerodynamics. With some modification the laws can be applied to global weather prediction. Here we have a good example to illustrate the nature of a dynamical theory in physical science: prediction of the future based on physical laws and initial conditions.

Whereas the physical laws involved are all known, the initial condition (at the very beginning) cannot be derived by reasoning and has to be given as a fact on an empirical basis. We have to go out to collect the data as they are given to us by nature.

We set up hundreds of weather stations across the nation. Data of temperature, pressure, humidity, precipitation, wind velocity are collected every hour at every station. They are telegraphed to the central bureau where they are analyzed for weather prediction. The initial condition is thus collected in the form of several hundred sets of weather data. In spite of the vast amount of work involved, the information is far from adequate. The initial condition should be specified by three-dimensional functions of temperature, pressure, and so on. Yet the data so collected are at best two-dimensional functions (on ground level). They help, but not very much. Weather prediction in the early years was often so poor as to be almost laughable.

To add information from the third dimension, balloons have been used, so that weather data can be collected at different heights over an observation station. Three-dimensional functions can then be constructed, albeit still very approximately. Prediction is much improved. Now, earth satellites provide the most efficient way of collecting weather data and prediction has been greatly improved (Fig. 24.4).

Thus the initial condition may comprise millions of pieces of data. To sift through them for weather prediction is such a time-consuming job that it is beyond human capacity. It may take a month to figure out the next day's weather. What use is a prediction that is one month late? At this juncture, the computer comes to our help. Electronic computers are designed exactly for such a job—to speed up computation. The differential equations of the weather system and the computer program to solve the equations for weather prediction were worked out by von Neumann (mathematician, physicist, meteorologist, economist, Atomic Energy Commissioner, and so on) in the 1950's. With all the help

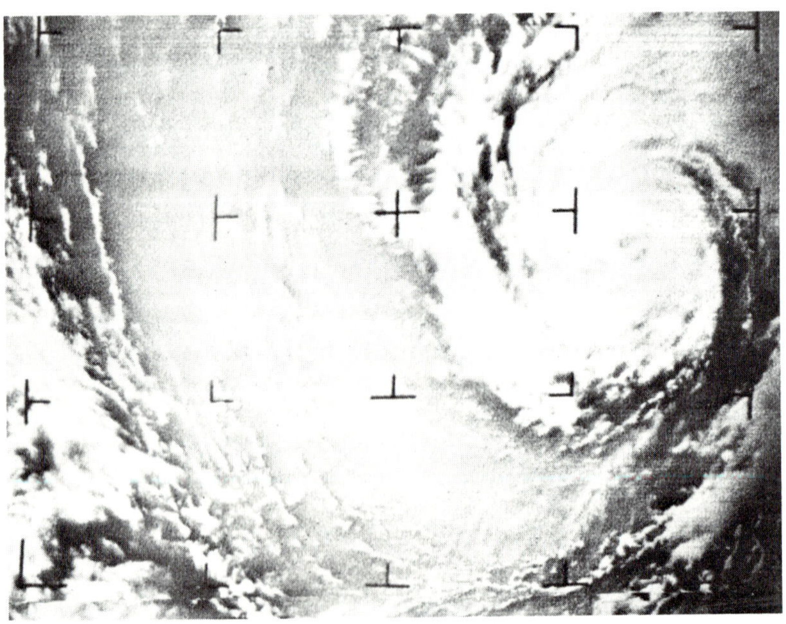

Figure 24.4 The weather satellite Nimbus-I photographed Hurricane Ethel on September 11, 1964, from an altitude of 490 miles. [NASA photograph]

of modern sciences, weather prediction has progressed in our lifetime from unreliable guesswork to verifiable forecast.

In previous dynamical problems we mentioned that once the initial condition is given, all in the future can be predicted. Why do we have to keep on collecting weather data? What is said is true in principle. In practice our mathematical capacity in the present case is still very limited. Even with thousands of weather stations our reconstructed initial condition is still a coarse approximation of the three-dimensional function. With inadequate input the prediction in the short run may not be far off, but in the long run it may be less reliable. Weather prediction within 24 hours is now fairly reliable but long-range forecasts are still uncertain. Thus we collect data every day and make a new prediction every day based on the fresh data. The situation is similar to that in astronomy. In principle, with the initial condition given at one time, we can predict all the eclipses to come. In practice, we observe the "present condition" of each eclipse and use it to correct the very small deviations of the theoretical predictions for the prediction of future eclipses.

This is also the reason why we do not make weather predictions from the initial condition at the beginning of the earth, far less at the beginning of the universe, even though the initial condition we use today for the prediction of tomorrow is rationally traceable to the earlier conditions. Aside from mathematical difficulties there are other factors affecting weather, such as sunspot activities, on which we do not have complete information. Thus we take the initial conditions of the weather of today as an empirical fact and use this fact for the prediction of tomorrow.

Weather prediction illustrates the importance of the initial condition in prediction by a dynamical theory. A dynamical law only tells how things *change,* not how things *were*. How things were must be given empirically as the initial condition. Then the law, specifying the rate of change, enables us to predict the future by adding the changes onto the initial condition. The bank can tell you only the interest rate. The amount of the initial deposit is up to you. The worth of the account at a future date can be predicted by adding the interest to the principal. Thus the prediction depends on both the law (interest rate) and the initial condition (principal).

24.4 The Lithosphere

The Geological History. The composition and structure of rocks reflect the conditions under which they were formed. Past geological history on the continents may be reconstructed by the rock remains. This study is aided greatly by fossils (Fig. 24.5) embedded in rocks. Thus the discussion of geological history is inseparably connected with the biological history of the earth. From the thin yet complex veneer of sedimentary rocks covering the surface of the continents we are able to reconstruct a history of the earth extending 2 billion years back in the past.

The earliest stage, the *Archeozoic era,* occurred more than 2 billion years ago. It was a time of great volcanic activity with some sedimentary deposition. On the biological side it left no recognizable fossils, but organic material has been found in rocks indicating the existence of simple forms of life. This period ended with the First Great Revolution, which presumably wiped out all fossil remains.

The next was the *Proterozoic era,* spanning a period from 2 to 0.5 billion years ago. This era was characterized by great sedimentation, with volcanic activity later. Repeated glaciations occurred. Primitive aquatic plants like algae and fungi are found in fossils. Marine protozoa occurred, and marine invertebrates appeared toward the end. This era ended with the Second Great Revolution, which also caused considerable loss of fossils.

Then followed the *Paleozoic era,* the age of ancient life, spanning the period from 505 million to 205 million years ago. This era is divided into seven *periods* (*Cambrian, Ordovician, Silurian, Devonian,*

Figure 24.5 Fossil footprint on the Dakota sandstone and sketch of the dinosaur that made the print. Print is 18 in. long. [U.S. Geological Survey]

Mississippian, Pennsylvanian, and *Permian*). The lands were low in the beginning, and then they submerged. Later they emerged from the waters and rose, resulting in the formation of the present continents. The Appalachian Mountains were formed at the close of the era. Climate also changed from mild to warm and humid, and then to cool and arid with glaciations. Plant life began with abundant marine algae, followed by the appearance of land plants, eventually culminating in the appearance of great forests of seed ferns that gave us the coal beds of today. Animal life developed from fishes to amphibians and then to reptiles. This era ended with the Appalachian Revolution, which involved some loss of fossils.

Then followed the *Mesozoic era,* the age of reptiles, spanning the period from 205 million to 75 million years ago. This era is divided into three periods (*Triassic, Jurassic,* and *Cretaceous*). Continents rose higher. The Andes, Alps, Rockies, and Himalaya mountains formed in the later part of the era. Seed ferns became extinct early; oak and maple forests appeared later. Dinosaurs dominated the animal kingdom. They became extinct toward the end; at the same time, modern birds and archaic mammals appeared. This period ended with the Rocky Mountain Revolution, which caused little destruction of fossils.

Last is the *Cenozoic era,* the age of mammals, spanning the last 75 million years. It is divided into the *Tertiary* and *Quaternary* periods; each is further divided into a number of *epochs* (*Paleocene, Eocene, Oligocene, Miocene, Pliocene* for the Tertiary; *Pleistocene* and *Recent* for the Quaternary). Old mountains eroded. The Sierra and Cascade mountains formed and rose. The climate became warmer at first, cooler later. The Pleistocene epoch, from about 1 million to 0.025 million years ago, was characterized by repeated glaciations; there were four *ice ages* in this epoch. The Recent epoch of the last 0.025 million years followed the end of the last ice age, with the climate becoming warmer. Forests spread to the maximum in the Oligocene epoch and declined in the Pliocene. Flowering plants developed. But many species became extinct during the Pleistocene. Archaic mammals spread and later became extinct. Modern mammals originated in the Oligocene and developed extensively in the following epochs, but the great mammals became extinct in the Pleistocene. Man perhaps evolved from manlike apes in the Pliocene, but records of man and his precursors can be traced only in the Pleistocene during the last 1 or 2 million years. The earliest species that

Figure 24.6 Taylor Glacier, Rocky Mountain National Park, Colorado, an example of a mountain glacier.

can still be identified with human beings is perhaps *Sinanthropus* (Peking man), who lived about 360,000 years ago. *Australopithecus,* discovered in East Africa, perhaps the earliest tool maker, was an ape-man of about a million years ago.

Glaciation. One important aspect of the earth's history is the change of climate, which is dramatized in the four ice ages of the Pleistocene. Within the past million years or so, which is a very short period in the life of the earth, the weather alternated between cold and warm in four cycles. During a cold period (*glacial period*) ice formed and covered a large part of the continents. As the glacial period advanced, more water from the oceans was converted into ice accumulated on land. At the peak, ocean level was down 300 ft and a huge ice cap covered the Northern Hemisphere as far south as the Ohio and Missouri rivers in America and northern England and Germany in Europe. In the center of the cap, ice piled up 10,000 ft high. The pressure caused the ice to flow southward. Glacial erosions resulted, leaving such geological features as the Great Lakes of North America. During the ensuing warm period (*interglacial period*), the ice melted, the glaciers receded northward, and the ocean level rose again. The cycle was repeated in the next ice age. The Recent epoch is in the early stage of a warm period following the last glaciation. On the high mountains, the glaciers (Fig. 24.6) at the present time are still receding, indicating the continued warming trend. Yet an ice cap remains in the polar region. If the past pattern should repeat itself, another glaciation will appear 50,000 to 100,000 years from now and a large part of the populated area of the world will be under the ice again. Before the ice ages of the Pleistocene, there were at least three recognizable earlier glaciations that took place 700 million, 600 million, and 230 million years ago.

It is to be recognized that the polar ice cap and the ocean water are in phase equilibrium, like an ice cube floating in a glass of water; so the changes of the water between glacial and interglacial periods are phase transitions represented by the horizontal line in Fig. 6.7. Phase transition takes place at a constant temperature, and therefore glaciation does not require large-scale temperature changes. No drastic astronomical changes are required. On the other hand, heat addition or removal is required for the conversion. Even a small amount may cause the transformation of a small amount of ocean water into polar ice, or vice versa, and the effect is cumulative year after year. A small amount of ice yearly can add up over many years to a huge ice cap covering the continents. Thus glaciation may be caused by a slight imbalance of the heat flow to the earth's surface. This point will be elaborated further in the next section.

24.5 Dynamical Processes of the Earth—Tectonics

What caused the geological changes? As the earth evolved in time, continents and oceans emerged, mountains and rivers formed, and glacial periods appeared. Some of the geological manifestations, such as the folding of the rock layers like ribbons and the appearance of marine fossils in high mountains, are so incredible that they seem to call for the supernatural. Yet all the changes may be considered the natural consequences of the dynamical processes of the physical system earth dictated by the laws of nature, evolved out of a specific initial condition.

Heat Balance of the Earth. The initial condition given to us from the study of the origination of the planetary system is that the earth was a globe in the process of cooling down after having been heated internally until molten. The following sequence of events is expected to happen according to the laws of physical science. When the surface has cooled sufficiently, a thin, solid crust will appear and will thicken gradually. Because the crust is solid, the total surface area is fixed. However, cooling makes the volume of the globe shrink. As a result, the surface becomes wrinkled and surface features are formed. The surface features change in time in response to changes on the surface and in the interior of the globe. The formation of continents and mountains will be discussed a little later.

The thickened crust provides a heat insulation to the hot earth and the rate of cooling slows down in time. Eventually the rate will be reduced to zero; the surface cools no more and the crust thickens no more (in the long run, of the order of a billion years, the earth will cool further because of the depletion of radioactive heat sources in the interior of the earth). The balance of energy flow puts the earth in an equilibrium state in a way similar to the equilibrium state in stellar evolution.

In working out the equilibrium condition of energy flow, we have to consider another source of energy input to the earth's surface, that is, the heat energy of the sun intercepted by the earth. Thus the surface of the earth receives two sources of energy input: geothermal energy from within and solar energy from without. The outflow of energy from the surface of the earth is the energy radiated to space by the earth according to the Stefan–Boltzmann law. When the outflow equals exactly the total input, the surface of the earth is neither heating up nor cooling down and a heat equilibrium is established.

If the equilibrium is upset because one of the three factors has changed, then there is a net energy input to (or outflow from) the surface of the earth. The earth will become heated up (or cool down) gradually, the ice in the ocean (polar ice cap) and on the land (glaciers) will melt (or grow) gradually and the global climate will become warmer (or cooler) gradually. If the factors causing such weather alterations change periodically, then the climate may change from warm to cool to warm alternately and we have an explanation of the glacial periods.

Thus the origin of glaciation may be sought in changes in the three factors affecting energy flow to the surface of the earth. The geothermal energy flow rate is not expected to suffer erratic changes. The solar radiation rate is quite constant. Yet many stars suffer from instabilities and there are pulsating stars and explosive stars. Erratic changes in the rate of solar energy production are not impossible. New nuclear reactions that may cause periodic changes in the rate have been sought as an explanation of glaciation. Another possibility is the change of the orbit of the earth or change in direction of the earth's axis of rotation. Glaciation might also be caused by factors on earth. Changes in air composition, variations in the CO_2 and water content of the atmosphere, may change the rate of solar heat absorption as well as the rate of heat emission of the earth. This is a point relevant to environmental considerations. Air pollution may have a long-range effect on global climate. Indeed, one of the fears concerning supersonic air transport is its pollution of the upper atmosphere, which might eventually lead to melting of the polar icecap and inundation of all coastal cities.

The important point is that the earth's surface is in a delicate balance of heat flow. Any slight disturbance, no matter how small, has an effect that will be accumulated over the years to an extent that will bring about such a calamity as a glacial period. If the glaciers advanced merely a foot a year, then in several thousand years a large part of the Northern Hemisphere would be covered by ice. This was so in the glacial periods of the Pleistocene.

Formation of the Continents. Theories of the formation of the continents are not conclusive. The following is a reasonable sketch of the process. As the thin crust of the earth becomes wrinkled, the wrinkled lines may develop cracks and become fracture lines. The pressure in fracture lines is reduced. As a result the mantle rock below becomes molten and changes into *magma,* the molten rock that changes into *lava* in a volcano eruption. This change is due to the fact that the melting point of mantle rock is lowered when the pressure is reduced, a general property of matter we studied in Sec. 7.3. The magma flows into the fissures and piles up to form an island arc in the primitive universal ocean that covered the whole earth more than 3 billion years ago. The process is similar to that taking place presently in the mid-ocean ridges.

Then a similar drifting process as discussed in Sec. 24.2 sets in; the arc moves away from the fracture line; later a new island arc is formed in the fracture line; and so on. Thus a series of parallel island arcs may be formed one after another, and they merge into a continent. According to this view the continent grows by *accretion* (stars, planets, and satellites all grow by accretion in contrast to catastrophic *creation*). This enables us to explain why the continental rocks are progressively younger toward the coastline.

As the continents rise the ocean basins sink because the total volume of the earth is a constant. Later the lands may sink and become submerged under water, and then rise again. There will be corresponding compensatory changes in the ocean basins. In fact, all parts of the land have become submerged at one time or another, though not at the same time. On the other hand, continents have always remained continents and ocean basins have always remained ocean basins.

Figure 24.7 The Grand Canyon.

When the continents were first formed, all rocks were *igneous,* that is, rocks solidified from molten material. When submerged in water, *sedimentary rocks* are formed on top of the original igneous rocks by the accumulation of sediments from erosion. When raised again, the sedimentary rocks may fold to form *fold mountains.* Most of the great mountain ranges, such as the Appalachians, the Rockies, the Andes, the Alps, and the Himalayas, were formed this way. The mountains undergo erosion, new sedimentary rocks are formed, topography changes again, newly formed rocks pile on the old and the surface features of the earth evolve gradually to the present state. The complicated past history of evolution is revealed when the sedimentary rock layers are exposed, such as by the erosion of a river gorge; from such exposures we reconstruct the geological history of the earth discussed earlier. The most spectacular of the river gorges is the Grand Canyon (Figs. 24.7 to 24.9), the geological history of which is shown in Fig. 24.8. The process of mountain formation and erosion will be discussed shortly.

Isostasy. The continental block floats on the plastic upper mantle, which is made of material of higher density. Just as an iceberg tip is accompanied by a large submerged part, the continent rising above the ocean basin has a lower bottom sinking deeper into the mantle. Horizontal compression may cause the surface to be folded and uplifted to form a mountain range. For a higher mountain, just as a taller iceberg, there must be a deeper mountain root sinking into the mantle. Weathering and erosion level the mountain and reduce its weight. The buoyant force will push up this part of the land and the mountain will be elevated again. Because the crust is solid and can withstand some stress, this adjustment may take place not instantaneously but later, when sufficient stress has built up. Then a drastic readjustment takes place and new mountains may be formed. Weathering and erosion then repeat the process. In the northern Appalachians this process has been repeated at least three times.

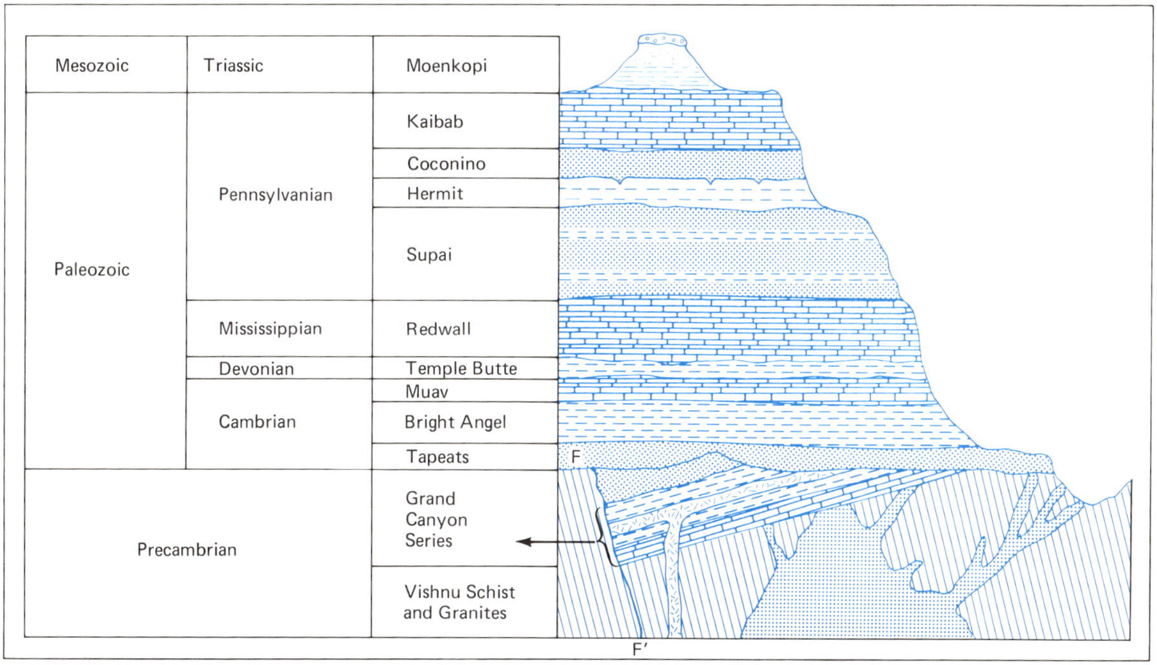

Figure 24.8 The geological history of the Grand Canyon as revealed by the south wall. Marine fossils are found in formations below Supai, indicating the presence of the sea in the area. The Supai and Hermit show evidence of stream deposition. The Coconino is a dune deposit of sand made by wind. The Kaibab indicates the return of the sea and the Moenkopi a retreat of the sea again. The cutting of the canyon started about 1 to 2 million years ago. [From Verne H. Booth and Mortimer L. Bloom, *Physical Science,* third edition, copyright © 1972 by The Macmillan Company]

Volcanism. One of the surface activities of the earth is volcanism. Volcano eruption is the most familiar manifestation (Figs. 24.10 and 24.11). Other forms are *fissure eruption* and *intrusive* activities beneath the surface forming *batholith* (rock volume), *sill* (rock sheet), and *dike* (rock finger), which are solidified magma filling cavities and crevices in the earth's crust (Figs. 24.12 to 24.14). We have mentioned the formation of magma at the weakened spots of the crust. The fluid seeps into any cracks it can find. When it reaches the surface, the eruption becomes a volcano. Magma contains a large amount of gas, largely water vapor, which causes violent explosions of the volcano. The release of gas in past volcanic activities contributed to the formation of the secondary atmosphere and surface water. The magma with the gas removed becomes *lava* and soon solidifies in air. Lava flow contributed to the formation of some of the surface features of the earth. Geysers are formed when underground water comes close to hot magma and becomes heated.

The state of Hawaii rests completely on volcanos. The island of Hawaii is a volcanic mountain rising 28,000 ft from the ocean floor, a height comparable to Mt. Everest. Citizens of Hawaii live on the slopes (albeit very gentle) of volcano cones.

Volcano activities over the world are concentrated in two *mobile belts,* which are also the regions of earthquake epicenters and young mountains. One of them runs along the west coasts of North and South

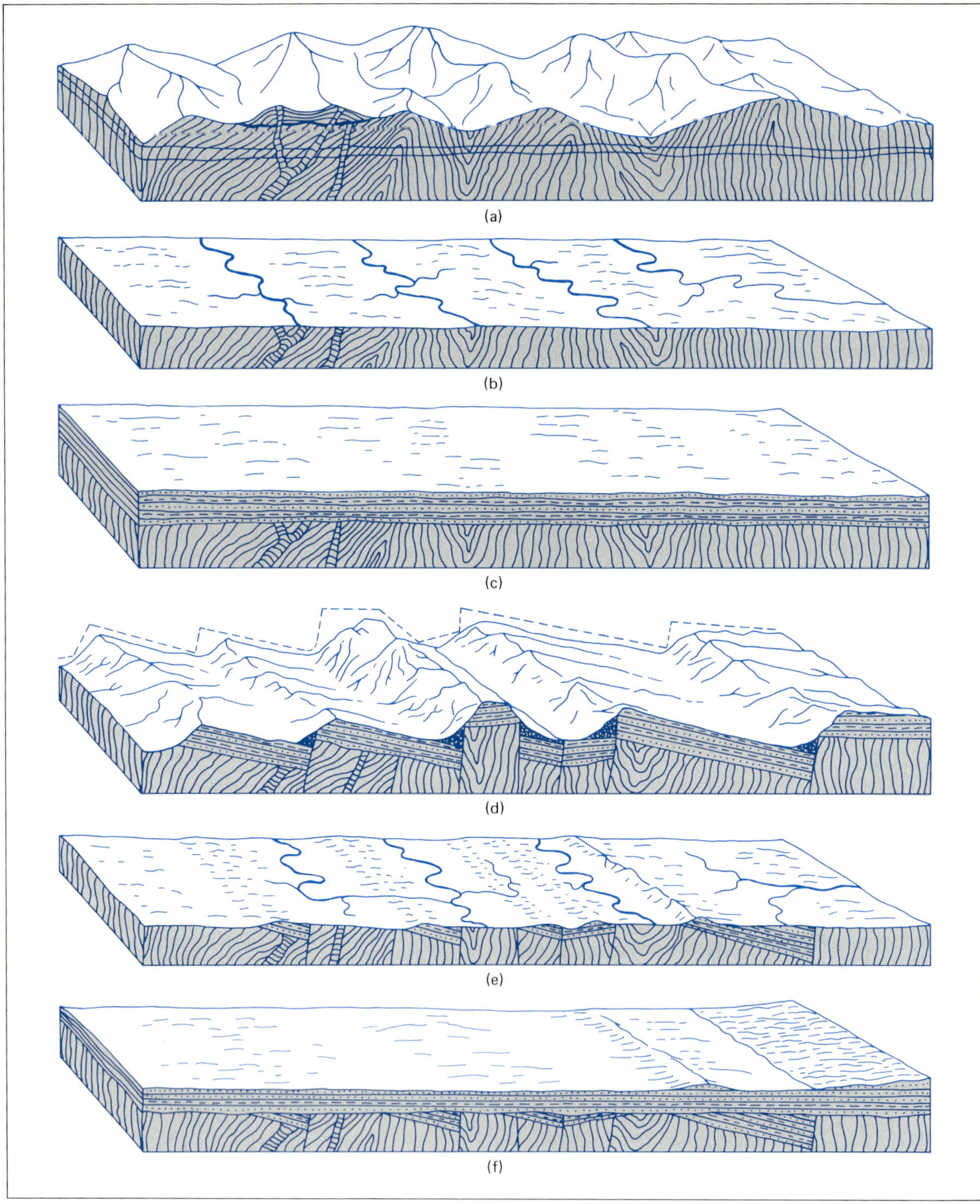
(a)
(b)
(c)
(d)
(e)
(f)

Figure 24.9 Six stages of the early history of the Grand Canyon area (after C. O. Dunbar). (a) The Vishnu mountains. (b) Erosion of the Vishnu to a plane. (c) Sediments spread over region. (d) Block-fault mountains form. (e) Leveling of the block mountains. (f) Younger rocks are formed. [From Bonner-Phillips, *Principles of Physical Science,* 1957, Addison-Wesley, Reading, Mass.]

Figure 24.10 Kilauea volcano, Hawaii Volcanoes National Park. [U.S. National Park Service photograph]

America from Chile to Alaska and extends to the Aleutians, the Kuriles, Japan, the Philippines, and Indonesia. The other belt includes the Alpine, Turkish, Persian, Himalayan, and Burmese mountains. In the ocean the mobile belt gives rise to island arcs (such as Kurile–Japan–Philippines islands) and deep sea trenches (such as the Japanese Trench) a short distance offshore. These trenches include the deepest parts of the oceans. The Japanese Trench reaches a depth of 37,000 ft, whereas the usual depth of the ocean floor is only about 12,000 ft.

Diastrophism. Another kind of activity on the surface of the earth, called *diastrophism,* is the large-scale movement of the earth's crust because of factors in the interior of the earth. A dramatic manifestation of it is the *earthquake,* which frequently occurs along a *fault,* such as the San Andreas Fault along the California coast. A fault is a crack in the earth's crust with lateral and/or vertical displacement of one side against the other. It may be hundreds of miles long and a few miles deep. The crust, floating on the plastic upper mantle like a raft,

Figure 24.11 A small volcano crater.

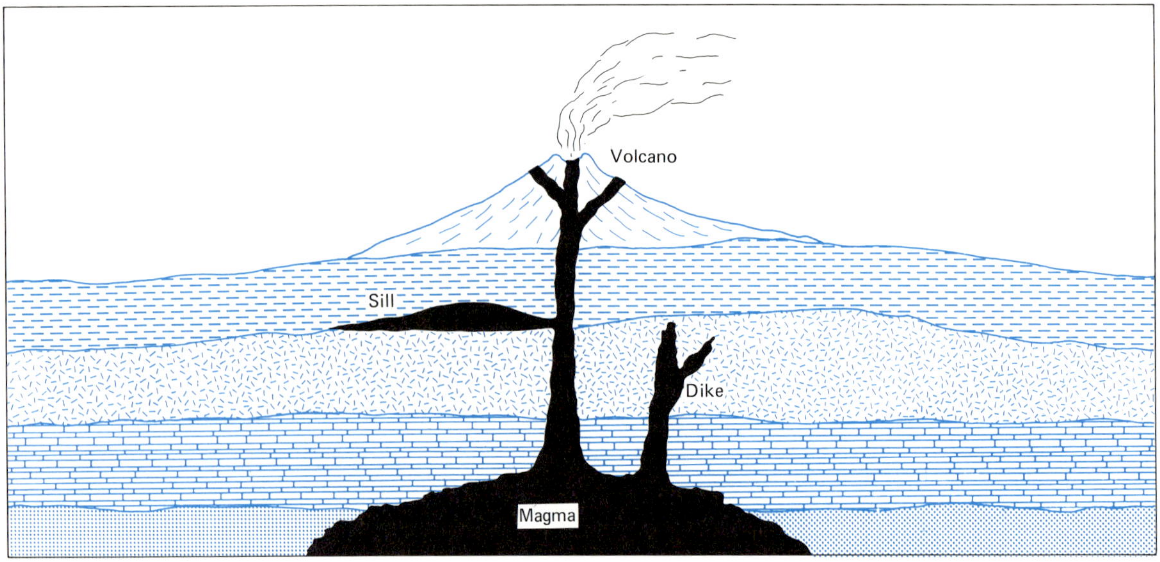

Figure 24.12 Volcanism includes volcano eruption and intrusive activities beneath the surface forming sill, dike, and so on.

Figure 24.13 Pegmatite dike in Precambrian gneiss in Bear Creek Canyon near Denver, Colorado. [U.S. Geological Survey]

Figure 24.14 Devil's Tower (Wyoming) is an ancient mass of intrusive rock formed by magma rising through the earth's crust that solidified without reaching the surface. After the surrounding rocks were removed by extensive erosion, this mass was exposed, giving us this spectacular view. [U.S. National Park Service photograph]

may undergo movement resulting from changes in the interior of the earth. The crust on the two sides of the fault may move differently. Because the crust is rigid and cannot readily move along the fault, stress is built up. When the stress is sufficiently large that the elasticity of the crust reaches its limit, a sudden movement, lateral and/or vertical, of one side of the fault against the other, on the order of a few inches to tens of feet, takes place. A violent shake accompanies the movement, resulting in an earthquake. After the movement the stress is relieved. Then stress is gradually built up again until a comparable earthquake takes place several years later. California has had major earthquakes once in half a century. The successive movements along a fault may cause a total displacement that is miles long.

Knowing the mechanism of earthquakes, we should be in a position to predict their occurrence (Prob. 4.10). We may monitor geological factors to detect incipient warning signals in time to alert the affected area, if not to prevent or avert the quakes. Work along this line has been progressing recently, and the first major prediction has been made.

Other manifestations of diastrophism besides *faulting* (slipping) are *warping* and *folding*. Warping is a large-scale uplifting or lowering, resulting in the formation of plateaus (such as the Colorado Plateau) and plains. In warping we find the origin of most plateaus and plains. As already mentioned, most of the great mountain ranges originated through folding. These mountains were originally sedimentary rocks formed in shallow seas. In fact, fossils of marine animals are found in them—an incredible feat for marine animals to land deep in high mountains. Horizontal squeezing of the earth's crust caused the layered strata of sedimentary rocks to fold in wavy contours. The upfold peaks are called *anticlines* and the downfold troughs are called *synclines*. The wavy surface topography may be leveled by weathering. In the process the inner strata of the sedimentary rocks are exposed. From such exposures we learn the history of folding that has taken place over geological history.

Origin of Mountains. How are the fold mountains (Fig. 24.15b) formed? What force, so tremendously large, squeezes the rock layers into folds? Could folding be due to the wrinkling of the earth crust because of the cooling of the interior? Although wrinkling may take place at the early stage of continental formation, it is unlikely to be the mechanism for the formation of the great mountain ranges that came into being in the last 200 million years. From the formation of the sediments to the folding into mountains, the time is not long enough for the earth to shrink sufficiently to form wrinkles. In fact, the crust of the present thickness has already been formed and heat flow balance already established so that no further shrinking is expected. Besides, there are other mountains, such as the Sierra Nevada, that are formed by stretching of the crust, instead of by compressing. The *thermal contraction theory* of mountain formation is giving way to other theories.

A new theory is the convection current theory. The convection current in the upper mantle has already been mentioned. Because the upper mantle is plastic (fluidlike), convection current flow can be expected, for the interior of the earth is hotter than the surface—just as in boiling water hot water at the bottom rises and cold water on the top sinks, forming a convection current. Its application to mountain formation goes like this. At the point the convection current turns down and goes into the interior of the mantle, the current drags down rocks from the crust and the crust at the point becomes thinner and depressed. A down-warping occurs, forming a large, gentle trough perhaps a thousand miles long and a hundred miles wide, which is called a *geosyncline*. Sediments pour in to fill the trough. By isostasy or further thinning of the crust the sediments sink and new sediments can be added. Therefore even a shallow sea hundreds of feet deep can create a sediment layer many thousand feet thick. The whole layer may later rise above sea level and then sink under again. The Appalachian geosyncline submerged under and emerged from the ocean two dozen times during the 300 million years of the Paleozoic era. When the original crust at the bottom wears thin, pressure on the mantle is reduced. Mantle rocks convert into liquid magma by a mechanism we discussed earlier. The magma rises and intrudes into the sedimentary rock layers, forming large-scale batholiths. The intense heat brought into the sedimentary rock layers by the magma causes the latter to expand, creating compression tension and resulting in folding and uplifting. The Appalachian Mountains were formed from the Appalachian geosyncline at the end of the Paleozoic era. This theory has the advantage of being able to explain the origination of the geosyncline, the formation of the fold mountains on geosynclines, and the formation of large-scale batholiths as the core of the fold mountain. The close folding of rigid, thick rock layers may be explained by the fact that the intense heat makes the rock layers plastic and thus able to be "folded."

In another way, mountains may be formed by a *fault block* (Figs. 24.15c and 24.16). The Sierra Nevada is formed by a single great fault block 400 miles long and 100 miles wide. The western side, during the faulting process, was tilted 20,000 ft higher than the

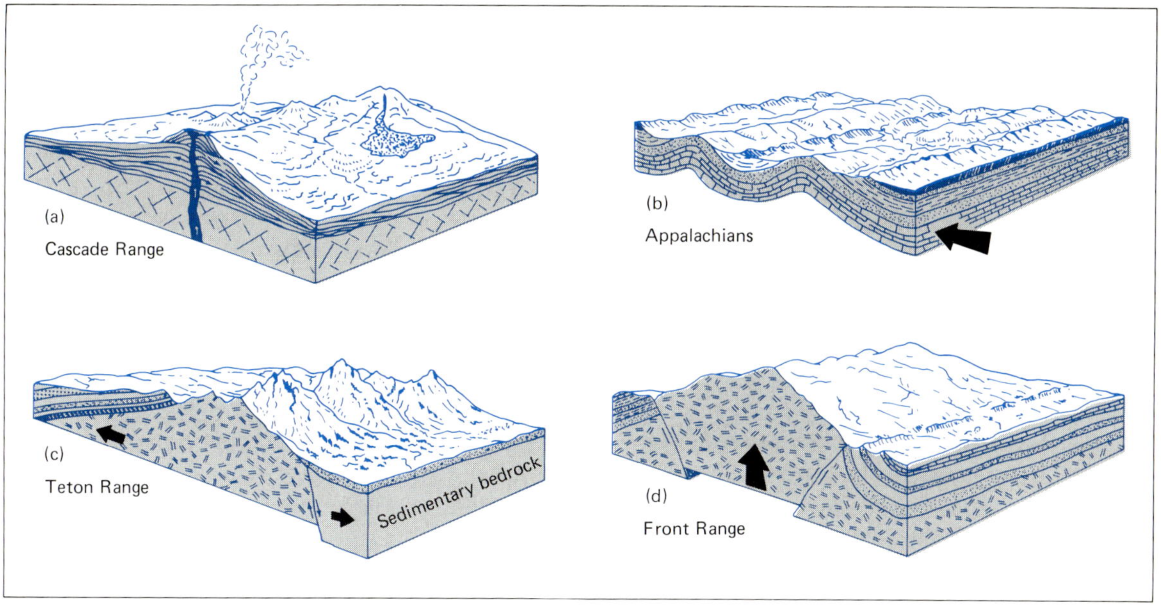

Figure 24.15 Origin of mountains. (a) Mountains formed by volcanic action; (b) mountains resulting from folded layers of rock; (c) mountains formed from fault blocks; (d) mountains formed as a result of vertical uplift. [U.S. Geological Survey]

eastern side. Erosion has reduced the height considerably. Along the fault the stress is a tension, rather than a compression. Mountains may also be formed by vertical uplift (Fig. 24.15d).

The conventional term *mountain* includes every earth formation that rises higher than the neighboring ground level. Thus it also includes such things as the erosion remnants of old highland areas such as the Catskill Mountains in New York State and the volcanic mountains formed by lava accretion (Fig. 24.15a). These, of course, are mountains of origins not included in the preceding discussion.

Figure 24.16 Teton Range. See Fig. 24.15C.

Weathering and Erosion. Whereas diastrophism refers to changes in the earth's surface resulting from factors in the interior of the earth, *weathering* and *erosion* refer to changes resulting from factors outside the earth. Diastrophism results in constructive

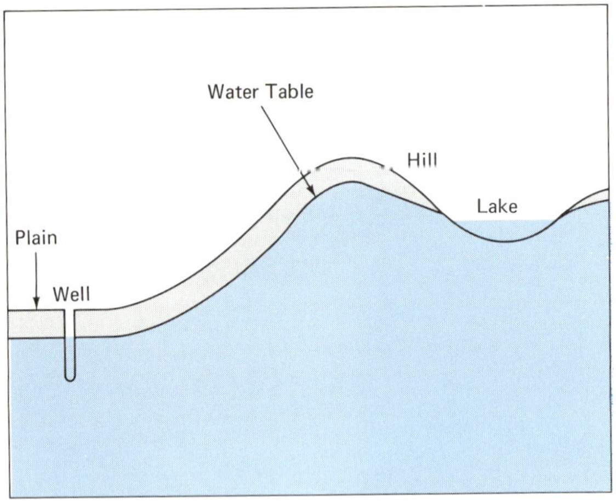

Figure 24.17 The water table illustrated.

activities such as mountain formation; the latter are invariably destructive, resulting in the gradation of elevated areas.

Figure 24.18 Royal Gorge, Colorado.

The agents of gradation are water and air in their various forms, such as wind, stream, and glacier. The process is largely mechanical, accompanied by some chemical changes. The energy required in these processes comes ultimately from the sun. Weathering and erosion may be considered as being caused by the sun, carried out with water and air as the agents.

It is the sun's heat that causes the evaporation of ocean water and raises it to a high altitude where it condenses to form rain clouds. The potential energy of the raindrops thus ultimately originated from the heat energy of the sun. It is this potential energy, after conversion into the kinetic energy of raindrops and running streams, that erodes the surface of the earth, carves out river valleys, and so on. When water is frozen, the potential energy becomes that of the ice in glaciers that also erode the earth's surface. Wind is produced because the sun heats up different parts of the surface of the earth unevenly and thus produces pressure differences. So the energy of wind erosion is also traceable to the sun.

The rainwater accumulated on the surface of the earth partly evaporates again and partly runs off on the surface of the earth into the rivers and lakes. Still another part enters the earth and, after saturating the soil, seeps into the underground water system. Surface and underground water eventually return to the ocean, completing the cycle. After a heavy rain all the soil is saturated with water. Unable to absorb any more water, the land lets additional rainfall run off on the surface. In the days afterward, the surface layer of the soil dries off first but the lower layers remain saturated. The boundary surface of the saturated part of the soil is called the *water table,* (Fig. 24.17), which lowers and rises according to the amount of rainfall in the course of time. The circulation of water on the surface of the earth is basic to discussions of many of the water pollution problems dealt with previously.

Weathering consists of the mechanical processes by which rocks disintegrate and the chemical processes by which rocks decompose. In a long and slow process, rocks break down and a mountain is leveled by weathering. Chemical weathering is instrumental in producing the soil, which has a different composition from the rocks. Mechanical weathering is responsible for the break-up of rocks into smaller and finer fragments. A major mechanism is *frost wedging.* Water seeps into rock crevices. When frozen in cold weather it expands and exerts pressure; the pressure breaks up rocks into smaller pieces. The sharp peaks of the Teton Mountains (Fig. 24.16) were formed by this process.

The water stream is a main source of erosion. As the water runs down the stream, it washes away loose material on the channel bed, carries it downstream and eventually deposits it in the river delta or on the continental shelf in the ocean. The process of erosion is also carried out by *abrasion;* the sand and pebbles carried by the stream rub against the channel bed and erode away bed material. Gradually a valley or a gorge is cut out of an otherwise level surface.

In the early stage the valley cut out by a stream is V shaped (Figs. 24.18 and 24.19) in cross section and straight lengthwise. The downward cutting has a limit—ultimately no stream can cut lower than the sea level. When the limit is reached, the cutting turns sidewise. Because it is the convex side of the curve of the stream that is doing the cutting, the curve tends to become more convex. The originally slight curvature becomes accentuated and the more or less straight stream develops meandering, widely swinging curves that wind back and forth. The original V-shaped valley thus gradually changes into a flat-floored valley which widens gradually.

The amount of erosion products that can be carried by a stream depend on its velocity—the faster the stream, the larger the load. When a stream is overloaded, the excess amount will be deposited out of the stream. A stream in a mountain flows rapidly and thus can carry a large load of rock debris. Once out of the mountain and into a plain the speed lowers and the stream becomes overloaded. The overloaded debris is deposited on the channel bed and soon fills the stream. The stream overflows and changes course. The new stream is filled again and

Figure 24.19 Black Canyon of Gunnison, Colorado.

the course changes again. The process repeats itself until the whole triangular area outside the mountain mouth of the stream is filled with erosion debris and becomes an *alluvial fan*—a delta on land.

Not all valleys are carved out by streams and are V shaped. Some are carved out by *mountain glaciers* and have a characteristic U shape (Figs. 24.20 to 24.23). Glacier erosion is another important mechanism of gradation. When the winter snowfall is more than can be melted in summer, snow accumulates year after year in a mountain valley and in time changes into a large body of ice filling the valley. A glacier is thus a "river of ice" (Fig. 24.24). This river, like a river of water, also flows but by a different mechanism, in spite of the fact that ice is solid. The flow speed is very low, such as a meter a day. The reason that the river of ice can flow is as follows.

Figure 24.20 Tyndall Glacier and Hallet's Peak in Rocky Mountain National Park. [U.S. National Park Service photograph]

The ice, under the action of gravity, tends to slide down the valley but is stopped by the friction force of the jagged surface of the valley wall. The tremendously large weight of the glacier exerts great pressure on the valley wall. The valley wall in turn exerts an equally great pressure on the glacier (Newton's third law). Now, as we learned in Chapter 7, ice melts under pressure. Therefore on the surface of contact ice melts into water, which in turn runs away. The melted ice creates a vacant space and the glacier under gravity immediately moves down to fill it. Melting starts again and the process repeats itself. As a result the river of ice *appears to flow* down the valley at a very slow speed.

The large pressure on the rocks of the valley wall causes them to break loose. The rock fragments are carried in the glacier and so "flow" down with the river. In the process the rock fragments are ground into finer pieces under the large pressure of the ice and the movement of the glacier. As the glacier flows year after year the valley walls are cut in deeply (Fig. 24.25). The result of this erosion process is a U-shaped valley, easily distinguishable from the V-shaped valley discussed earlier.

The flow of the glacier moves the ice mass down the valley. The melting of ice at the head of the glacier (the ice front) moves the position of the head backward. The two tendencies compete with each

Figure 24.21 Kawuneeche Valley, a glacially sculptured valley and headwaters of the Colorado River. [U.S. National Park Service photograph]

other in opposite directions. They reach an equilibrium when the head moves to a certain point of the valley where the river "stops"—the river of ice terminates there. Actually the river of ice flows to this point at which the ice on the front disappears by melting.

The rock debris carried by the glacier also flows down the valley to this point and stops but rocks cannot melt and disappear. As a result rock debris piles up continually at this point. Eventually the pile forms a ridgelike hill called a *terminal moraine.* It contains fertile soil and yields lush vegetation. Thus it stands out of the landscape and can be easily identified. Glacier debris piled up on the side of a glacier forms a *lateral moraine* (Fig. 24.26).

When the climate becomes warmer, ice melts faster. The equilibrium between the flow and the melting is upset and the glacier disappears before

Figure 24.22 (Left) V-shaped stream valley. (Right) U-shaped glaciated valley. [U.S. Geological Survey]

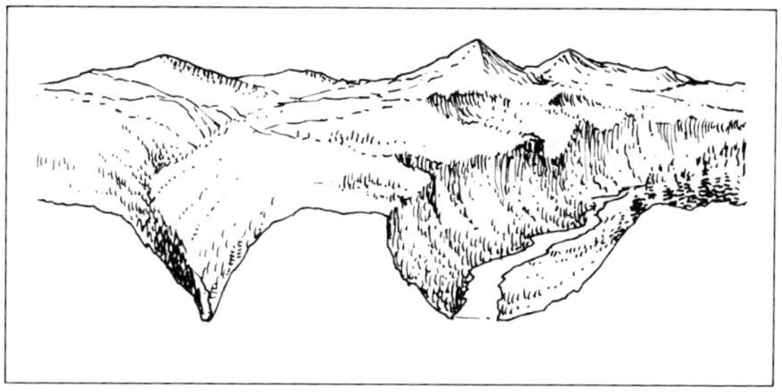

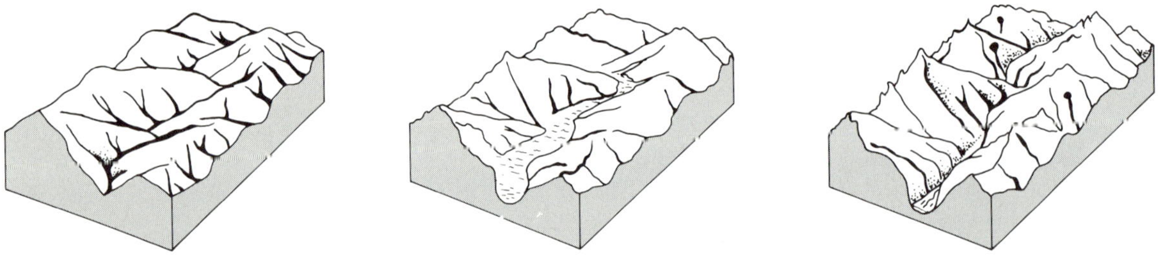

Figure 24.23 Before, during, and after alpine glaciation. The V-shaped stream valley is changed to the U-shaped glacier valley. [From Verne H. Booth and Mortimer L. Bloom, *Physical Science,* third edition, copyright © 1972 by The Macmillan Company]

reaching the moraine. The ice front of the glacier moves backward, and we say the glacier is receding. If the climate becomes colder the glacier advances. The moraine previously formed cannot move and remains there. We are now in a warming stage after the last ice age and glaciers are receding slowly. On the other hand, the moraines formed in the previous ice ages stand out in the plain without any ice in their vicinity. It is from the moraines that we learned of the existence of previous glacial ages and the extent to which the glaciers had advanced.

After a glacier has receded, the valley may find itself blocked by a moraine like a dam (Figs. 24.27 to 24.29). Water accumulates behind it and forms a *glacial lake,* another earmark of the previous glacial period. In Rocky Mountain National Park the moraines, glacial lakes, and U-shaped valleys are easily recognizable.

Figure 24.24 Muir Glacier, Glacier Bay National Monument, Alaska. [U.S. National Park Service photograph]

Figure 24.25 Yosemite Valley and Half Dome, Yosemite National Park. [U.S. National Park Service photograph]

Figure 24.26 Horseshoe Park (left) and lower Hidden Valley (with beaver ponds) separated by lateral moraine, Pinedale Glaciation, Rocky Mountain National Park. [U.S. National Park Service photograph]

Figure 24.27 Cameron Lake, Glacier National Park. [U.S. National Park Service photograph]

During the last ice age, a large ice sheet covered a large part of the Northern Hemisphere. This ice sheet, under the pressure of the ice cap, also moved in various directions. This movement also produced great erosion. The terminal moraines produced stretches from Long Island to the Rocky Mountains. Below the terminal moraine there is no evidence of glaciation. The *ice sheet glacier* smoothed out topography, removing material from high areas and depositing it in low-lying regions. However, mountain glaciers made the topography more jagged. The ice sheet glacier also carved out most of the lake basins, such as the Great Lakes of North America.

Wind is another agent of erosion. It can remove sand and soil and produce a *blowout.* It may accumulate sand to produce a *dune* (Fig. 24.30). Wind produces *ocean waves* and *longshore currents,* which are agents of erosion on the shorelines. Erosion is the process that produced many spectacular scenes on the earth (Fig. 24.31).

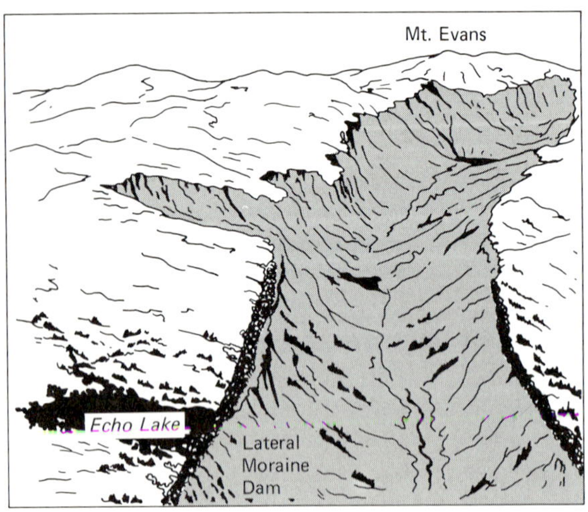

Figure 24.28 Echo Lake (near Denver, Colorado), dammed by lateral moraine along the rim of the glaciated valley of Chicago Creek. [U.S. Geological Survey]

Figure 24.29 Echo Lake and Mt. Evans, near Denver, Colorado. See Fig. 24.32.

Figure 24.30 Great Sand Dunes. [U.S. National Park Service photograph]

Figure 24.31 Natural wonder produced by erosion. Bryce Canyon National Park.

Life. The evolution of the living species is a topic belonging to biology. Even so, physical factors may still play an important role. For example, one novel explanation advanced recently for the extinction of the dinosaurs attributes it to the appearance of a supernova very close to earth. The intense radiation sterilized the unprotected dinosaur eggs.

Conclusion. In this chapter we are actually studying earth sciences by considering the earth as a dynamical system evolving out of a specific initial condition dictated by the laws of physics and chemistry. The initial condition, given to us from the study of planetary systems, is that the earth began as a globe with a given mass, a given composition, and a given temperature. The sequence of processes we discussed, i.e., internal heating, melting, fractionating, cooling, wrinkling, volcanic eruptions, diastrophism, weathering, and erosion, are likely or even destined to happen. As a result, continents and oceans formed, mountains emerged, valleys were carved out, lakes came into being, and the earth's surface gradually developed to the present state with a complex and ramified history. Of course, our theory here is far from complete. We cannot predict exactly the happening of each event at the exact time in the exact place to the exact extent, such as the formation of the Appalachian Mountains 200 million years ago and their exact size then. From this standpoint the theory is far less complete than the astronomical theory predicting eclipses. Yet all major events can be understood to a large extent and there is none contradicting science and none that science cannot explain (in other words, there is no compelling reason to invoke the supernatural). It is hoped that when all necessary information is available, a complete dynamical description will be worked out, and that is the ultimate aim of the earth sciences.

Problems

24.1 Why is the glacial valley U shaped and the stream valley V shaped?

24.2 If the continent were made of lighter material, what would happen to the root of the continent?

24.3 In a geosyncline if the sedimentary rock is lighter, will the thickness of the sedimentary rock layer be thinner or thicker than otherwise?

24.4 Why is there no erosion on the moon?

24.5 The surface of the earth is in a delicate heat balance. Would the human consumption of energy upset this balance now? In the future? In which direction? With what consequences?

24.6 In the preceding problem does the kind of energy consumed make any difference? Nuclear energy? Geothermal energy? Solar energy? Fossil fuel energy, which is ultimately of solar origin?

24.7 How is the Mid-Atlantic Ridge related to the drifting apart of the Americas from the Old World?

25

The Origin of Life

25.1 Life and the Inanimate Material World

The origin of life and man, like the origin of the universe, was a problem outside the scope of science until recently. There are varied motivations for attempts to set life, man, or even one kind of man apart from all other material things. The phenomena of life and man are so subtle and so complex that for a long time they seemed to be beyond scientific comprehension. But recent scientific advances have shown that some of the seemingly unanswerable questions can be and have been answered; the explanation of genetics by molecular biology is a typical example. Together with other advances, the study of the origin of life has gradually become a respectable scientific pursuit.

Life and Matter. Living organisms are made of material substances that are not different from substances in the physical world. They are made of the same 92 chemical elements; there are no new elements peculiar to life or to man. Nevertheless, it was speculated that there were compounds peculiar to life that set the living apart from the nonliving. This view was based on the earlier observation that all organic compounds originate in living systems and cannot be made from nonliving material. Then in 1820, Woehler first synthesized an organic compound, urea, from an inorganic compound, ammonium cyanate. The barrier between organic and inorganic material thus was torn down. Nowadays, synthesis of organic materials from inorganic substances is carried out routinely.

Life and Force. Even though the material substances involved in life are the same as those of the physical world, there was speculation that a different kind of force, called the *vital force,* was involved in living systems. The Brownian movement, first observed with pollen particles, was at one time thought to be due to the vital force of the living pollen, which was thought to make the particles move incessantly.

Then experiments were carried out with dead pollens many years old and even with pollen taken from Egyptian mummies that had been dead for several thousand years; the Brownian movement appeared just the same. Finally, small particles of mineral materials like glass were shown to exhibit Brownian movement in just the same way. The motion is thus obviously not related to life. No force peculiar to life was ever found.

For the same material and the same forces, the same conclusions from physical science should follow. There is no way life can be set apart within the framework of physical science. If life is to be different from the inanimate in nature and origin, the additional element will have to be supernatural, something that science can neither prove nor disprove. No further scientific discussion is possible beyond this point.

Within the framework of physical science we can investigate and explain as fully as possible the phenomena of life in terms of known physical laws. Molecular biology has been particularly successful in this pursuit, and now it appears that there is no problem in lower forms of life that cannot be solved in this manner. On the lowest level, a virus particle can be taken apart and put back again even with some parts replaced, just as an automobile is overhauled (resurrection?). There are even claims that life (a virus) has been made in a test tube out of nonliving substances (creation?). However, in higher forms of life the central nervous systems of man (soul?) and primates remain an intriguing problem. Science still does not have any idea of how the brain works; this is the last frontier of biological science. Nevertheless, recent studies on primates show that many of the physical, mental, and social activities of man are already present in primates. Some biologists are of the opinion that the only thing that can separate man from the primates is the extraordinary ability to use language for communication.

Life and Information. Now that we know all this we may ask whether the phenomena of life and man can be made accountable in terms of physical science. A significant feature of life is the existence of a body of information, genetic and acquired. The information is not supernatural, even though it is not simply matter or energy. It is the relationship between the material entities, and as such it is subject to analysis by science. There is a possibility that those aspects of life that cannot be reduced to physics and chemistry can be analyzed and understood as information.[1] The one characteristic peculiar to man, the use of language, is but a sophisticated device for transmission of acquired information. The problem of the origin of life and man may be approached from such a standpoint, but we still do not have a complete solution. Up to now, most considerations of the origin of life have been limited to biochemical aspects, particularly in relation to the early physical environment of the earth.

25.2 Abiogenesis of Organic Compounds and Organisms

For life to appear, we need first the basic biochemical subunits: amino acids, nitrogenous bases, sugar, and so on. We have already discussed the Urey-Miller experiment and similar ones by which these building blocks can be synthesized by natural processes from a primitive reducing atmosphere composed of methane, ammonia, and water such as was believed to have existed on earth in its early stages. From these basic units, the proteins and nucleic acids may be formed that eventually lead to living organisms.

The primitive atmosphere, rich in CH_4 and NH_3, is maintained as a reducing atmosphere by the pres-

[1] P. Fong, "Phenomenological Theory of Life," *J. Theoret. Biol.*, **21**, (1968), 133; P. Fong, "Thermodynamic and Statistical Theory of Life," *Biogenesis, Evolution and Homeostasis, a Symposium* (New York: Springer-Verlag, 1973), pp. 93–106.

ence of hydrogen. Yet the earth's gravity was too small to hold hydrogen, so the primordial hydrogen gradually escaped and finally disappeared from earth 2 billion years ago. CH_4 and NH_3 then became unstable, and they were gradually replaced in the atmosphere by N_2 and CO_2. By that time, the abiogenesis of organic compounds should have been completed. The products were collected in the primitive oceans, forming the "primordial soup" from which organic macromolecules and finally organisms were to be formed.

The First Organism. After the formation of proteins and nucleic acids in the primordial soup, the next step was the formation of the primitive self-replicating molecular assembly that may be considered the earliest form of life. This step is the crux of the matter, the watershed separating the living from the nonliving. Yet not much attention has been paid to this step by both physical and biological scientists. How this step came about is not likely to be solved like a standard physics and chemistry problem, i.e., as a problem of the mere transformation of matter and energy. For the first time, the element of information comes into play; self-replication involves the preservation, reproduction, and utilization of genetic information. This goes beyond the mere transformation of matter and energy, and thus beyond routine physics and chemistry. We shall merely point out the existence of the problem without further elaboration here.

25.3 Life and Energy

Once a living organism has originated, an interesting problem closely related to physical science arises. All living organisms use energy, and thus energy metabolism is a major biochemical problem in all forms of life. Here we have independent *biological* evidence that the primitive atmosphere did not contain free oxygen. In living organisms the biochemical process of using energy by breaking down a fuel molecule—say, glucose—is basically anaerobic, requiring no help from free oxygen. The particular step involving the use of free oxygen comes last in the sequence, apparently a later addition in the more recent oxygen-using era, incorporated into an already complete, anaerobic system that had developed earlier under reducing conditions. This evolutional change reflects the environmental change from an oxygen-free atmosphere to an oxygen-laden one.

Early organisms are thus anaerobic and derive their energy from *fermentation*. Fermentation leads to the product CO_2, which is not abundant in the primitive atmosphere. The increase of CO_2 paved the way for photosynthesis to arise later. Photosynthesis led to the production of free oxygen, which was released into the atmosphere. Thus the oxygen content of the atmosphere gradually increased. Then a crucial change took place—fermentation was replaced by oxidation, corresponding to the last step in glucose metabolism discussed earlier. Oxidation is much more efficient than fermentation in energy utilization, and life acquired a more powerful energy source. At the same time, the lethal pollutant oxygen was made into a vital nutrient, truly a tour de force in the ascent of life.

The fuel molecule—say, glucose—presumably existed in the primordial soup, having been synthesized abiogenetically from the primitive atmosphere. The primitive organisms only had to break down the glucose for energy utilization, but did not have to synthesize it. When the glucose was used up and the primitive atmosphere (from which glucose was synthesized) disappeared from the earth, the organisms were faced with a crisis of "famine." Another tour de force took place; living things learned to synthesize glucose from its immediate precursor. An enzyme was developed for this purpose, and the genetic information for making it was incorporated into the genome. Once this precursor in the primordial soup had been used up, organisms were forced to synthesize it from its precursor. Another enzyme was developed and incorporated into the genome. In this way a series of enzymes were developed, eventually enabling organisms to synthesize glucose from the raw material CO_2 and H_2O. The whole body of genetic information thus developed formed the glucose *operon* (a number of related genes, one for each enzyme, correlated for a single mission, the synthesis of glucose); the complex process of photosynthesis thus came into being. The preceding discussion shows how such a complex, coherent system may be brought forth by evolution, in spite of the fact that evolution does one thing at a time without correlation with the past and the future. The random chance of developing many correlated enzymes at the same time is extremely small; thus the appearance of a whole operon and the development of life would be difficult to understand without such a mechanism.

25.4 Life and Matter[2]

By photosynthesis and respiration (oxidation), living organisms constantly exchange material with the natural environment. At the present time, it is estimated that all the oxygen in the atmosphere passes through living organisms once in 2,000 years. All the carbon dioxide circulates once in 300 years. All the water on the earth recycles every 2 million years.

The present biochemistry in living organisms is very complicated. One reason for this is that there are many redundant parts. The bare essentials are reasonably simple. In fact, it can be shown that life can start business with only 29 molecules as the basic units: 20 amino acids to form proteins, five nitrogenous bases plus ribose and phosphoric acid to form nucleic acids DNA and RNA, glucose as the fuel molecule, fat as the fuel storage, and phosphatides for structural purposes. The origination of such a simple system does not seem to be overly difficult. Presumably life began with a simple system somewhat like this and later developed into more complex systems.

As far as chemical elements are concerned, only 16 to 21 elements are used in life. They are divided into four groups: water (H, O), organic elements (H, C, N, O, P, S), ions (Na^+, K^+, Mg^{++}, Ca^{++}, Cl^-), and trace elements (Mn, Fe, Co, Cu, Zn, and a few others for special organisms). The ion concentrations (take the composition of blood, for example) seem to be the same as those in the water of ancient seas. That water was where the primitive organism lived and was later enclosed inside the organism, disconnected from the environment. There in the body fluid, the ion concentrations of the ancient seas were preserved to this day. This is another example of the influence of the ancient environment, this time the hydrosphere instead of the atmosphere, on the evolution of life.

Exobiology. Is there life beyond earth? The other planets in the solar system have surface conditions (temperature, pressure, chemical composition of the atmosphere, hydrosphere, and lithosphere) so different from ours that life of our kind is not likely to exist. Outside the solar system, the question hinges on the way planetary systems are formed. According to the collision theory, the formation of a planetary system is a rare, singular event and the number of planetary systems in the universe is likely to be extremely small. The chance of finding life beyond our solar system would be extremely small and we are truly unique. On the other hand, according to the condensation theory planet formation is the rule rather than the exception, making the chance of life existing beyond our solar system much greater. The current view follows the condensation theory, and the general outline is not likely to be changed in the future. According to one estimate, there are 1 billion planets in our galaxy alone that are similar to ours and thus may harbor life.

If life should be a widespread phenomenon in the universe, how would the other forms of life compare with ours? Although chemical elements are all the same throughout the universe, we may still wonder

[2]George Wald, "The Origin of Life," *Proc. Nat. Acad. Sci.* **52** (1964), 595.

if life can be based on a different set of elements and another group of reactions. For example, since silicon and carbon are similar to each other, could another kind of life be formed with silicon as a substitute for carbon in our biochemistry? Could ammonia be a substitute for water? And so on. George Wald considered the problem and found that each major element (H, C, N, O) and compound (such as water) used in life plays such a unique role in our biochemistry that they can hardly be substituted. He concluded that life beyond earth would still be made largely of H, C, N, O, and based on water. In other words, it would be based on essentially the same biochemistry as ours. It must be recognized that this only means that our kind of life is one that allows little room for substitution. Whether there exist completely different kinds of life elsewhere depends on the definition of life. Dissociated from a particular biochemistry, if life is to be defined by a set of characteristics such as reproduction, growth, and so on, then the possibilities are far from limited.

Problems

25.1 Concerning the origin of life there is a microsphere theory: the complex organic compounds formed in the primordial sea aggregate to form a small, spherical structure which is to be the beginning of simple organisms. What is wrong with this theory?

25.2 The chicken–egg problem: Which one comes first, the individual organism or the genetic system?

25.3 We already have a great variety of life forms on the earth. What other forms of life could be possible elsewhere in the universe? Would a tri-sex form of life be possible and probable?

Epilogue

The laws of physics and chemistry we studied in this book comprise the *basic laws* of dynamics and forces and the *secondary laws* of properties of material systems. Simply stated, the basic laws include the three dynamical laws of Newton (augmented by the field, relativity, and quantum theories), the four laws of the four basic forces (possibly two more at most at the present), and one probability law of statistical physics (the second law of thermodynamics). The principles of conservation of mass and energy are already contained in the dynamical laws. By applying these basic laws to deal with the various material systems, that is, solids, liquids, gases, plasmas, molecules, atoms, and nuclei, we deduced a set of secondary laws concerning the properties of these specific material systems, such as the Boyle law and the law of mass action. The secondary laws do not represent new, independent information and thus are not fundamental. Nevertheless, they are very useful and are the workhorses of day-to-day applications in science and technology.

This body of scientific knowledge is used to study the origin and evolution of the physical universe in Chapters 21 to 25, and we find the application quite successful. Though there are still unsettled questions, we do have a reasonably good understanding and a reliable outline of the origin and evolution of the universe. Our knowledge is thus far better than a mere inventory of the heavenly bodies and earthly objects. Chapters 21 to 23 deal with astronomy, and Chapter 24 touches upon the earth sciences of geology, oceanography, and meteorology. In every topic the scientific problem is considered as a dynamic process, starting from an initial condition (either assumed at the very beginning of the universe or derived from the dynamical study of the previous stage of evolution) and developing in the course of time according to the basic laws of dynamics and forces, supplemented by the secondary laws of properties of matter. In the process we have also explained the existence and the properties of many astronomical and geological systems and established a set of secondary laws concerning these systems (such as the color-luminosity correlation of the stars) that emerged in the course of evolution.

These systems, main sequence stars, red giants, white dwarfs, supernovae, planets, satellites, continents, mountains, all have peculiar properties, not shared by other systems, so that at first they seem unrelated to the basic laws of dynamics and forces and thus governed by a set of independent laws. That they are explicable at all is due to the fact that we are able to analyze the complex manifestations of these systems as the results of the accumulation of a long series of small changes, each one of which is simple enough that it can be explained by the basic laws we learned before. The stars are formed in a process of accretion. The planets are formed in a process of accretion. The continents are formed in a process of accretion. Each step of accretion, a material addition as well as an evolutionary change, is small and simple and can be explained without great difficulty. Thus we may formulate a *principle of accretion* as a general guide in explicating evolutionary processes and in explaining complex, mysterious systems in terms of the basic principles of science. Without its help we require a *principle of creation,* which is supernatural, such as the early doctrine of catastrophism in geology, to account for the sudden appearance in toto of complex systems.

•

The evolutionary process did not stop at the formation of mountains. In time emerged life, man, and civilization. These are even more complex systems. Can these new systems be analyzed and their origination be accounted for on a scientific basis like all the other new systems that previously emerged in natural history, such as the supernovae, the planets, or the secondary atmosphere?

Inasmuch as the new systems did not exist previously and emerged at a certain stage of natural history, they either evolved out of the previous natural system dictated by the laws of that system as a natural consequence of these laws and as a natural extension of the on-going, unending transformation process or they resulted from the unexpected intervention of an element previously unknown and inactive.

The previous new systems in astronomy and geology can be accounted for according to the first alternative. Many mysteries of old times, such as the burning of the sun, the cooling of the earth, the folding of the mountains, and so on, have been satisfactorily answered by modern science. There is nothing in these areas that definitely cannot be explained by science.

The emergence of life is more complicated. There are many obvious manifestations of life that seem to go beyond or even to be contradictory to the laws of physics and chemistry. Some problems of life still remain unsolved. Inasmuch as life is new, not existent before, and different from all that went before, it seems as if a new element transcending nature had come into play, the laws of nature were interceded, and a new set of laws of biology was suddenly imposed. This new element, beyond matter and energy, has been called life, vital force, vis viva, entelechy, and so on if it is not ascribed to divine action.

However, because the emergence of these new systems is a part, indeed a very small part, of the whole evolutionary process of which we have a reasonably good account, we cannot help but ask if the same methodology may be applied here. As mentioned before, a scientific analysis of life must include the new element of information in addition to the old elements of matter and energy, which we are already familiar with in physical science (see references cited on p. 419). Rather than pursuing the problem along this line we will satisfy ourselves here by looking into how the principle of accretion may help in this pursuit.

In the development of life the principle of accretion is manifested in the *Haeckel law* in embryology, which states that ontogeny recapitulates phylogeny. The human embryo goes through stages of development characterized by gill, tail, body hair, and so on. If the human being came into existence by catastrophic creation, there is no reason to re-enact stages characteristic of lowly animals. These stages indicate a process of accretion. A skyscraper is built one floor on top of another; the lower floors, even if not used, cannot be removed. A complex system, once identified with the result of a process of accretion, may be analyzed and understood by studying each of the small steps one by one. The mystery can thus be expelled.

The brain is still a mystery in science today. However, the process of accretion is also evident. Different parts of the brain developed at different stages of evolution. The impulsive, bestial behaviors (food- and sex-related) are governed by the primitive parts of the brain. The rational, social behaviors are governed by the modern parts of the brain, essentially the cerebrum. The old bestial behaviors remain, but are suppressed by the new rational behaviors. When control is relaxed or lost, such as in a dream or in psychosis, antisocial activities emerge, and we ascribe them phenomenologically to the subconscious. If the emergence of the brain follows the doctrine of catastrophism, there is no reason to incorporate into it a troublesome subconscious. By the process of accretion some brain activities may be explicated.

Civilization is perhaps the most wonderful thing of all on earth. The social genome also expands by accretion just as the biological genome. The United States Constitution expands by adding one amendment after another, just as DNA expands by adding one operon after another. Some accretion processes may be just aimless wandering, rather than purposeful seeking of a definite goal. The Appalachian geosyncline has been wet (submerged in water) and dry and wet again two dozen times in the Paleozoic era. Some animals, like the whale, have been wet (marine animals) and dry (land animals) and wet again in biological evolution. The United States has been wet and dry (18th Amendment introducing Prohibition) and wet again (21st Amendment abolishing Prohibition) in the past century.

•

The evolution of the universe did not stop at the emergence of life, man, and civilization. In recent years the evolution of natural history has taken a strange and drastic turn. The changes are brought about by human activities, and the processes take place much faster than any of the natural changes that happened before, and they affect our well-being and even survival in a significant way. These changes bring forth the environmental issues, including the pollution of the environment, the depletion of resources, and other problems concerning the relation between man and nature.

Pollution of the air means the change of the chemical composition of the atmosphere. This is the latest phase in the history of the atmosphere, which has undergone many changes already. Inasmuch as these changes involve changes in the physical world, they are naturally a part of physical science. They represent an extension of the natural evolution process starting from the very beginning of the universe. Their relevance to our well-being makes them significant and timely topics of discussion. Thus we have discussed them in a number of chapters. This part of the evolutionary process represents the artificial transformation of the physical world.

In addressing environmental problems, we use the term *pollution,* meaning that an originally good and pure state is spoiled and degraded by some specific human activities that must then be regarded as undesirable. Although we all suffer from a degraded environment and deplore all the violations leading to it, abhorrence of pollution can easily develop into a kind of religious fanaticism regarding nature as originally pure and perfect, thus sacred and inviolable, and all human activities modifying natural environment, such as the use of natural resources, as deplorable. One of the arguments against the Alaska oil pipeline is that it disturbs nature—meaning caribous, permafrost, and so on.

Such a pious view is scientifically not meaningful. Nature has been undergoing modification all along; the only original pure state of nature is that of the primordial hydrogen gas. All other 91 elements are "waste products" of nuclear activities in the stellar interior. These stars exploded and polluted the virgin interstellar hydrogen gas with the 91 pollutants. Ironically, our earth and all of us are made of such "waste products." As discussed earlier, the original atmosphere of the earth was a reducing atmosphere of methane and ammonia. It was polluted by oxygen from living species, which was lethal to living species then. Yet life executed a grand tour de force, converting the lethal pollutant to a vital nutrient, and most living things now are dependent on oxygen. We cannot expect our genetic system to mutate so that we not only are immune to automobile exhaust but cannot live without it. Our air pollution problem cannot be expected to be solved this way. Nevertheless, the discussion does point out that modifications of nature are not necessarily undesirable.

The emergence of life, of the animal kingdom, and of man may also be looked upon in the same light. Life is parasitic to nature, the animal kingdom is parasitic to the plant kingdom, and man is the greatest parasite of all. There are nine amino acids and about a dozen vitamins man cannot synthesize in his own body. These have to be robbed from other living species. As far as the animal kingdom is concerned, it would be far better if man did not exist. As far as the plant kingdom is concerned, it would be better if animals did not exist. Many changes in nature have taken place, often to the disadvantage of the existing species, yet frequently these changes have brought forth more interesting, more sophisticated, and more exquisite things, including man. And man can even synthesize vitamins in factories!

Thus there is no perfect, pure state of nature that should be preserved at all costs. Modification of nature, such as that resulting from the utilization of natural resources, by itself is not necessarily sinful. The final judgment of all environmental considerations must be based on the well-being of mankind.

A similar situation exists concerning the depletion of natural resources. The first energy crisis happened 2 billion years ago when the primitive organisms in the primordial sea used up the glucose in their environment. It was truly a crisis, for the glucose was formed by abiogenetic processes in the primitive reducing atmosphere, which by then had disappeared. The glucose in the primordial sea, once used up, could never be replenished. Life was on the brink of extinction. Yet life executed a tour de force by developing an enzyme to synthesize glucose from its immediate precursor, which still existed in the primordial sea and which now became a new fuel molecule. This process continued until all substitute fuels were used up. But in the process organisms had acquired the ability to synthesize glucose from an inexhaustible source, inorganic CO_2 and H_2O. The energy crisis was then completely solved and did not recur for 2 billion years. The way we use up our irreplaceable fossil fuels and fission fuels may appear as alarming as the depletion of glucose and its precursors in the primordial sea. Yet the previous example gives us hope that before these fuels are depleted we shall have mastered the control of fusion energy, solving the energy crisis "forever."

Thus environmental and energy issues are not sacred issues, but secular issues. As such they should be considered together with other secular issues in their proper priority. If all of us are rich men who have everything, then indeed better environment is the first priority. If some of us still suffer from hunger and cold, then we should not be so callous as to ignore these needs.

•

As far as energy is concerned, fusion energy is practically inexhaustible. Fusion fuel may last 5×10^{10} years at the current rate of consumption. The age of the earth is only 4.6×10^9 years. The sun will burn out in 7×10^9 years. The fate of "heat death" is written by the second law of thermodynamics. It is completely meaningless to plan a future to infinity. In fact, before we were frozen to death by the extinguishing of the sun, other more "pressing" catastrophes would befall upon us. Before the sun burns out, it will become a red giant. Its size will be so large that it will engulf the orbit of Mercury. The earth will be so hot that all water in the oceans will be boiled dry. All life will be wiped out. In hundreds of millions of years continents may become submerged again. And in as short a time as 50,000 years the next glacial age may have covered much of the Northern Hemisphere with ice. All buildings will be ground into pebbles. And before this happens the present interglacial age may become so extended that the ice cap melts, partly inundating all coastal cities. The history of man is only about a million years old. Any planning beyond 1 million years is of dubious significance. The fusion fuel age, lasting 50 billion years, is much longer than the time between the development of bacteria and man. It would have been preposterous for the bacteria to plan for the energy needs of man 2 billion years later. With an inexhaustible source of energy the problems of environment and resources can be solved (Sec. 20.4). Two obstacles against economic development can be removed.

This discussion is also related to future population development. It is absurd to think that population can expand indefinitely or even perpetuate itself indefinitely. Flexible population control geared to the environmental capacity is required. When the sun becomes a red giant, a small human population with a high technology may still manage an underground existence for a period of time, even though not forever. Then the age of the DNA-protein-based life on the earth will draw to an end. The study of astronomy and geology reminds us of the ephemeral nature of human existence on this planet and helps us shed the pre-Copernican egotism that considered man as the grand master of the universe. We still tend to appoint ourselves the grand custodians of the earth. Let us remember no monument on earth, no matter how solid, will remain when the polar ice cap extends over it in a glacial age. We should be content with the more humble role of making the best use of the resources for ourselves and our posterity during our fleeting existence on the earth.

For example, if we knew that the next glacial age is about to commence, would we preserve nature, letting the glaciers advance and wipe out the northern cities or would we try to modify nature to stop the advance of the glaciers? The latter option is not impossible. Suppose it takes 10,000 years, a minimal estimate, for the polar ice cap to extend to its southmost reaches, with a lowering of the ocean level by 300 ft; then the ocean level is lowered by 0.03 ft (about 1 cm) per year. Based on the surface area of the oceans, 3.6×10^{14} m^2, we calculate the amount of ocean water annually converted into ice to be 3.6×10^{15} kg. The latent heat of fusion for this transformation totals 2.9×10^{20} cal/yr (still small compared with the total solar radiation intercepted by earth, 1.3×10^{24} cal/yr). On the other hand, current world energy consumption amounts to 6×10^{19} cal/yr. If the current energy consumption rate is increased five times, then, remembering that all energy consumed eventually becomes waste heat dumped into the environment, there will be enough heat to melt that much ice and thus to stop the advance of the glaciers. This rate of energy consumption may be reached in a few decades if we do not change our energy policy and is well within the capability of the nuclear energy resources of the earth to supply. Moreover, for this purpose we can only depend on nuclear energy (fossil fuel energy would have already been exhausted), not on direct solar energy, which is the only other energy source, because the heat needed is an amount in addition to the solar heat input to the earth. Therefore, even after we have developed the ideal solar battery to tap solar energy, nuclear energy is still needed to stop the advance of the glaciers.

If the current interglacial age is prolonged, human utilization of energy may accelerate the melting of the polar ice cap to inundate the coastal cities. We need a comparable amount of energy to refrigerate the earth to avert the catastrophe. Again this is within the capability of nuclear energy (this time the solar battery will be a great help). It is assumed, of course, that our current energy consumption rate will not expand without bound.

•

In the long run, the energy and material resource problems as well as cosmic weather modification can be dealt with by harnessing fusion energy. Therefore, scientific research on and technological development of fusion energy should be given high priority. Yet these problems are centuries away and fission energy is a decent interim substitute. Even fusion research need not be our most urgent pursuit.

What is the most urgent, then? No matter how rosy the scientific prospect of the future, the road ahead is still full of dangers. A nuclear world war not only would be a great disaster for the present population, but would also produce intense radiation pollution, making the earth uninhabitable for future generations. Prevention of a nuclear war is definitely one of the most urgent concerns at the present time.

Although there is a possibility that a nuclear war may start by accident, the most important measure to prevent such a catastrophe is to keep a stable world peace. International war sometimes breaks out because of domestic social problems. World peace and domestic tranquility are inseparable. These must be considered the most important social objectives, not only for their own sake but also for sheer survival.

How can we maintain world peace and domestic tranquility? Searching through all government systems, social institutions, economic practices, and political strategies, we find that the most practical way perhaps is through *economic development*. We may add many volumes of research but the conclusion is not likely to be changed substantially.

We need only consider recent history to convince ourselves. World War II broke out because of Hitler's demand for *Lebensraum* (living space), the cause of war for ages. Now, only 30 years later, Germans and Frenchmen have become good neighbors; Americans and Russians are enjoying detente. No one is fighting for *Lebensraum* any more. Human nature has not and cannot be changed in thirty years. What has happened is that the world economy expanded from \$1 trillion to \$3 trillion annually in the past quarter century (1950–1975). Let us *not* consider economic growth as serving no other purpose than fulfilling human greed. It is actually a life-sustaining medicine. The environmental and energy crises are but the undesirable side effects of this potent medicine.

The rate of economic growth of the past quarter century cannot be sustained indefinitely because the earth is finite. However, what is necessary to maintain world peace and domestic tranquility is the growth of per capita income, not necessarily gross income. This objective may be attained and the burden of the earth kept at a tolerable limit if population is stabilized or even allowed to decline. In fact, in some developed nations population is already stabilized or even beginning to decline. Thus, this objective is not wishful thinking but is practically realizable. Because population control is not a problem in developed nations but is still a problem in developing nations, economic development is even more imperative.

Because the energy and environmental crises are brought about by recent economic development, can we afford further development from that view? The answer is yes. Actually, we need economic development to undo some of the misuses of nature in the past quarter century. Cleaning up the environment requires manpower, and this increases employment opportunities. To use cleaner and scarcer fuels costs more. This means more production efforts (for the fuels) and more employment. The demands from these sources lay the foundation for economic growth in the next quarter century.

•

The rash economic development of the recent past, with all its abuses, deplorable and condemnable, nevertheless fulfilled a historical mission, that of pulling much of mankind out of hunger, poverty, natural calamities, and struggle for *Lebensraum*. It also elevated mankind to such a height that we are now in a position to be able to rehabilitate the abuses. The issue is somewhat parallel to that of child labor in the early years of industrial development. Inhuman and unjust as it was, child labor nevertheless played a part in helping the industrial revolution move ahead by providing cheap labor. The child came to the labor force not by bond of slavery, but by his own choice—a choice at least better than other alternatives. Once developed to a certain stage, the industrial society is in a position to ban child labor because industry can afford higher wages and the child can afford no employment. The rapid economic development of recent years was greatly helped by disregarding the hidden environmental costs and by the availability of cheap liquid fuel. But now we are developed enough to be able to pay for the environmental costs and more expensive fuels. Economic development proceeds by stages, one leading into another, that cannot be skipped—

just as a star cannot go into the helium-burning stage without passing through the hydrogen-burning stage.

The evolution of the technological-industrial-economic system to the present state is similar to the evolution of a biological system. Once an endangered species becomes extinct, it can never be resurrected again. The same is true for the present technological-industrial-economic system. Its genesis and its precariousness have been discussed in Sec. 9.6. It should be treated as an endangered species.

The present system is based on the concepts of growth, expansion, and progress. These concepts originated in the seventeenth-century scientific revolution. In the medieval times, ancient teachings, institutions, and practices were considered to be ideal, perfect, and authoritative. Solutions of problems were sought in ancient classics. Aristotle was impeccable. The scientific revolution of Galileo and Newton changed all this. It was demonstrated that ancient authority could be wrong and present knowledge obtained by a process of scientific investigation could excel ancient teachings. If modern knowledge could be better, why not modern institutions, practices, and ways of life? Thus the emulation of the established was abandoned and an endless search ensued for the new, the novel, and the expanded. The worship of growth, expansion, and progress followed.

Development of the American nation, more than all others, was based on the concept of growth. Our social institutions and mentality are geared to the idea of growth. This manifested itself in the rapid pushing back of the frontier in the nineteenth century. After the vanishing of the geographical frontier, momentum was perpetuated on the industrial frontier. To stop economic growth is an "un-American" concept and is certain to change the character of the nation. Without growth many of our noble ideals and cherished institutions will fade and we run the risk of relapsing into such practices as dictatorship and genocide. No amount of moral preaching can stop the resurrection of these skeletons. They are only submerged and superseded by the system characterized by economic growth, just as the bestial behavior in the subconscious is submerged and superseded by the rational brain.

The recent attack on science, technology, and economic development, largely spawned by environmental considerations, left the scientific community shaken. With the misunderstanding removed, it is time for all of us to devote ourselves to the necessary tasks before us so that we will have a better world for ourselves and our posterity.

Bibliography for Special Topics

Environment

Federal Water Pollution Control Administration, *The Cost of Clean Water,* 1968, 4 vols.

P. F. Krenkel and F. L. Parker, *Biological Aspects of Thermal Pollution.* Nashville, Tenn.: Vanderbilt University Press, 1969.

William W. Murdoch, ed., *Environment.* Stamford, Conn.: Sinauer Associates, Inc., 1971.

National Academy of Science–National Research Council, *Eutrophication: Causes, Consequences and Corrections.* Washington, D.C., 1969.

National Academy of Science–National Research Council, *Waste Management and Control,* NAS-NRC Pub. 1400, Washington, D.C., 1966.

President's Science Advisory Committee, Environmental Pollution Board, *Restoring the Quality of Our Environment.* Washington, D.C.: Government Printing Office, 1965.

A. C. Stern, ed., *Air Pollution,* 2nd ed. New York: Academic Press, 1968, 3 vols.

Richard H. Wagner, *Environment and Man.* New York: W. W. Norton, 1971.

G. F. White, *Strategies of American Water Management.* Ann Arbor, Michigan: University of Michigan Press, 1969.

Energy and Resources

Bulletin of the Atomic Scientists, *The Energy Crisis,* **27,** 7–9 (September, October, and November, 1971).

Bureau of Census, *Statistical Abstracts of the United States.* Washington, D.C.: Government Printing Office, annual volumes.

P. R. Ehrlich and A. H. Ehrlich, *Population, Resources and Environment.* San Francisco: W. H. Freeman, 1970.

M. D. Fiske and W. W. Havens, Jr., eds., *Physics and the Energy Problem—1974.* New York: American Institute of Physics, 1974.

Jerry B. Marion, *Energy in Perspective.* New York: Academic Press, 1974.

Appendix

National Academy of Science, *Resources and Man.* San Francisco: W. H. Freeman, 1969.

Howard T. Odum, *Environment, Power and Society.* New York: Wiley, 1971.

Science, special issue on energy, **184,** 4134 (April 19, 1974).

Science, special issue on food, **188,** 4188 (May 9, 1975).

Scientific American, special issue on energy and power, **244,** 3 (Sept. 1971).

U.S. Geological Survey, *United States Mineral Resources.* Washington, D.C.: Government Printing Office, 1973. (Cat. No. I19.16:820.)

Richard Wilson and William J. Jones, *Energy, Ecology and the Environment.* New York: Academic Press, 1974.

Nuclear Power Safety

American Physical Society, "Study on Light-Water Reactor Safety," *Rev. Mod. Phys.* **47,** Supplement 1 (Summer 1975).

Atomic Energy Commission, WASH-1238 (on transportation accidents of spent nuclear fuels).

Atomic Energy Commission, ORNL-4762; ORNL-TM-4639; BNWL-1764; WASH-1297 (on radioactive waste disposal).

B. L. Cohen, *Nuclear Science and Society.* Garden City, N.Y.: Doubleday, 1974.

Environmental Protection Agency, report 520/9-73-003 B, C, D (on reactor routine emission).

David Inglis, *Nuclear Energy: Its Physics and Its Social Challenge.* Reading, Mass.: Addison-Wesley, 1973.

National Academy of Science–National Research Council, Committee Report on Radioactive Waste Management (1970).

National Academy of Science–National Research Council, *The Effects on Populations of Exposure to Low Level of Ionizing Radiation.* Washington, D.C., Nov. 1972. (This is commonly referred to as the BEIR report.)

G. T. Seaborg, "Fast Breeder Reactor," *Sci. Am.,* **223,** 5, 13–21 (Nov. 1970).

U.S. Atomic Energy Commission, WASH-1400, *Reactor Safety Study.* Washington, D.C., 1974. (This is commonly referred to as the Rasmussen Report.)

Units and Their Conversion

Length

1 meter (m) = 100 centimeters (cm) = 10^6 microns (μ) = 10^{10} angstroms (A) = 3.28 ft

1 kilometer (km) = 1,000 m = 0.62137 miles

1 foot (ft) = 12 inches (in.) = 0.3048 m

1 mile = 1,760 yards = 5,280 ft = 1.6093 km

Area

1 km^2 = 0.3861 square miles

1 square mile = 640 acres

1 acre = 43,560 ft^2

Volume

1 m^3 = 10^3 liters (l) = 10^6 cm^3 (cc) = 35.315 ft^3 = 264.18 gallons (gal)

1 bushel = 35.2383 l (dry measure, grains)

1 gallon = 4 quarts (qt) = 8 pints (pt) = 3.7853 l (liquid measure)

1 barrel = 42 gal (for oil only)

1 gallon (c.) = 8 pints (o.) = 128 fluid ounces = 3.8016 l (Apothecaries' fluid measure, drugs)

Time

1 year = 365 days = 8,760 hours = 525,600 minutes = 31,536,000 seconds

Mass

1 metric ton = 10^3 kilograms (kg) = 10^6 grams (g) = 10^9 milligrams (mg)

1 metric ton = 1.1023 short tons (each short ton is 2,000 lb)
= 0.9843 long tons (each long ton is 2,240 lb)

1 metric ton of grain (wheat, corn, peanuts) in weight $\simeq$ 40 bushels in volume

1 slug (English system) = 14.6 kg (mks system)

Force

1 newton (mks system) = 10^5 dynes (cgs system) = 0.224 lb (English system)

Energy

1 joule (mks system) = 10^7 ergs (cgs system) = 0.7376 foot-pounds (English system)

1 kilowatt-hour (kwh) = 3,600,000 joules = 3,412 Btu

1 MeV = 10^6 eV = 1.60206×10^{-13} joules

Heat

1 calorie (cal) = 4.186 joules

1 British thermal unit (Btu) = 252 calories = 1,055 joules

1 therm = 10^5 Btu

Power

1 megawatt (Mw) = 10^3 kilowatt (kw) = 10^6 watts (w)

1 horsepower (hp) = 550 foot-pounds/sec = 0.746 kw

Electric units (practical system): coulomb, volt, ampere, ohm, joule, watt

$P = VI$ watt = volt × ampere

$V = IR$ volt = ampere × ohm

Universal Constants

Gravitational constant
$G = 6.670 \times 10^{-11}$ newton · m^2/kg^2

Speed of light in vacuum
$c = 2.997930 \times 10^8$ m/sec

Avogadro number
$N_o = 6.02486 \times 10^{23}$/mole

Electronic charge
$e = 1.60206 \times 10^{-19}$ coulomb

Electron rest mass
$m_e = 9.1083 \times 10^{-31}$ kg

Proton rest mass
$m_p = 1.67239 \times 10^{-27}$ kg

Planck's constant
$h = 6.62517 \times 10^{-34}$ joule · sec

Boltzmann's constant
$k = 1.38044 \times 10^{-23}$ joule/°K

Index

A

D

I

J

K

L

M

N

S

T

U

V

W

X

Y